BOSTON STUDIES IN THE PHILOSOPHY OF SCIENCE

VOLUME XI

PHILOSOPHICAL FOUNDATIONS OF SCIENCE

SYNTHESE LIBRARY

MONOGRAPHS ON EPISTEMOLOGY,

LOGIC, METHODOLOGY, PHILOSOPHY OF SCIENCE,

SOCIOLOGY OF SCIENCE AND OF KNOWLEDGE,

AND ON THE MATHEMATICAL METHODS OF

SOCIAL AND BEHAVIORAL SCIENCES

VOLUME 58

BOSTON STUDIES IN THE PHILOSOPHY OF SCIENCE

EDITED BY ROBERT S. COHEN AND MARX W. WARTOFSKY

VOLUME XI

PHILOSOPHICAL FOUNDATIONS OF SCIENCE

PROCEEDINGS OF SECTION L, 1969,
AMERICAN ASSOCIATION FOR THE ADVANCEMENT OF SCIENCE

Edited by

RAYMOND J. SEEGER AND ROBERT S. COHEN

D. REIDEL PUBLISHING COMPANY

DORDRECHT-HOLLAND / BOSTON-U.S.A.

Library of Congress Catalog Card Number 73–83555

Cloth edition: ISBN 90 277 0390 6
Paperback edition: ISBN 90 277 0376 0

Published by D. Reidel Publishing Company,
P.O. Box 17, Dordrecht, Holland

Sold and distributed in the U.S.A., Canada and Mexico
by D. Reidel Publishing Company, Inc.
306 Dartmouth Street, Boston,
Mass. 02116, U.S.A.

Printed in The Netherlands by D. Reidel, Dordrecht

PREFACE

At the 1969 annual meeting of the American Association for the Advancement of Science, held in Boston on December 27–29, a sequence of symposia on the philosophical foundations of science was organized jointly by Section L of the Association and the Boston Colloquium for the Philosophy of Science. Section L is devoted to the history, philosophy, logic and sociology of science, with broad connotations extended both to 'science' and to 'philosophy'. With collaboration generously extended by other and more specialized Sections of the AAAS, the Section L program took an unusually rich range of topics, and indeed the audiences were large, and the discussions lively.

This book, regrettably delayed in publication, contains the major papers from those symposia of 1969. In addition, it contains the distinguished George Sarton Memorial Lecture of that meeting, 'Boltzmann, Monocycles and Mechanical Explanation' by Martin J. Klein.

Some additions and omissions should be noted:

In Part I, dedicated to the 450th anniversary of the birth of Leonardo da Vinci, we have been unable to include a contribution by Elmer Belt who was prevented by storms from participating.

In Part II, on physics and the explanation of life, we were unable to persuade Isaac Asimov to overcome his modesty about the historical remarks he made under the title 'Arrhenius Revisited'.

Part IV, on cosmology, benefits now from two supplementary papers, those of Fred Hoyle and G. de Vaucouleurs, but suffers the omission of the contribution of E. R. Harrison, which he feels to be too dated for inclusion.

Part V, concerned with objectivity in anthropology, is complete.

In Part VI, planned as a survey of issues in the comparative history and sociology of science, we have been unable to include the paper of Benjamin Schwartz on 'Science in China' which he felt to be too tentative, and that of one of the editors (R.S.C.) which is to appear elsewhere. But we have been fortunate to obtain the commentary on

Toulmin's paper by Ernan McMullin, and also the introductory remarks of the session chairman, Joseph Agassi.

Part VII, on the unity of science, does not contain 'A Non-Unitary Approach to Theories', the paper by Peter Achinstein who has substantially revised his materials for presentation at another academic occasion. Further, we have not been able to include Herbert Bohnert's paper on 'The Unity of Language'.

Finally, the entire session on 'Brain and Language' with presentations by Noam Chomsky ('Linguistics as Theoretical Psychology'), Norman Geschwind ('The Organization of Language and the Brain'), and Eric Lenneberg ('Neurological Correlates of Language'), and with Stephen Toulmin as commentator and Jerome Lettvin as the lively chairman, proved impossible to include here. The contributors have published their materials fully elsewhere, and indeed regarded their session as a discussion and debate concerning established contributions.

We are of course indebted to the speakers, but also to the scholars who agreed to help organize the meeting and then to participate as chairmen of the several symposia, and to the hundreds of students, natural scientists, philosophers, social scientists, physicians and 'educated laymen' who joined in a memorable three days. We hope the program, as preserved in this book, will be interesting and useful to a similarly wide range of readers.

RAYMOND J. SEEGER

U.S. National Science Foundation

ROBERT S. COHEN

Boston University Center for the Philosophy and History of Science

TABLE OF CONTENTS

PART IV / CURRENT PROBLEMS OF COSMOLOGY

(Chairman: John Stachel)

PART V / OBJECTIVITY AND ANTHROPOLOGY

(Chairman: John M. Roberts)

PART VI / COMPARATIVE HISTORY AND SOCIOLOGY OF SCIENCE

(Chairman: Joseph Agassi)

PART VII / UNITY OF SCIENCE

(Chairman: Ernest Nagel)

PART I

450th ANNIVERSARY OF THE DEATH OF LEONARDO DA VINCI

(*Chairman:* LOREN C. EISELEY)

JOHN SHAPLEY

AN ASPECT OF LEONARDO'S PAINTING*

The 67 years of Leonardo da Vinci's lifetime were fraught with epoch-making events. One was the burst of exploration, including the discovery of America, of which, as far as we know, Leonardo was quite unmindful – in astonishing contrast to his almost exact contemporary and fellow Florentine (also son of a notary) Amerigo Vespucci. Another great event of the time was the rise and spread of printing, something Leonardo could not be ignorant of, in which, however, his own interest can be assumed to have been peripheral. A third event extending over the years of Leonardo's lifetime, and one that concerned him very nearly, was the shift in Italy of easel painting from tempera to oil. The reaction of Leonardo to this momentous change of painting medium is the occasion of this paper.

We all recall the dazzling metallic brilliancy of Italian tempera pictures common for centuries before Leonardo's time. Shortly before his birth Fra Angelico's studio was still turning out such jewel-like objects as the *Madonna of the Star*, in which the old generic relationship between enamelwork and tempera painting is evident. And about the year of Leonardo's birth Filippo Lippi painted his best-known and most widely reproduced *Madonna*, with its crystalline gleaming surfaces, hard and smooth as metal, a picture which, partly because of its roguish angels, remains a household favorite to the present day. Then, a dozen years later, about the time Leonardo was ready to begin his apprenticeship, Piero della Francesca was painting, in what was still popular taste, his firm and polished tempera portraits, among them the *Federigo da Montefeltro*, which has come to be for us a symbol of the Early Renaissance.

Meanwhile, however, the oil medium was gradually beginning to make its triumphal entry into Italy from Northern Europe. Roger van der Weyden came at the middle of the century (perhaps in some connection with the Papal Jubilee of 1450), and aroused great and profitable admiration with his oils. For Florence, more particularly, the key work was the monumental *Portinari* altarpiece by Hugo van der Goes, dating

R. J. Seeger and R. S. Cohen (eds.), *Philosophical Foundations of Science*, 3–16.

from the middle 1470's, and opening Florentine eyes to the revolutionary possibilities of the oil medium formerly unwonted to them. Elsewhere in Italy, say at Ferrara, despite the advent of Roger van der Weyden, a Cosimo Tura might go on undisturbed to paint in perfect tempera his panel of *St. Christopher* of ten years or so later, still pursuing the traditional enamel-like technique and its metallic rigidity. Yet, as to the progressive artists' workshops of Florence, a city in touch with the wide world through its banking and mercantile connections, and one whose citizens had been acquainted for decades with the oil painters of the Low countries, these ambitious Florentine studios were not in a mood to ignore blindly whatever of advantage might be gained by cautious and judicious experimentation with the unfamiliar medium of oil. For a generation or more, in the later fifteenth century and beyond, Florentine painters were using varying mixtures of media, of tempera and oil, the exact analyses of which have not been, and seem unlikely to be, fully accomplished (pending the uncertain development and exploitation of safe scientific techniques, such as, for example, autoradiography is thought to have some prospect of becoming). For even in the hopeful pursuit of knowledge there are restraints to be observed with precious Renaissance pictures as with medical patients.

Into this period of indecision, of the ambiguity of painting medium, was precipitated the young painter Leonardo. The inherited tempera medium was not lightly to be given up. It had its own unequaled merit of *claritas*, as St. Thomas would have said; it accorded with the undying association of light with beauty, indeed with things divine; it had centuries of prestige. As a hallowed concept, how better could the Crucifixion be painted, physically speaking, than in the tempera technique of such a painter as Francesco del Cossa (Figure 1)? Even in progressive Florence the Umbrian Perugino maintained for decades a flourishing traditional workshop, down past the end of the fifteenth century, in which (whether he mixed his media or not) the ideals and appearance of tempera painting were kept alive for the sake of a prosperous, if conservative, clientele.

Leonardo did not wish to give up entirely these attractions of tempera, yet, if we can trust the observations of modern scholars, which are admittedly not, in the nature of the case, at present capable of complete scientific control, Leonardo's early works are already of mixed media.

An essential quality of tempera painting is the smooth surface. It was achieved by a very laborious and precise technique, sometimes described even as niggling, much like that of a miniaturist, its purpose being the production of a beautifully worked, gleaming *objet d'art*, like a piece of

Fig. 1. Francesco del Cossa, *Crucifixion*.

enamel or of jewelry. Gold ornaments or, earlier, gold backgrounds were entirely in the spirit of such painting. As the early printers endeavored to make their work acceptable by producing books that retained many characteristics of manuscripts, and as the early automobile manufacturers imitated horse-drawn carriages, so in the Italian painters' transition to oil painting it was felt important not to lose the smooth and finished, enamel-like quality of tempera, such as we may see in a painting by Tura or Cossa (Figure 1).

Leonardo's *St. Jerome* in the Vatican Gallery is unfinished, and in imperfect preservation; but, partly because unfinished, it is a most instructive indication of Leonardo's reaction to the new utilization of oil. Though the medium is no longer pure tempera – again, that is, if our not scientifically fully controlled observations are correct – Leonardo has aspired to the sleek surface of tempera. Recent X-ray photographs published by Brachert[1] of the Swiss Institute for Art Studies at Zürich show greatly magnified segments of the *St. Jerome* landscape. In these (cf. Brachert's Figures 4–7) we see how Leonardo has manually rubbed the surface to give it an enamel smoothness. Brachert's magnification, which shows the traces of the papillae of the painter's thumb or finger tips, gives in the reproduction a false impression of roughness, but in actuality this rubbing is a clear indication of the effort to achieve a smoother surface than that left by brushwork in the mixed media. Brachert reports similarly rubbed passages from the Uffizi *Adoration* and from the London *Madonna of the Rocks*, but it is to the *Ginevra dei Benci* in the National Gallery at Washington, where I have been able to check out Brachert's X-ray photographs with others taken independently, that I should like to call especial attention. The passages reproduced in Brachert's Figures 13–14 are from the landscape to left and right. Of the actual smoothness of the surface of the picture it is hard to convey an impression by way of illustration of the landscape surface. But illustration of the face may serve better to convey this smoothness in a reproduction (Figure 2). Looking at the face we get the effect of a considerable measure of survival of the quality of enamel, of the egg-shell limpidity admired in tempera painting. This seemingly impenetrable hardness of surface gives a strange metallic appearance to the curls; they look as if made of steel tape.

After Leonardo's time it remained characteristic of the Florentine painters to continue to strive for blond tempera-like effects though their use of oil increased. Deep into the sixteenth century we find the most representative Florentine Mannerist, Pontormo, still obtaining a metallic brilliancy almost unbelievable, as in the recently damaged painting in Sta. Felicita.

There was indeed another way in which a uniform surface could be

Fig. 2. Leonardo da Vinci, *Ginevra dei Benci* (detail).

attained in oil painting. This was by the use of a very thin oil medium of minimal viscosity, sacrificing precision of contour to softness of effect. It was briefly adopted in Venetian circles, as by Antonello da Messina and Giovanni Bellini. Judging from a brief written comment of Leonardo's it seems unlikely that he would have found the slick lustre of the glistening surfaces of such pictures desirable; their surfaces are quite different from the metallic surfaces of tempera paintings. Returning to Pontormo, we see that the surface he (and Leonardo) favored was a different one (Figure 3). It was to remain standard for the Florentine Mannerists of the sixteenth century.

Fig. 3. Pontormo, *Monsignor della Casa* (detail).

Fig. 4. Titian, *Venus with a Mirror* (detail).

In Venice, however, things soon took a different course. While retardataire painters, such as Crivelli, tried long and laboriously to maintain the well-recognized glories of tempera, other Venetians developed, in contrast to the Florentine mode, what we call the Venetian mode in painting. This mode (some aspects of which are sometimes loosely called tonal painting) had a long championship by Titian. Compare a detail of the subtle, but in places rather rough, complexion of Titian's painting of the head of a *Venus* in the National Gallery, Washington, with the face of Leonardo's *Ginevra* (Figures 4 and 2). Or compare the hand of the same *Venus* by Titian with a hand by Pontormo (Figures 5 and 3). The smooth surface achieved by Pontormo and representing the Florentine mode can be more fully appreciated if compared with a section of the surface, shown under raking light, of Titian's contribution to Giovanni Bellini's *Feast of the Gods* in the same National Gallery (Figure 6). This rough surface, calculated to refract, not merely to reflect, light is the sign

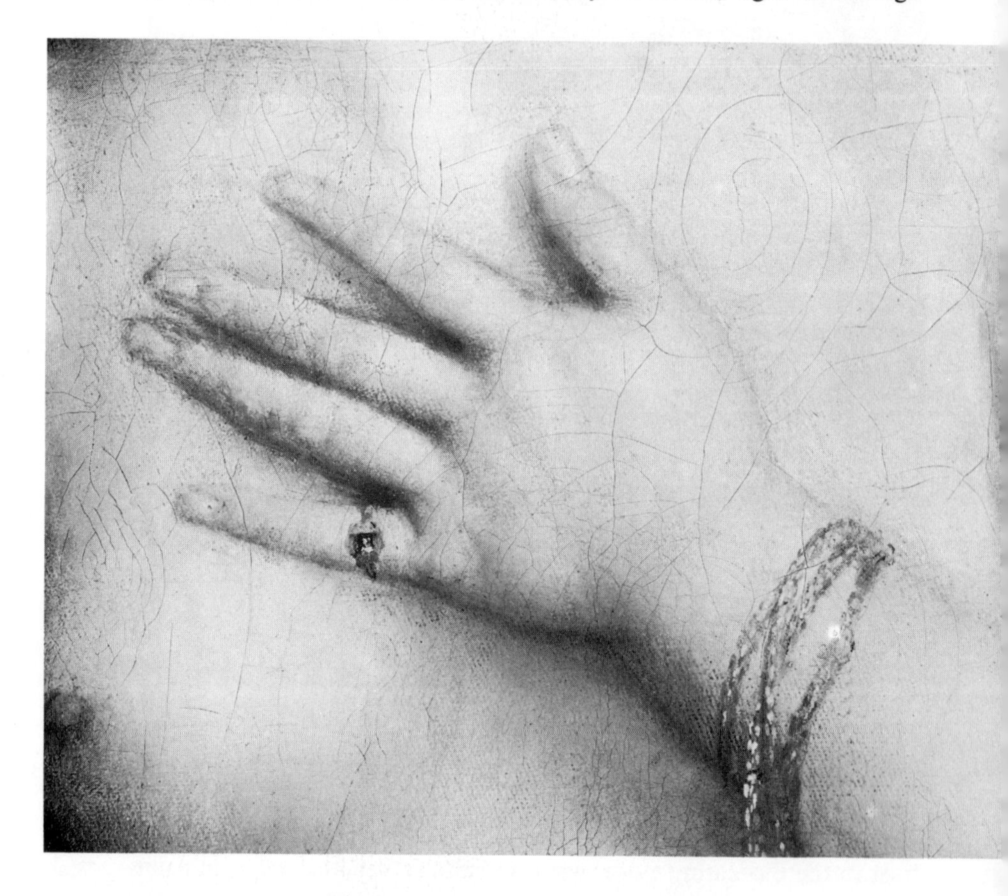

Fig. 5. Titian, *Venus with a Mirror* (detail).

Fig. 6. Titian's contribution to Giovanni Bellini's *Feast of the Gods* (detail).

manual of the Venetian mode in painting. Tintoretto, with still broader brushwork, carried on after Titian. It is interesting to compare one of Tintoretto's portraits, so utterly unlike a tempera painting, with an almost coeval portrait by the Florentine Bronzino, in whom still dwelt an aspiration to cling to something of the hard brilliancy of the tempera tradition. Details of the two portraits, one of the man's head from the Tintoretto portrait, the other of the child's head from the Bronzino, serve well to illustrate the contrast between the two modes (Figures 7 and 8).

It was Leonardo, of course, who was mainly instrumental in carrying the Florentine tradition of painting to Milan and the surrounding territory of Northern Italy. There, largely under his influence, painters strove for the smooth surfaces he valued, as is exemplified strikingly by the head of the *Venus* of Luini, that, in this respect, reminds us forcibly of the head of the *Ginevra*, with which it is available for comparison in the same gallery. Even more striking are the hard and brilliant surfaces, with unnaturally metal-like coloring, of the accompanying landscape in Luini's picture. Such a style long survived in the Lombard and adjacent region, even to the point of influencing the young Caravaggio, whose famous *Bacchus* looks entirely at home in Florence in the Uffizi Gallery, where it now hangs along with the works of the Florentine Mannerists, who were more directly heirs of Leonardo. In view of Leonardo's influence radiating from his adopted home, Milan, it is not at all strange that the works of North-Italian Mannerism resemble in finish those of Florentine Mannerism. Thus, for example, the Parmigianino (Parmese) *Madonna of the Long Neck* is likewise quite at home, stylistically, in the Uffizi, though perhaps this artist's *Cupid* at Vienna is an even more vivid illustration of the continuity of style from Leonardo.

It was the Florentine mode, rather than the Venetian, which was carried to France, not so much by Leonardo himself as by his Florentine (and related) Mannerist successors who went to constitute the School of Fontainebleau, of whom Primaticcio is a representative example. Once there, these immigrant artists succeeded in establishing an enduring French tradition. This runs on astonishingly true to type in the works of Clouet, as we can see in the so-called *Diane de Poitiers* of the National Gallery, Washington; and when we look at her face we immediately feel it to be familiar (Figure 9). In the seventeenth century Poussin further

Fig. 7. Tintoretto, *A Procurator of St. Mark's* (detail).

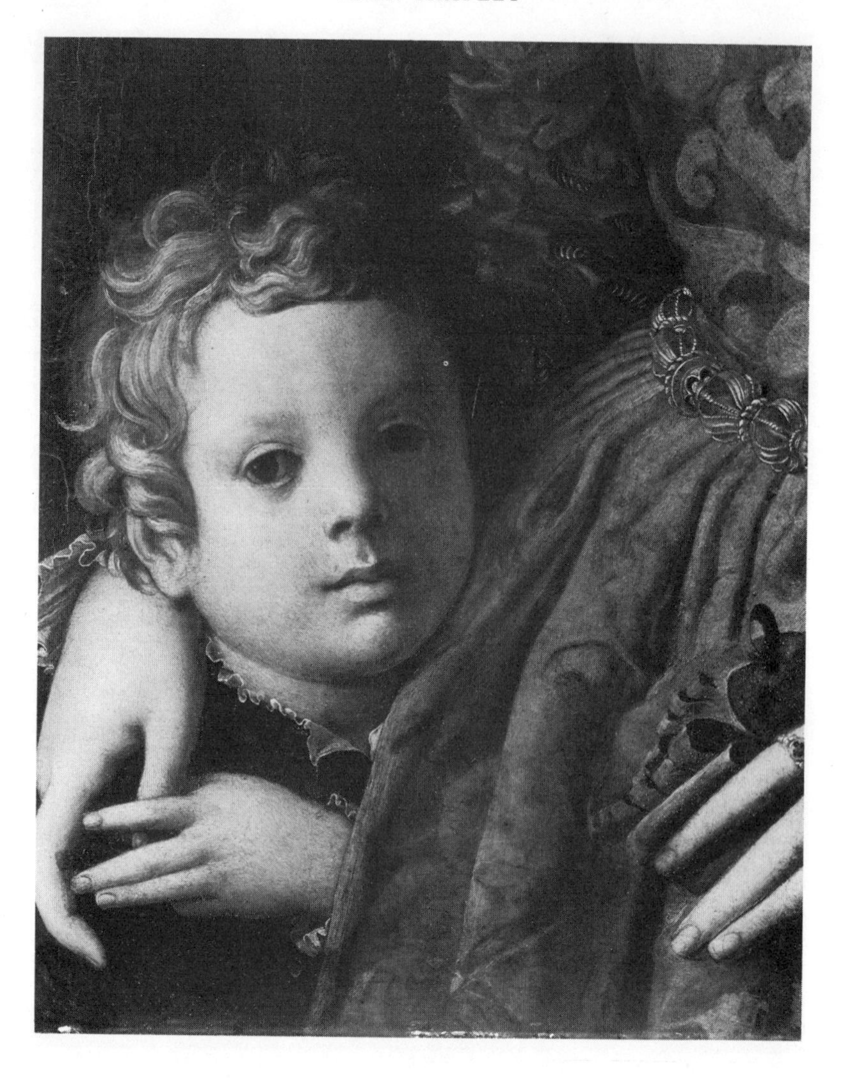

Fig. 8. Bronzino, *A Young Woman and Her Little Boy* (detail).

Fig. 9. Clouet, *Diane de Poitiers* (detail).

represents the same tradition. His *Inspiration of the Poet* in the Louvre provides an effective demonstration, especially if we look at the female head to the left. We are reminded at this point of the bitter debate between the Poussinistes and the Rubensistes, which we can now see in terms of this paper as a kind of sequel to the contrast of Florentine and Venetian modes. Even in the nineteenth century the reflection of clinging to the old tempera ideal of painting is still apparent in Ingres, as, for instance, in the Chantilly portrait of *Mme. de Vaucay*, a picture having the smooth complexion that we have been tracing through the centuries. Nor did this ideal disappear: witness some of the early pictures by Renoir, whose beginnings in China painting must in the light of history seem perfectly natural.

George Washington University

NOTES

* All accompanying illustrations are from pictures in the National Gallery of Art, Washington, D.C., which has courteously given permission for their reproduction.

[1] Brachert, T., in *Maltechnik*, 1969, 33ff.

RAYMOND S. STITES

LEONARDO DA VINCI AND THE SUBLIMATORY PROCESS

In 1907 Sigmund Freud gave a lecture to his group of students and colleagues on the subject of Leonardo da Vinci. In this talk he presented the hypothesis that Leonardo was mentally disturbed and that the disturbance could be related somehow to a dream Leonardo had recorded from his childhood [1]. Also, Freud assumed that Leonardo had an unnatural affection for his mother, who had been deserted by her lover, Piero da Vinci, in the year Leonardo was born [2]. Among other conclusions, in his diagnosis of what he considered Leonardo's almost neurotic constitution, Freud made the statement that "this man, otherwise so universal in his many interests, avoided all investigations in the realm of psychology" [3]. This statement is particularly intriguing, for to anyone acquainted with Leonardo's art and particularly his portraits, it must be apparent that he was capable of very deep studies of the human personality. For example, in Manuscript G Leonardo wrote

> Words which fail to satisfy the ear of the listener always either fatigue or weary him; and you may often see a sign of this when such listeners are frequently yawning,... And if you would see in what a man takes pleasure without hearing him speak, talk to him and change the subject of your discourse several times, and when it comes about that you see him stand fixedly without either yawning or knitting his brows or making any other movement, then be sure that the subject of which you are speaking is the one in which he takes pleasure [4].

In folio 43 r. of the *Codex Trivulzianus*, while observing auditory and visual effects, Leonardo wrote: "The eye keeps in itself the image of luminous bodies for some time" [5].

Furthermore, Leonardo mentioned in his notebooks that he had used a very interesting (almost 20th century) method of composition. For example, *The Madonna of the Grotto* in the Louvre, as I have discovered, contained a childhood memory of a beautiful grotto a few hundred yards away from Leonardo's birthplace in Vinci. Another realistic image within the composition which, with the dolomitic rocks and the stream, formed part of the topological elements in Leonardo's mind, is the portrait of that young woman *Cecelia Gallerani*, with whom he had fallen in love at the time. There she may be seen as the angel looking almost

R. J. Seeger and R. S. Cohen (eds.), Philosophical Foundations of Science, 17–40.

Fig. 1. Doodle, inception of a Madonna study. *Codex Atlanticus* folio 249 v.

exactly like her portrait. In the *Madonna of the Grotto* we have then one element, the grotto, from Leonardo's subconscious and another, the face of his beloved, from his conscious mind. There is, however, one more factor, the pyramidal composition form or the archaic heritage (as the psychiatrists Jung or Fromm might call it). This inner symbolic form can be traced to Leonardo's training in religious art under Verrocchio. This triangle appears first in Verrocchio's *Baptism of Christ*, where the angel at the left was painted by Leonardo. That angel, as I can now demonstrate, is Leonardo's self-portrait.

But to these images, some conscious and others subliminal, we must add a third type coming from a deeper level of the subconscious, represented by doodles like Figure 1, basic to the initial composition of the *Madonna and Child with St. Anne*.

Another one, a doodle of agonizing complexity (Figure 2 [6]), became the initial part of his battle-scene where it joins with memory images of members of his family like the grotesque head at the left. Of phenomena such as these Leonardo was quite aware. Indeed he tells us in several places in his notes, that he could not only see battles and other compositions in cloud-formations and blots on old walls, but that he noticed related phenomena in word-sounds emerging from the clanging of bells. Now of these psychological phenomena as recorded by Leonardo, Freud knew nothing.

For the past forty years I have been recovering these and much more such related material in Leonardo's notes, in his artistic creations and

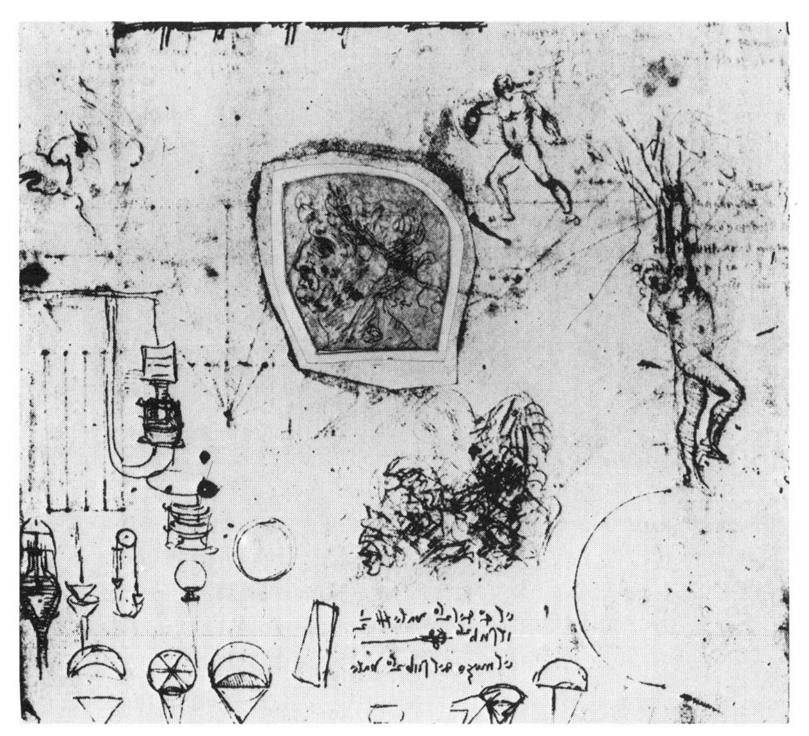

Fig. 2. Doodle with genesis of the battle-scene (*Codex Atlanticus* folio 299 r.a. with head in center inserted from Windsor Library cut-out No. 12485) (after Pedretti [6]).

his scientific inventions. In this report I present a few of my findings which have finally been incorporated in a book entitled *Sublimations of Leonardo da Vinci, with a Translation of the Codex Trivulzianus.* [5]

In the *Trivulzian Codex* of Leonardo, among lists of words set down in a rather freely associated order, I found at the end of one list the sentence, "It seems to be a voice which comes from the air", and this, to my mind, suggests that Leonardo's subconscious not only yielded images which may have almost been eidetic in character, but also that he had a strong auditory recall. This, had he been a religious ecstasist, he might have considered as being messages from the realm of the spirits.

To repeat, some of Leonardo's paintings arose from observations of natural phenomena and other compositions were based upon archaic schemata taken from earlier orthodox religious paintings. Still a third kind of composition arose from the stimulating effects of blots on walls.

A fourth type, rather surrealistic, came from doodles which could only have come from the subconscious or subliminal self.

Much that Freud had to say in his lecture of 1907 he then began to enlarge into a book which was first published in 1910. There it appeared that his information had come from rather unreliable sources such as the novel by the Russian Merejkowski and from other rather literary legends which had grown up about the character of Leonardo. For example, to prove Leonardo neurotic Freud used a false drawing with its many errors in his book [7]. It is, as any Leonardo scholar would see at once, a poor copy of the famous coition sheet in the royal collection of Windsor Castle. A dream which Leonardo had noted (the only one in fact), and Leonardo's own very brief interpretation of the same, also was enlarged upon and interpreted by Freud in terms of his own prepossession with sex. However, there is evidence from Leonardo's life and writings that leads to a more satisfactory interpretation for this dream.

Kenneth Clark, the well-known English scholar dealing with Leonardo's life as a painter, quoted Freud's interpretation of this dream, but in the 1961 edition of his book *Leonardo da Vinci: An Account of His Development as an Artist*, Clark has written

> We are still too ignorant of psychology to interpret such a memory with any finality, but it is not surprising that Freud has taken this passage as the starting point for the study of Leonardo. His conclusions have been rejected with horror by the majority of Leonardo scholars. And no doubt the workings of a powerful and complex mind cannot be deduced from a single sentence explained by a rather one-sided system of psychology. Freud's study, though it contains some passages of fine intuition, is perhaps as over-simplified as that of Vasari. Yet it helps our conception of Leonardo's character by insisting that he was abnormal [8].

With both Clark and Freud I must disagree; – Leonardo, as I can demonstrate, was one of the most normal geniuses who ever lived.

Since Freud's hypothesis concerning Leonardo's psyche, arising from what Clark calls his "one-sided system of psychology", has had considerable impact upon those scientists interested in explaining their own mental struggles, as well as artists and art historians interested in drawing from Leonardo's works some knowledge of their own creative processes, the earliest attempt at a psychoanalytic interpretation calls for a stringent re-examination sixty-three years after Freud first made it. This examination must be made in terms of what may actually be found in Leonardo's inventions in science and art, and in his notes as they reflect his manner of working. After a forty-year exploration of this problem, at last I am prepared to demonstrate that Leonardo's reactions to the struggles within his own nature, and those brought about by the vicissitudes and frustrations of life around him, were, in the main, quite normal.

Moreover it is now clear that he wrote many notes which show an intimate experimental approach to the problems of psychology. In this he was far ahead of the other scholars of his time, for he observed and used what we today call chain-association, thereby inventing for himself a kind of *self-psychoanalysis* with art-therapeutic overtones.

Much of the evidence herewith presented stems from an extended study which includes a translation into English of the *Codex Trivulzianus*. This manuscript, consisting of many notes on architecture, bronze-founding, philosophy and health, with drawings and doodles, also contains 45 pages of word lists with strongly affective or emotional overtones. The original is today in the Trivulzian Library of the Sforza Castle in Milan. The manuscript was first recorded as having belonged to one Galeazzo Arconati, who, in 1637, gave it to the Ambrosian Library in Milan. In 1570, Prince Trivulzio acquired it, and in 1891 the Italian scholar Luca Beltrami published a facsimile edition of only 200 copies, one of which I first saw in the Albertina Library in Vienna in 1925.

The initial folio of that manuscript which contains notes, consists of some caricatures and four sets of words. The first note, above on the page, reads

> Ammianus Marcellinus reports that 700000 volumes of books were burned in the attack on Alexandria in the time of Julius Caesar.

This suggests that Leonardo, who probably wrote this page around 1484, was disturbed by the destruction of the Library of Alexandria. He himself had read or heard of this holocaust in a book by a Roman historian. We may today, following this lead, begin to examine the books Leonardo read for the source of some of his ideas on war and peace.

The next note, of more psychological import, reads

> If Petrarch loved so stoutly the laurel, perhaps it was because it was between the sausage and the sage – for me it is impossible to go so far as to get the treasure.

Now these lines may have a double meaning, for, as I have learned, Leonardo was a master of *double entendre*. Pierina Castiglione, an Italian scholar well-versed in late medieval Tuscan, who carefully checked my translation, assures me that the sage and the laurel are used in a sausage sandwich still much enjoyed by Florentine gourmets. But Petrarch, the poet, had been in love with Laura and so might have desired the laurel crown, which Leonardo also pictured on certain heads drawn at this time.

And our artist, at the age of 32 or thereabouts, had fallen in love with a young woman. She, his beloved Cecelia as he named her, was skilled as a poetess writing verse both in Italian and Latin. So, at this time, Leonardo apparently began to study Latin in earnest.

Indeed, four of the pages in this very notebook consist of lists of Latin words with their equivalents in Tuscan. Two Italian scholars, Govi and Marinoni [9], on seeing these lists concluded that the *Trivulzian Codex* was little more than Leonardo's attempt to garner words he was lacking for a dictionary of scientific terms. The third note on the initial page reads – "Two principles need to be defined" – and then that line peters out. But the fourth line of that introduction seems to have been one of these principles. It reads *Saluticho e quel che si salva*. Now this may be translated in three different ways. First, as Richter translated it many years ago, it might run "Savage is he who saves himself." However, farther on in the *Codex* are lines which suggest that a closer translation might be either, "Saved is he", or "healthy is he"... "who saves himself." I had to use the third meaning "healthy is he" because of still further definitions in Leonardo's notes found later in this manuscript. This was also the interpretation of the psychiatrist James Sagebiel who studied the word lists with me in 1934. My discoveries since that time indicate that indeed Leonardo had already begun to use the means of self-liberation found in other word lists which followed still earlier in his career, i.e. around 1480. So this page may have actually been one of the later pages in a manuscript whose pagination as left by Leonardo is still not known exactly (although scholars such as Marinoni and Pedretti have tried to place it more accurately).

Other caricatures which accompany the notes later, such as that in Figure 3, show the head of an old man, like one seen also in folio 1, in two aspects, surrounded by four grotesques which may stand either for the four temperaments or four vices. That Leonardo believed in the Hippocratic doctrine of the temperaments is shown at other places in his manuscripts and in his anatomical notes.

When I first saw the *Codex Trivulzianus* in 1925 I had been studying

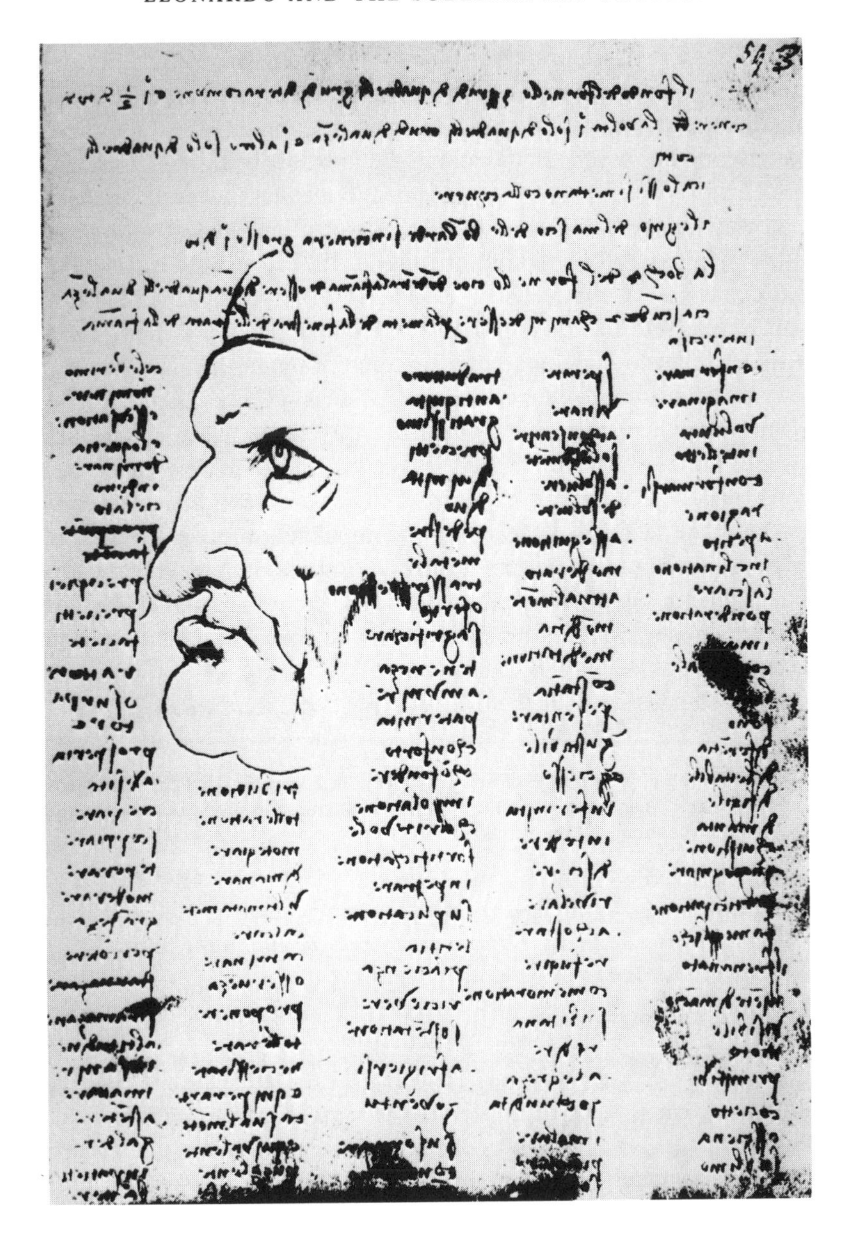

Fig. 3. *Codex Trivulzianus* folio 30 r. with head of Leonardo's father; compare with head in Figure 2.

psychology for six years. In that time I had become acquainted not only with the book about Leonardo by Freud but had also studied the association word list tests of Jung [10], used by Jung in perfecting a more objective kind of psychoanalysis than the one then being used by Freud. On glancing at Leonardo's lists it seemed to me that they were related in some way to the discoveries of chain-association made by James of Harvard [11], as well as to those of Jung. In 1927 I began to translate the *Codex* and had it completed by 1934. At that time the word lists in English were sent to three reputable psychiatrists and all three assured me that Leonardo had apparently been interested in the phenomena of the unconscious mind. One of these psychiatrists, Oskar Diethelm, then at Johns Hopkins University, and since, until recently, head of the Harry Payne Whitney Clinic at Cornell Medical College in New York, cautioned that any interpretation, in order to be reasonably accurate, could only be made in light of the thought-compelling events in Leonardo's life. His reaction to these events could have inspired him to write the lists. With Diethelm's good advice in mind I have spent the intervening years relating the lists to Leonardo's inventions in science and art and to his life course.

In 1970 I first read in Diethelm's book entitled *Treatment in Psychiatry* [12], that

> Psychiatry based on the principle of integration cannot be merely a science of abnormal 'mental functions'. The whole personality, with due attention to physical and mental aspects, has to be considered if one wants to understand and treat these disturbances.

Further he wrote

> Of fundamental importance for the understanding of the working of a personality, and therefore the base for treatment, are the concepts of integration and individuation.

And finally, again from Diethelm come these salutary words for the investigation of a personality

> The most essential characteristic of the psychobiologic personality is that, with the aid of symbols, it makes use of reality and imagination, past, present, and future, personal and general, as if everything existed at this particular time and place and only with this particular object.

My research, undertaken over the last forty years in the spirit of Diethelm's original advice, has discovered a new concept of Leonardo's life style which almost exactly answers to this picture of the mentally sound individual.

One of the most important points to be made is that Leonardo, in the interrelationship of all his artistic and scientific *Invenzioni* or discoveries, as he and his contemporaries called them, proceeded by a method of free chain-association as described by our great psychologist William James several years prior to Freud's publications.

For those physical scientists who hitherto may have had little need to explore the concepts in this field of psychological research, I would suggest the reading of an essay on mathematical creation by Henri Poincaré in a book entitled *The Creative Process* [13]. There, on pages 36–39, he describes the creation of his first memoir on the Fuchsian functions. He ends with the words

> The unconscious, or, as we say, the subliminal self, plays an important role in mathematical creations; this follows from what we have said. Usually the subliminal self is considered as purely automatic... but the subliminal self is in no way inferior to the conscious self; it is not purely automatic; it is capable of discernment...

He continues by telling us that

> only the interesting combinations of numbers break out of the subliminal self into the field of consciousness.

Then he asks "why do some pass the threshold of the subliminal self while others remain below?" Finally he concludes (on page 40) that

> emotional sensibility is evoked in the process apropos of mathematical demonstrations, which, it would seem, should interest the intellect alone.

He ends by invoking the sense of mathematical beauty and the harmony of numbers; that is, of geometric elegance, and he tells us that this is a true aesthetic feeling known to all mathematicians, concluding "and surely it belongs to emotional sensibility."

Since Leonardo several times mentions in his notes on geometry that all true science rests not only on experience but must be capable of geometrical demonstration, we may begin our research by seeking out his notes on geometry for evidence of this faculty of the human mind. He records, for example, on a folio of the *Codex Atlanticus* "Have Benedetto d'Abbacco teach you how to square the circle." The problem of the squaring of the circle, which has intrigued mathematicians for centuries, was one on which he worked for many years. He started with a study of the lunules of Hippocrates and ended finally around 1510 as this, one of his most famous drawings of human proportions, shows (Figure 4a, b), by solving the problem of squaring the circle with a dia-

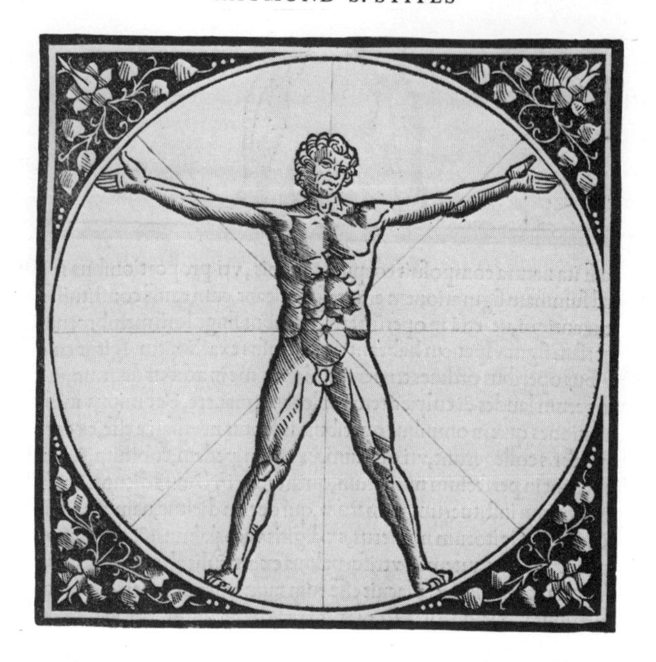

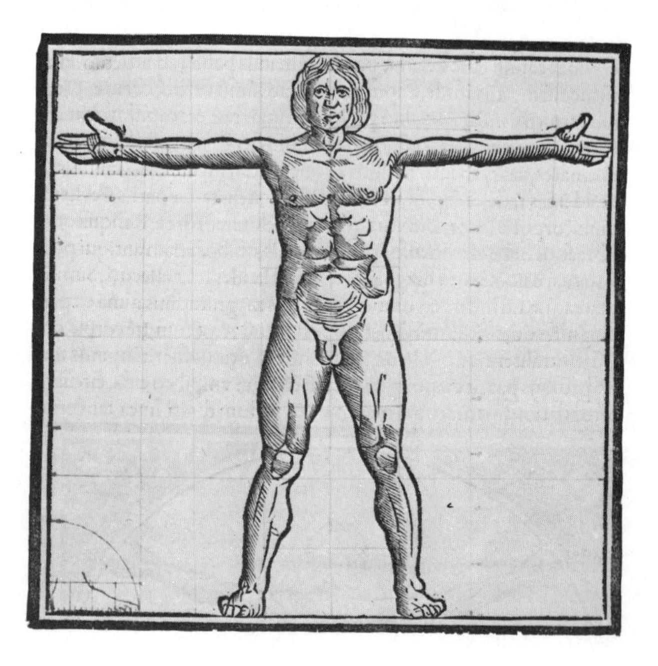

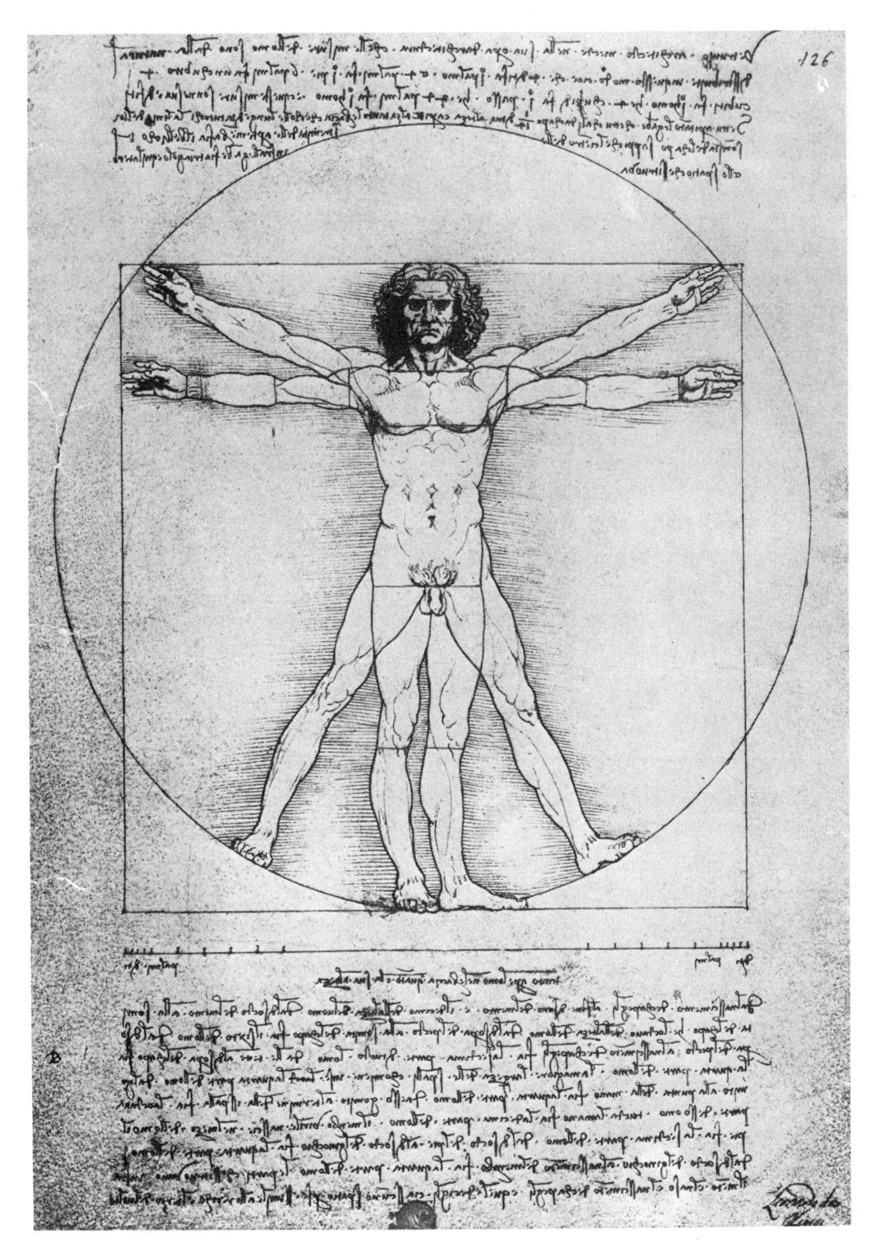

Figs. 4a–b. (a) Vitruvius, *De Architectura Liber Tertius* page 22, Venice 1511 (Study of human proportion). (b) Leonardo's solution of problem of squaring the circle with study of human proportion with quotation from Vitruvius' *Liber Tertius*, Academy, Venice.

gram which has definite psychological overtones. Earlier illustrations of this concept of human proportion tending toward the beautiful, as shown in contemporary manuscripts of Vitruvius, show the symmetrical man placed in the circle so that his center is the umbilicus, sign of his earlier feminine connection with his mother. Leonardo, placing his two men, one in the circle, the other in a square, has the second with his center at the testis, thus signifying his masculinity. In his coition sheet, (Windsor Drawing Guaderni III, 3 v.) and in a number of notes, Leonardo shows that he acknowledges both the masculine and the feminine side of his nature and in this he was far ahead of his time as, before that, ancient anatomists saw only the masculine side in the creation of the new child.

Throughout Leonardo's pages there are literally hundreds of drawings which are a part of his attempts at solutions of what are called (by Leonardo scholars) his *'ludi mathematici'*. In one page of *lunules*, the 'personal sensibility' element of Poincaré appears in two heads identifiable as those of Leonardo and his father. One is that of a young person, the other of an old man. A psychiatrist, finding similar associations of these two head types on many other pages of Leonardo's manuscripts might well consider that this is one of those cases of emotional sensibility mentioned by Poincaré.

In a drawing from the Uffizi Gallery (Figure 5) the same two heads appear again. Here the page is dated 1478. By that time Leonardo was twenty-six years of age. Here the old and young man confront each other. The older man looks very stern, the younger seems almost to be pleading. The related notes and doodles on both sides of the sheet tell us what Leonardo's activities were at the time this confrontation was made. First of all, in the line below, we read "I commenced two virgins Mary" i.e. two Madonna pictures. Secondly there are a number of mechanical devices, six in all, on both sides of the folio. At the top of the page there is mentioned a trip to Pistoia. This is followed by two Latin words 'love' and 'beloved'. Next comes the Latin phrase *In Dei Nomine*. This last gives us a clue to the inner or subconscious background for this recording of a creative experience.

As Clark and other scholars have noted, Leonardo wrote this line, *In Dei Nomine*, many times on various pages of his notebooks. Sometimes it is accompanied by part of a sentence, "tell me if ever" (*dimmi se mai*). Clark assumed, correctly, I believe, that these words came to mind

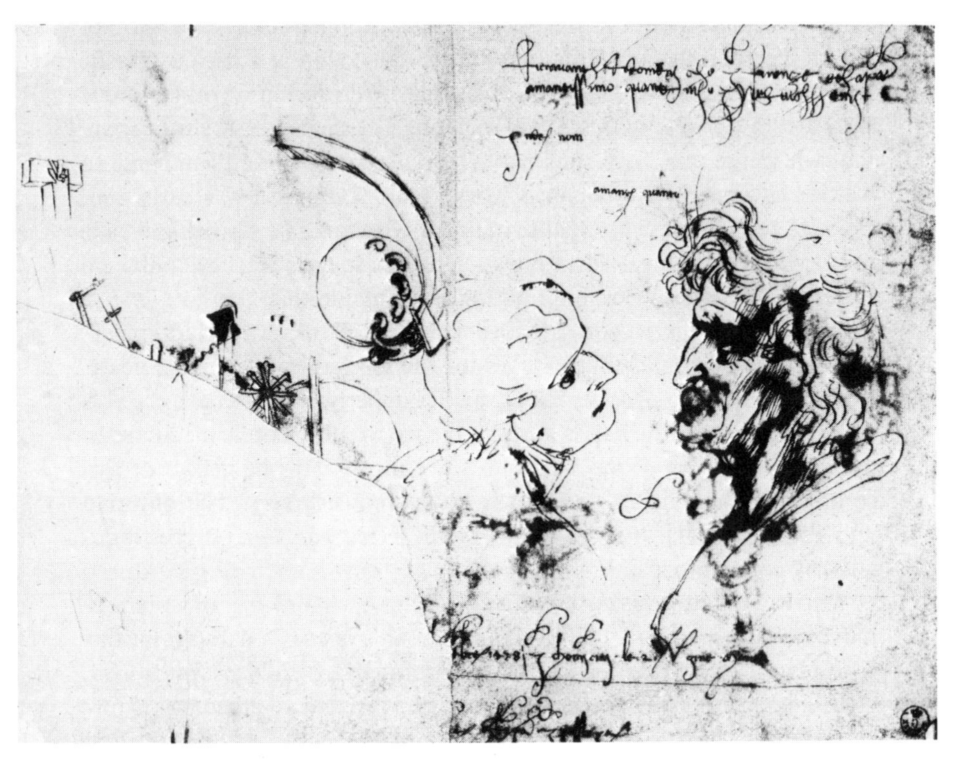

Fig. 5. Leonardo and father, dated 1478 (reversed). Courtesy of the Uffizi Gallery, Florence.

whenever Leonardo sharpened and tried out his quill pen. What do these two almost compulsive symbolic phrases suggest about Leonardo's unconscious mind? We may well imagine that many times the young Leonardo had read these words "In God's name, Amen", in wills drawn up by his lawyer father. Since they are written with the finality of a will, Leonardo later probably used them to signify a conclusion reached as an event was consummated. So the abbreviation *In Dei* might be taken as a kind of personal symbol or signature. The other phrase, *dimme se mai* (tell me if ever), as Herzfeld has shown [14], may well refer to the fact that in Verrocchio's workshop many works of art remained unfinished. So the young Leonardo probably became used to his master exclaiming, in frustration, "tell me if ever anything is finished." This I assume because the very same words appear in the notebook of another one of Verrocchio's students (now in the collection at Chantilly Castle in France). And so we may conclude, since Verrocchio or others in his shop also used these words, that the young Leonardo already was aware that where the mind is fertile in ideas and schemes not all can come to fruition at once in a completed work. Thus, Leonardo was early conditioned not to expect to be able to finish everything he started and gave vent to his feelings of frustration when he had to try out a new quill pen. Contrary to Freud, I can find nothing abnormal in this.

On this Uffizi sheet Leonardo began, as this tells us, "two Madonnas". He had also begun, as the sketches for the mechanical devices indicate, to work upon or to observe the principles underlying these devices. This is the evidence that he began early in life to invent scientific machines or devices; again Freud and Clark to the contrary. In the center of this complex of thought are the words 'love' and 'beloved'. The old man and the young also have with them the word *compare*, suggesting a fatherly companion.

When Leonardo was still relatively young (his handwriting would indicate a dating around 1480) this older man's head appears again, this time placed in a square, with the features geometrically proportioned. Here it seems that the older man was being reduced to a geometric formula relating to the square – while the young man, at the right first looking like the older, then below, with flowing locks, finally riding free upon his racing steed. It seems to me that at this time Leonardo already perceived that, although some humans may be static and square (in terms of modern adolescent slang) that others, like himself, being more romantic, were free, with unbounded energy and infinite possibilities for speed and for fleeing the restrictions of their elders. At first the young man's head is shown completely bald, and with exactly the same shaped cranium as that of the older man – much as though Leonardo was still identifying with his father. Later, his abundant hair flies free!

Another later sketch (in the British museum) (Figure 6), this time for Leonardo's painting of the *Virgin and Child with St. Anne and the Infant St. John*, was probably done around 1498. The same old man's head reappears below the women, strongly drawn in black ink and upside down as though it might have been a geometric *point d'appui*; an early concept of the painting's composition.

Fig. 6. Father's head connected with early studies of *Virgin and Child with St. Anne and Infant John the Baptist*. Courtesy of the British Museum.

On the reverse of this particular folio one may find the same composition reversed. Probably this was the earlier of the two sketches. It has bled through to the other side and has perhaps inspired the creation of the old man's image so that the two appear to be related in Leonardo's subliminal mind. Two compositions below on the reverse side also show a still earlier stage in the development of this picture which is usually dated around 1498, that is, just after Leonardo had finished the *Last Supper* and just before he left Milan to return to Florence. Since there are also drawings on this sheet for a mill wheel and for a revetment, which might have been used in connection with some canalization project, we become aware that Leonardo certainly had by this time entered the engineering phase of his development. The relationship of the old man's head to Leonardo himself and to the developing composition of the *Virgin and St. Anne* is there obvious. I have counted this particular type of old man's head at least 32 times in the Windsor Drawings alone, and I have found it 19 times more throughout the course of Leonardo's manuscript notes. To some psychiatrists it might well seem that Leonardo was 'obsessed' with this image of his father.

Fortunately, two pages of the *Codex Trivulzianus* (Figure 3) give us the answer to that problem. One folio has the profile of the old man characterized by the scholar *Gombrich* as "old nutcracker face". Note that the column begins with the word *traduttoro* – traitor. Incidentally, in my book I have had printed each folio reversed so that all may read Leonardo's otherwise mirror-script from the original, thus checking the translation. Gombrich relates this type of face to the origin of many of Leonardo's caricatures. Some of them, certainly based upon this face, are almost horrendous. For example, on Windsor folio 12495 r. this same face has become quite stern, and the four caricature heads there are usually interpreted by Leonardo scholars as standing for the four temperaments.

I believe from the associations in the *Codex* that the ugly ones may actually stand for various forms of vice. The lists of associated words begin on the left of page v with the words *intercisa*, i.e., killed, then "to conform, to imagine, will, intellect, confusing, the reason, appetite", etc. The appearance of the head here, on folio 30 r., with these words, serves to remind us that about this time or perhaps a little earlier, the same profile of Leonardo's father appears again on Windsor 12626 as that of a much younger man. He is also found on a sheet of dissections along with the list of the various faculties: *memoria*, *senso*, *commune* (or common sense), *volunta* or will, *impresiva*, *intellecto*. It also appears on Windsor 12604 r., with rays of light falling on various parts. There the older man looks upward toward the light.

The second list on this folio (Figure 3), reading from the right, begins with the word *sperare*, that is, to hope, and continues in a kind of dialogue in which Leonardo seems to be considering occasions for hope and despair (ending with the word *sospirare*). Finally, in a dramatic and intense confrontation of words, we have what I believe to be the undoubted identification of the person we have been considering.

Here is the translation of the whole of this list: traduttoro (from traditoro, traitor, or from tradurre, to transfer), antiquity, grateful, precepts, divinity, divine, pedestrian, mental, transgression, offered, to sacrifice, tenderness, *amunire*, to warn, paternity, comfort, confuse, to sacrifice, charitable, fortification, *inpetrare* (?), supplication, gladness, pleaşure or pleasing, to receive, maintenance, to be saddened, youth, to whisper, contemporary, fellow countryman. As we continue down the list we come to the word 'paternity' or fatherhood. So this undoubtedly is Leonardo's father. Leonardo's feelings, as shown here in these very affective or emotionally loaded terms, were ambivalent, presumably because the older man, certainly a tyrant as he appears here, had at times been tender and charitable. Or else we may read here that Leonardo, taking into account the older man's faults, is telling himself that he must be tender or charitable with respect to his father. The final sense in which I have taken these two varying interpretations must be considered further in light of what appears on several other pages of the *Codex* and in many notes and designs from other places in the notebooks. This much I can say for certain, the old man was definitely Leonardo's father. In my research, once this discovery was made, it was not too difficult to identify others of Leonardo's relatives and to see how they appeared in Leonardo's paintings to suggest what they must have symbolized in Leonardo's thoughts.

For example, the father type head, in its most malevolent aspect, may be recognized again in sketches of Judas for the *Last Supper*. Leonardo was probably working on that picture as he wrote this list. We must remember that the paternal word list began with the term *traduttoro* or traitor, and Judas was the traitor among the disciples!

Reading backward into Leonardo's earlier life we find the father's head again in the profile of a warrior wearing an elaborate helmet which is in the British Museum. That drawing is usually dated around 1480. We note that the face is younger than that of the one done in 1494 in the *Codex*. Then we come to what must have been the earliest version of the profile. Here the father's head appears on what (following the theories of Alfred Adler) we may well call Leonardo's 'family constellation' sheet (Figure 7). Although this has been dated by Clark as having been done around 1478 and by Möller around 1490, it is my opinion that it should be dated as early as 1468, when Leonardo was 16 years of age. Here, it seems to me, we have to deal with his earliest drawing, recording the images of his family as they had appeared to him up until that time.

Fig. 7. Constellation sheet of Leonardo and family, Windsor No. 12276 v. Courtesy of Her Majesty Queen Elizabeth II.

Leonardo himself, we recognize as the fine-looking youth near the signature *I dei* (*nomine*). He has the largest sized profile. Next to him in size is the profile of what I take to be that of his Uncle Francesco, looking like a young Greek hero. The father's face appears then, next in size, very strongly drawn and looking like the profile on a coin of a Roman emperor. (Indeed, Clark in his catalogue of the Windsor drawings characterizes several similar faces as being of "the Roman emperor type".)

The very old man, above, with a profile similar to that of Leonardo's father, must have been the grandfather Antonio da Vinci, who died around 1465. Next to that face is one of a very old woman, presumably Leonardo's grandmother Lucia. Next to her, is a younger woman, of hateful mien, who appears to be shouting. This could have been Leonardo's Aunt Violante, who may well have shouted at him when he was a child – thus inspiring fear (inducing a trauma). Leonardo, in a note in Manuscript 2038 *Bibliothèque Nationale* 17 v. tells us

> old women should be represented as bold with swift passionate movements like the infernal furies, and these movements should seem quicker in the arms and hands than in the legs.

If I had the space I could follow the further development of these varied familial features through several other passages in Leonardo's creative phantasy. The hateful face of Violante, for example, (Figure 8) becomes that of Nemesis, the Fury. Circumstances came to force this face back into Leonardo's conscious mind in connection with the assassination of Giuliano de'Medici in 1478. Later on, as I have shown you, it appeared again in the doodle (Figure 2) for the battle scene on a sheet done around 1505, just after Leonardo's father's death. In this final sheet the father has at last been crowned with a wreath, as the Renaissance Florentines of the time crowned their funeral wax masks. Leonardo's father dead, the artist was at last legally emancipated after a long period of familial servitude. At that time Leonardo also had been sued by four of his half brothers concerning the inheritance from the family estates of which they intended to deprive him. This was a time of great frustration for him, due to the lack of love on the part of some of his relatives, as well perhaps as to some feeling of hurt pride.

The sublimations of Leonardo's self-preservative aggressive instincts, as demonstrated here, were about to be objectified in his great battle scene. Here then is found a close-up of the doodle with the same trau-

Fig. 8. left: Nemesis or a fury, from Scipio relief in Louvre; Right: Nemesis from bust of Giuliano de Medici. Courtesy of the National Gallery of Art, Washington, D.C.

matic image, in back of the horse derived from his Aunt Violante of his early childhood.

Still earlier it had appeared as Nemesis, goddess of revenge, on the cuirass of the bust of Giuliano de Medici. Giuliano's face was done by Verrocchio I believe from a death mask presumably made when Giuliano had just been murdered. This bust is now in the National Gallery of Art in Washington.

In the final rendition of the central action in the Battle mural, the *Fight for the Standard*, as it is called, it appeared for the last time as a surrealistic demonic face in the very center of the composition. Thus we may identify Nemesis, the goddess of revenge, whose ancient Tuscan name, which must have been known to Leonardo, was *Mania*. At that time also, returning to his earlier unfinished picture of the *Adoration of*

Fig. 9. Right background of *Adoration of the Magi with Battle-scene* (here dated ca. 1504) with pointing hand. Courtesy of the Uffizi Gallery.

the Magi (Figure 9) Leonardo inserted a record of the central action of his battle scene in the background, using a pointing hand to indicate the importance of the fact that killing and warfare draw us away from the loving reverence for the blessed Mother and her Divine Child. From that time on Leonardo seems to have lost his interest in creating new weapons, having seen warfare at its worst while in the service of Cesare Borgia.

In one place in Leonardo's word lists we have a grouping of words suggesting that insanity is derived from having a spell or bewitching (a 'hexing') placed upon one. Could Leonardo have felt that his Aunt Violante had bewitched him? Certainly there could hardly be a better primitive explanation for a harmful neurotic complex derived from some traumatic event than that one had been placed under a spell. That Leonardo escaped being bewitched or damaged by this subliminal image, to my mind, is due to the fact that he was able to objectify himself both in his word lists and artistic creations. The most progressive practice of psychoanalysis today employs (following a suggestion made by Pastor Pfister of Zurich [15]) objectification through art. This is that form of art therapy where words fail but images reveal. This type of art therapy is that exercise in which, I believe, Leonardo first showed the way. It has been possible to show that Leonardo probably gained some idea of the process of psychotherapy from reading the works of certain philosophers

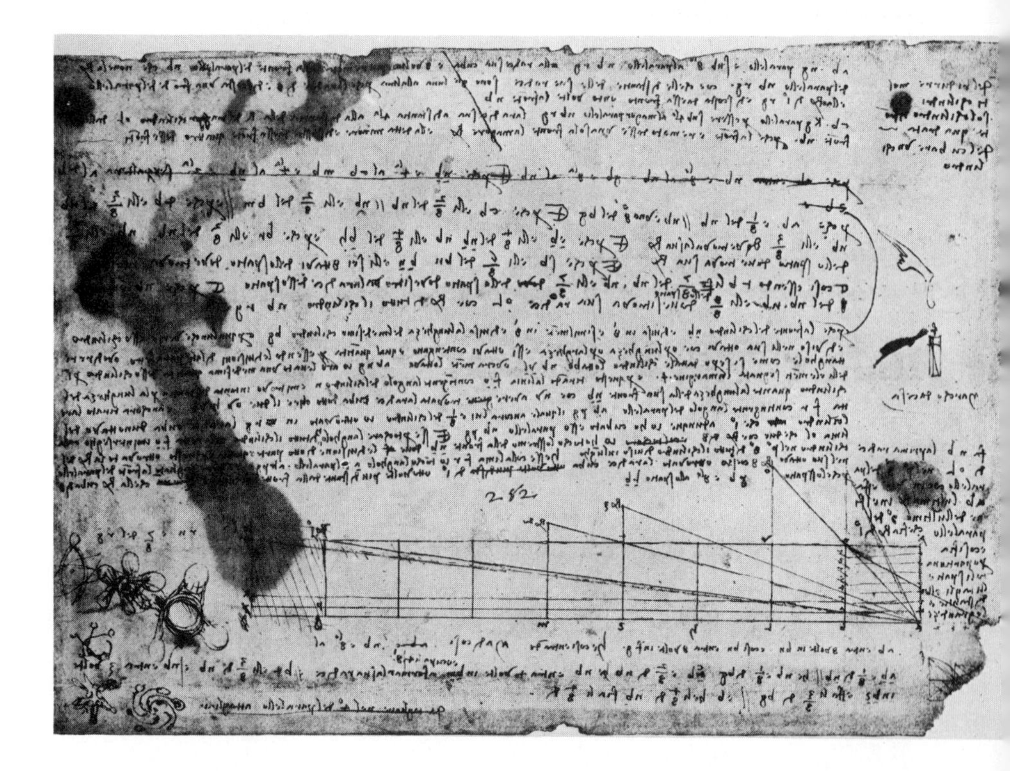

Fig. 10. Hand pointing to Leonardo's solution of the problem of cubing the cylinder from *Codex Atlanticus* folio 303 r.b. Reproduced from Collection of the Library of Congress.

and medical writers such as Celsus, then being revived from antiquity; but that is another story.

In conclusion, I believe that I have been able to demonstrate, with many of his notes and some 325 illustrations, how Leonardo's stream-of-consciousness functioned as he brought up his subliminal images in the form of innumerable scientific and artistic *invenzioni* and so sublimated not only his procreative but also his self-preservative instincts or drives, thus turning his aggressive and amatory desires into sicially acceptable acts. If there were place here I could take you farther along the road to what I have called 'psychosynthesis' (a term borrowed from the Swedish psychiatrist Poul Bjerre), which is what Jung called 'individuation'. By this I mean that Leonardo discovered, in his search for the definition of his own soul, how to integrate the egoistic and social factors in his own make-up. This process I have demonstrated by charting his use of the pointing hands and other indicators. These he employed throughout his works whenever he found a solution to his problems in both art and science, like this one from folio 303 r.b. of the *Codex Atlanticus* (Figure 10). Arranged in chronological order upon a single chart these indicators provide, I believe, a reasonably accurate picture of the essential characteristics of Leonardo's life-style such as, I hope, he would find fairly acceptable if he were here to criticize my efforts.

Maryland

BIBLIOGRAPHY

[1] S. Freud, *Eine Kindheitserinnerung des Leonardo da Vinci*, Leipzig/Wien 1923, p. 53ff.
[2] *Ibid.*, p. 26ff.
[3] *Ibid.*, p. 16.
[4] E. M. McCurdy (ed.), *Notebooks of Leonardo da Vinci*, New York 1955.
[5] R. S. Stites and M. E. Stites, *Sublimations of Leonardo da Vinci, With a Translation of the Codex Trivulzianus* (co-transl. by P. B. Castiglione), Smithsonian Institution Press, Washington, 1970, plate Folio 43 recto.
[6] C. Pedretti, *Leonardo da Vinci: Fragments at Windsor Castle*, London 1957, plate 19.
[7] S. Freud, op cit., p. 9.
[8] K. Clark, *Leonardo da Vinci. An Account of his Development as an Artist*, Baltimore 1961, p. 20.
[9] A. Marinoni, *Gli Appunti Grammaticali e Lessicali di Leonardo da Vinci*, 2 vols, Milan 1944–1955.

[10] C. G. Jung, *Studies in Word Association. Experiments in the Diagnosis of Psychopathological Conditions Carried on at the Psychiatric Clinic of the University of Zurich* (transl. by M. D. Eder), New York 1919.
[11] W. James, *Principles of Psychology*, Chapter 2, New York 1893.
[12] O. Diethelm, *Treatment in Psychiatry*, 3rd ed., 1955.
[13] H. Poincaré, *The Creative Process*, New York 1955.
[14] M. Herzfeld, 'Bemerkungen zu einem Skizzenblatt Leonardo's', *Mitteilungen der Gesellschaft fuer vervielfaeltigende Kunst* **1** (1928) 1–7.
[15] O. R. Pfister, *Expressionism in Art. Its Psychological and Biological Basis* (transl. by B. Low), London 1922.

RAYMOND J. SEEGER

ON THE PHYSICAL INSIGHTS OF LEONARDO DA VINCI

Leonardo da Vinci can be regarded as a person from several points of view: his vocation as an artist, his profession as an engineer (essentially military, including civil engineering in those days), his avocation as an amateur of nature. The resulting views, however, are not independent. Leonardo, therefore, must be looked upon as a whole person; indeed, he must be continually re-viewed by each generation, individually by each person.

Nevertheless, my concern here is to consider Leonardo primarily as a physical scientist. Unfortunately, considerable confusion exists among the various writers commenting on this phase of his activity. At one end of the spectrum of opinion, Clark, an art critic, claims, "He [Leonardo] was essentially a scientist and a mathematician" [1]. Hart, a science historian, also boasts that he was "a truly great man of science" [2]. At the other extreme Randall, a philosopher, maintains, "Leonardo was not himself a scientist in the sense in which he and his contemporaries understood science – or in any sense in which anybody else has understood science" [3]. Truesdell, a mathematician, notes that "for the historical initiate he [Leonardo] must be dismissed as a curious artist who learned a bit of science" [4a]. I agree with his conservative judgment that "Leonardo did not discover *everything*,... he did discover *something*" [4b].

Some of the confusion stems from differing popular notions as to the very meaning of the term science. The question is how to translate the Latin word *scientia*, which means generally knowledge, in the context, to be sure, of knowledge as understood in the period under consideration. I cannot subscribe to interpretations of certain art historians such as Ackerman, who surmises, "Leonardo meant theory", and Gombrich, who regards Leonardo "as a theoretical physicist" [5], and certainly not to Hart's dictum that Leonardo was "completely scientific as in the modern sense of the term" [2]. Such a priori misconceptions give a false credibility to logical deductions from them. For example, Ackerman concludes, "It [perspective] was invented primarily to rationalize visual

R. J. Seeger and R. S. Cohen (eds.), Philosophical Foundations of Science, 41–53.

data rather than to achieve realism.... Every object within the space is relative to every other object and to the observer by a fixed proportion" [5]. Nowadays, of course, we identify three kinds of perspective, dependent upon size, color, and distinctness which is a function of the ambient atmosphere. Clark reminds us that it was, indeed, "luminosity, the feeling for atmosphere, which distinguishes all Leonardo's genuine works from that of his pupils" [1] (cf. chiaroscuro).

The confusion arises not only in translating the word *scientia* but also in current conceptions, or rather misconceptions, of science itself. Randall, for instance, is cited as saying that "science is systematic and methodical thought" [5], hence it is "a cooperative inquiry" [3]. On the basis of such a definition, Marinoni rightly concludes that Leonardo is obviously not a scientist. The conception itself, however, is hardly adequate, inasmuch as science is necessarily based upon the bedrock of experience; it may, indeed, be quite individualistic. Gombrich speaks of "Leonardo's original plan to make the science of water a deductive science, starting from first principles like Euclid's Elements. These principles from which Leonardo started he found, of course, in the traditions of Aristotelian science" viz., "systematic inventory of the world" [5]. Here we have a rather common attempt to fit everyone into either a Platonic or Aristotelian mold. It would, indeed, be strange if an unschooled person like Leonardo, who was a persona non grata to the Platonic Florentines, as well as to the clerical Aristotelians, should become imbued with the traditional philosophy of that period.

As a theoretical physicist, I shall confine my present remarks to those areas of physical science in which Leonardo was apparently interested, viz., mechanics and hydrodynamics. The complementary biological aspects of his broad interests will be considered by Belt, a physician.

Let us begin with some popular misconceptions about motion itself. I do not refer here to an obvious misstatement like that of Clark, namely, "under the influence of Archimedes, questions about the first principles of dynamics..." [1]. Archimedes contributed only to statics, nothing at all to dynamics. I have in mind rather an interpretation like Hart's:

> Nothing whatever can be moved by itself but its motion is effected by another. This other is the force.... We have here, at one and the same time, a conception of the term force which is completely valid today – and, implicit in it, we have also the principle of inertia [2].

The idea expressed here is essentially Aristotelian – not at all Newtonian.

Keele claims further, "He [Leonardo] finds that action equals reaction, Newton's third law of motion. This law Leonardo demonstrates by experiments" [5]. He cites as an illustration the rebound of a ball bouncing from a wall being almost the same as its initial fall. According to Newton's third law, however, the reaction at the wall would be equal to the action there even if the rebound were less. Gombrich goes so far as to suggest: "This imaginary law [oblique rebound] which you might compare with the preservation of energy..." [5]. Leonardo actually failed to comprehend even so simple a phenomenon as tension. Consider a four-pound weight on the end of the rope balanced by another four-pound weight on the other end, the rope itself passing over two pulleys at the same height; Leonardo wrongly indicates the tension at the middle of the rope to be eight pounds. The confusion, therefore, is owing both to indefiniteness of the traditional terms translated and to misconceptions about the scientific concepts used.

With respect to dynamic phenomena, Leonardo, I believe, can be credited with at least three significant contributions: (1) He was the first person to study a body moving down an inclined plane; he noted that its speed at the bottom is independent of the height of the plane, and that its time of fall is proportional to the plane's length. (2) What is more scientifically important was his insight as to the independence of simultaneous motions. Observing the zig-zag path of a bouncing ball he realized that the time of its passage was the same as if the ball had been thrown directly with the same initial speed. (He noted, too, that the angle of reflection of a ball is equal to its angle of incidence.) (3) Finally, Leonardo approached the phenomenon of friction empirically:

> The friction made by the same weight will be of equal resistance as the beginning of its movement, although the contact may be of different breadths or lengths... Friction produces double the amount of effort if the weight be doubled. [6]

He also noted reduction of friction by round objects inserted between sliding surfaces. What we today call the coefficient of friction he indicated to be 0.25, which is true only for wood sliding on wood. These unexpected conclusions are not obvious; they surprised the French when announced independently in 1696 by Amontons. Despite the absence of any recorded measurements by Leonardo, they could only have resulted from quantitative experience.

Leonardo had a life-long interest in statics, both practical and theoretical. For example, he understood the physical characteristics of a lever, including the behavior of the so-called 'potential arm' for a bent lever. I do not believe, however, that he had any conception of the role of a product like force times lever arm – what we nowadays call the moment of force. For a balanced lever (Figure 1) with a load L, effort E,

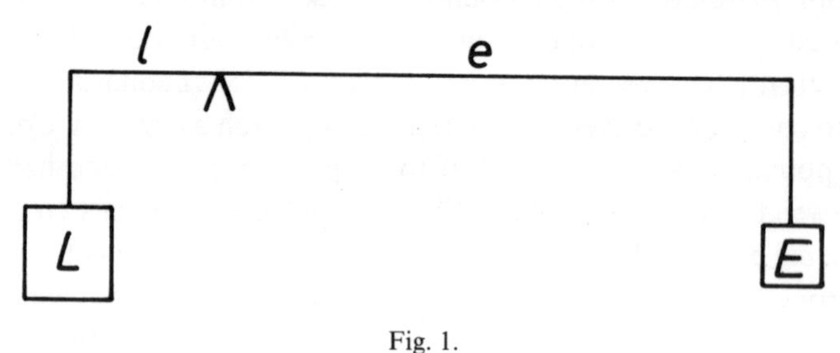

Fig. 1.

and lever arms l and e, respectively, there had long been experientially known the analogical relation that

$$\frac{L}{E}=\frac{e}{l}.$$

Nevertheless the fact that $Ll=Ee$ was not yet a commonplace for any such a proportion.

Leonardo studied also the forces needed to support a uniform beam at any two different positions. Moreover, he utilized this understanding of balancing to determine the center of gravity of a pyramid (about one-fourth up from the base to the vertex) – probably by balancing the pyramid in different positions, at least, in his mind or in his drawings. He was also the first to describe a catenary curve and to propose a model of it for study; he described the phenomenon of buckling. His interest in balance per se is shown further in his study of the equilibrium of different liquids in connecting vessels. He was quite familiar with the operation of the Archimedean screw.

Professional historians of science are generally wont to stress the so-called principle of virtual velocities. Hart, for example, speaks of it as

> the Aristotelian principle that tells us that the same motive power that moves a body slowly must move a lighter body more quickly, the velocities produced being inversely proportional to the weights. Implied in this is the principle of work [2].

For an unbalanced lever (see Figure 2)

$$\frac{L}{E}=\frac{D_E}{D_L}.$$

Such a relationship, however, by no means implies that scholars deduced the relation (obvious to us):

$$LD_L=ED_E.$$

This product, the physical concept of work, indeed, was not introduced formally until about the middle of the 19th century, after engineers had become concerned with its use as the measure of thermodynamic efficiency. In the 18th century the natural philosophers had been primarily interested in celestial mechanics where the concept of work at best can be regarded possibly as a mathematical function convenient for analysis.

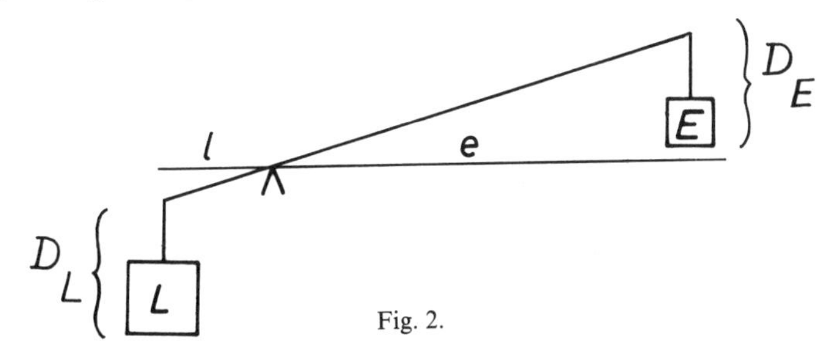

Fig. 2.

These considerations of statics raise questions as to the roles of mathematics and of measurement. In no sense can Leonardo be regarded as a mathematician. In this connection I am not bothered about his errors in adding fractions. More significantly, Leonardo showed no interest whatever in the exciting development of algebra in his own age. What is more, in geometry he disregarded the traditional limitation of construction by straight edge and compass; he was quite content with numerical approximations. His celebrated boast, "Let no man who is not a mathematician read the elements of my work" is a warning for only a very restricted class of so-called 'mathematicians'. There is only one mathematical aspect that truly fascinated him, viz., proportion. As Clark argues,

> If man was the measure of all things, physically perfect man was surely the measure of all beauty, and his proportion must in some way be reducible to mathematical terms and correspond to those abstracts perfections, the square, the circle, and the golden section [1].

In this connection, we are all familiar with Vitruvius' man [7], which became Leonardo's trade mark after his Venice sojourn in 1485–90. Luca Pacioli, the link between Piero della Francesca and Leonardo in this matter, instructed him about Euclid's summary statements about proportion. Their joint interest is acknowledged in the *Divina Proportione* (written in 1498 and published in 1509). Truesdell, it should be noted, regards Leonardo's proportionality as a term covering increase generally – not strictly proportional increments. Clark reminds us that "This union of art and mathematics is far from our own way of thinking, but it was fundamental to the Renaissance. It was the basis of perspective," i.e., "scientific representation of receding figures in space" [1].

The general public is often prone to look upon the handling of any quantities as being essentially mathematical. In many cases, however, it is the quantitative (measurement) aspect itself alone which is the more significant. Hart emphasizes Leonardo's recommendation that "in order best to understand a happening or phenomenon or observation, it should be capable of measurement" [2]. Clark refers to Leonardo's "predominant interest in the human figure" [1], including its measurement. Truesdell, however, cautions that Leonardo usually tells what to measure but has left no record of actual measurements. Nevertheless, to me, Leonardo's pragmatic interest in concrete measurements and in the mathematics associated with them is of paramount importance. Certainly he should not be identified in any sense with the abstract Neoplatonists of his time.

Leonardo's investigations in hydrodynamics are of much greater scientific significance than his interests in mechanics. First of all, we note his definitely experiential approach. As suggested by Clark, it probably arose out of an aesthetic appeal; his "love of curves was instinctive" [1], beginning with the hair of the Angel in Andrea del Verrocchio's *Baptism*. Clark concludes, "Of all Leonardo's interests the most continuous and obsessive was that of water" [1]. Notes Gombrich, "None of these subjects... reoccurs with greater persistence than the subject of movement in water and air"; indeed, the "sheer bulk of his writings on that subject testifies to the key position it held in his mind" [5]. Pater reflects, "Two ideas were especially fixed in him, as reflected of things that touched his brain in childhood beyond the measure of other impressions – the smiling of women and the motion of great waters" [8]. You may recall the 1513 Windsor drawing of a melancholy old man meditating on studies of water (alongside Leonardo's notes on plaited hair and swirling water). Even *La Gioconda* has a network of dividing streams in the background cirque of fantastic rocks.

Coupled with Leonardo's keen interest in phenomena themselves was his accuracy of observation. Clark notes, "His superhuman quickness of eye has allowed him to fasten on the decorative aspects of the subject, since confirmed by spark photography" (he does not cite evidence of the latter); "he had been drawing horses all his life with a matchless power of observation" [1].

Clark further remarks, "The quality of his drawings [water] shows a passion with no relation to practical life.... Some of his studies of swirling water are amongst the most direct expressions of his sense of form, springing from the same mysterious source of his love of knots and tendrils" [1].

We immediately think of the portrait of *Genevra de' Cenci* and of the appearance of his assistant Andrea Salaino with his curled and wavy hair. Randall also emphasizes, "His artist's interest in the particular and the concrete which implied his careful, precise, and accurate observation is carried further by his inordinate curiosity into an analytic study of the factors involved" [3].

Gombrich, on the other hand, believes, "Leonardo's drawing is not a snapshot of water falling on water but a very elaborate diagram of his ideas on the subject" [5]. Ackerman, too, would have us believe that "Leonardo's drawings of hydraulic action, which for all their apparent naturalism, are logical deductions, not landscape vignettes.... Just as the artist sought to produce not a particular man but man himself, the paragon, so the scientist sought to describe not a particular instance of nature, but her general laws. This was *scientia*; it was concerned with the ideal more than the actual" [5]. This view, I believe, is somewhat anachronistic, inherent in the author's own conception of science. He further comments, "His extraordinary studies of drapery have never been surpassed as a demonstration of the power of light to suggest form." On the other hand, he admits, "The botanical sketches represent particular plants, rather than species... a sense of the object's biological function as well as its form." I am more inclined to agree with Truesdell's judicious remark about Leonardo's famous waterfall sketch; "Other artists have drawn waterfalls as beautiful, but few as accurately as this, which like Leonardo's anatomical studies, must be the result of many observations" [6]. Leonardo's persistent search for a typical (average) natural form, I believe, may account for the apparent lack of precision which modern photographers obtain automatically. Leonardo's own advice in this regard is noteworthy, "Remember when discoursing about water to use first experience and then reason." He concluded, "Experience is not at fault; it is only our judgment that is in error promising itself from experience things which are not within her power" [6a].

Not only did Leonardo observe but he recorded his observations. Says Randall, "This talent for scientific illustration is indeed impressive" [3] (cf. Dürer and Vesalius in this regard). Clark speaks of Leonardo "throughout life an untiring draftsman." One may recall his bread-and-butter occupation of map making for Cesare Borgia. The art critic Panofsky concludes, "The invention of a method of recording observations through revealing drawings, an invention made by the great universal artists at the turn of the 15th century, deserves to rank with the invention of the telescope and the microscope in the 17th century and the camera in the 19th century" [3]. Truesdell comments, "Graphic art indeed bears close upon the rise of science" [4b]. I agree with the engineer Reti, "There can be no doubt, however, that most of Leonardo's technical ideas are grounded in firm and accurate experience, even if the historical records are meager" [5].

Now, to be sure, experiential observation and recording are an important part of science, but by no means the whole of science. Science, it seems to me, involves four essential elements: I – and you – experience something with nature as a source; I – and you – view these findings somehow with imagination as our inspiration; I – and you – deduce something else with reason as a guide; I – and you – check these findings with nature as a re-source. Leonardo's activity does not comprehend the whole gamut of these characteristics. He exhibits little interest in generalized laws and in imaginative theories. In hydrodynamics, however, he possibly showed genuine insights, e.g., with respect to flow continuity, vortices, and waves.

Leonardo noted leaves being carried along by streams, thereby tracing streamlines. He himself suspended small grass seeds as flow indicators in a glass-walled tank of water. He watched the progress of atmospheric currents in heavy rain, driven like dust by the wind. As Truesdell notes, Leonardo was probably the first individual to call attention to the similarity of the flow of water (hydrodynamics) and that of air (aerodynamics) [6a]. We can agree with Clark's conclusion that Leonardo was fascinated by the grace of continuity in "flowing gestures, flowing draperies, curling or rippling hair" [1]. Leonardo, however, went further; he found that "a river of uniform depth will have a more rapid flow at the narrower section than at the wider, to the extent that the greater width surpasses the lesser" [5]. As we say today, the speed of flow is inversely proportional to the cross-sectional area; in other words, the product of the speed and the cross-sectional area remains constant – an expression of the so-called principle of material continuity. In this connection Gombrich emphasizes the power of Leonardo's thought as opposed to his mere observation or measurement. Leonardo's physical insight, I believe, is probably a combination of both, with the caveat that measurement in his case does not necessarily imply precision but rather a rough estimate of magnitudes. In his observation of flowing streams he noticed also their geological erosion.

Furthermore, the whole question of turbulence genuinely interested Leonardo. He observed that eddies form with any abrupt expansion in the flow, as in wakes; that entrapped air in foam retards the flow of water. By placing sticks in the water Leonardo observed that eddies decrease in size as they are farther from the shore. What is more, he recognized that in vortical fluid motion the speed of the circulating fluid is inversely proportional to the length of its path, or, as we would now say, the product of the speed and that length is constant – essentially the modern principle of circulation (cf. the contrast of a spinning wheel which has the speed at any point proportional to the length of the circular path).

Leonardo investigated also the discharge of fluids from orifices of different shapes. He noted that the efflux is proportional to the speed. In this connection the hydraulic engineer Rouse points out an error in Leonardo's conclusion that the efflux is proportional to the head of flow. Nevertheless, he concludes, "Leonardo was still the country's first hydraulician" [9].

From the standpoint of physical theory probably Leonardo's most notable contribution was his insight as to the nature of a wave. Noting that straws remain in their place on disturbed fluid surfaces, that ripples move forward while the water does not do so, Leonardo concluded, "It is motion, not matter that they transport" [4b]. He watched the reflection of water waves; he listened to sound echoes, the multiple echoes from steps. He noted that the angle of reflection here, too, is the same as that of incidence, as in the case of a ball bouncing from a wall. He concluded that sound and light behave similarly. He observed the motion of a straw on a lute string when a similar string of a nearby lute was plucked – the physically important phenomenon now termed resonance.

Most important, Leonardo appreciated the phenomenon that we nowadays call the principle of superposition of waves. For example, he sketched ripples spreading from two stones dropped simultaneously in water, he observed how the wavelets penetrate each other without at all disturbing the progressing waves. Leonardo noted the pattern produced likewise by the reflection of wave incident upon a wall. As Truesdell notes, Leonardo was probably the first to conclude that "a small disturbance of the surface of the water pursues its course unaffected by other small disturbances" [4b]. Leonardo referred even to the dust collected in heaps on vibrating boards, in other words, the phenomenon of so-called standing waves. All in all, Leonardo exhibited broad and perceptive observations of the flow of water.

What then is the legacy of Leonardo? Merely a list of unfinished works? Is it true, as some moderns would have us believe, that he was a universal failure? Great stock is taken in Vasari's emphasis that on his deathbed Leonardo protested to the king of France that "he had offended God and mankind by not working at his art as he should have done." We are all familiar with the reputed derogatory comment of Leo X, "Alas, this man will never do anything for he begins thinking about finishing the work before he even starts it" [10], i.e., when Leonardo was worrying about how to distill oils and herbs in order to prepare a varnish. The "recent view that Leonardo was the victim of psychological frustrations, which prevented him from concentrating on a single theme" [1] is a modern variation of an old subject.

Then, too, what about his unpublished writings? Truesdell concludes regretfully, "His greatest achievement, the record of his observations of fluid motion, lay unfinished and unpublished in his disorderly notes and remained unknown – a legend of what he might have done" [4b]. I cast my eyes over my own scattered notes, some of which were made at random, many of them copied, most of them unfinished. In the future, if someone were to come upon these private notes (and hopes) of mine, would he make the same judgment about me as is nowadays common about Leonardo? There is no question that Leonardo was neither a writer, nor a teacher – indeed, he did not even have an academic appointment. On the other hand, he was a successful courtier; he is reputed even to have had a courtier's charm. Certainly he was not at all a recluse. I am, therefore, inclined to agree with Hart that "it seems most inconceivable that during Leonardo's active lifetime there was no informed knowledge of what he was doing scientifically" [2]. Certainly, we know that Leonardo influenced scholars such as Cardano, Benedetti, and Baldi.

For me, however, what is more important is Leonardo's unwavering enthusiasm. He had a determined curiosity; first of all, curiosity about nature itself. Clark remarks, "We are aware of landscapes as something full of movement, light moving over the hills, wind stirring the leaves of trees, water flowing and falling in cascades" [1]. His curiosity comprehended not only what he observed, but also how it could possibly be explained. Clark comments, "He was not content to record how a thing worked: he wished to find out why" [1]. Accordingly, we find Leonardo making a study of light striking a sphere in order to satisfy his empirical passion for chiaroscuro. (Anatomy, indeed, was secondary in importance to perspective during his studious sojourn in Florence.) Ackerman surmises, "His observations had a way of leading him to more observations rather than to principles or laws of nature, and he seldom arrived at the mathematical description of phenomena" [5]. As Randall notes, "In practice, Leonardo always became fascinated by some particular problem – he had no interest in working any systematic body of knowledge" [3]. Clark concludes, "In general Leonardo's scientific [?] researches were undertaken for their own sakes" [1]. I myself am of the opinion that Leonardo exhibited only a limited serendipity. He did not behave quite like Leacock's Lord Ronald who "flung himself upon his horse and rode madly off in all directions." On the contrary, serendipity throughout his whole life was restricted to following only certain kinds of bypaths.

Leonardo's enthusiasm was illustrated by his imaginative use. Reti proclaims "He sought to make the mechanics of man into a useful science" [5]. Hart admires his technical skills: "He had an extraordinary flair for the design of mechanical contrivances" [2]. Randall admits his technological concern: "Leonardo can fairly claim to belong not to the line of scientists but to the noble line of inventors" [3]. We recall various practical projects with which he was associated – canals such as those at Vigevano, Limellini, Ivrea, Milan, Arno, as well as drainage projects like the Pontine marshes at Rome and the one at Piombino. Most of us, of course, regard despairingly Leonardo's futile attempts to conquer flight through the air, his dream obsession which turned into a nightmare – or, to change the metaphor, his Achilles heel – in view of his wrong emphasis upon the birdlike role to be played by the flapping wings of a flyer's harness. Incidentally, Leonardo's notebooks, "the stone which the builders refused is become the head stone," is a veritable treasure house of knowledge about the technology of his age.

Leonardo's enthusiasm was not confined to his determined curiosity about phenomena, nor even to his imaginative use of them, but he himself was entrapped by mysterious beauty. Pater, indeed, cautions us, "How great for the artist may be the danger of overmuch science" [8]. Gombrich notes that Leonardo was "convinced that a painter must be more than 'merely an eye'" [5]. Clark emphasizes the vital relation Leonardo saw between the grace of flowing hair and the energy of flowing water. Pater goes so far as to speak of Leonardo's emotional response to "the solemn effects of moving water" [8]. The pointing finger and the enigmatic smile are Leonardo's signature, indicating a "sense of mystery, that disturbing quality which comes first to mind at the mention of Leonardo's name" [1]. In his later life Leonardo became more and more concerned about nature beyond human control – and man's comprehension. "His studies of hydrodynamics suggest a power of water beyond human control" [1]. "Leonardo was fascinated and obsessed with catastrophic destruction" ... "some interfusion of the extremes of beauty and terror" [8] – due possibly to "contemporaneous emotions" [1] associated with current predictions of another Deluge. Visions of such a Deluge became related in Leonardo's mind to his studies of moving water – undoubtedly "the most personal in the whole range of his work" [1] (Dürer, too, was influenced even by the thought of a coming Deluge.)

In reviewing Leonardo's enthusiastic explorings, one is impressed not so much by his scientific shortcomings or even by his physical insights, but rather by his integrated personality – his artistic vocation and his engineering profession and his avocation as an amateur of nature.

U.S. National Science Foundation

BIBLIOGRAPHY

[1] Kenneth Clark, *Leonardo da Vinci*, Penguin, Harmondsworth, 1968.
[2] Ivor B. Hart, *The World of Leonardo da Vinci*, Kelley, Clifton, N.J., 1970.
[3] John Herman Randall, Jr., 'The Place of Leonardo da Vinci in the Emergence of Modern Thought', in *Roots of Scientific Thought* (ed. by P. P. Wiener and A. Noland), Basic Books, New York, 1957.
[4] C. Truesdell, 'Leonardo da Vinci', *The Johns Hopkins Magazine* **XVIII** (1967) 29–41 (a); *Essays in the History of Mechanics*, Springer-Verlag, Berlin, 1968 (b).
[5] C. D. O'Malley (ed.), *Leonardo's Legacy: An International Symposium*, University of California Press, Berkeley, Calif., 1968.
[6] Edward MacCurdy (ed.), *Notebooks of Leonardo da Vinci*, George Braziller, New York, 1955 (a); *The Mind of Leonardo da Vinci*, Dodd Mead, New York, 1939 (b).
[7] Pollio Vitruvius, *On Architecture* (transl. by F. Granger), 2 vols, Heinemann, London, 1931 and 1934.
[8] Walter Pater, *Renaissance*, New American Library, New York, 1959.
[9] Hunter Rouse and Simon Ince, *History of Hydraulics*, Dover, New York, 1957.
[10] Giorgio Vasari, *Lives of the Artists* (transl. by G. Bull), Penguin, Harmondsworth, 1970.

BERN DIBNER

LEONARDO AS MILITARY ENGINEER

The recent discovery of two codexes written in the hand of Leonardo da Vinci, comprising some 700 pages, was a most welcome event to the historian of technology and to the expanding number of Vincian scholars. The 'loss' of these important texts was, in a way, a blessing because when exposed, they showed Leonardo's notes and drawings in a state of freshness no longer evident [1] in most of the nearly 6000 pages that have survived of the four or more times that number that Leonardo entrusted to his favorite pupil, Francesco Melzi, at the time of his death in 1519. The delicate drawings in pencil and chalk would have been obliterated by casual handling during a period of less concern with the conservation of such material. Since their rediscovery, the Madrid drawings have been copied in color and their life and original brightness extended by photography and printing.

When one is given the opportunity of examining the new find, one hunts among the hundreds of illustrations for items in one's chosen area of interest. Only about a dozen references are made in the *Madrid Codexes* to items of a military nature. Of these, two describe Leonardo's efforts to increase the range of a cannon ball by linking it with rockets, and three show match-lock guns. Leonardo's match-lock had already been known from a drawing in the *Codex Atlanticus*.[2] This large gathering (like the Atlantic!) of Leonardo's designs and writings, now in Milan, contains most of the notes and investigations that he made in military matters. In the many fields of interest that drew Leonardo, he became a specialist in a wider range of the physical world around him than any other man on record. Among these interests was his concern with the weapons of war. Leonardo hated war as strongly as any rational person has ever hated it, calling it 'bestial madness', but living in the unsettled times in northern Italy at the end of the 1400s, he was exposed to the demands of war and the sweep of invading armies against his people in Tuscany and Lombardy. In 1494 Leonardo watched the invasion under Charles VIII of France, followed in a few years by invading armies from

R. J. Seeger and R. S. Cohen (eds.), Philosophical Foundations of Science, 55–95.

Germany and Spain, and twice again from France. Papal armies joined one force or another or acted on their own in the continued contest between the cities that formed the political structure of what is now Italy.

Added to the unsettling times caused by the many wars, we must remember that Leonardo lived in a most restless period in the history of Europe. His birth coincided with the Gutenberg invention, one of the most revolutionary developments in all human history. Further, Vasco da Gama had sailed around Africa connecting Europe with India by ship, and Columbus and Leonardo's fellow Florentine, Amerigo Vespucci, had opened new and vast continents to the adventurous sailors and soldiers of Europe.

1. His military service

Leonardo's interest in things military predated the wars of invasion. Having finished his apprenticeship with the master Verrocchio in Florence, Leonardo was ready for larger undertakings than the painting, sculpting, architecture and casting of bronze on which he had trained. He therefore prepared a letter which was sent to Ludovico Sforza, Duke of Milan. Leonardo was then about 30 years old and whether he was retained as a painter, musician or military engineer is not as important as the contents of his letter of application. It reads:

Having, My Most Illustrious Lord, seen and now sufficiently considered the proofs of those who consider themselves masters and designers of instruments of war, and that the design and operation of said instruments is not different from those in common use, I will endeavor without injury to anyone to make myself understood by your Excellency, making known my own secrets, and offering thereafter at your pleasure, and at the proper time, to put into effect all those things which for brevity are in part noted below – and many more, according to the exigencies of the different cases.

I can construct bridges very light and strong, and capable of easy transportation, and with them pursue or on occasion flee from the enemy, and still others safe and capable of resisting fire and attack, and easy and convenient to place and remove; and methods of burning and destroying those of the enemy.

I know how, in a place under siege, to remove the water from the moats and make infinite bridges, trellis work, ladders and other instruments suitable to the said purposes.

Also, if on account of the height of the ditches, or of the strength of the position and the situation, it is impossible in the siege to make use of bombardment, I have means of destroying every fortress or other fortification if it is not built on stone.

I have also means of making cannon easy and convenient to carry, and with them throw out stones similar to a tempest; and with the smoke from them cause great fear to the enemy, to his grave damage and confusion.

And if it should happen at sea, I have the means of constructing many instruments capable of offense and defense, and vessels which will offer resistance to the attack of the largest cannon, powder and fumes.

Also, I have means by tunnels and secret and tortuous passages, made without any noise, to reach a certain and designated point; even if it is necessary to pass under ditches or some river.

Also, I will make covered wagons, secure and indestructible, which entering with their artillery among the enemy will break up the largest body of armed men. And behind these can follow infantry unharmed and without any opposition.

Also, if the necessity occurs, I will make cannon, mortars and field-pieces of beautiful and useful shapes, different from those in common use.

Where cannon cannot be used, I will contrive mangonels, dart throwers and machines for throwing fire, and other instruments of admirable efficiency and not in common use; and in short, according as the case may be, I will contrive various and infinite apparatus for offense and defense.

In times of peace I believe that I can give satisfaction equal to any other in architecture, in designing public and private edifices, and in conducting water from one place to another.

Also, I can undertake sculpture in marble, in bronze or in terra cotta; similarly in painting, that which it is possible to do I can do as well as any other, whoever it may be.

Furthermore, it will be possible to start work on the bronze horse, which will be to the immortal glory and eternal honor of the happy memory of your father, My Lord, and of the illustrious House of Sforza.

And if to anyone the above mentioned things seem impossible or impracticable, I offer myself in readiness to make a trial of them in your park or in such place as may please your Excellency; to whom, as humbly as I possibly can, I commend myself.[3]

This recital of some 36 abilities which were put at the disposal of the Duke can also be used as a measure of Leonardo's interests at that time. Thirty of the claims are of technical nature and six are in the field of art. The studio of Verrocchio was decidedly non-military; one therefore concludes that Leonardo was drawn by the mechanical and engineering stimulus that military devices offered – the extension of power in firearms, the inventive challenge of new combinations of weapon elements for both offensive and defensive application. One sees this dependence on increased effectiveness of the machine in his note:

Instrumental and mechanical science is the noblest and above all others the most useful, seeing that by means of it all animated bodies which have movement perform all their actions; and the origin of these movements is at the center of their gravity...[4]

Leonardo continued as military engineer to the court of Sforza for some 17 years. In this time he was commissioned to design, cast and erect a colossal statue of the founder of the house of Sforza, Francesco, an equestrian statue to be 22 feet high, twice as large as any other equestrian statue then attempted. After many years of labor, a clay model had been erected in one of the squares in Milan, but it fell victim to the Gascon archers of the invading French armies. It was only from the Madrid manuscript that we finally learned what the actual shape and posture of this great undertaking was to have been.

Leonardo's second task as a military engineer was for Cesare Borgia, the adventurous son of Pope Alexander VI, whose armies swept through Faenza, Urbino, Cesena, Imola, Forli and the other cities of Central Italy. The military maps that Leonardo prepared in this campaign continue as the most faithful records of the area of that time (Figure 1). Borgia's appointment of Leonardo and the high regard in which he was held, are indicated by the commission of appointment:

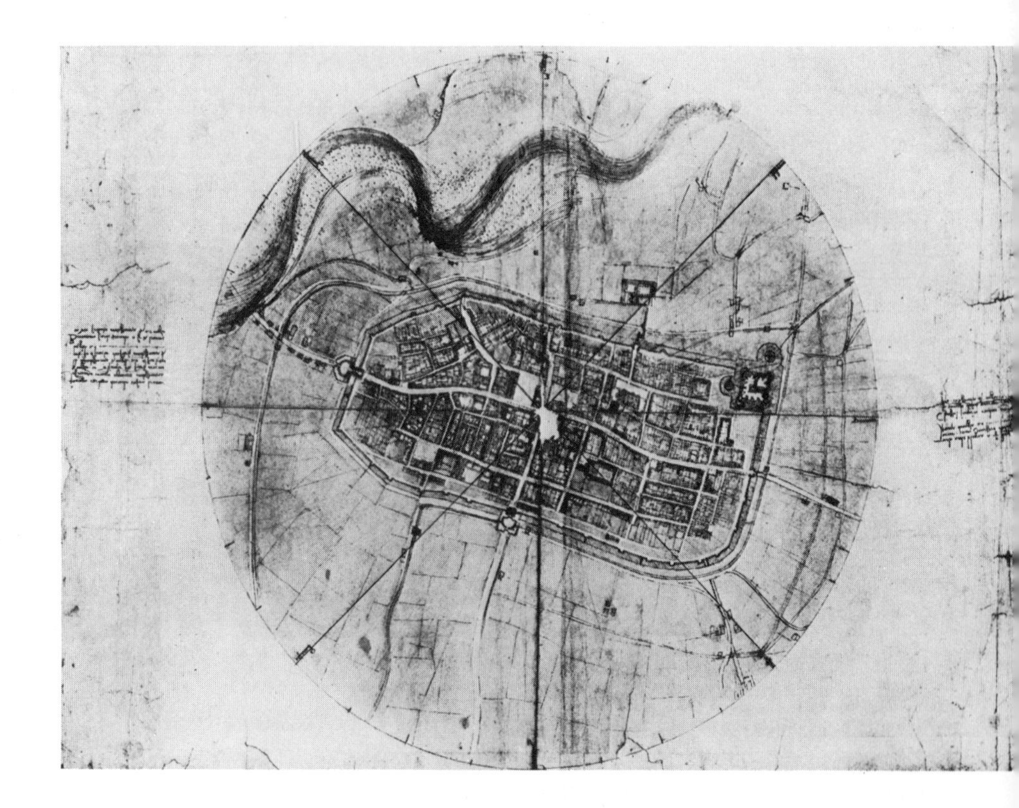

Fig. 1.

To all our Lieutenants, Castellans, Captains and Condottieri, officers and subjects, who receive this notice, it is commanded and ordered, in favor of our very excellent and favorite private Architect and Engineer General Leonardo Vinci, bearer of these letters present, whom we have charged to visit the places and the fortresses of our estates, in order to have them put back into condition according to the exigencies and in his own decision, to guarantee him freedom of movement, all exemption from payment of public taxes; to give a friendly welcome to him and his companies; to help him in visiting, measuring and estimating the localities as much as he pleases; and to this end to place at his disposal the men he requires; to give him all aid and assistance that he requires and finally carry out our wish, that for the work to be done in our domains all engineers shall confer with him and conform to his orders.[5]

Borgia's campaigns were short-lived and Leonardo, in 1503, returned to Florence where he was with the troops at the siege before Pisa.

2. Military engines

The military engineering works of Leonardo can be divided into three categories: forts and structures for defense, engines for offense, and general equipment for construction, excavation, mining, ordnance and for the services of supply. Forts and military architecture will not be covered in this study. Engines for offensive action, greater fury, firepower and speed form the subject of many drawings by Leonardo. He designed engines of greater mobility, increased the caliber of cannon, improved the firing means and was mindful of more rapid fire such as from multiple-barrel guns. He introduced mechanical means to replace manual ones wherever possible, especially in repetitive motion. The roster of classic authors on military subjects which he compiled in his *Madrid Codex* included Vitruvius, Vegetius, and his contemporaries, Roberto Valturio and Francesco di Giorgio Martini.

One of his better known military devices that quickly catches the imagination for its combination of cleverness and horror is a chariot with whirling scythes drawn in several variations and shown in the Paris manuscripts (Figure 2). Horse-drawn vehicles have their wheels geared to whirling scythes that moved through a field of men, mowing them down. Leonardo derived this idea possibly from Lucretius, who described scythes attached to chariot wheels. To these Leonardo added improvements such as means to raise the blades when the machine moved among friendly troops.

On the same page showing the scythed chariots, one finds what has

Fig. 2.

often been called a tank. It is really intended as an armored, self-propelled vehicle but it lacks the basic feature of the modern tank, its tread. However, for its time, a highly ingenious device is shown. The left view shows the top removed. The four drive-wheels are rotated by two cranks, each crank turning two lantern pinions which engage the gear-points on the wheel faces. The whole vehicle is topped by a protective cover of planks, with a cap on top. Two horizontal firing slots are provided, so that men inside may fire as the 'tank' moves, like a kind of land-ship reminiscent of the *Monitor*.[6]

These take the place of the elephants. One may tilt with them. One may hold bellows in them to spread terror among the horses of the enemy, and one may put carabineers in them to break up every company.[7]

Leonardo derived his military information from reading the references in technological texts in contemporary military literature and through his contact with military men, who, like himself, were in the employ of the Duke of Milan. Through his father, Piero d'Antonio, notary to the Signoria in Florence, he had contact with the military men of that strong city-state. From reading the works of Archimedes, Pliny and, especially, Vitruvius, he learned the military methods and technology of the Roman period. He transcribed notes from his readings into his own notebooks. Among his contemporaries, he was influenced by Francesco di Giorgio Martini (1425–1506)[8] and by Roberto Valturio (1413–1483). Valturio was instructor to Malatesta, Tyrant of Rimini, and his book *De Re Militari* was first published in 1472 and again in 1483. In it are shown the traditional weapons with variations of the pike and crossbow, the scaling ladders, battering rams, wheeled shelters for protection against dropped stones and burning pitch. It all represents much power expenditure, but not too much ingenuity. An older variation on his work is that of Flavius Vegetius, a Roman engineer, whose treatise on military tactics was reissued through the centuries and constantly modernized. Both books were illustrated by crude, naive woodcuts, in contrast to the clear, striking draftsmanship of Leonardo.

In Leonardo's time, as in our own, canals were used to divert rivers, flood valleys, drown out armies and destroy communications and supplies. Leonardo, in his notes, has several recommendations on methods of using canals for the diversion of rivers as an offensive weapon. He also made several proposals for mechanizing canal-digging.

A test of Leonardo's inclination to use mechanical power instead of muscle power, even in an age of cheap and abundant labor, is shown by his design for a canal-cutting engine (Figure 3). The engine moves upon three beams; the main central beam is fitted with a bolt and nut arrangement that causes the entire engine to be dragged forward as the work progresses and the two outrigger rails balance the engine. In the center, a vertical mast supports two swinging booms of unequal length; hooks and ropes hang from the end of each boom. Digging proceeds on two levels, an arc of buckets being filled by men with picks and shovels. As the lower buckets fill, the lower boom picks them up and swings them to either side where they are dumped. As the upper tier of buckets fills, the upper boom swings the buckets out to the side forming a second row of

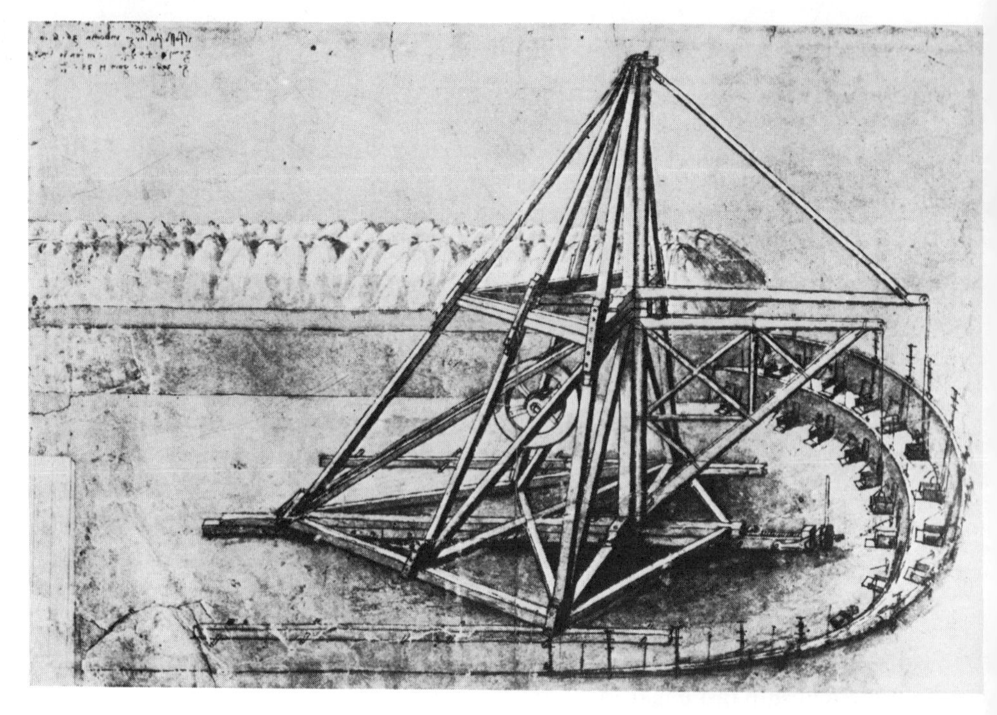

Fig. 3.

mounds shown in the drawing's background. When the arcs of excavation have been completed, the engine is drawn forward and the excavating operation is repeated. Among the interesting things to note is the method Leonardo uses to brace the booms. The upper brace is in tension, therefore no additional support is provided. The lower boom is under compression; it therefore is provided with elaborate bracing.

Another example of Leonardo's use of intricate mechanical devices in order to obtain some desired end is the machine shown in Figure 4, for translating reciprocating into rotating motion. This is shown in a drawing so clear and instructive as to hardly require detailed explanation. An assembly view is shown at the left, an exploded view of all the elements is shown at the right. Motion begins with the vertical reciprocating lever at the right which imparts its swinging motion to the

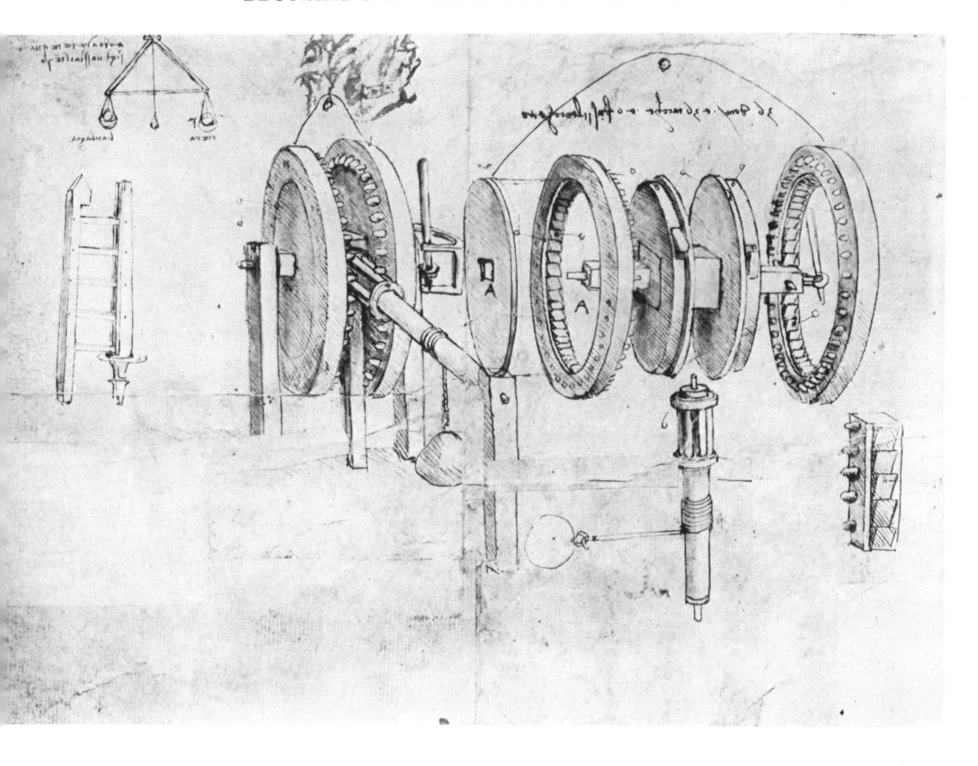

Fig. 4.

square shaft. Upon this shaft are fixed two wheels which have pawls set into their perimeters, one facing up, the other down. These pawls engage ratchets fixed into the inner circumferences of two rings that rotate on the pawl-bearing discs. The ratchets, like the pawls, face up and down. The opposite sides of the rings are studded with gear teeth and both gear faces engage a common lantern gear that connects with a final shaft. When the initial lever is rocked, one pawl moves one gear, the other becoming passive. When the lever moves back, the other pawl grips the gear and turns the shaft, the first becoming passive. No matter which gear is activated, the lantern gear still will be moved counterclockwise and, as shown, the suspended weight will rise or equivalent work will be done. This mechanism can be used as a windlass, or a hoist, or in the several schemes Leonardo had for automotive vehicles. One application of this ingenious mechanism is for the propulsion of a boat as sketched by Leonardo in Figure 5. We see the ship's hull and the motive paddles. To

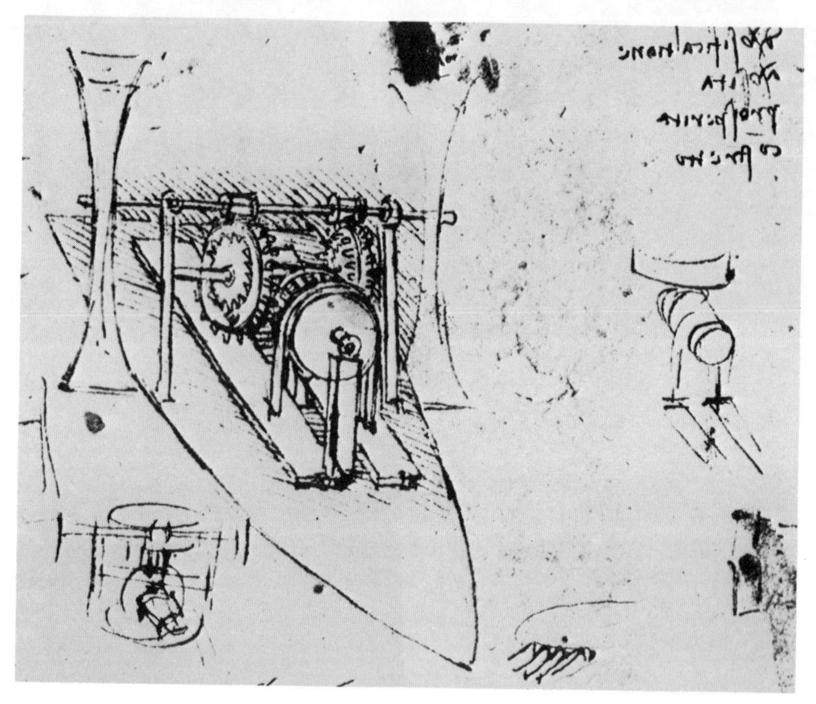

Fig. 5.

the deck are hinged two treadles connected by a belt riding over a lantern gear. The teeth of this engage the teeth of a crown gear forming the inner disc of such a translating mechanism. The outer rim is ratcheted and a pawl engages the ratchet; it gives rotary motion to a ring having teeth on its outer rim. The teeth engage a gear on the paddle-wheel shaft. There is a matching mechanism on the ship's port side; the two in combination give the paddle shaft and paddles continuous rotary motion as the treadles move up and down. The germ of the idea is seen rapidly sketched by Leonardo at the right.[9]

For assault troops, Leonardo designed light portable bridges that were to be moved in elements and then assembled at the river's edge or at the moat. Similarly hinged, swivelled and cantilever bridges, quickly constructed and easily swung across the river or canal, are sketched in his notebooks.[10]

Fortifications are massive, of polygonal design, with high lookout towers and bastioned angles. Leonardo advocated the use of brushwork matting to absorb the shock of oncoming shot. In another sketch (Figure 6), Leonardo devised a means for dislodging scaling ladders leaned

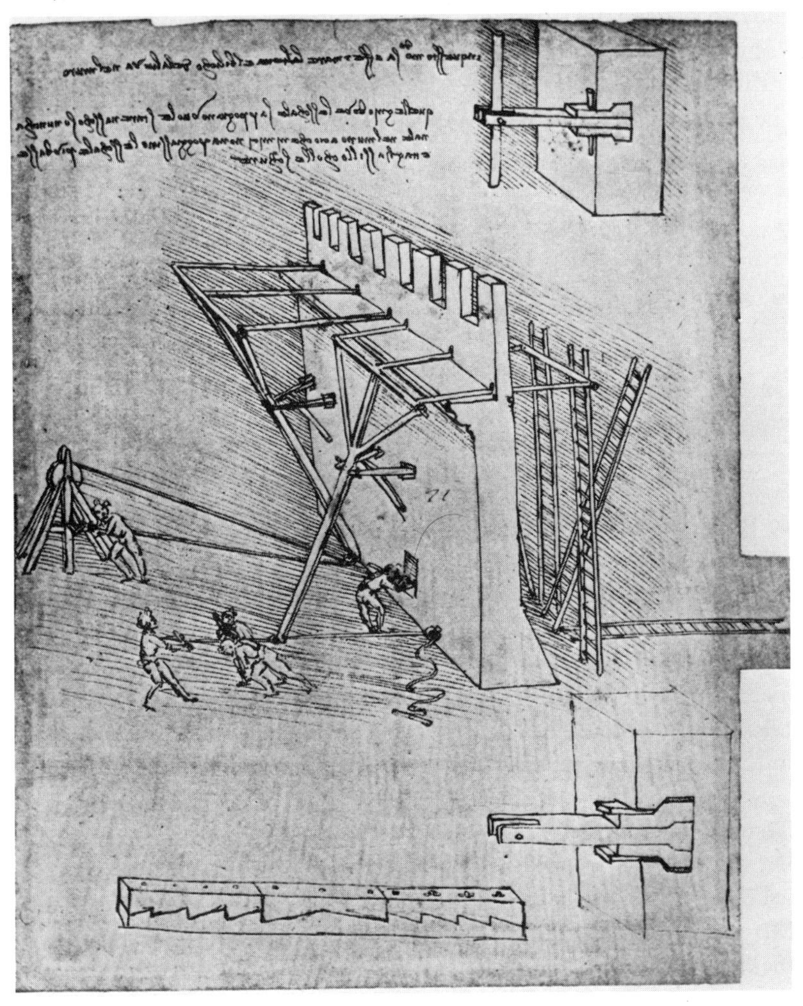

Fig. 6.

by an enemy against the top parapet. Vertical pivoted bars are pulled away from the inner wall by defense troops. In so doing, the tops of the bars push outwardly and thereby force out a hidden horizontal bar set flush with the escarpment. This bar moves out, forcing the ladders away and pitching them beyond their footing.

As indicated in his letter to the Duke of Milan, Leonardo had advanced notions on mining and countermining. This was one way of breeching a moat-protected fort without direct assault. Against such mining, he recommended detection of the sound of the sappers at work by placing a drum above the area where the enemy is suspected of mining. Upon the drum are placed a pair of dice,

> and when your are near the place where the mining is, the dice will jump up a little on the drum, through the blow given underground in digging out the earth.[11]

Further,

> if you do not wish to make the test with the rod in order to discover a mine, go every hour with a light above each hole, and when you come to the hole which is above the mine the light will be immediately extinguished.[12]

Once located, Leonardo proposed methods of burning out the tunnel supports.

3. Ordnance

In Renaissance editions of Vegetius and Valturio, reference to artillery is less then expected, although cannon had been in use for more than a century. That was so, because the methods of cannon manufacture, laying and firing were all quite primitive; their combined result was noisy but not very destructive. The imaginative Leonardo perceived a method of refining the manufacturing methods of boring by reducing the tolerances in order to increase the fire-power and range (Figure 7). Advice to the founders of bronze cannon is contained in Manuscript M, 54 v.,

> That part of the bronze is most compressed within its mould which is most liquid. And that is most liquid which is hottest, and that is hottest which comes first out of the furnace. One ought therefore always to make first in the casting that part of the cannon which has to receive the powder before that which has to contain the muzzle.

Mortars or bombards of short-barrel and firing high-trajectories, rather then low-trajectory long-barrel cannon, were more frequently used at that time because they were more easily made. Not that they were

Fig. 7.

necessarily small, for the cannon founders competed in outdoing one another in cannon size. At the siege of Constantinople in 1453, cannon balls weighing up to 1800 pounds were fired, and the Venetians used guns firing even heavier shot. Projectiles were mostly stone, of rough calibration and wearing on the bore. Leonardo proposed many refinements (Figure 8).

In artillery, Leonardo's drawings and notes indicate several classes of equipment: mortars, cannon, great slings, catapults and crossbows, field guns and several types of rapid-fire guns. Some pieces of artillery were conceived of such great dimensions as to be impossible to transport over the terrain and roads of his day, but they did stir the imagination and sketching such monsters must have been pleasant to his fertile imagination.

One great crossbow, using a laminated bow section for maximum flexibility, has some interesting features, as shown in Figure 9. The bow string is drawn back by the worm and gear shown in the lower right hand corner. Two releasing trips are shown, lower left. The upper is spring-pivoted and released by a hammerblow; the lower is tripped by lever action. The bowman, as shown also in the main view, bears down on the bar, and by lifting the encased lever, the string is released. The wheels are mounted to give them a wider, more stable, base and also to reduce road shock. A similar means was later used in canting the spokes of most artillery wheels.

Many of Lenardo's sketches show slings and catapults, using the energy stored in bent and twisted wooden arms, in metal springs and in the torque of twisted ropes by the use of worms and gears, or racks and pinions (Figure 10). Many of these designs appear in older sources, for Leonardo transferred to his notebooks any that were of interest to him.

It was 150 years from that fateful day at Crécy in 1346, when two cannon, attended by twelve English cannoneers, first fired a shot propelled by gunpowder at an enemy in battle. Yet a casual examination of some of Leonardo's sketches would indicate that his concerns were those of an artillerist of the mid-1800s. In some respects he was ahead of even those days. He was, for instance, concerned with good shrapnel design,[13] with breech instead of muzzle loading, with ease and speed of fire, with multiple fire and with advanced gun construction methods.

Shrapnel shot is contained in a leather cover, and a fuse ignited at

Fig. 8.

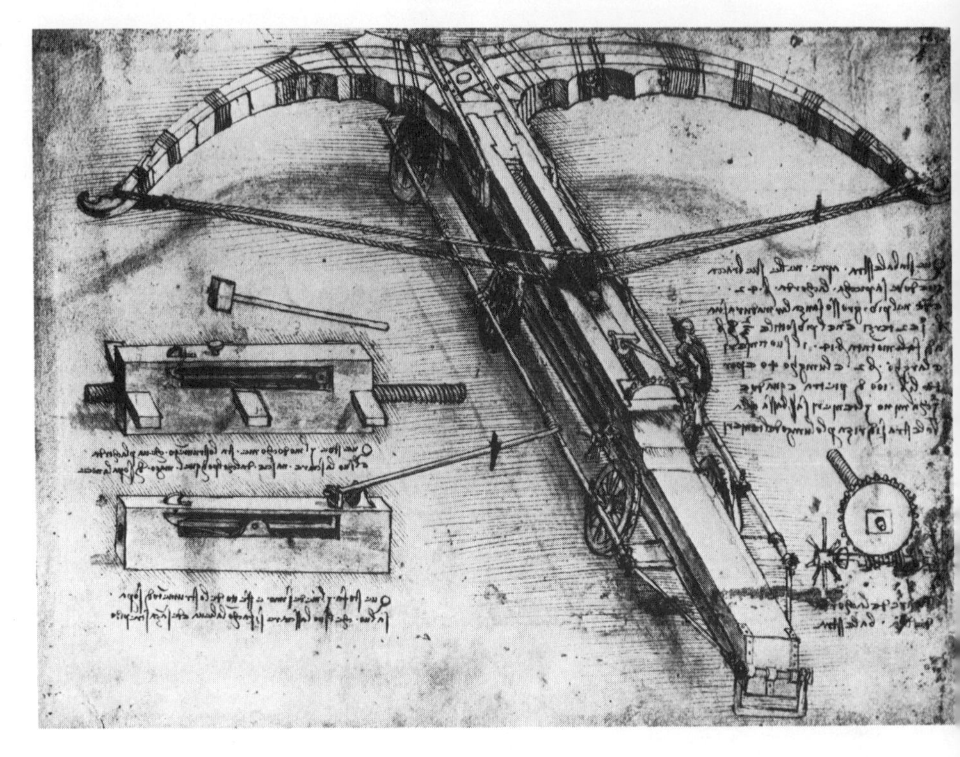

Fig. 9.

firing causes the shell to burst in the air (Figure 11). Another drawing formalizes the barrage to be laid down on a given area outside a defense wall. Figure 12 shows means of assembling gunbarrel sections by means of spanner wrenches and levers. One of the earliest known attempts to use the breech as the loading end of a gun is shown in Figure 13. Note the chamber in the breech block, intended to hold the firing charge.

Although Leonardo was constantly solving problems in weights and balances and in the mechanical advantage of various combinations of pulleys, yet he was not the mathematician to develop mathematical ballistics. That remained for Nicolò Tartaglia (1501–1559) of Brescia and Galileo Galilei (1564–1642) of Florence. His interest in the theory of ballistics tended toward general solutions as can be seen from the following problems he set for himself:

If a bombard hits a mark in a straight line at ten braccia how far will it fire at its greatest distance? And so conversely if it fires three miles at its greatest distance how far will it carry in a straight line? If a bombard fires at different distances with different curves of motion, I ask in what section of its course will the curve attain its greatest height?[14]

If a bombard with four pounds of powder throws a ball weighing four pounds two miles with its maximum power, by how much ought I increase the powder for it to carry four miles?[15]

What difference there is between the movements made upwards or crosswise, or in damp or dry weather, or when it is windy or rainy or with snow falling, either against or across or in the direction of the course of the ball. Where the ball makes most rebounds – upon stones, earth or water. How the smooth ball is swifter than the rough one. Whether the ball revolves in the air or no.[16]

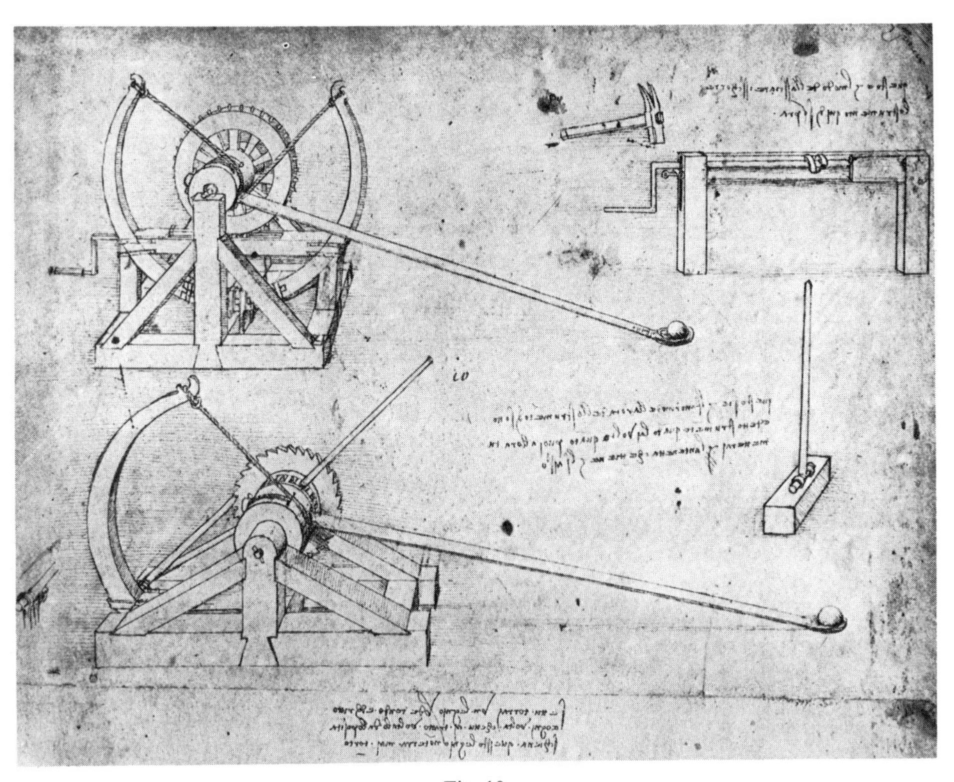

Fig. 10.

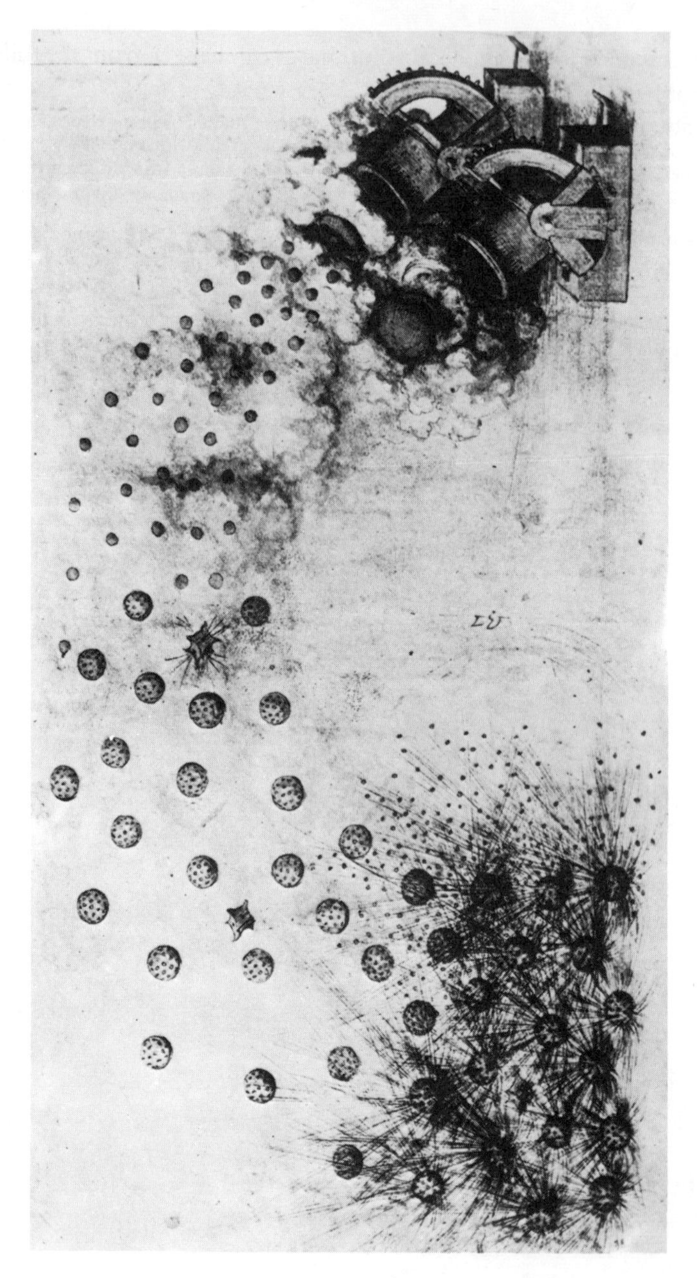

Fig. 11.

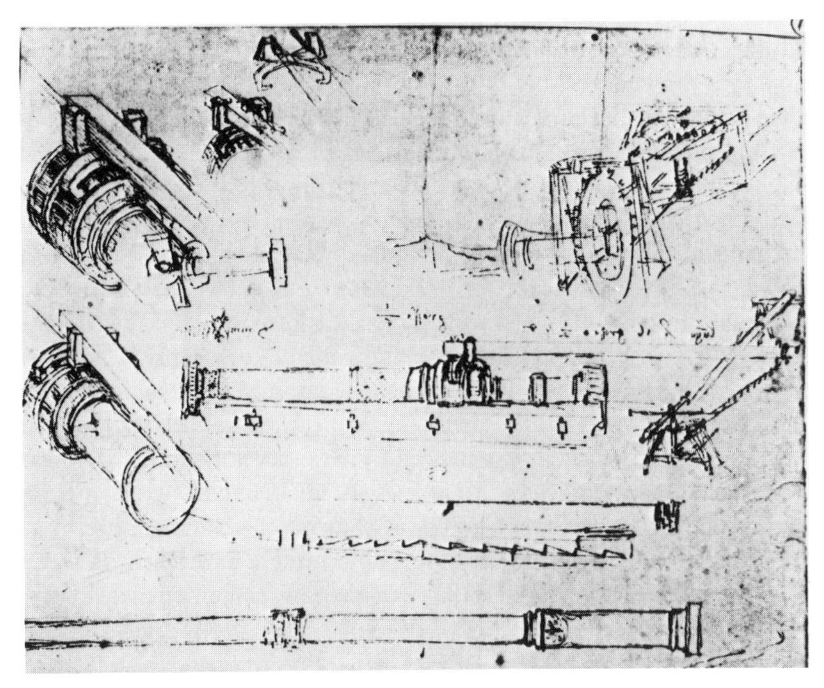

Fig. 12.

Leonardo explained dynamic phenomena on the theory of 'impetus' which meant to him that a moving body continued motion in a straight line. In the matter of a projectile fired from a position of rest in the breech of a gun, he held to the view of Aristotle that the subsequent acceleration after leaving the muzzle was due to the push of air. He also adhered to the view of Albert of Saxony that the trajectory of a projectile was divided into three periods: the initial violent impetus; the secondary, in which natural gravity begins to act; the tertiary, in which gravity and air resistance dominate over the impressed impetus.[17] Cannon, in Leonardo's day, were cast in iron and in bronze. He sensed the increase of air resistance to a projectile with the increase of the projectile's velocity, without determining the exact ratio of this increase. "The air becomes denser before bodies that penetrate it swiftly, acquiring so more or less density as the speed is greater or less." (Manuscript E, 70 v.). In spite of his inadequacies in analytic equipment, Leonardo perceived the parabolic nature (without naming it as such) of ballistic trajectories and made numerous drawings of them.[18]

He recognized the important role of air in influencing the curve, whereas Galileo considered air resistance negligible. Leonardo studied the manageable curves of jets of water issuing from an orifice under varying heads and related these to projectile curves. The mathematical solution of the ballistic curve involving air resistance was provided by Newton in 1687.

The effectiveness of the hand-held gun was recognized in the century 1470–1570. The *arquebus* with its clumsy match-lock evolved into the musket with its various types of firing mechanisms, including the flintlock. To Leonardo goes the credit for the earliest representation of a functioning wheel-lock, as can be seen from the several alternative forms in the sketches in the *Codex Atlanticus*, where a connecting chain from the mainspring to the wheel spindle is shown.[19] This design, however, drew the attention of historians of technology more for its clear showing of a chain-and-sprocket linkage than as a gunpowder ignition mechanism.

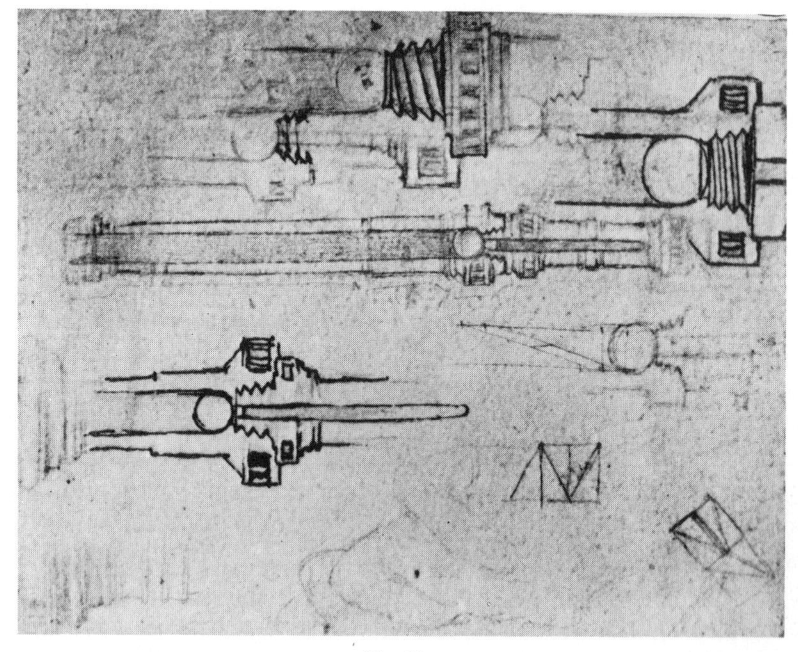

Fig. 13.

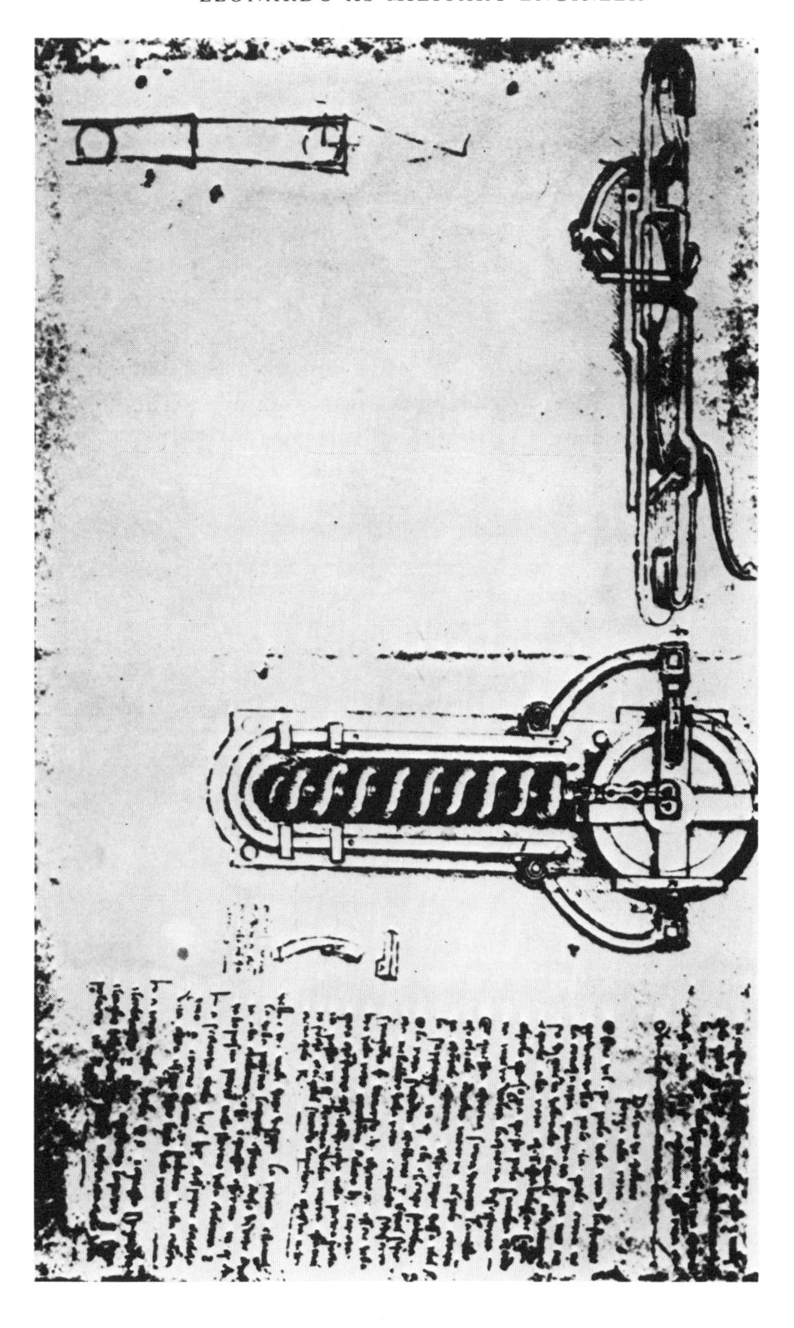

Fig. 14.

Leonardo's sketch (Figure 14) is believed to have been made ca. 1505, but the earliest wheel-lock mechanism is otherwise credited to Nuremberg a decade later. The complex wheel-lock continued in use for two centuries and although more efficient than the flint-lock, complexity and economics lost out to the simpler and cheaper flint-lock until it also was replaced by the percussion musket with the fulminating cap in the mid-1830s and by the magazine rifle of 1890. The wheel-lock principle carries over into the modern flint-and-wheel cigarette lighter.

The two left drawings in Figure 15 show an arrangement for joining gun-barrel sections applied to a threaded breech block. This breech section is boxed into a pivoted quandrant rotated by a crank and pinion.

Fig. 15.

When the breech block is unscrewed, the crank is turned and the breech block tilts, giving the cannoneer access to the chamber.

The use of a complete prefabricated cartridge was to Leonardo as obvious and expedient as was the advantage of breech loading. A combination of both is seen in Figure 16. The cartridge complete with ball,

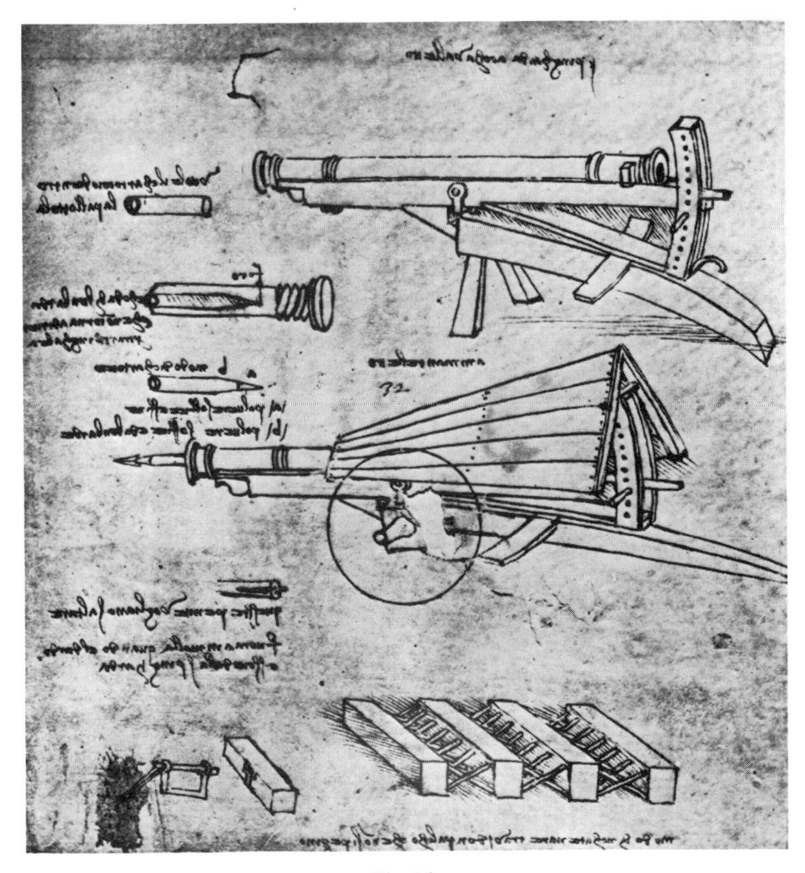

Fig. 16.

powder charge and fine powder primer is shown assembled before a screw-type breech block. A plank cover protects the artillerists as shown in the assembly view. Similar cartridges are recommended for use by mounted carabineers.

See that they are well supplied with guns with a thin single fold of paper filled with powder and the ball within, so that they have only to put it in and set alight. Being thus ready, they will have no need to turn as have the cross-bowmen when they are preparing to load.[20]

Leonardo also has sketches of long projectiles with ogive heads and tail sections formed into longitudinal fins (Figure 17). He intended these fins to give aerial stability to the projectiles; of these four variations are shown in his manuscripts of that period (Figure 18). The stabilizing fins resemble modern air missiles and were intended to extend their range and improve their accuracy. The more intricate dart-shaped projectile had a fuse arrangement with horns extending from the nose. Simpler but sophisticated projectiles showed the ogive nose with streamline body gradually formed into cross guiding vanes. A similar but less practical type of projectile showed interrupted smoothness and was therefore of probably shorter range, induced by turbulence behind the sharp break where body and fins join.

For firing encampments or wooden structures, fireballs are to be shot from bombards.

Fireball worked up: take tow smeared with pitch and turpentine of the second distilling. And when you have made the ball make four or six holes in it as large as the thickness of your arm, and fill these with fine hemp soaked in turpentine of the second distilling and powder for the bombard; then place the ball in the bombard.[21]

Shrapnel is built up so that

this is the most deadly device that exists: when the ball in the centre drops it sets fire to the edges of the other balls, and the ball in the centre bursts and scatters the others which catch fire in such time as is needed to say an Ave Maria, and there is a shell outside which covers everything.[22]

An unusual piece of artillery was designed and termed 'architonitro' by Leonardo (Figure 19). This consists of a cannon that depended on the sudden generation of steam to drive the shot out of the barrel. The breech was built into a basket-like brazier, containing the burning coals. With the shot rammed back, and the breech section sufficiently heated, a small amount of water is injected into what would normally be the powder chamber.

And when consequently the water has fallen out it will descend into the heated part of the machine, and there it will instantly become changed into so much steam that it will seem marvelous, and especially when one sees its fury and hears its roar. This machine has driven a ball weighing one talent six stadia.[23]

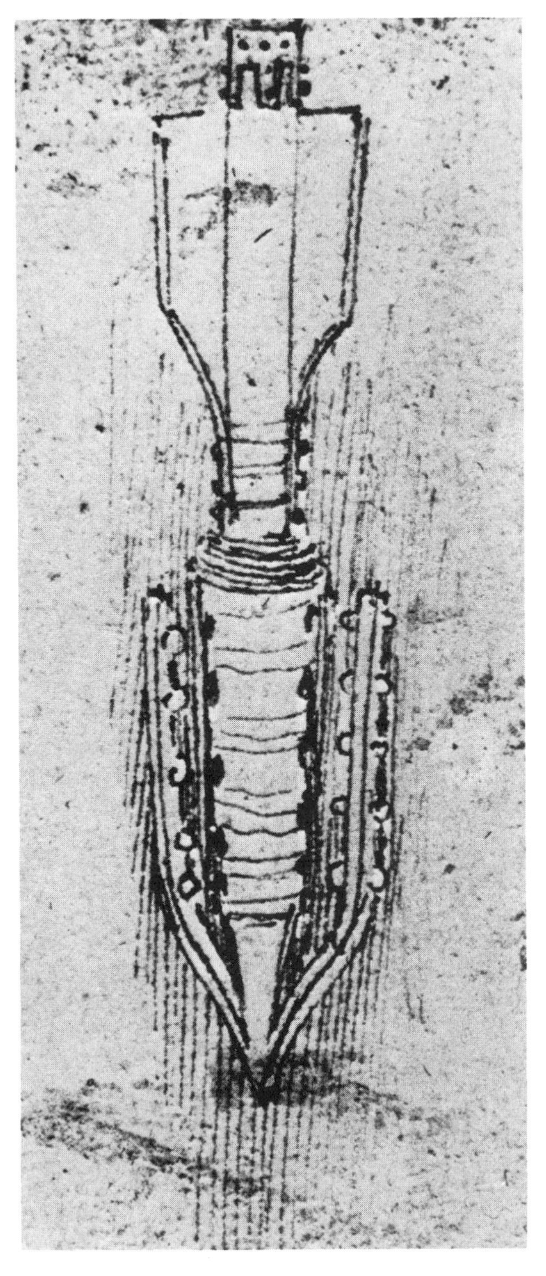

Fig. 17.

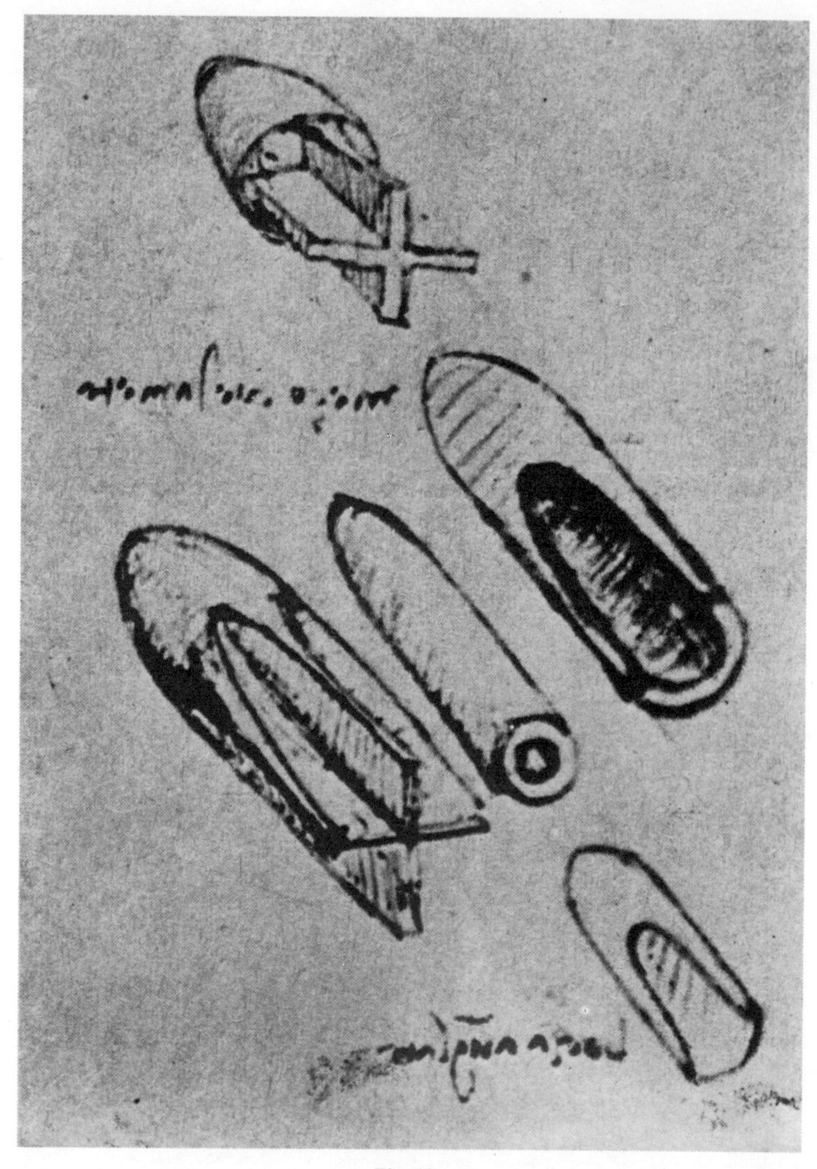

Fig. 18.

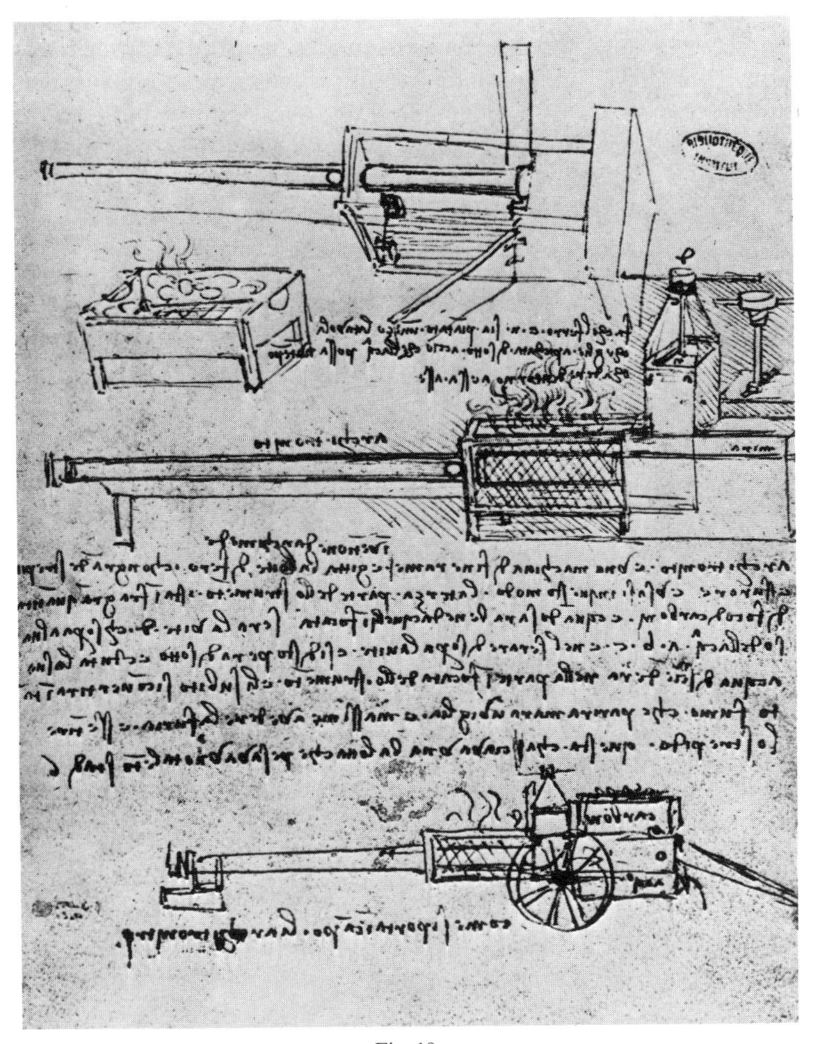

Fig. 19.

This would indicate that such a gun had actually been made. Evidently the rate of fire was a lesser consideration. Putting steam's expansive power to work was novel indeed, in Leonardo's time.

Large cannon were built up by combining forged tapered segmental sections to form the barrel, then binding the segments together by driving metal rings upon the barrel segments, much like driving hoops upon barrel staves, or by winding multiple layers of steel wires. Leonardo's scheme for drawing gun-barrel segments is shown in Figure 20, one of several such proposed methods. Starting with the prime mover, a

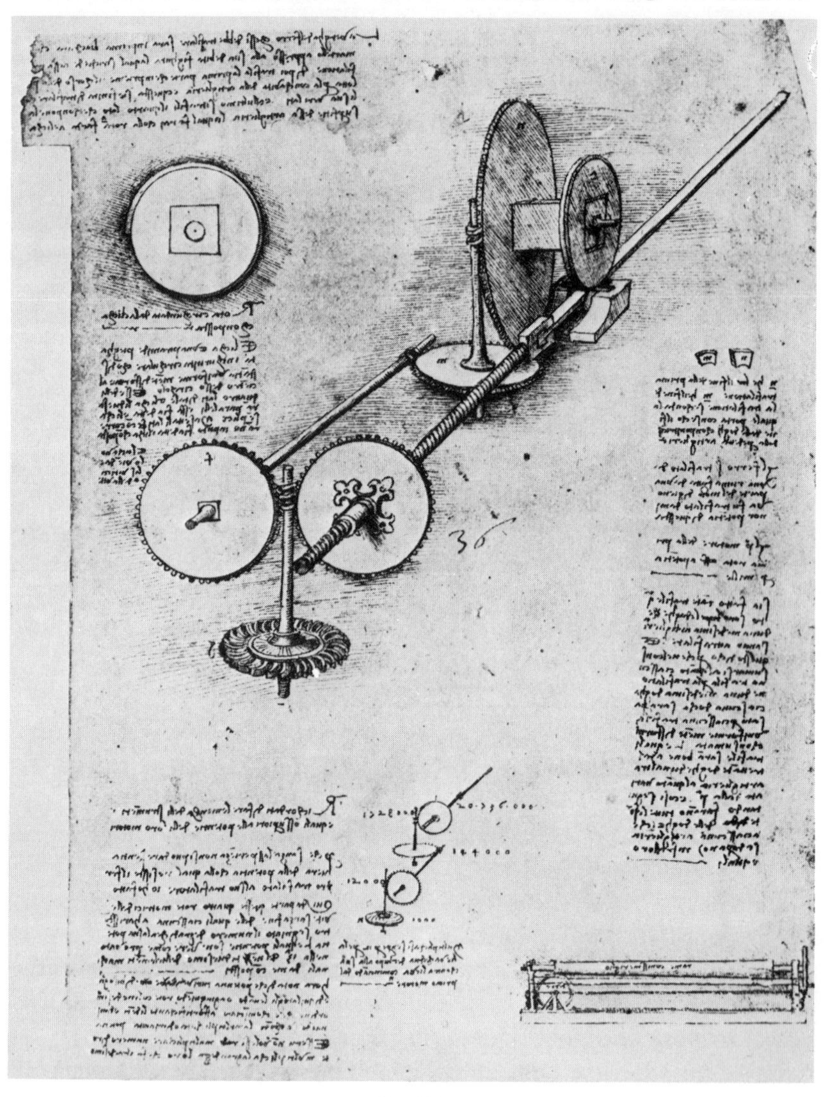

Fig. 20.

water-driven reaction turbine shown at the lower left of the assembly, the vertical power shaft drives the two geared wheels through a common worm. This causes the shaft attached to the left wheel to turn and give power to its own worm. The gear set horizontally is caused to turn, thereby powering another vertical shaft and yet another worm. The engaged large vertical gear in the back is thereby rotated and its power, now greatly multiplied (at a slower turning speed), is passed through the square shaft to the wheel on the right. This wheel is sturdily built and heavily armored by a helicoidal cam shown in the upper left sketch. Its purpose can be seen by following the sequence of operations passing from the prime mover through the gear wheel engaged on the right of its worm. This gear wheel has fastened to it an internally threaded shaft, through which rides a long threaded rod.

As the right gear wheel is turned, the threaded rod is drawn forward, thereby drawing a blank rod after it. This blank rod is forced into a segmental shape, as shown above the column of writing at the right. In addition, because of the heavy cam pressure, the breech end is drawn heavy, the muzzle end light; their inner arc radii must remain the same. In the diagram below the drawing appears the schematic power ratio due to the gearing. Starting with the figure 1000 at the turbine, there appears 12000 at the first gear; 144000 at the horizontal gear; 1228000 at the top big gear, and 20736000 at its shaft. A mechanical advantage ratio of 1:12 is created at each step. Leonardo certainly knew the reduction of forces by friction, even though he does not seem to apply it here. Roller and ball bearings for friction reduction is one of his frequently used design elements.[24]

Some elaborate applications of ball and roller bearings are shown in several devices in the *Madrid Codex* where also the general subject of friction and friction-bearing materials is discussed. One also observes in these new notes his striving for optimum mechanical advantage and with a minimum of friction. The *Madrid Codexes* feature quite a number of mechanisms depending on ball and roller bearings and, to the surprise of many modern designers, with conical bearings. He experimentally determined an average coefficient of friction of one-quarter the weight of a flat-surfaced object.[25] He scoffed at perpetual motion and moved towards the elements of automation that one expects of modern mechanisms. In military affairs he aimed at replacing the vulnerable horse in battle by protected muscle-propelled vehicles. He was conscious of the lack of a motive prime mover (which came in our time with steam, the gas engine and the electric motor).

4. Naval warfare

Living his life inland, as at Florence and Milan, and dying inland (at Amboise in France in 1519) Leonardo yet shared with his contemporaries a strong interest in things pertaining to water and the sea. So valued were his notes and drawings on the motion of water and the flow of rivers, that these were gathered and published more than a century ago.[26] One of his ideas – how a man can remain under water – was used by several writers of military texts in the 1500s and 1600s, and they credited Leonardo with the invention.[27] On the other hand, an even more effective idea was withheld by Leonardo from his own notebook, for the reason he gives:

> How by an appliance many are able to remain for some time under water. How and why I do not describe my method of remaining under water for as long a time as I can remain without food; and this I do not publish or divulge on account of the evil nature of men, who would practice assassinations at the bottom of the sea by breaking the ships in their lowest parts and sinking them together with the crews who are in them.[28]

For the annihilation of enemy shipping Leonardo proposed the use of Greek fire, a weapon first used in the seventh century. This was a combination of sulfur and quicklime, which on touching water was ignited. Another proposal was to

> Throw among the enemy ships with small catapults, chalk, pulverized arsenic and verdigris. All who inhale this powder will be asphyxiated by breathing it, but be careful that the wind be such as not to blow back the fumes, or else cover your nose and mouth with a moist cloth so that the powder fumes cannot penetrate.[29]

Leonardo knew the methods and devices developed by fighting seamen during the several thousand years of naval warfare on the busy Mediterranean. Throwing of burning sulfur and pitch, of mixtures to poison the air, to induce asphyxiation, throwing grease and soap to cause slipping and sliding, catapulting many-pointed caltrops to pierce the feet, these were studied and recommended by him. Then there were the swinging arms ending in cutting scythes to slash enemy rigging, rend sails, and destroy shrouds and stays. Among his own contributions are sub-surface pivoted hooks whose powerful thrust into the under side of an enemy ship is intended to cleave or stave in a ship's bottom. Another device was to cut its way through planks of any thickness and then open the seams. One of the countermeasures that Leonardo proposed for such tactics by an enemy was to build naval ships with double bottoms.

An unmanned attack vessel is described as

> This zepata is good for setting fire to ships which have kept a blockade after having besieged some harbour or other ships in the harbour, and it should be made thus: first, wood a braccio[30] above the water, then tow, then powder as used for a bombard, then tiny faggots and so gradually larger ones; and put iron wires and burning rags on the top; and when you have the wind as you want it direct the rudder. And as the fire... spreads in the ship the bent wires will set fire to the powder and it will do what is necessary. It is also useful for setting fire to bridges at night, but make its sail black.[31]

The camoufleur's touch is here included.

In the *Madrid Codex II*, several pages are given over to navigation and sailing problems in which the complex forces of wind direction and force, motion of the waters and target are stressed.

We learn from Leonardo that the use of poison gas in World War I was not as novel as some of us were once led to believe:

> The Germans, in order to asphyxiate a garrison, use the smoke of feathers, sulfur and realgar, and they make the fumes last seven and eight hours. The chaff of corn also makes fumes which are thick and lasting, as does also dry dung; but cause it to be mixed with sanza, that is with the pulp of crushed olives, or, if you prefer, with the dregs of the oil.[32]

5. Aviation

Leonardo did not, like Franklin, envision the air as a sphere of combat but he was modern enough to make flight a very intense area of thought and design. To him air was a ponderable element to be used for transport, as was land or water. Aviation being so fundamental an element in today's weaponry, the design and thoughts of Leonardo on flight and on rockets become pertinent. This is especially so when it is realized that any reality of flight was so completely non-existent in his day.

Leonardo's notes on the flight of birds and of his proposed flying machines are scattered in many of his notebooks and on odd pages of notes, for the thought of mechanical flight was one of his most persistent. In the *Codice Atlantico* (161 r.), we find:

> A bird is an instrument working according to mathematical law, an instrument which it is within the capacity of man to reproduce with all its movements, but not with a corre-

sponding degree of strength, though it is deficient only in the power of maintaining equilibrium. We may, therefore, say that such an instrument constructed by man is lacking in nothing except the life of the bird, and this life must needs be supplied from that of man.

That his studies and notes on the flight of birds were, like his other studies, eventually to find their orderly way into 120 encyclopedic books, is shown by his note,

I have divided the *Treatise on Birds* into four books; of which the first treats of their flight by beating their wings; the second of flight without beating their wings and with the help of the wind; the third of flight in general, such as that of birds, bats, fishes, animals and insects; the last of the mechanics of this movement.

As the function of each component part of a bird is broken down to show its part in flight, so the kinds of flight are grouped into straight, curved, spiral, circular, falling and the complex combination of these.

In reading Leonardo's notes on the construction of the various versions of his flying machine, and those on the flight of birds, it is difficult to tell to which drawing he refers. His machine, an ornithopter, was intended to follow closely the successful flight of birds he saw constantly about him (Figure 21). Like his studies of different kinds of flight, so his keen sight and patient analysis broke each flight cycle into its elements of motion of beak, head, body, wings, tail and feet. The same detailed breakdown of motion that each one of his airplane elements was to make was similarly studied under conditions of ascent, soaring, in turbulent air and in descent.

The movement of the bird ought always to be above the clouds so that the wing may not be wetted, and in order to survey more country and to escape the danger caused by the turbulence of the winds among the mountain defiles which are always full of gusts and eddies of winds. And if, moreover, the bird should be overturned, you will have plenty of time to turn it back again following the instructions I have given, before it falls down again to the ground.

Clearly, 'the bird' was Leonardo's handiwork. Two sentences from the same notebook:

The descent of the bird will always be by that extremity which is nearest to its center of gravity,

and

When without the help of the wind the bird is stationary in the air without beating its wings in a position of equilibrium, this shows that its center of gravity is identical with the center of its bulk,

indicate an appreciation of the concept of the roving center of pressure

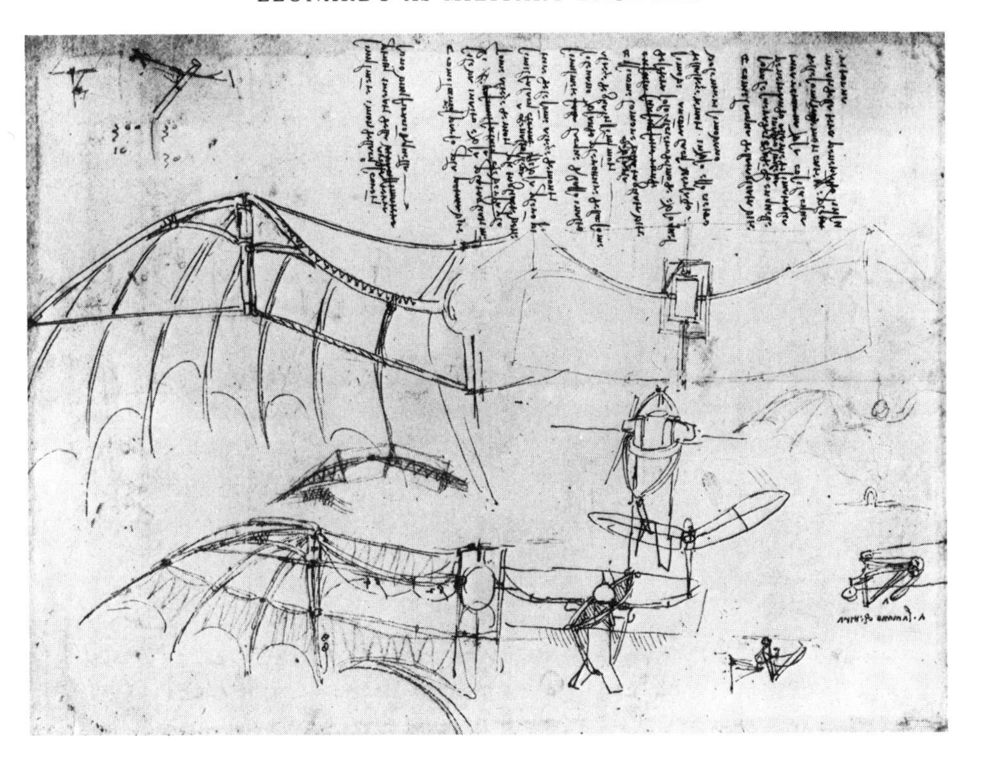

Fig. 21.

or center of position, and its couple, the center of gravity, as well as the resulting interaction as the wind blows or the wings move. This shift is shown further in:

The man in a flying machine must be free from the waist upward, in order to be able to balance himself as he does in a boat, so that his center of gravity and that of his machine may oscillate and change where necessity requires, through a change in the center of its resistance (*Cod. sul Volo*).

Ascent results,

since the wings are swifter to press the air than the air to escape from beneath the wings (then) the air becomes condensed and resists the movement of the wings; and the motive power of these wings by overcoming the resistance of the air (and) raises in a contrary movement to the movement of the wings.

Reduced to this simple process, it remained only to construct wings of sufficient structural strength and lightness. With the necessary harness to provide mobility, a device capable of rising in the air was certain. Such wings and harness were designed and evidently constructed in experimental full-scale models. No record of actual flight exists, but there was much promise.

The great bird will take its first flight upon the back of the Great Swan [33] filling the world with amazement and filling all records with its fame; and it will bring eternal glory to the nest where it was born.

Four hundred years were to pass before the realization of this great plan, and the place where the great bird flew was not from Swan Mountain, but rather over the sand dunes of Kitty Hawk. That Leonardo undertook to fly his machine was known to Girolamo Cardano, who, in his book *De Subtilitate*, 1550, wrote:

It has turned out badly for the two who have recently made a trial of it (flying). Leonardo da Vinci, of whom I have spoken, has attempted to fly but he was not successful; he is a great painter.

The powerful chest muscles that Leonardo knew to be the basis of a bird's or a bat's ability to fly must have an equivalent in man if he also were to be capable of flying with the aid of artificial wings. Did man have the required strength to equal that of the bird? Leonardo believed that if he could put man's powerful leg muscles to operating the wings, a sufficient source of power would be made available. The bird's muscles, he believed, were not only to sustain flight but they also had a triple factor of safety, in order to acquire a speed of flight necessary to attack its prey. Man, not requiring this, surely had sufficient power for only sustained flight. Leonardo, therefore, designed his wings to be operated by the leg muscles rather than depending on the chest muscles alone.

That actual flight was in Leonardo's mind is reflected by his notes on safety. There is the parachute, the first known reference to such a device (Figure 22).

If a man have a tent roof of which the pores have all been filled, and it be twelve braccia square and twelve in depth, he will be able to let himself drop down from any great height, without suffering any injury (Codex Atlanticus, 381 v.).

Then the fear of falling from land into water is covered by a note recommending a double chain of leather bags, presumably air-filled, to be

Fig. 22.

tied underneath, you so manage that these are what first strike the ground.

Soaring upon a wind is different from flight, because the source of the energy required to counter gravity has changed.

But when the bird finds itself within the wind, it can sustain itself upon it without beating its wings, because the function which the wing performs against the air when the air is motionless is the same as that of the air moved against the wings when these are without motion.[34]

Also, as written in another notebook (Manuscript E, 38 v.),

it takes as much to move the air against the immovable object as to move the object against the immovable air,

– clearly, Newton's third law of motion.

With the behavior and properties of air clearly in his mind, Leonardo's imagination extends itself to conceiving and designing a helicopter (Figure 23).

I find that if this machine made with a helix is well-constructed, that is to say made out of linen of which the pores are closed with starch and it is turned with great speed, the said helix is able to make its spiral in the air, and it will rise high.[35]

The helix was to be eight braccia in radius, and the linen was to be stretched on stout cane. He recommends that a small pasteboard model,

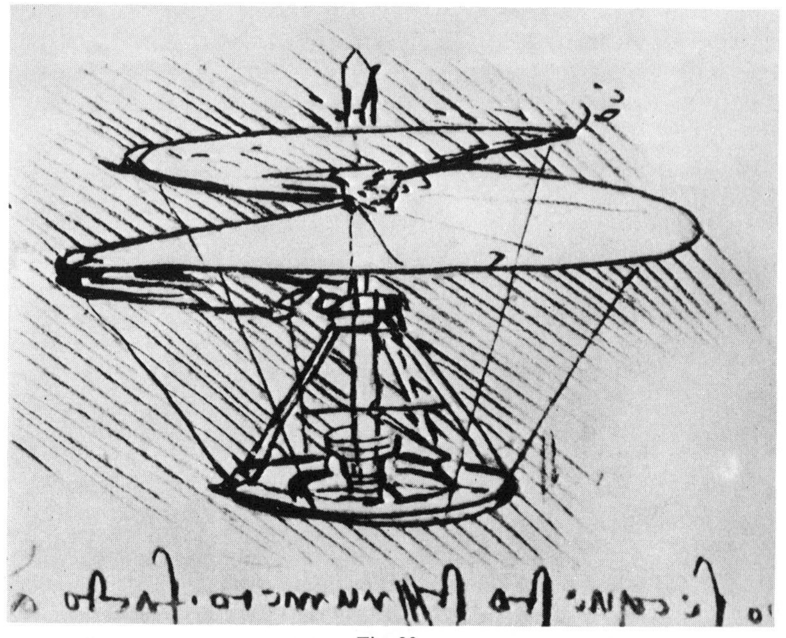

Fig. 23.

if dropped, will demonstrate the reverse action of helix and the air, in that in falling the helix will turn. He seems to have made no provision in the small sketches for means to provide counter-rotation to the air-screw. Yet he was thoroughly aware that, at least in other forms, such counter-thrust must be present for

> no movable thing moves of itself, unless its members exert a force in other bodies outside of itself,

proceeding to give the example of the man vainly trying to move a boat by pulling the rope attached to the boat's stern instead of to a pier (Leic., 29 v.).

Many are the schemes that Leonardo devised to make the flying machine work. In some, the flyer rests prone; in others, upright, the latter facilitating take-off by leaving the feet free to run for the start. For strength and pliability, structural members were to be of laminated hemlock or fir, springs of laminated steel, or of cow's horn and hinges of leather. Wings were to be made of netting with shutters twenty braccia in length and breadth to permit wing-lift with reduced resistance. These were to open on the up-beat and shut on the down-beat. Having designed and made the wing, its effectiveness must be tested. Therefore, a testing block (Figure 24) is rigged up to determine the difference in behavior between up-beat and down-beat. Leonardo notes,

> And if you wish to ascertain what weight will support this wing place yourself upon one side of a pair of balances and on the other place a corresponding weight so that the two scales are level in the air; then if you fasten yourself to the lever where the wing is and cut the rope which keeps it up, you will see it suddenly fall; and if it required two units of time to fall of itself you will cause it to fall in one by taking hold of the lever with your hands; and you lend so much weight to the opposite arm of the balance that the two become equal in respect of that force; and whatever is the weight of the other balance so much will support the wing as it flies; and so much the more as it presses the air more vigorously.[36]

Mobility to navigate in the air is accomplished by the relative forces applied on each of the four main wings or lifting surfaces. If moved with equal force, the flying machine will proceed on straight and level flight, and

> if they are used unequally, as in a constant proportion, the flying body will rise in a circling movement.[37]

His intense interest in mechanical flight is indicated by the large number of studies which have survived – over 500 drawings and sketches and

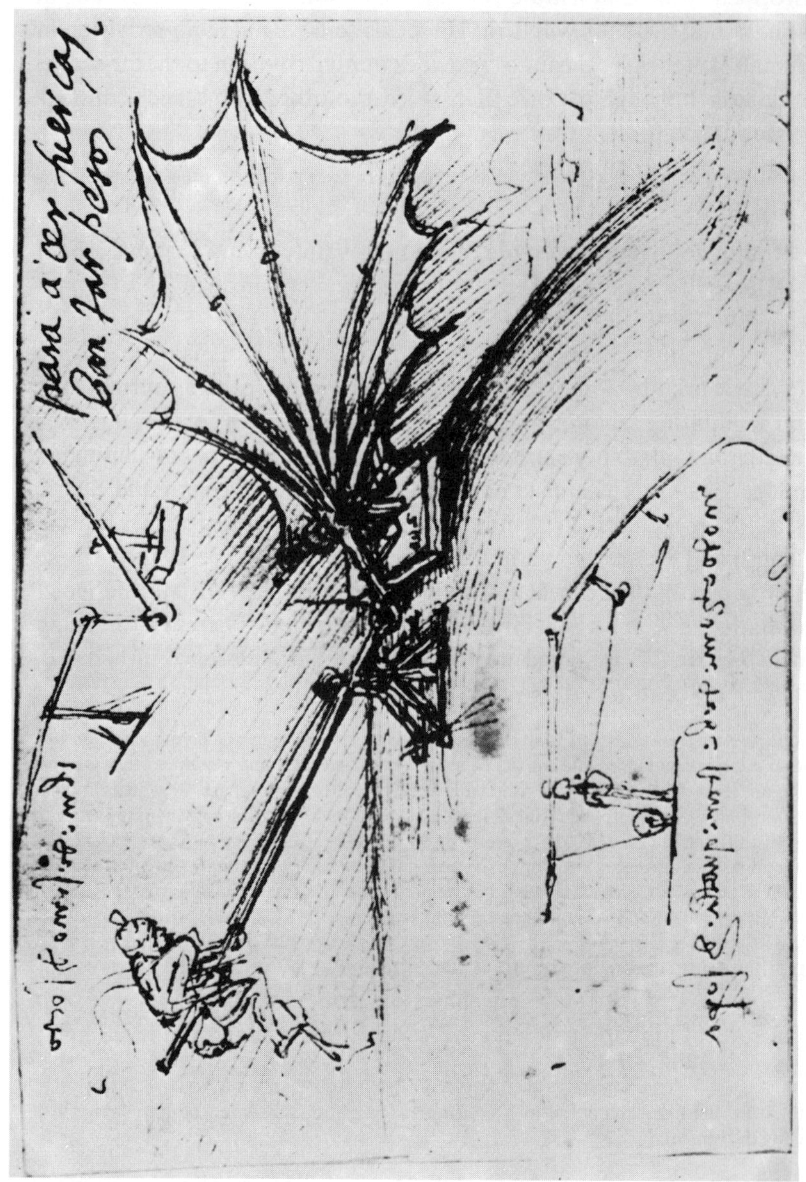

Fig. 24.

more than 3500 descriptive words.[38] His interest in flight was most intense during his service with Sforza, especially in the years 1486 to 1490 when his concern with armaments and military equipment was most intense.

The final military appointment in Leonardo's career was from the young and powerful monarch of France, Francis I, who designated Leonardo as his 'ingénieur, architecte, mécanicien'. In the summer of 1515, Pope Leo X felt threatened by an invasion under Francis I leading a fresh French army in a move to reestablish his rights in Italy. Leonardo joined the Papal force under Giuliano de'Medici, continuing with the headquarters staff in Piacenza and returning to Florence to meet the young French king at Bologna. Francis greatly admired all of Leonardo's talents and had not only planned to bring the aging master to France for retirement and peace, but also hoped to have all of his work to accompany him. The wall in the church in Milan on which Leonardo had painted the *Last Supper* was also to be moved, but this proved to be an impossible task. In 1516 the frail old man moved in convoy over the Alpine passes along with many of his paintings and these have remained in France ever since. For three years, from 1516 to 1519, Leonardo was made comfortable at the chateau of St. Cloux in Amboise where he worked on several canalization schemes and was consulted on some architectural projects in the area. There he also attempted to bring order into the vast accumulation of notes, designs and observations of his long, busy and productive life. In 1519, at age 67, Leonardo finished his labors and left the world to its historic evolution.

One of the large halls in the Palace of Discovery at the world exposition held in Paris in 1937 honored the great physicists of all time. Large operating models demonstrated some basic mechanical principles and held the attention of the visiting hundred thousands. Upon the walls were framed the portraits of six outstanding men in the mechanical sciences – Aristotle, Archimedes, Leonardo, Galileo, Newton, Lorenz – a noble company. This is one more token among many of Leonardo's work in the field of mechanics becoming recognized as of a highly creative order.

On the other hand, one wanders wide-eyed through the cobbled streets of the small town of Vinci nestling among the same Tuscan hills that held Leonardo's gaze. One enters the small but ancient cathedral and notes the bronze plaque on the sanctum's rear wall honoring the soldiers of Vinci who fell in World War I. The names are read, typical Tuscan names, but there is not one Leonardo among the more than score of surnames engraved there.

In our own day, the name of Leonardo da Vinci appears among the first 25 Immortals of Science engraved in perpetuity on the wall of the Hall of Science at the University of Bridgeport, Connecticut. And so, in our time, Leonardo has come into his own and the ways of science and mechanics which he envisioned are being realized.

This figure, who thought so intensely on matters of war and peace and worked with equal intensity in so many other areas of human knowledge, continues to astonish the most scholarly among us. Further, that this technologist, scientist and engineer should also have painted the *Mona Lisa*, most admired of all secular paintings, and the *Last Supper*, most famous religious painting, assuredly qualifies him for our deepest admiration.

Burndy Library (Connecticut)

NOTES

[1] A recent viewing of the famous self-portrait of Leonardo in Turin shows it to be much faded and foxed in the few decades since its earlier publication at the turn of the century.

[2] *Codex Atlanticus*, folio 56 v.

[3] *Codex Atlanticus*, folio 391 r. Leonardo's codexes here referred to, other than the *Codex Atlanticus* (C.A.), are the *Leicester Codex* (Leic.) in London, the *Codice sul Volo degli Uccelli* (Cod. sul Volo) in Turin, and the Manuscripts 2037 and 2038 in the Bibliothèque Nationale in Paris (B.N.). Most quotations were from MacCurdy's *The Notebooks of Leonardo da Vinci*.

[4] *Cod. sul Volo*, folio 3 r.

[5] Original formerly in the archives of Duke Melzi, Milan.

[6] As Leonardo placed his pinions, the front and rear wheels would rotate in opposite directions, producing no vehicle motion. The pinions should be on the same side of each wheel.

[7] Manuscript B, folio 83 v.

[8] Chief engineer to the Duke of Urbino. He served with Leonardo on the commissions to erect the spire of the Cathedral of Milan and construct a cathedral at Pavia. His *Treatise on Civil and Military Architecture*, now in the Laurenziana in Florence, written and illustrated by the author, has manuscript notes by Leonardo.

[9] C. D. O'Malley, *Leonardo's Legacy: An International Symposium*, University of California Press, 1968, p. 118.

[10] Examples are shown in Manuscript B, folio 23 r., and *Codex Atlanticus*, folios 22 r., 312 r.v., and 391 v.
[11] Manuscript 2037 B.N., folio 1 r.
[12] Manuscript 2037 B.N., folio 8 v.
[13] British Brigade General Henry Shrapnel lived 1761–1842.
[14] Manuscript I, folio 128 v.
[15] Manuscript I, folio 130 r.
[16] Manuscript I, folio 134 r. Further concern with this problem is contained in *Madrid Codex* I, folio 58 v.
[17] See A. C. Crombie, *Augustine to Galileo*, London 1952, pp. 254, 280; also A. R. Hall, *Ballistics in the Seventeenth Century*, Harper and Row, 1969. L. Reti, *Il moto dei Projetti e del Pendolo secondo Leonardo e Galileo*, Le Machine, Milan, Dec. 1968, treats Leonardo's ballistics with scholarly thoroughness.
[18] *Madrid Codex* I, folio 147 v.; Manuscript I, folios 128 v. and 129 r. Water jet curves are shown in Manuscript C, folio 7 r.
[19] See H. L. Peterson, *Pageant of the Gun*, New York, Doubleday, 1967, p. 18; and T. Lenk, *Flintlock: Its Origin and Development*, Saifer, West Orange, N.J., 1965, p. 13.
[20] Manuscript B, folio 46 v.
[21] Manuscript B, folio 5 v.
[22] Manuscript B, folio 50 r.
[23] Manuscript B, folio 33 r.
[24] *Madrid Codex* I, folio 26 r.
[25] *Codex Atlanticus*, folio 198 v.
[26] Cardinali, *Trattato del Moto e Misura dell'Acque*, Bologna 1826.
[27] Vallo, edition of 1535.
[28] *Leicester Codex*, folio 22 v.
[29] Manuscript B, folio 69 v.
[30] A Florentine braccio equals about 23 inches.
[31] Manuscript B, folio 39 v.
[32] Manuscript B, folio 63 v.
[33] Monte Ceceri, near Fiesole.
[34] *Codex Atlanticus*, folio 77 r.
[35] Manuscript B, folio 83 v.
[36] *Codex Atlanticus*, folio 381 v.
[37] Manuscript L, folio 60 v.
[38] C. H. Gibbs-Smith, *Leonardo da Vinci's Aeronautics*, London, H.M.S.O., 1967, p. 3.

LADISLAO RETI

LEONARDO DA VINCI AND THE BEGINNINGS OF FACTORIES WITH A CENTRAL SOURCE OF POWER

Today's factories are powered by a large number of independent electric motors, each directly connected to the operating machine. Only elderly people and scholars are aware of the times when, in factories and workshops, power was delivered from a central source, a steam engine or a hydraulic turbine, and the power distributed by means of shafting.

The advent of electric power, distributed from large central stations, changed this picture, though the change came slowly and by steps. For a long time after power plants were in operation, the working machines of a factory were driven by a single electric motor, and even today small workshops that operate according to this scheme can be found.

In the past, shafting was the principal method for the transmission of power. A shaft may transmit its power by means of gear-wheels or through the use of pulley and belt drives. The velocity ratio may be modified by shifting the belt over stepped pulleys.

Power distribution from a central source, small or large, is one of the basic characteristics of the modern factory; its origin, therefore, is of legitimate concern for the historian of technology.

Printed sources offer a satisfactory amount of information about power distribution from steam engines, even in the early stages of their development. Such information becomes scanty if we search for the method of apportionment of water and animal power, and disappears altogether when we pursue our investigation before the 16th century. One explanation for this lack of evidence might have been the recognition that primitive prime movers, such as hydraulic wheels, treadmills, or horse whims, were barely sufficient to give motion to a single machine – be it a mill, a pump, or a hoisting device – excluding, thus, the practicability of using a single power source to move several machines. But this would only be a half truth when, in fact, numerous manufacturing operations which required no more than a fraction of the output of the available power-source existed even in the pre-industrial age. Nor should we underestimate the efficiency of some of these old prime movers, one interesting example

R. J. Seeger and R. S. Cohen (eds.), Philosophical Foundations of Science, 97–115.

which I should like to remember being the well known projects of Leonardo da Vinci for rolling-mills powered by a horizontal water-wheel.[1] In spite of the fact that these projects (*Codex Atlanticus*, 2 r.a., 2 r.b.) may be dated to a time when Leonardo was in his sixties, far past juvenile enthusiasms and master of a technological knowledge accumulated through relentless experience and experimentation, some historians rejected these admirable projects as fantastic because it was not generally believed that a horizontal water-wheel could deliver sufficient power to operate such heavy and exacting machinery.[2] Today, we know better, and the horizontal water-wheel may be considered as the most powerful prime mover that the old mechanicians had at their disposal. The humble flour mills of Persia, moved by free-jet horizontal water-wheels, are capable of delivering a power output of 8–10 horse power with a coefficient of efficiency of 75 per cent, as found by Wulff.[3] The analogous water-wheels of 16th century Spain must have been even more powerful, if we accept the testimony of the unpublished manuscript of Turriano at the Biblioteca Nacional in Madrid.[4]

If we wish to systematically analyze the principle involved in the multiple application of a single power source, the following practical possibilities may be considered:

(1) A single power source may impart motion to different, though coordinated, parts of one machine. Such would be the case in an astronomical clock or a water-powered sawmill where, aside from the reciprocating motion of sawing, there is an additional mechanism for keeping the log pressed against the saw. According to White, the sketch of a sawmill of this type, depicted in the notebook of Villard de Honnecourt (circa 1235), represents the earliest industrial automatic power-machine to involve two motions.[5]

(2) A single power source may move two or more identical process machines, such as a water-wheel which activates two mills (Figure 1) at the same time or a silk-throwing mill in which hundreds of spindles revolve. Machines of this type have been known since at least the 13th century. Series of machine tools operated by a central source of power are often described by Leonardo da Vinci. Other interesting early examples are to be found in Biringuccio's *Pirotechnia* (1540), in the chapter which discusses devices for working bellows. Here, several methods are discribed and depicted for the activation of a pair of bellows that alter-

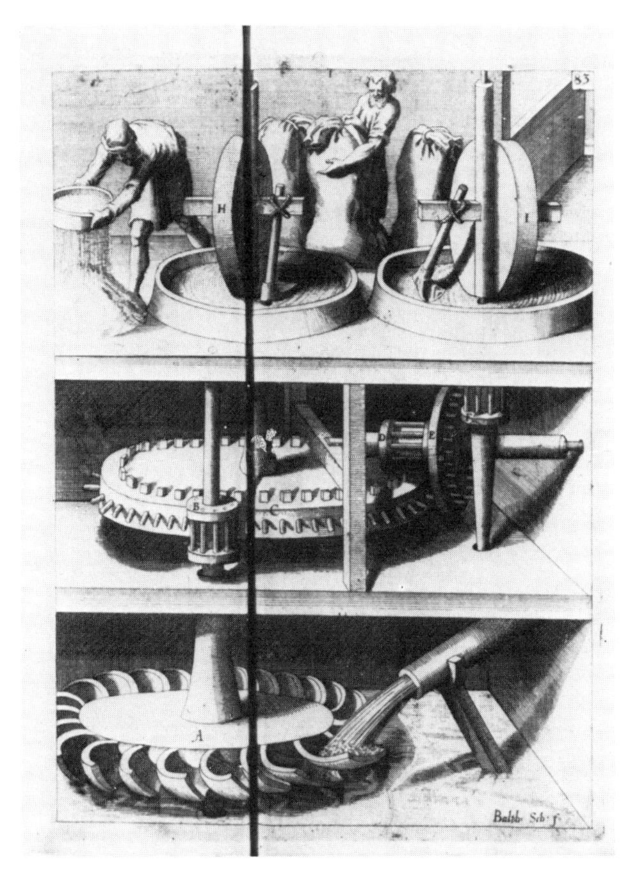

Fig. 1.

nate their blasts to insure a continuous stream of air. However, when Biringuccio mentions a multiple bellow-station, he is obliged to confess: "I cannot show you this in drawing because it is too difficult a thing for me to draw." Figure 2 demonstrates how Beck attempted to interprete Biringuccio's somewhat foggy description of a central powered system of four pairs of bellows.[7]

(3) The single prime mover activates a number of independent and different machines. Each performs a specific task in a complex and coordinated technological process. According to Beck, the earliest industrial application of this type is to be found in Agricola's famous *De Re Metal-*

lica Libri XII (1550).[8] Agricola's descriptions are extremely valuable for the historian of technology interested in how things were really done at that time. Fortunately for us, Agricola was a physician and not a practicing engineer. His purpose in writing *De Re Metallica* was to describe

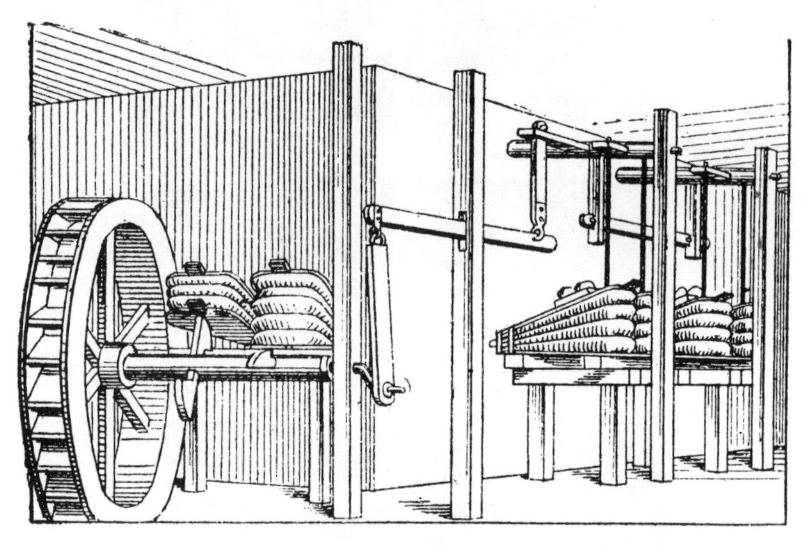

Fig. 2.

the mining and manufacturing processes he had seen, and not to suggest new, fantastic, or ingenious methods of constructing working devices, as was done later by the various authors who composed 'Theaters of Machines' (Besson, De Caus, Ramelli, Strada, Branca, Verantius, etc.).

Figure 3, from the VIIIth Book of *De Re Metallica*, represents an industrial complex, designed for the mercury amalgamation treatment of gold ores. A single water-wheel turned by a stream activates an ore-crusher visible on the left of the driving wheel. The main axle of the motor on this side is provided with cams which raise the stemps that crush the dry ore. Then, the crushed material is thrown into the hopper of the grinding mill, the first device visible at the right of the water-wheel. From here, the powdered ore falls continuously into the first amalgamation tub along with water, and from there runs into a second tub, then out of the second and into a third, which is the lowest.

Fig. 3.

The amalgamation tubs contain quicksilver and are provided with agitating devices that keep ore powder, quicksilver, and water in permanent motion. The gold contained in the crushed ore, being in contact with the quicksilver, is amalgamated and extracted; the exhausted mineral is then discarded while gold is recovered by first separating the amalgam from the excess quicksilver and then by distilling it.

Beck rightly calls our attention to the fact that in this unique machine [9] a single water-wheel imparts motion to an ore-crusher, a mill, and three mixing vats through a transmission device applied to the main shaft of the water-wheel, which consists of a drive shaft with crown-wheels and lantern-pinions. The concept of the transmission of power to a variety of operating machines is, according to Beck, one of the most important for the development of industrial machinery.

Is this remarkable installation described by Agricola truly the first example of central distribution of industrial power? The answer is no. More than 50 years before Agricola wrote his book, Leonardo da Vinci proposed similar solutions. The most significant ones shall herewith be illustrated and discussed.

One of them is depicted on folios 21 v. and 22 r. of *Codex I* at the Bibliotheca Nacional of Madrid (Figure 4). There is no explanatory text, but then one is needed. On the right, we see the complete assembly of a corn-mill combined with a bolting machine; on the left, the pictorial explanation of the bolting mechanism. The mill is of a type well known. The motor is not shown by Leonardo, indicating that the mill could be driven by water, wind, or animal power.

The important thing is the bolter, as confirmed by the drawings on the facing page: a cylindrical sieve suspended in its case, agitated by a lever that receives a periodical impulse from pegs fastened to the lower trundle of the lantern-pinion of the mill-gear. Both the mill and the bolter are activated by the same prime mover.

Until now, Ramelli (1588) has been credited with having described a cylindrical bolter moved by the main mill-drive for the first time.[10] However, a similar machine was known to Turriano, and its description can be found in his as yet unpublished manuscript at the Biblioteca Nacional of Madrid, datable circa 1565.[11]

The bolting machine discussed by Cardano in 1550 has similar features, but is was moved by hand and not by the mill-drive.[12] Schmidten's

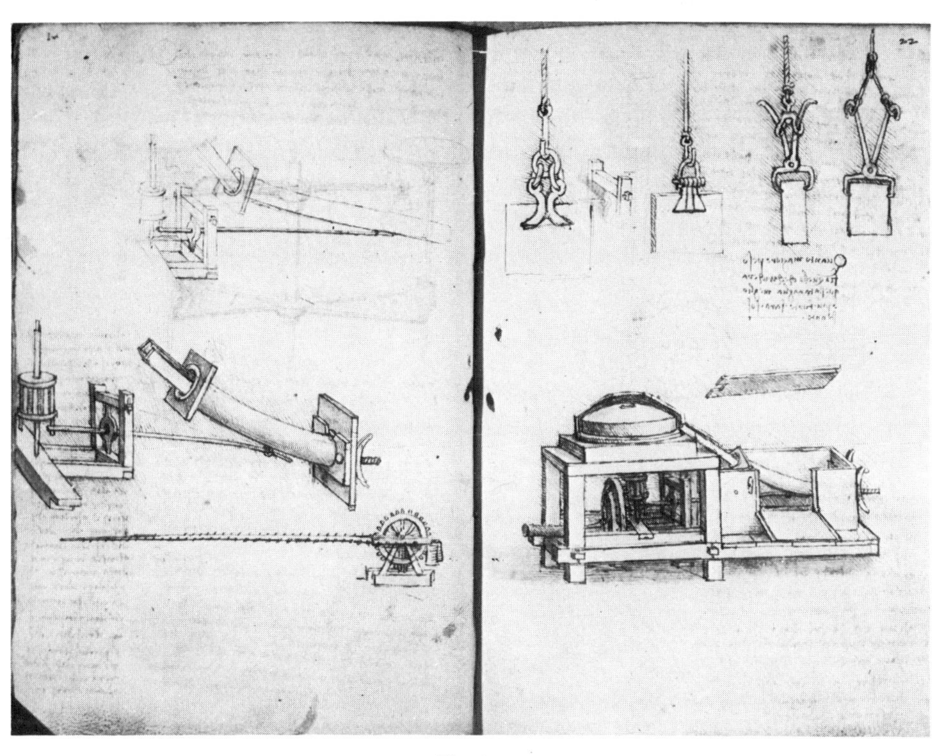

Fig. 4.

Chronica Cygnea contains a notice about one Nicholas Boller who proposed to use mill-power to activate the bolter as early as 1502.[13] The information entered the history of technology through Beckmann, but the late date of the source makes it suspect, however.[14] Nevertheless, here is the completely developed combination, lovingly outlined by Leonardo, according to internal evidence, around 1495.

This original conception was bound to survive for almost four centuries. In Leupold's well known *Theatrum Machinarium Molarium* (1735),[15] the milling-bolting combine still looks exactly like the design of Leonardo (Figure 5) and the same device was in use until the late 19th century.

Next, I would like to discuss another surprising development, one which is to be found in the *Madrid Codex I* on folios 46 v. and 47 r. (Figure 6). Here, Leonardo examines the possibility of running a complete oil factory through the use of a single power source, in this case a horse whim. The power source is only hinted at in the drawing, but it is clearly described in the accompanying text which runs as follows:

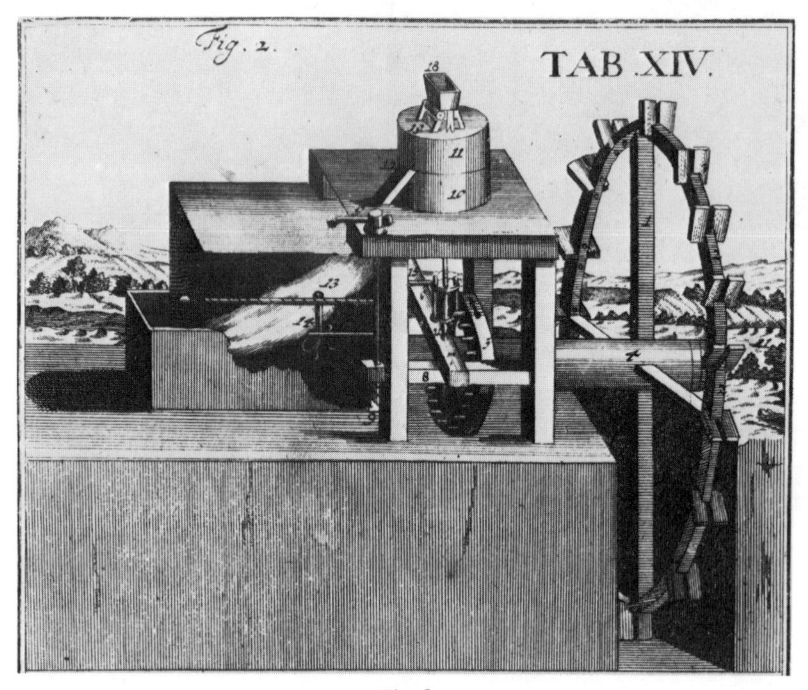

Fig. 5.

Here a horse grinds the nuts and rakes them under the rollers, extracts the oil with the press, mixes them thoroughly on the fire, and produces six barrels of oil a day, which is 4 kegs.

And it is done this way: the striking ram *m* goes up and down, hitting upon wedge *m*. At the same time, the horse turns the millstones crushing the nuts, and the nuts are mixed on the fire in heating-pan *S*, as you can see. When the wedge does not penetrate any more, hang ram *m* upon its hanger to avoid its being touched by the motor. Now, unfasten the second ram, connect it with the motor, and let wedge *n* be struck as many times as is needed for releasing the wedge and slackening the entire instrument. Said instrument is depicted above. The aforementioned pressing wedge is *a* and the reverse wedge, which opens the instrument, is wedge *b*.

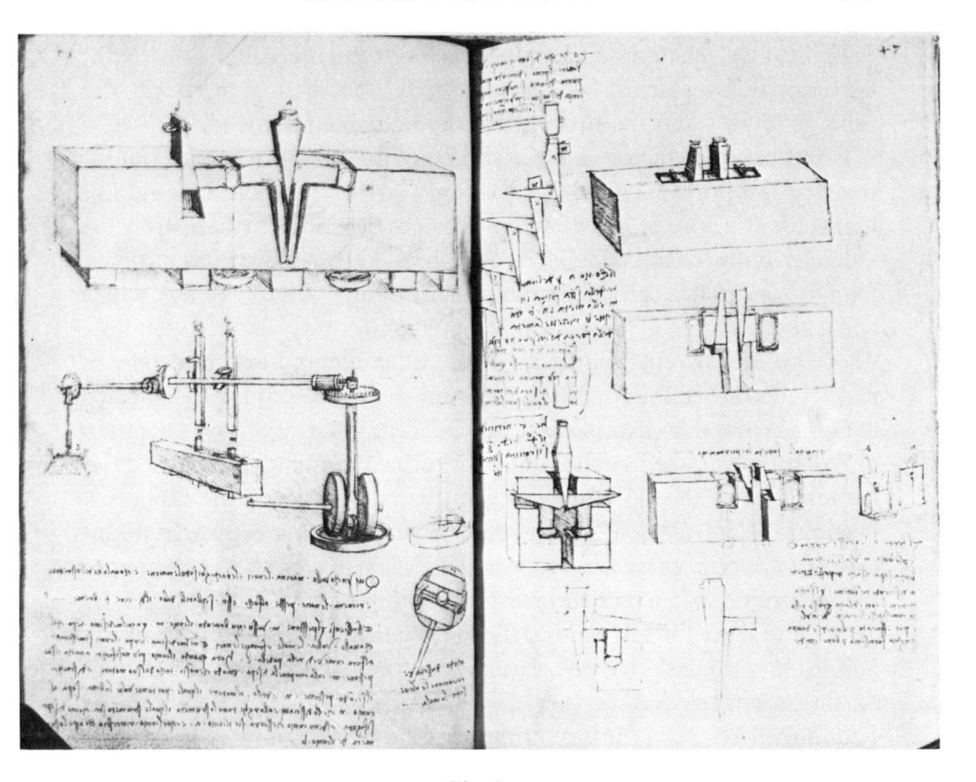

Fig. 6.

Beneath the small drawing, at the right margin, there is an added notation: "Method for rearranging and driving the olives under the roller." Very likely, this is the oldest known description of an integrated factory with a central source of power that utilizes shafting for its transmission. The following operations are activated from the single motor:

(a) Crushing of the nuts or olives with the aid of a roller or edge-mill.

(b) Rearranging of the crushed material with the aid of a pair of scrapers or sweepers.

(c) Pressing out the oil in a wedge press, using a striking ram.

(d) Opening the press by a releasing ram and wedge.

(e) Mixing of the material in a heating-pan.

The central shaft transmits the power from the roller mill to the heating-pan by gearing. At the same time, the shaft is provided with cams in order to lift the striking and the releasing rams.

Is this complex factory a fancy of Leonardo, or was it carried out for practical purposes? Before trying to answer this question, I would like to discuss the novelty of the various devices described by Leonardo.

Roller mills for crushing seeds have been known since ancient times. I am, however, less convinced about the antiquity of the sweepers, which add substantially to the continuity and uniformity of the operation. There are no sweepers in other older, or contemporary, drawings of roller mills as are found, for example, in the manuscripts of Francesco di Giorgio (datable circa 1485).[16] It appears as if the important operation of redistributing the material under the crushing rollers was carried out manually, with the aid of a shovel. In the manuscript at the Biblioteca Nazionale Centrale of Florence under *Palatino 767*, a copy after Francesco di Giorgio, there is a roller mill with a spade or shovel secured to the horizontal axle of the instrument. A similar figure in the Ms. *II–III 314* of the same library, containing materials of the same source, has a hoe leaning against the axle, clearly pointing to a manual operation. At the end of the 16th century, Jacob de Strada only knew about hoes bound with a rope at the axle of the edge-roller[17] (Figure 1). In Zonca's book we see how a laborer actually shovels material under the stone of a roller mill.[18]

The automatic sweepers delineated by Leonardo are here to stay, and will be found in every modern roller-mill project. In a book dealing with modern Dutch windmills and their industrial applications, we are able to recognize a late descendant of the Leonardine scheme (Figure 7).[19]

I must, however, call the attention of the historians of technology to the main feature of the oil factory sketched by Leonardo: the wedge press. This instrument had an important place in the oil industry before the advent of the hydraulic press, that is before 1800, and its popularity did not wane until long after the introduction of the hydraulic devices. In describing the 'Dutch mills' that operate with wedge presses, Appleton, writing in 1852, remarks: "They are still in very general use, and are, by many persons, supposed to be preferable to the hydraulic presses."[20]

How old is the wedge press? There are no classical literary references to it, but we know from Pompeian frescoes how a Roman wedge press may have looked. The wedges were driven horizontally into a frame which held the material to be pressed between square beams that slid up and down. It was a crude affair, and was used for the rather small scale preparation of pharmaceutical products. The large oil presses of Antiquity operated by means of levers and screws.[21]

In China, according to Needham, the most important type of press was one which used wedges that were driven home vertically.[22] Even if

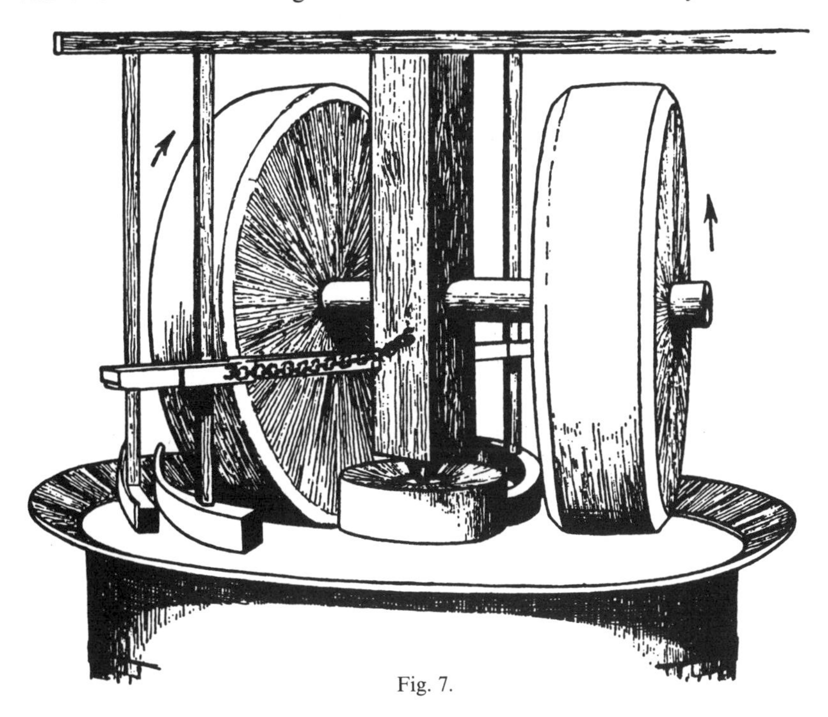

Fig. 7.

such presses precede the European models by centuries, they differ from them in several essential details. A direct oriental derivation, in this case, is not likely to have occurred.

Beck, whose book was first printed in 1899 and is still one of the best aids for the historian of technology, could not find a wedge press among the documentary and literary evidence available to him prior to 1612.[23] He found it in the *Theatrum Machinarum* of Zeising where a wedge press was described.[24] Zeising's *Theatrum* was most popular and went through many printings, the last one being in 1708. Though the book contains very few original projects, Beck succeeded in identifying most of the plates as being derived from earlier authors, especially Vitruvius, Besson, Rivius, Cardano, Ramelli, and Zonca. Among the few plates that might be attributed to Zeising, there is one representing an oil factory where the material is exhausted by the aid of a press that had wedges rammed in horizontally. A wedge press with horizontal wedges is also described in the aforementioned manuscript of Turriano.[25] I have, however, succeeded in finding a wedge press in a printed source earlier than Zeising's, in the little known, rare book of Bachot, published in 1598[26] (Figure 8).

Fig. 8.

It forms part of a central powered oil factory where, to our greatest surprise, we identify the same sequence of operations and the same mechanical arrangement as found in Leonardo's scheme, which precedes it by more than a century. We might feel compelled to suspect a direct derivation were it not for the fact that it could represent a standard manufacturing complex, handed down by tradition.[27] Three hundred years after Leonardo, oil factories were schematically presented the same way, as can be seen in Figure 9 from Knight's excellent Dictionary.[28]

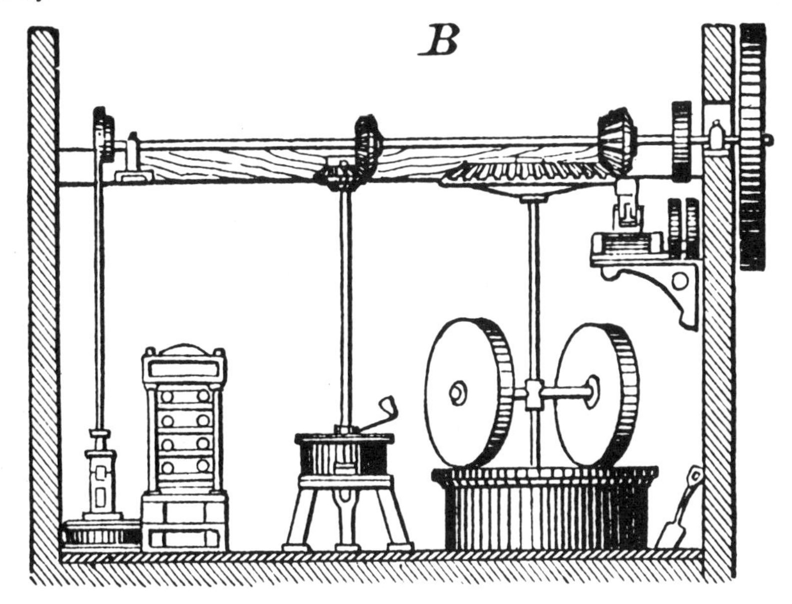

Fig. 9.

Let us, however, examine the wedge press of Leonardo on folio 47 v. of *Codex Madrid I*. There, facing the sketch of the integrated oil factory, there are several designs for wedge-press arrangements. Over the last one, at the center, there is a significant note: "Original made by Bre Pierantonio." First of all, the note disclaims once more the customary criticism about the machine projects of Leonardo as pertaining to a theoretical 'armchair' technology. This is an actual instrument, made by a contemporary craftsman. Unfortunately, Leonardo is not clear enough; we are not instructed as to the novelty of the machine: was the press made to order for Leonardo, or is he dealing with a device in general use at his time? We would search in vain for wedge presses in the writings of the engineers preceding Leonardo (Brunelleschi, B. Ghiberti, Taccola, Fontana, Francesco di Giorgio, etc.), nor are such presses mentioned in contemporary documents. As far as we can tell, based on the admittedly scarce testimonies of the history of engineering, Leonardo's is the first record of a device which will hold its own, unchanged, for at least four hundred years.

There are also sketches on the same page that represent alternate solutions to the basic problem. Those most interesting are at the left, where Leonardo examines the possible advantages of compounded wedges.

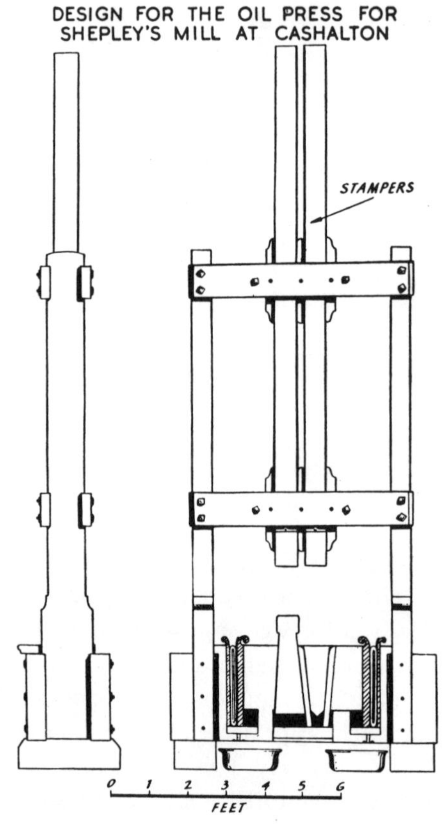

Fig. 10. Details of Stamper Press 1778. Reproduced by courtesy of the Royal Society.

The press on the facing page, over the complete factory, is undoubtedly a new conception of his, as he seeks here to eliminate the unequal pressure to which the two packages of materials to be pressed are subjected as a consequence of constructive asymmetry. Be as it may, wedge presses of the type described by Leonardo on the page facing folio 47 v. were to enjoy widespread popularity until the early 20th century. Curiously enough, graphic documents about its construction can be found only from the 18th century onward. One of the best examples is offered by a design of Smeaton, the famous English mechanician[29] (Figure 10).

The technical literature of the 16th century offers several other examples of multiple utilization of a central power source aside from those found in Agricola's great book, which points to a growing awareness for the necessity of mechanization. One of the most appealing, even if some of its features appear premature and difficult to carry into effect, is the project for a central powered, automated cane sugar factory found in the unpublished Turriano manuscript[30] (Figure 11). Here, a single horizontal water-wheel takes care of four distinct operations:

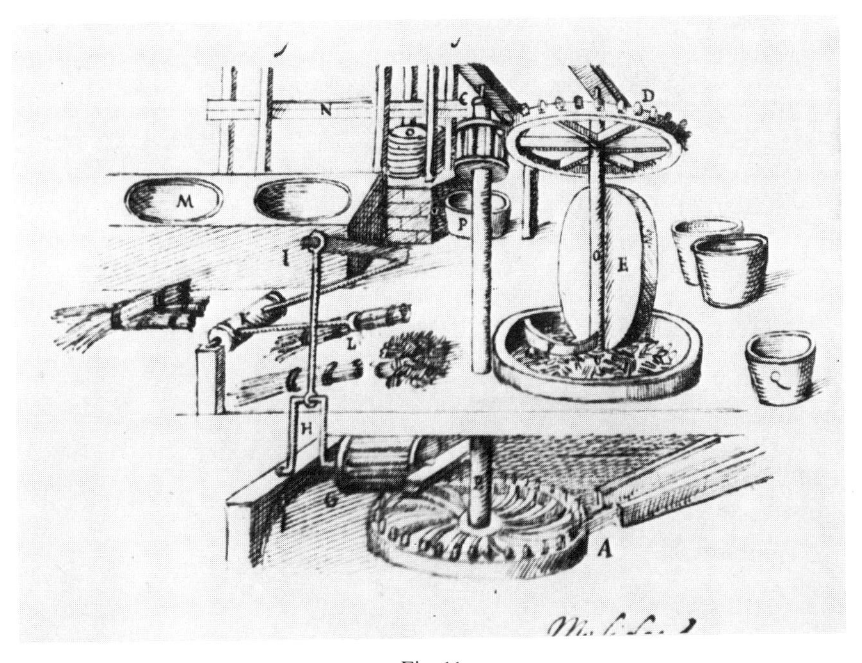

Fig. 11.

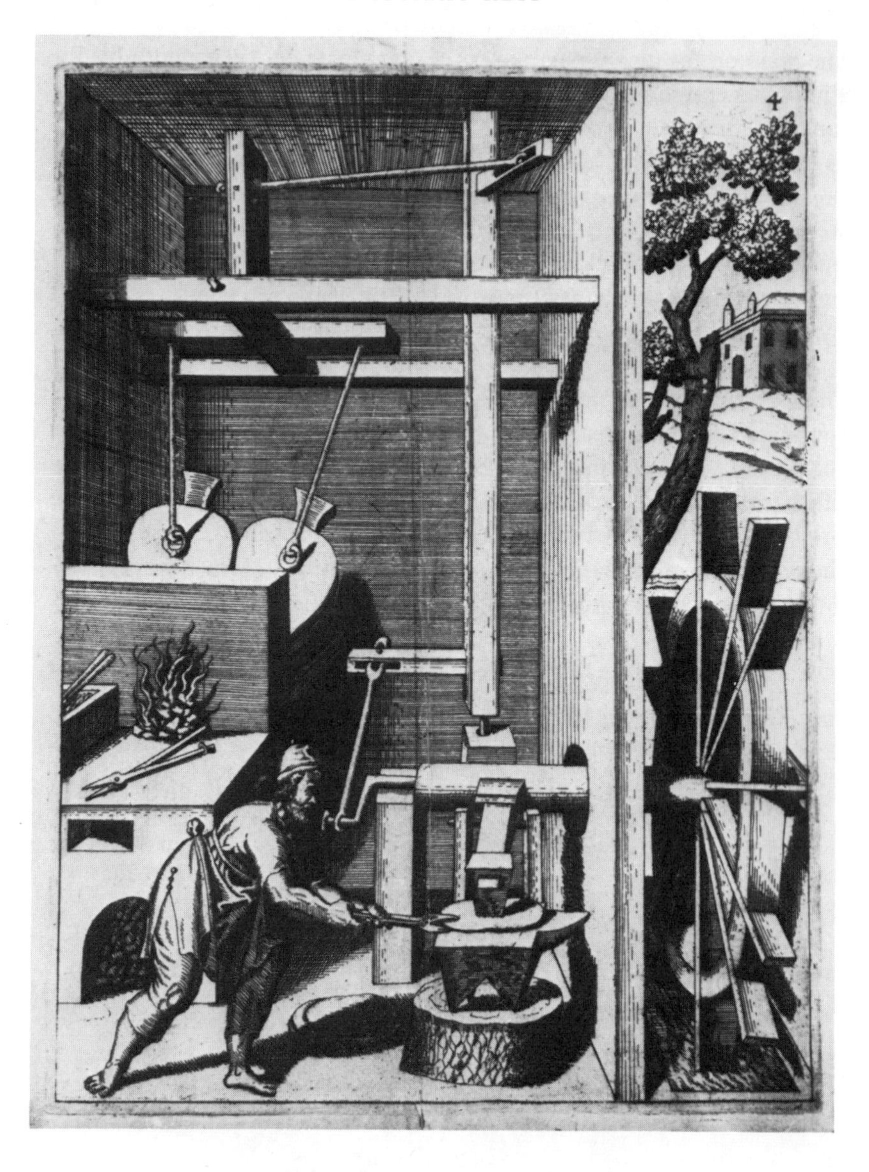

Fig. 12.

(a) Translation of the cane bundles.
(b) Cutting the cane.
(c) Crushing the cut cane.
(d) Rearrangement of the crushed cane by a sweeper.
In the background a cane press and the cauldrons for the evaporation of the cane juice are visible.

Kinematically, Turriano's mechanized sugar factory is more primitive than the oil mill of Leonardo. Movement and power are transmitted by lantern and spur-gear wheels and by the use of the antiquated rocking roller. The handling of the cane press is not included in this combine. There is no evidence, and little likelihood, that such a mechanized factory ever worked.

The previously cited book by Strada contains a fair number of examples of two machines, and sometimes three, powered by a single motor. The association is not always functional, though, as when for example Strada combines a millwork with a waterpump or a grinding wheel. But in other cases the pairing is reasonable, such as when a waterwheel moves a forge hammer together with a pair of bellows, even if the power transmission is clumsy (Figure 12). The repeated appearance of such combined operations in the technical literature of the 16th century points to a new spirit animating theorists and practitioners alike: the mechanization of technological operations, the increasing awareness of the possibility of substituting manual labor with mechanized devices. As in many other fields of human endeavor, Leonardo da Vinci was the first representative of this new breed of man in the march towards the conquest of a still hostile world. They had not the prescience to see that at the end of their quest yet another challenge would confront humanity; the problem of man's social insufficiency.

University of California, Los Angeles

NOTES

[1] Cf. Bern Dibner, 'Leonardo: Prophet of Automation', in *Leonardo's Legacy: An International Symposium* (ed. by C. D. O'Malley), University of California Press, Berkeley and Los Angeles, 1968, p. 111.

[2] H. Ch. von Seherr-Thoss, *Die Entwicklung der Zahnrad-Technik, Zahnformen und Tragfähigkeitsberechnung*, Springer, Berlin-Heidelberg-New York, 1965, p. 45. The author attributes this project to 1495, stating that "it would have been impossible to draw or to

roll iron in structural shapes because of the lack of sufficiently powerful prime movers at that time." But if the rolling mill, installed in Paris by Fayolle in 1729, could do the job by means of a whim powered by six horses, the horizontal water-wheel used by Leonardo in his rolling mills could do a still better job, according to new evidence.

[3] Hans E. Wulff, 'A postscript to Reti's Notes on Juanelo Turriano's Water Mills', *Technology and Culture* **7** (1966) 398–401; also *Traditional Crafts of Persia*, MIT Press, London, 1966, pp. 280–283.

[4] Ladislao Reti, 'A postscript to the Filarete Discussion: On Horizontal Waterwheels and Smelter Blowers in the Writings of Leonardo da Vinci and Juanelo Turriano', *Technology and Culture* **6** (1965) 428–441; 'On the Efficiency of Early Horizontal Waterwheels', *Technology and Culture* **8** (1967) 388–394.

[5] Lynn White, *Medieval Technology and Social Change*, Oxford University Press, 1962, p. 118.

[6] Vannoccio Biringuccio, *Pirotechnia* (orig. published in 1540; translated and annotated by M. T. Gnudi and C. S. Smith), MIT Press, London, 1966, pp. 300–306.

[7] Theodor Beck, *Beiträge zur Geschichte des Maschinenbaues*, Springer, Berlin, 1899, p. 120.

[8] Georgius Agricola, *De Re Metallica Libri XII* (orig. published in 1556); (translated and annotated by H. C. and L. H. Hoover); Dover, New York, 1912.

[9] Beck, *op. cit.*, p. 152. The 'uniqueness' of the machine is, however, not emphasized by Agricola, as Beck thought, translating *Machinae unica est rota, quam rivi impetus eius pinnas percutiens versat* with "Diese *einzig dastehende* Maschine hat ein Rad, welches vom Stosse des Wassers getroffen und umgedreht wird." *Unica* refers to the singleness of the wheel, and Hoover's translation is the correct one: "This machine has one waterwheel, which is turned by a stream striking its buckets." We find the same interpretation in the German translation of Agricola's book of 1557. The central-powered amalgamation plant may not have therefore been so sensationally new in Agricola's time, as supposed by Beck.

[10] Agostino Ramelli, *Le Diverse e Artificiose Machine*, Paris 1588, Plate 99.

[11] Ladislao Reti, 'The Codex of Juanelo Turriano (1500–1585)', *Technology and Culture* **8** (1967) 53–66.

[12] G. Cardano, *De Subtilitate Libri XXI*, Nürnberg 1550.

[13] T. Schmidten, *Chronica Cygnea*, Vol. 4, Zwickau 1656, p. 249.

[14] John Beckmann, *A History of Inventions, Discoveries and Origins* (translated from the German by William Johnston), London 1846, 4th ed., Vol. I., p. 161.

[15] Jacob Leupold, *Theatrum Machinarum Molarium*, Leipzig and Rudolstadt 1735.

[16] Francesco di Giorgio Martini, *Trattati di Architettura Ingegneria e Arte Militare.* A cura di Corrado Maltese (trascr. di Livia Maltese Degrassi), Polifilo, Milano, 1867, Vol. I, f. 34, Tav. 63.

[17] Jacob de Strada a Rosberg, *Künstlicher Abriss allerhand Wasser-, Wind-, Ross- und Handmühlen*, Frankfurt 1629, but written around 1590.

[18] Vittorio Zonca, *Novo Teatro di Machine et Edificii*, Padova 1607, p. 30.

[19] Frederick Stokhuyzen, *The Dutch Windmill*, van Dishoeck, Bussum 1965, 2nd ed., p. 79.

[20] Appleton's *Dictionary of Machines, Mechanics, Engine-work and Engineering*, New York, 1852, Vol. II, p. 421.

[21] A. G. Drachmann, *The Mechanical Technology of Greek and Roman Antiquity*, Munksgaard, Copenhagen, 1963.

[22] Joseph Needham, *Science and Civilization in China*, Vol. IV, Physics and Physical Technology, Part 2, Cambridge University Press.

[23] Beck, *op. cit.*, p. 407.

[24] H. Zeising, *Theatrum Machinarum*, Leipzig, 1612–14, Vol. III, No. 20.
[25] See note 11.
[26] Ambroise Bachot, *Le Govvernail*, Melun 1598.
[27] To give only a single example of the obstinate survival of technological prototypes, I should like to recall the history of the silk-throwing mill. Invented in Lucca, according to an early tradition, this highly developed instrument was brought to Bologna around 1272 by an exile, and its secret strictly kept, even if it was impossible to avoid its spreading to Florence and Venice in the mid-fourteenth century. The design was little improved until the 19th century, and its early stage is best illustrated in Zonca's book (1607). In spite of this publicity, silk machinery was so relatively undeveloped in England that the supply of thrown silk had to come chiefly from Italy. John Lambe, returned from Italy in 1717 with drawings of the silk-throwing mill obtained by bribery, applied for, and received, English patents for his 'improved' mill in 1718. The silk factory erected by Lambe in 1719 in Derwent at Derby was considered one of the wonders of the age. Comparing Lambe's designs with Zonca's pictures, and even with those of a 15th century Florentine manuscript, the constructive differences become irrelevant. Needless to say, the basic design of such mills, as they are represented in the famous Diderot-D'Alembert Encyclopedia, are still found unchanged in the France of 1750.
[28] Knight's *American Mechanical Dictionary*, Boston 1876, Vol. II, p. 1554.
[29] From Harold W. Brace, *History of Seed Crushing in Great Britain*, Land Books, London, 1960, p. 35.
[30] See note 4. *Technology and Culture* **6** (1965).

PART II

PHYSICS AND THE EXPLANATION OF LIFE

(*Chairman:* George Wald)

EUGENE P. WIGNER

PHYSICS AND THE EXPLANATION OF LIFE*

1. Preamble

I am a physicist, but the problem on which I wish to present some thoughts is not a problem of physics. It is, at present, a problem of philosophy, and I may well be told 'ne sutor ultra crepidam' – the shoemaker should stick to his last. I do have, however, several excuses for venturing into this difficult field. The first is that, if no solid knowledge is available in a field, it is good if representatives of neighbor sciences put forward the views which appear most natural from their own vantage point. The second reason for my speaking here today is that since I started to think and also to write on the subject, I have received many letters and verbal comments from colleagues, agreeing, on the whole, with my point of view. This means, I hope, that there is some interest in the subject among physicists and some consensus on it. It also means that much of what I will have to say will not be original but must have been conceived, at least in part, before me. My third excuse for putting forward views which do not have the solid foundation which one is used to expect from a physicist is that many others before me have done likewise and my fourth excuse is simply that the subject is of overwhelming interest and I like to speculate about it.

2. A bit of history

It would be difficult to review in this session even that small part of the philosophers' thinking on the problem of body and mind, physics and consciousness, with which I am familiar. Let me confine my attention to the ideas of three schools, all of which had a profound effect on our thinking.

Descartes seems to have been the first in modern times to have devoted a great deal of thought to our question. Descartes is, of course, well-known as the originator of the rectangular coordinate system, and for

R. J. Seeger and R. S. Cohen (eds.), Philosophical Foundations of Science, 119–132.

his pronouncement, 'cogito ergo sum'. This saying indicates that he recognized the thought, an evidence of the consciousness, as the primary concept. Descartes was also the first to recognize the nerves as transmitters of sensations and the brain as the depository of our emotions and our memory. He said that the brain is the body's link to the soul.[1]

Let me comment only on three characteristics of Descartes' thoughts. The first is the truly mechanistic picture which he used throughout. He considered the nerve impulses as a flow of a liquid through the nerves, which he imagined to be tubes. He imagined that the memory consists of an expansion of pores in the brain. I mention this because it shows how easily even a truly great thinker succumbs to the ideology of the contemporary state of science and makes too detailed images in accordance with that state of science. Of course, in Descartes' times nobody could dream of the travelling electric impulses which do constitute nerve action. The second point to which I wish to call attention is that Descartes considered mind and body to be two separate entities, the body acting on the mind. He did not think of body and mind as fused into an entity. Lastly, he maintained that only man has a soul, animals are mere machines or automata, devoid not only of thoughts and emotions, but also of sensations. One of his successors, Malebranche,[2] said:

> Thus dogs, cats, and the other animals, have no intelligence, no soul in the sense in which this concept is usually understood. They eat without pleasure, cry without pain, grow without knowing this. They have no desires and no knowledge.

This sound fantastic to us, pupils of Darwin's recognition that man is an animal species. It sounds more fantastic than it should: some insects lack sensations to a surprising extent. Wasps, the thorax of which was cut off suddenly, did not appear to notice this but continued to eat – the food dropping out of the channel leading to the thorax. I'll return later to this illustration of the enormous differences between the inner lives of different animals.

The next philosopher whose ideas I wish to mention is Thomas Huxley.[3] He recognized the near absurdity of Descartes' view as far as the sharp and absolute difference between the states of consciousness of animals and man is concerned. He did accept, however, Descartes' view that animals are pure automata, and extended this to man. According to Huxley, and many other philosophers who followed him, man's and animals' volitions, their intentions, are consequences rather than causes of the physical circumstances. At first hearing, this appears absurd but, as will appear later on, in the deterministic framework of these philosophers it is more nearly meaningless than absurd. Causation is not a well-defined concept in a deterministic picture of the world – it may not have an unambiguous meaning in any known picture. This statement will be made more explicit and concrete later on; it will also be expressed in the physicist's language. However, you probably have heard the story of two Eskimos watching a water-skier. "Why boat go so fast?" asked the first. "Chased by fool on end of string," was the reply.

The last body of thought that I wish to refer to is that of the Gestaltslehre of Wertheimer, Köhler, and others. They point out that a steam engine, for instance, would be very inadequately described as a steel cylinder, covered at one end, having a closely fitting but movable disc on the inside and a rod, attached to this disc, protruding on the other side. Rather, in order to describe a steam engine, its purpose and the cooperation of its parts should be given. Similarly, an explanation of an animal in terms of the physical functioning of its parts will be inadequate; it is a whole, much more than the sum of its parts.[4]

The point of the Gestalt-theoreticians is undoubtedly correct and it is a valuable observation. However, it seems to me to be more a pedagogical than an ontological observation. Surely, we do not obtain a vivid picture of man by just describing his bones and muscles and how they are attached to each other. However, it is possible to describe the functioning of man's organs without answering the question of the relation between his body and soul, his emotions and his physical constitution, his volition and his movements. Hence, it appears to me that the statements of the Gestaltlehre, though both true and relevant, really avoid the principal issue which confronts us.[5]

Let me now attack our problem from the point of view of the physicist, the physicist familiar with the fundamental changes which quantum mechanics initiated in the physicist's picture of the world. I shall begin with the part of our title with which I should be familiar – with physics.

3. WHAT IS PHYSICS? WILL IT FORM A UNION WITH THE LIFE SCIENCES?

One often hears the statement that the purpose of physics is the explanation of the behavior of inanimate objects. To most of us physicists this does not appear to be a very incisive statement. What our science is after is, rather, an exploration of the regularities which obtain between the phenomena, and an incorporation of these regularities – the laws of nature – into increasingly general principles (the theories of physics), thus establishing more and more encompassing points of view. I like to quote David Bohm[6] in this connection, who said that "science may be regarded as a means of establishing new kinds of contacts with the world, in new domains, on new levels." No ultimate explanation is possible and our science is rather a constant striving for more encompassing points of view than the provider of an explanation for one or another phenomenon. Furthermore, as Einstein often emphasized, the more encompassing point of view for which we strive must have a conceptual simplicity – otherwise it will not be credible. And, as Polanyi emphasized, it must be interesting.

If all this is accepted, it follows that the phenomena of life and mind will form a unit with our regularity-seeking physical sciences if regularities in the behavior of the thought processes can be discovered, and an encompassing point of view developed which embraces both the phenomena of the mind and those of matter. Surely, psychology has pointed to many regularities in our thought processes and has made many, many interesting observations. These are, however, at present entirely divorced from the regularities in the behavior of matter which are the subjects of present day physics.

My discussion of life and consciousness will be based on the assumption that these phenomena will become, along with ordinary physical phenomena, the subjects of a regularity-seeking science. It will be based on the assumption that a picture will be discovered which will provide us with a view encompassing both mental and physical phenomena and describes regularities in both domains from a unified point of view. Clearly, these are assumptions for which a proof is lacking at present.

Are there tendencies in the sciences of life to expand in the direction of physics and are there, conversely, tendencies in physics to consider the phenomena of life and consciousness? If the disciplines in question are considered broadly enough, both tendencies are present. It is true that physics, in the true sense of the word, is foreign to basic psychology. However, the life sciences, particularly those dealing with the lowest organisms, are endeavoring to acquire a base in chemistry and physics. They also hope to extend their interest, eventually, to organisms of which consciousness is an essential characteristic. Conversely, the basic principles of physics, embodied in quantum mechanical theory, are dealing with connections between observations, that is contents of consciousness. This is a difficult statement to accept at first hearing, and I must hope that most are familiar with it. In essence, it recalls that quantum mechanics is not a deterministic theory. The formulation of its laws in terms of our successive perceptions, between which it gives probability connections, is a necessity. Classical physics, of course, also can be formulated in terms of, this time deterministic, connections between perceptions and the true positivist may prefer such a formulation. However, it can also be formulated in terms of absolute reality; the *necessity* of the formulation in terms of perceptions, and hence the reference to consciousness, is characteristic only of quantum mechanics. In fact, the principal objection which your present speaker is inclined to raise against the epistemology of quantum mechanics is that it uses a picture of consciousness which is unrealistically schematized and barren. Nevertheless, there is a tendency in both physics, which we consider as the most basic science dealing with inanimate objects, and in the life sciences, to expand toward each other. Furthermore, the tendency is strongest in the modern parts of the two disciplines: in quantum mechanics on the one, in microbiology on the other, hand. Both feel that they cannot get along by relying solely on their own concepts.

Nevertheless, that the tendencies to which I just alluded will ultimately lead to a merger of the disciplines can be only a hope at present. Neither of them proposes in its present form a more encompassing point of view; both wish to use the concepts of the other only as the basic concepts in terms of which their own regularities can be formulated. That some such basic concepts are unavoidable is, I believe, clear: there must be some things which are the subjects of regularities. It is not clear, however, that these subjects must be either entirely in the realm of orthodox physical theory, or entirely in psychological subjects. Nevertheless, it is encouraging that there is a tendency on both sides of the chasm to take cognizance of the other side – even if both sides are forced to do so, or perhaps *because* both sides are forced to do so.

Let me now come to my last subject: the physicist's view of the relation between body and mind. I will try to give a rational discussion of the two alternative roles which present-day physics can play in a future regularity-seeking science the realm of which extends to the phenomena of mind as well as to those of present-day physics.

4. A PHYSICIST'S VIEW ON THE MIND-BODY PROBLEM – THE FIRST ALTERNATIVE

As particularly the historical discussion indicates, there is a possibility that the laws of physics, formulated originally only for inanimate matter, are valid also for the physical substance of living beings. To put it in a somewhat vulgar fashion, even most physicists, if unexpectedly presented with the question of the validity of the laws of physics for organic matter, would affirm that validity. On the other side of the chasm, many, if not most, microbiologists would concur in this view. The view does not lead automatically to Huxley's view that we are automata because the present laws of physics are not deterministic but have a probabilistic character. Furthermore, if we are honest about it, we cannot now formulate laws of physics valid for inanimate objects under all conditions. Hence, the statement which we are considering should be formulated somewhat more cautiously: that laws of nature for the formulation of which observations on inanimate matter suffice, are valid also for living beings. In other words, physical laws, obtained by studying the traditional subjects of physics, and perhaps not very different from those that physicists are trying to formulate now, will form the basis from which the behavior of living matter can be derived – derived perhaps with a great deal of effort and computing, but still correctly derived.

The assumption just formulated is surely logically possible. It is very close to Huxley's views which were mentioned before. Would it mean that, eventually, the whole science of the mind will become applied physics? In my opinion, this would not be the case even if the assumption which we are discussing is correct. What we are interested in is not only, and not principally, the motion of the molecules in a brain but, to use Descartes' terminology, the sensations which are experienced by the soul which is linked to that brain, whether it is pain or pleasure, stimulation or anxiety, whether it thinks of love or prime numbers. In order to obtain an answer to these questions, the physical characterization of the state of the brain would have to be translated into psychological-emotional terms.

It may be useful to give an example from purely physical theory for the need for such a translation. The example which I most like to present derives from the classical theory of the electromagnetic field in vacuum, that is, the simplest form of Maxwell's equations. These give the time derivative of the electric field in terms of the magnetic field, and the time derivative of the magnetic field in terms of the electric field. Both fields are free of sources. Although the actual form of the equations is not very relevant for our discussion, it may render this more concrete if I write down the equations in question for the electric and magnetic field, E and H:

$$\frac{\partial H}{\partial t} = -\, c \operatorname{curl} E \qquad \frac{\partial E}{\partial t} = c \operatorname{curl} H.$$

I shall refer to these equations, briefly, as Maxwell's equations; actually, they are Maxwell's equations for empty space; c is the velocity of light. If E and H are given at one instant of time, these equations permit their calculation for all later times, and for all earlier times. They will serve as model equations for the discussion which follows – they give both sides of the picture, the electric and the magnetic side, and do not prefer one over the other.

It is possible, however, to formulate an equation for the magnetic field alone. This is again, and should remain, free of sources and its time-dependence is regulated by the equation

$$\frac{\partial^2 H}{\partial t^2} = c^2 \left(\frac{\partial^2 H}{\partial x^2} + \frac{\partial^2 H}{\partial y^2} + \frac{\partial^2 H}{\partial z^2} \right).$$

I shall refer to this equation, briefly and somewhat incorrectly, as Laplace's equation. One can observe now that, if H and $\partial H/\partial t$ are given at one instant of time, this equation permits the calculation of H for all later times – and for all earlier times. Is now this equation, which is just as valid as Maxwell's original equations, a full substitute for the latter? The answer is no. If we want to obtain the force on a small charge at rest, the original form of the equations furnishes this directly: it is the electric field at the place where the charge is, multiplied by the magnitude of the charge. In order to obtain the force from the second, that is

Laplace's equation, referring only to the magnetic field H, one has to calculate first the electric field in terms of H. This can be done, though the formula is quite involved. The formula gives the translation of the magnetic field into the electric one and this is, in the case considered, more relevant than the magnetic field itself. We have, therefore, an example before us in which a theory – Laplace's equation for H alone – is completely valid but is not very useful without the translation which should go with it.

The example also shows that the translation into the more relevant quantity can be quite complicated – more complicated than the underlying theory, that is Laplace's equation. The translation equation is also more complicated than the set of equations, in this case Maxwell's equations, which uses both concepts: the one which turns out to be the more relevant one, that is E, along with the other, H, which does suffice for the formulation of the time-dependence. It is unnecessary to remark that, in the preceding illustration of a future theory of life, H plays the role of the purely physical variables, E plays the role of the psychological variables. In this illustration, the use of both types of variables in the basic equations is much preferable to the use of only one of them – the problem of translation does not arise in that case.

The example just given illustrates also the observation on the meaningless nature of the concept of causation in a deterministic theory – such as Maxwell's theory of the electromagnetic field in vacuum. Looking only at Laplace's equation, and the translation thereof, one will conclude that the magnetic field is the prime quantity, its development is determined by its magnitude in the past. The electric field will appear as a derived quantity, caused and generated by the magnetic field. Maxwell's original form of the equations shows, on the other hand, the possibility (and in the opinion of the physicists, the desirability) to consider the two to have equal rank and primitivity. One can also go to the other extreme and derive an equation similar to the last one, but involving only the electric field E and then claim the E is the primitive quantity, the magnetic field H the deribed one, the product of E.

I believe I have discussed the assumption that the laws of physics, in the sense described, are valid also for living matter. We also saw that this assumption need not imply, as is often postulated, that the mind and the consciousness are only unimportant derived concepts which need not enter the theory at all. It may be even possible to give them the privileged status. Let me now discuss the assumption opposite to the 'first alternative' considered so far: that the laws of physics will have to be modified drastically if they are to account for the phenomena of life. Actually, I believe that this second assumption is the correct one.

5. THE SECOND ALTERNATIVE: LIFE MODIFIES THE LAWS VALID FOR INANIMATE NATURE

I wish to begin this discussion by recalling how wonderfully actual situations in the world have helped us to discover laws of nature. The story may well begin with Newton and his law of gravitation. It is hard to imagine how he could have discovered this, had he not had the solar system before himself in which only gravitational forces play a significant role. The discovery, also due to Newton, that these forces also determine the motion of the Moon around the Earth, and the motion of freely falling bodies too heavy to be much affected by air resistance, was a wonderful example for science's power to create a unified point of view for phenomena which had, originally, widely differing characters. Newton, of course, recognized that there must be other forces in addition to the gravitational ones – forces which, however, remain of negligible importance as far as the motion of the planets is concerned.

Newton's discovery was followed by the discovery of most laws of macroscopic physics. Maxwell's laws of electromagnetism – the ones which we just considered in the special case of absence of matter – are perhaps the most remarkable among these. Again, these laws – those of macroscopic physics – could not have been discovered were not all the common objects which surround us of macroscopic nature, containing many millions of atoms, so that quantum effects, for instance, play no role in their gross behavior. Again, the unifying power of science manifested itself in a spectacular way: it turned out that Maxwell's equations also describe light and, as we now know, all electromagnetic radiation from radio waves to X-rays.

The next step of comparable, perhaps even greater, importance was the development of microscopic physics, starting with the theory of heat and soon leading to quantum theory. Most of this development took place in the first half of our century but, in some regards, the development is still incomplete. If we assume that it can and will be completed – most of us believe this – the question which we should face is whether our present microscopic theories also presuppose some special situation, the absence of certain forces or circumstances. The point of view which we are discussing maintains that this is the case. Just as gravitational theory can describe only the situation in which no other but gravitational forces play a role, and macroscopic physics describes only situations in which all bodies present consist of many millions of atoms, present microscopic theory describes only situations in which life and consciousness play no active role.

Similarly, just as macroscopic physics contains gravitational theory as a special case, applicable whenever only gravitational forces play a significant role, and just as microscopic physics contains macroscopic physics as a special case, valid for bodies which contain millions of atoms, in the same way the theory which is here anticipated should contain present microscopic physics as a special case, valid for inanimate objects. Thus, each successive theory is expected to be a generalization of the preceding one, to recognize the regulatities which its antecedent postulated, but to recognize them as valid only under special conditions. This should apply also to the theory foreseen here, in the form of the 'second alternative'.

Naturally, the preceding story does not *prove* that the present, microscopic, physics will also have to be generalized, that the laws of nature as we now know them, or try to establish them, are only limiting cases, just as the planetary system, macroscopic physics, were limiting cases. In other words, it does not prove that our second alternative, rather than the first one, is correct. Can arguments be adduced to show the need for such modification? I know of two such arguments.

The first is that if one entity is influenced by another entity, in all known cases the latter one is also influenced by the former. The most striking and originally least expected example for this is the influence of light on matter, most obviously in the form of light pressure. That matter influences light is an obvious fact – if it were not so, we could not see the moon. We see it because it scatters the light emitted by the sun. The influence of light on matter is, however, a more subtle effect and is virtually unobservable under the conditions which surround us. Light pressure is, however, by now a well-demonstrated phenomenon and it plays a decisive role in the interior of stars. More generally, we do not know any case in which the influence is entirely one-sided. Since matter clearly influences the content of our consciousness, it is natural to assume that the opposite influence also exists, thus demanding a modification of the presently accepted laws of nature which disregard this influence.

The second argument which I like to put forward is that all extensions of physics to new sets of phenomena were accompanied by drastic changes in the theory. In fact, most were accompanied by drastic changes of the entities for which the laws of physics were supposed to establish regularities. These were the positions of bodies in Newton's theory and the developments which soon followed his theory. They were the intensities of fields as functions of position and time in Maxwell's theory. These were replaced then by the outcomes of observations (the perceptions referred to before) in modern microscopic physics, that is, quantum mechanics. In the development which we are trying to envisage, leading to the incorporation of life, consciousness, and mind into physical theory, the change of the basic entities indeed appears unavoidable: the observation, being the entity which plays the primitive role in the theory, cannot be further analyzed within that theory. Similarly, Newtonian theory did not further analyze the meaning of the position of an object, field-theory did not analyze further the concept of the field. If the concept of observation is to be further analyzed, it cannot play the primitive role it now plays in the theory and this will have to establish regularities between entities different from the outcomes of observations. An alteration of the basic concepts of the theory is necessary.[7]

These are the two arguments in favor of what I called the second alternative, that the laws of physics which result from the study of inanimate objects only are not adequate for formulating the laws for situations in which life and consciousness are relevant parts of the picture.

6. Conclusion and Summary

I realize that the hope expressed in the last two sections, that man shall acquire deeper insights into mental processes, into the character of our consciousness, is only a hope. The intellectual capabilities of man may have their limits just as the capabilities of other animals have. The hope does imply, though, that the mental and emotional processes of man and animals will be the subjects of scrutiny just as processes in inanimate matter are subjects of scrutiny now. The knowledge of mind and consciousness may be less sharp and detailed than is the knowledge given by present day physics on the behavior of inanimate objects. The expectation is, nevertheless, that we can view mind and consciousness – at least those of other living beings – from the outside so that their perceptions will not be the primitive concepts in terms of which all laws and correlations are formulated. As to the loss in the sharpness and detail of the laws, this is probably unavoidable. It has taken place throughout the history of physics. Newton could determine all the initial conditions of the system of his interest and could foresee its behavior into the indefinite future. Maxwell's and his contemporaries' theories can be verified only by creating conditions artificially under which a verification is possible. Even then, it is possible only for limited periods of time. The laws of quantum mechanics, finally, neither make definite predictions under all conditions, nor have its equations of motion been verified in any detail similar to those of macroscopic theories. A further retrenchment of our demands for detail of verification is probably in the offing whenever we extend our interest to a wider variety of phenomena.

You will want to ask me, I believe, at least two questions. First, whether other physicists would agree with me, and second what good all this does, considering that I do not even specify the basic entities the behavior of which is subject to the new regularities to be established. My answer to the first question is that most physicists do not concern themselves too much with the questions I discussed. Their reason may well be given by the answer I'll give to the second question. However, Bohr, in his inimitable, profound, and somewhat ambiguous way, concurred in the view which I am embracing.[8] Also, I was just a few days ago reminded by Dr. Hartshorne, of the University of Texas, that Heisenberg spoke, in his *Philosophical Problems of Nuclear Science*, of the limited applicability of our present physics, of the necessity of broadening its laws if they are to apply to life.[9] Pascual Jordan, another founder of quantum mechanics, made a similar statement.

As to the usefulness of the considerations, I must admit that I do not see much of it. This may well be the reason for the lack of a more general interest on the part of physicists in the questions discussed. What I spoke about is philosophy and it would be presumptuous on my part to voice an opinion whether it shares the usefulness of the newborn child on which Abraham Lincoln commented. Even if not useful, I would like to summarize it when concluding my address.

I believe that the present laws of physics are at least imcomplete without a translation into terms of mental phenomena. More likely, they are inaccurate, the inaccuracy increasing with the increase of the role which life plays in the phenomena considered. The example of the wasp which does not seem to have sensations may indicate that even animals of considerable complexity are not far from being automata, largely subject to the present ideas of physics. On the other hand, the fact that the laws of physics are formulated in terms of observations is strong evidence that these laws become invalid for the description of observations whenever consciousness plays a decisive role. This also constitutes the difference between the view here represented and the views of traditional philosophers. They considered body and soul as two different and separate entities, though interacting with each other. The view given here considers inanimate matter as a limiting case in which the phenomena of life and consciousness play as little a role as the non-gravitational forces play in planetary motion, as fluctuations play in macroscopic physics. It is argued that, as we consider situations in which consciousness is more and more relevant, the necessity for modifications of the regularities obtained for inanimate objects will be more and more apparent.

ACKNOWLEDGMENT

I am much indebted to Dr. A. Shimony for his critical review of this article.

Princeton University

NOTES

* An article fully based on the address here presented appeared in *Foundations of Physics* **1** (1970).

[1] R. Descartes, *Oeuvres* (ed. by C. Adam and P. Tannery), Librairie Philosophique, Paris, 1967, Vol. XI, p. 119ff.

[2] Quoted by Thomas Huxley, ref. 3, p. 218.

[3] Thomas H. Huxley, *Selected Works*, Vol. 1: *Method and Results*, Appleton and Co., New York, 1902, p. 199ff. I am greatly indebted to Dr. W. Schroebel for calling my attention

to this essay. Ideas similar to those of Huxley were held by many others. P. B. Medawar, in *The Art of the Soluble* (Methuen, London, 1967) mentions, with approval, D'Arcy Thompson's very similar convictions.

[4] See, for instance, B. Petermann's *Gestaltslehre* (J. A. Barth, Leipzig, 1929), or W. Köhler's *The Task of Gestalt Psychology*, (Princeton University Press, 1969).

[5] An interesting account of the views of many philosophers, physicists, and biologists is presented in Chapter VII of S. L. Jaki's *The Relevance of Physics* (University of Chicago Press, Chicago, 1966).

[6] D. Bohm, *Special Theory of Relativity*, W. A. Benjamin, New York, 1965, p. 230.

[7] This is a point which was also brought out by G. G. Harris.

[8] N. Bohr, *Atomic Theory and the Description of Nature*, Cambridge University Press, 1934.

[9] W. Heisenberg, *The Philosophical Problems of Nuclear Science*, Faber and Faber, London, 1952. See also his *Physics and Philosophy*, Harper, New York, 1958, p. 155.

J. BRONOWSKI

NEW CONCEPTS IN THE EVOLUTION OF COMPLEXITY

Stratified Stability and Unbounded Plans

I. INTRODUCTION

Vitalism is a traditional and persistent belief that the laws of physics that hold in the inanimate world will not suffice to explain the phenomena of life. Of course it is not suggested, either by those who share the belief or by those like me who reject it, that we know all the laws of physics now, or will know them soon. Rather what is silently supposed by both sides is that we know what kind of laws physics is made up of and will continue to discover in inanimate matter; and although that is a vague description to serve as a premise, it is what inspires vitalists to claim (and their opponents to deny) that some phenomena of life cannot be explained by laws of this kind.

The phenomena that are said to be inaccessible to physics are of two different kinds. One school of vitalists stresses the *complexity of the individual* organism. The other school of vitalists asserts that physical laws are insufficient to explain the direction of evolution in time: that is, the *increase in complexity in new species*, such as man, when compared to old species from which they derive, such as the tree-shrews. The two grounds for finding physics to fall short are therefore quite distinct, and I shall discuss them separately. I begin with a summary sketch of each.

The first ground, then, is that the *individual* organism (even a single cell) functions in a way which transcends what physics can explain, and implies the existence of laws of another kind – what Walter Elsasser calls 'biotonic laws'. Elsasser argues that the development of an organism is too complex to be coded in the genes, and that there must be larger laws of biological organization that guide it overall. Eugene Wigner argues that development and reproduction is subject to so many statistical variations that there can be no certainty that the organism will survive them unless it is controlled by higher laws.

These arguments do not differ in principle from the classical argument, put forward (for example) by Bolingbroke early in the 18th century, that an

R. J. Seeger and R. S. Cohen (eds.), Philosophical Foundations of Science, 133–151.

organism is at least as complicated as a clock, and that we cannot imagine a clock to have come into being by accident. True, neither Elsasser nor Wigner speak of origins, but both imply that the configuration of parts and the sequence of functions in the cell requires a higher coordination than is provided by the laws of physics – by what one might call the simple engineering rules between the parts of the clock. Bolingbroke ascribes this higher coordination to God, and Elsasser and Wigner to biotonic laws, but this is only a difference in nomenclature.

The second school of vitalists finds another ground for claiming that the laws of physics are biologically incomplete, namely in questions about the *evolution* of organisms. Michael Polanyi asks questions of this kind, though he lumps all levels together – origins, functioning of individuals, and the sequence of species. He claims, as vitalists have always done, that there must be an overall plan which directs them all, and I shall criticize the confusion of meanings in his idea of plan or purpose. I shall distinguish between two concepts, the usual concept of a closed or bounded plan (that is, a tactic or solution for a defined problem), and a new concept of an *open or unbounded plan*, that is, a general strategy.

But beyond these concepts, there remains the crucial question raised by Polanyi – and others before him, of course, in earlier forms. Evolution has the direction, speaking roughly, from simple to more and more complex: more and more complex functions of higher organisms, mediated by more and more complex structures, which are themselves made of more and more complex molecules. How has this come about? How can it be explained if there is no overall plan to create more complex creatures – which means at least, if there is no overall law (other than evolution as a mechanism) to generate complexity? In particular, how do we square this direction with the Second Law of Thermodynamics, which (as a general description subsuming ordinary physical laws) predicts the breakdown of complex structures into simple ones? This is the constellation of questions to which I shall give most attention.

The course of my argument will incidentally reveal what additional physical laws we expect to discover as we continue to unfold the chemistry of life. In essence, they can be expected to be laws of specific relations between a few kinds of atoms which govern the *stability* of the structures that can be assembled from them. These are indeed laws of cooperative phenomena or ensembles, but they are highly particular and empirical, being simply accounts of the stability to be found in different conjunctions of matter under the conditions we know on earth.

II. MACHINERY OF THE CELL

I shall assume that we are all familiar with the way in which heredity is mediated by genes, which are molecules made up from four fairly small chemical bases that are strung out on two paired strands of DNA. Since there are many varieties of living creatures, and many genes in each, there are many different forms of DNA, in each of which the sequence of bases is different and is characteristic for directing some chemical process in that creature. The sequence of bases in a molecule of DNA spells out the twenty amino-acids which in their turn make the proteins. We have a simple hierarchy: the four bases are the four letters of the alphabet, each set of three letters makes up a word which is a fundamental amino-acid, and the twenty words in their turn are assembled into different sentences which are the different proteins.

The book of heredity is not the whole book of life. It records only those instructions which make a species breed true, so that the child is revealed as a copy of the parents. Yet this is far-reaching, because the living child, the living cell, is not a static copy, but is a dynamic process in which one action follows another in a characteristic pattern. We only dimly understand how the process of maturation unfolds this inborn ability. Nevertheless, we have made a beginning by seeing how the processes of making one protein after another are programed from the vocabulary of life so that they develop a stepwise, coordinated sequence.

Elsasser has argued that the development of living creatures (some of which consist of only a single cell) is too complex and too closely integrated to be directed by the genetic machinery. To this fundamental and, so to speak, primitive claim a biologist can only reply that there is absolutely no evidence to support it. No counting of constants, no calculation of the content of information in a set of chromosomes, can give any ground for it, because we simply do not know what the inner relations and restrictions between the parts of a complex molecule are. We do not even understand yet why a long protein molecule folds into the specific geometrical configuration which is its own, and not into any other. But we expect to find these laws, and we expect them to be no more esoteric or biotonic than, say, the laws which inform us that some assemblies of fundamental particles in physics make up stable atomic nuclei, and others do not.

In the same way, there is no evidence at all that the interaction of genes, either on the same or on different chromosomes, requires any kind of master law. In general it is mediated by local relations in the organ that is being shaped, just as the growth of a set of normal cells on a microscope slide is controlled in a regular array by chemical contact between the walls of neighboring cells. It may be that there are some places on the chromosomes where master controls reside, but if so, they can be expected simply to have the character of special genes. We already know that there are some master genes which control groups of other genes, for example by making them all more mutable or more stable.

A single-cell bacterium goes through its life cycle on an exact schedule, and every step in the sequence is a rearrangement of and within the molecules which compose it. So the machinery of the cell, the clockwork that drives and demonstrates its life, is a constant shaping and reshaping of its molecular material. It is suggested by vitalists that these cycles are so matched to its environment that they have the manifest plan or purpose, they are patently *designed*, to preserve the life of the bacterium. Of course this cannot be simply on the ground that the processes of life are cyclic: for there are plenty of physical actions, say in gravitation, which are cyclic – from the tides to the seasons. Nor can it be on the ground that there is something mysterious about the resistance of a cycle against disturbing forces: for that is displayed by any cycle – for example, a spinning top. The fact that a living cell is geared to go on living in the face of a disturbance is no more supernatural than the fact that a falling stone is geared to go on falling, and a stone in free space is geared to go on moving in a straight line. This *is* its nature, and does not require explanation any more (or in any other sense) than does the behavior of a ray of light or the complex structure of an atom of uranium.

Therefore the vitalist must have some more sophisticated idea of a plan than the mere persistence of a cycle, or even of a linked group of cycles. Usually what he does is to propose a distinction between different levels of explanation (and, by implication, of action). He says that we may well explain the mechanics of each cycle, but that this still misses the point of what they achieve as a totality; and that this achievement, namely the perpetuation of life in general, cannot be understood except as an overall plan or purpose.

I shall return to the discussion of the concept of plans later. Here I will content myself with repeating again that though there is much about life that we do not understand, there is no evidence at all that this is because there is a mystery in its basic processes. Living matter is different from non-living, but not because it follows different rules. The rules of organization by which the parts of a cell work together, the sequence of procedures which make it live, are understandable in the same terms as any other molecular process. The basic structures and sequences of life follow from those of dead nature without the intervention of any special powers or acts. There is no evidence for vitalism in the analysis of the cell, or of any simple assembly of cells that we know: a micro-organism, a limb, or a cancer.

III. THE ROLE OF ERRORS

A different and deeper question has been raised by Eugene Wigner. He remarks that living cells go through their life cycles by taking nourishment from the environment, and incorporating it either into their own structure or into that of daughter cells. During this transformation, he argues, quantum effects make it impossible to ensure with certainty that the cell will make an accurate copy of itself (or of some specified modification of itself); and therefore the laws of physics cannot suffice to explain how living matter perpetuates itself. Wigner calculates in detail how the manufacture in the cell of exact or similar, specified copies can be shown to be (in his phrase) 'infinitely unlikely'. Even if we treat the calculation as only indicative (as Wigner himself does at the end of his paper) the indication is that copying of any specific molecule (a strand of DNA, for instance) must produce an unacceptable error, as a result of unpredictable quantum events. And this, Wigner implies, flies in the face of our daily observation of the process of reproduction.

The argument is strange because it uses the same quantum effects that Max Delbrück and Erwin Schrödinger used long ago, to arrive at exactly the opposite conclusion. Schrödinger reasoned that quantum effects are essential to explain the uniqueness of a living form. Wigner turns this reasoning upside-down, and concludes that no living form can maintain its identity in the face of quantum disturbances.

When we examine the argument which leads Wigner to his conclusion, its limitations are evident (and are acknowledged by him). Wigner's procedure is to transform the quantum state vectors of the cell and its nourishment taken together, into the state vectors of the two similar cells that are to result (and of the rejected part of the nourishment). It appears that the number of equations to be satisfied is far larger than the number of unknowns at our disposal, if it is assumed (as in the absence of any more specific knowledge Wigner has to assume) that the Hamiltonian matrix which represents the transformation consists, apart from its symmetry, of *random* elements. But of course this begs the whole question, because it necessarily disregards all relations within the matrix of transformation – that is, the organized inner structure of the process of ingestion and cell division. So long as we do not know the relations between the elements of the matrix, the counting of unknowns and variables at our disposal is quite inconclusive, and can be wholly misleading.[1]

But in fact Wigner's procedure is (as he admits) even less realistic than this first criticism implies. For it is the nature of any argument that proposes a count of unknowns and variables that it can only assert or deny the existence of a solution that makes the outcome of the process *certain*. Even if we were to accept Wigner's assumptions, therefore, we could only conclude that the process of cell division as he idealizes it cannot be guaranteed to yield a second similar cell with *certainty*. But nothing is asserted, or could be concluded, about any process of cell division which has only a *probability* of producing viable offspring – even if the probability is as high as 0.99. The reasoning[2] is not applicable to probable outcomes, however high or low.

However, experience shows that no biological process works with certainty, and that few organisms produce similar and viable offspring with as high a probability as 0.99. We must therefore recognize that Wigner's argument misses the essence of biological processes, namely that they do *not* function with certainty. Indeed, the evolution of more highly organized forms of life than the cell would not have been possible if they had done so.

A cell in its task of simply living makes proteins over and over again from the same blueprint, and now and again it also replicates the blueprint when it divides in two. No conceivable machinery, within the known laws of nature or any that we might think of, can carry on this endless work of copying with zero tolerance – which means, without making individual mistakes from time to time. We put these mistakes down to quantum effects, and that is right; yet in a sense quantum physics is simply a formalization of our well-founded and much wider conviction that no natural process can work with zero tolerance and so be immune from error. Just as we know that perpetual motion is impossible, and can derive a great part of classical mechanics from this Law of the Impossible, so we know that perpetual accuracy of reproduction is inconceivable, and we can regard quantum physics as a specification of that knowledge: the specification of a non-zero tolerance, namely Planck's quantum constant, below which we cannot press.

The discrepancy arises because life has two separate components, and Wigner ignores just that creative component which is characteristic of life and is absent in dead matter. Life is not only a process of accurate copying: that is carried out quite as neatly in the geometrical scaffolding of a dead crystal. Life is also and essentially an evolutionary process, which moves forward only because there are errors in the copy, and every so often one of these errors is successful enough to be incorporated as another step or threshold in its progression. It is important to understand that the living creature combines both procedures, and to see how it does so.

The accumulation of individual errors is certainly a handicap to a cell: but we have to distinguish between two kinds, or better between two places, of error. Errors made now and again in producing a molecule of a protein which is merely a momentary step in the metabolism of the cell are not likely to be important, for the next molecule of the protein can be expected to be normal again. But there are some proteins which play a basic part in the productive machinery of the cell, and act as jigs or machine-tools to help make copies of other proteins. When such a master molecule is wrongly made, it will in its turn cause errors in the making of other copies, and as a result the error will be cumulative. Leslie Orgel has suggested that errors in such master molecules may be the cause for cells breaking down, and the suggestion is supported by recent experiments. Moreover, if Orgel's picture is right, every cell must break down sooner or later when it accumulates errors in some master molecule. There can be no immortal cells. Indeed, Leonard Hayflick has found in careful experiments that by the time a cell has divided about 50 times the clone of cells formed from it all fail to divide. *The machinery of life ensures the death of individuals.*

But exactly this machinery also ensures the evolution of new forms. *The errors which destroy the individual are also the origin of species.* Without these errors, there would be no evolution, because there would be no raw material of genetic mutants for natural selection to work on. There would only be one universal form of life, and however well adapted that might have been to the environment in which it was formed, it would have perished long ago in the first sharp change of climate. When Wigner and Walter Elsasser say that there must be some biological law different from the laws of physics in order that copying in the organism shall be free from error, and the storage of the instructions which govern its exact form and development from the cell shall be perfectly accurate, they are asking for the immortality of the individual but ensuring the destruction of species. We have only to look about us to see that the evidence is against them, on both counts.

IV. EVOLUTION IS CRUCIAL

Evolution is crucial to any discussion of biology, because this and only this makes biology a different kind of subject from physics. A recent survey by Desmond Bernal of what is known about biological molecules begins with a chapter on 'The Nature of Biology', and he opens it with a section whose title bluntly asks 'Does Biology Exist?'. The question is meant to remind us abruptly that biology is a different kind of study from other sciences, because it studies a very specialized and, as it were, accidental phenomenon. Bernal puts the distinction precisely:

> I believe there is a radical difference, fundamentally a philosophical difference, between biology and the so-called exact or inorganic sciences, particularly physics. In the latter we postulate elementary particles which are necessary to the structure of the universe and that the laws controlling their movements and transformations are intrinsically necessary and in general hold over the whole universe.
>
> Biology, however, deals with descriptions and ordering of very special parts of the universe which we call life – even more particularly in these days, terrestrial life. It is primarily a descriptive science, more like geography, dealing with the structure and working of a number of peculiarly organized entities, at a particular moment of time on a particular planet.

Perhaps Bernal is too narrow when he compares biology with geography, which is mainly a description of space. The comparison that comes to my mind is rather with geology, which like biology deals with configurations in space and traces their behavior in time. But essentially the distinction that he makes is well grounded; life has a more accidental and local character than the other phenomena of the physical world.

I would go further, and say that life has a more *open and unbounded* character than the other phenomena of the physical world: it is incomplete and unfinished in a way that they are not. That is, biology has a different character from physics at every point in evolutionary time. The biological universe that we are discussing today is different from that which we could have discussed 3 million years ago – when indeed there was no *homo sapiens* to discuss it. And we must expect that the biological universe of 3 million years hence may be quite as different again.

Let me make this distinction explicitly. The development of life from one form to another is unlike that of the rest of the physical world, because it is triggered by accidents, and they give each new form its unique character. Life is not an orderly continuum like the growing of a crystal; it creates new expressions, and remains constantly open to them, as a succession of errors which can only occur *because life is accident-prone*. The nature of life is only expressed in its perpetual evolution, which is another name for the succession (and the success) of its errors.

The molecules that make up a cell or an individual form a physical system with many states. If we map all possible states (disorderly as well as orderly) by the points of an abstract space, there is a narrow sequence of points which maps the sequence of steps in the life cycle of the cell, or the sequence of states of the individual before he returns to (approximately) the same state – say, waking in the morning. Since the cycle of states returns on itself, the sequence of points forms (virtually) a closed loop. Thus on the scale of the cell and of the individual, life is a process which is topologically closed. It runs over more or less the same loop again and again, and in time it runs down.

Yet life does not run down in the way in which natural processes run down, by the general leveling of energy peaks which is called the Second Law of Thermodynamics. The death of a cell or of an individual is not a leveling out, a falling apart of the architecture, as decay after death is. Instead, death is a failure of the metabolism to continue its cycles, and it begins in a failure to repeat the cycles accurately. It seems likely that the cell or individual is clogged by errors which are inevitable and become cumulative: whatever the underlying cause proves to be, the topological loop wavers and comes to a stop.

But life as an evolutionary sequence is not a closed loop. On the contrary, life as evolution is topologically open, for it has no cycle in time. Yet it derives this openness from just such accidents or errors, at least in kind, as kill the individual. The mechanism of survival for a species is its evolution, and evolution is that quantum resonator or multiplier, the exploitation of an accident to create a new and unique form, for which Max Delbrück was looking when he came into biology.

In summary, the closed loop of an individual life and the open path of evolution are dual aspects of life. The common mainspring of quantum accidents (that is, of errors in the copying of biological molecules) may be responsible for individual death, and is certainly responsible for evolution. Both are only properly understood when they are put side by side as complementary parts or processes of life.

V. BOUNDED AND UNBOUNDED PLANS

I will make this distinction in another form by examining the different arguments for vitalism which have been advanced by Michael Polanyi, first at the level of the cell, and then at the level of evolution. At the first or lower level, Polanyi simply says that to explain the *machinery* of a cell is like explaining the machinery of a watch while missing the most important thing about it, which is that a watch is planned *for a purpose* – to tell the time.

The design of a watch is the classical illustration for God's design in man that deists introduced in the 18th century. Henry St. John, Viscount Bolingbroke, and William Paley in the *Evidences of Christianity*, used it to claim that man is a more ingenious machine than is a watch, and must therefore be supposed to have been designed by a more ingenious creator. Polanyi now gives this argument a new look by saying that just as the design of the watch points to and is only understood in its purpose, so the design of the machinery of life points to and is only understood at a higher level of explanation by purpose. He calls this the boundary conditions for any mechanism, but these words only restate the requirement which he proposes, namely that it must fit into and serve some overall plan outside itself. In essence, the argument remains in the

18th century: we *know* that the watch is ingenious because we know the purpose for which it is planned.

Perhaps it is easiest to see what is wrong with this argument by deriving a paradox from it, as follows. The argument is intended to show that man (and any other living form) is not simply a machine. In order to show this, he is compared with a typical machine, namely a watch; and it is concluded that he is more sophisticated, namely more purposeful, than the mechanism that drives the watch. How is this concluded? By showing that the watch itself, as a machine, is more sophisticated, namely more purposeful, than the mechanism that drives it. In short, even a machine is not merely a mechanism, or what we usually call a machine. Man therefore is not a machine because he is a machine, and it has already been shown that a machine is not a machine.

How does this paradox come about? Evidently, by confusing the *external function* for which the machine has been made (by the watchmaker, for example) with the *inner plan* which the living creature follows as its natural and species-specific sequence of operations. To claim that this inner plan means anything more can only be justified by appealing to a quite abstract, classical tenet of philosophy, that the reduction of a sequence to its parts is not a sufficient explanation of their totality. But the implication is out of place here, where it merely restates the analogy of the watch which is designed to tell the time. There are indeed contexts in philosophy in which reductionism is not enough. But reductionism is valid and sufficient when it is an *historical* explanation, so that it presents a temporal and logical sequence of steps by which the result has been reached. (Indeed, all causal explanations are of this kind, and can only be challenged if we challenge the first cause.) To reduce a whole to its parts is a valid exposition of its plan if in fact the parts have come together in time, step by step, in building up a sequence of lesser wholes. So it is valid to regard an organism as an historical creation whose plan is explained by its evolution. But the plan of life in this sense is *unbounded*. Only unbounded plans can be creative; and evolution is such a plan, which has created what is radically new in life, the dynamic of time.

So it is timely now to consider *evolution as an open and unbounded plan*, and to ask what additional principles are needed to make it capable of creating the new living forms that we know. For it is essential that we recognize these forms as genuine creations, which have not been formed on a bounded plan like that which runs its rigid course from the seed to the full-grown plant.

The distinction here is between a sequence of actions which is fixed in advance by the end state that it must reach, and a train of events which is open and unbounded to the future because its specific outcome is not foreseen. *Any bounded plan is in essence the solution to a problem*, and life as a mechanism has this character. By contrast, the sequence of events that constitutes an unbounded plan is invented moment by moment from what has gone before, and the outcome is not solved but created. Life as an evolution is a creation of this kind.

In this analysis, the vitalist's question becomes directed to a different issue: the relation between the direction of evolution and the direction of time. In a history of 3000 million years, evolution has not run backward – at least, by and large, and in a definable statistical sense, it has not run backward. (The existence of some lines of regression, such as those which have produced the viruses, does not change this general characterization.) Why is this? Why does evolution not run at random hither and thither in time? What is the screw that moves it forward, or at least, what is the ratchet that keeps it from slipping back? Is it possible to have such a mechanism which is not planned? What is the relation that ties evolution to the arrow of time, and makes it a barbed arrow?

The paradox to be resolved here is classical in science: how can disorder on the small scale be consonant with order on the large scale, in time or in space. If this question is asked of the molecules in a stream of gas, the answer is easy – the motion imposed on the stream swamps the random motions of the individual molecules. But this picture will not help to explain evolution, because there is no imposed motion there. On the contrary, if we were to assume an imposed motion we would be accepting the postulate of vitalism. Evolution must have a different statistical form, in which there is *an inherent potential for large-scale order to act as a sieve* or selector on the individual chance events. There are such cooperative phenomena in physics: for example, in the structure of crystals, which we understand, and perhaps in the structure of liquids, which we do not. The existence of a potential of order in the selection of chance events is clear, but what is never clear in advance is how it will express itself. It is here that we need two additional principles in evolution to give it a natural order in time.

VI. STRATIFIED STABILITY

There are five distinct principles which make up the concept of evolution, as I interpret it. They are: (a) family descent; (b) natural selection; (c) Mendelian inheritance; (d) fitness for change; and (e) stratified stability.

The first three are familiar, and I need not elaborate them; they make up the standard account of the mechanism of evolution that has been accepted since R. A. Fisher first formalized it in *The Genetical Theory of Natural Selection.*

But in my view, it is now necessary to add the two further principles which I propose, namely (d) fitness for change, and (e) stratified stability.

They are concerned, the one with the *variability* of living forms, and the other with their *stability*; and between them they explain how it comes about that biological evolution has a direction in time – and has a direction in the same sense as time. The direction of evolution is an important and indeed crucial phenomenon, which singles it out among statistical processes. For in so far as statistical processes have a direction at all, it is usually a movement towards the average – and that is exactly what evolution is *not.*

Of the two new principles that I propose, the first is only peripheral to my theme here, and I shall deal with it quite briefly. In order that a species shall be capable of changing to fit its environment tomorrow, it must maintain its *fitness for change* today. The dormant genes that may be promoted tomorrow when they become useful must be preserved today when they are useless. And for this they must be held now in a setting of other genes which makes it possible to promote them rapidly. If this is to be done in the present, without some mysterious plan for the future, it must be by *natural selection* not for this or that variant but *for variability itself.*

It is evident that there is natural selection in favor of genetic variability. The selection is made by the small changes, up and down and up again and down again, by which the environment flutters about its mean. A long-term trend in the environment which lasts a hundred generations or so will in that time select a new adaptation. But the short-term fluctuation which goes one way for a few generations and then the other way for a few generations will meanwhile select for adaptability. That is, the short-term fluctuation favors the establishment of an arrangement of genes that will help mutant genes to express themselves.

Indeed, we know now that there are single genes which function specifically to enhance variability. For example, there are single genes which increase the rate of mutation in several other genes at the same time. Their action could explain the tendency for genetic change in one part of an organism (particularly, and in the first place, a haploid organism) to keep pace with change in other parts. A master gene of this kind, which increases mutation, is a mechanism that opens up the future, not by foreseeing it but by promoting the capacity for change.

I turn now to the crucial part of my argument. It is evident that we cannot discuss the variability of organisms and species without also examining their stability. We have therefore also to trace a mechanism for stability, as the second of the two balanced mechanisms that are needed to complete our understanding of evolution. I call this, the fifth and last principle in my analysis of evolution, the concept of *stratified stability.*

Evolution is commonly presented, even now, as if it required nothing but natural selection to explain its action, one minute step after another, as it were gene by gene. But an organism is an integrated system, and that implies that its coordination is easily disturbed. This is true of every gene: normal or mutant, it has to be incorporated into the ordered totality of the gene complex like a piece in a jigsaw puzzle.

Yet the analogy of the jigsaw is too rigid: we need a geometrical model of stability in living processes (and in the structures that carry them out) which is not so land-locked against change. Moreover, the model must express the way in which the more complex forms of life arise from the simpler forms, and arise later in time. This is the model of stratified stability.

There are evolutionary processes in nature which do not demand the intervention of selective forces. Characteristic is the evolution of the chemical elements, which are built up in different stars step by step, first hydrogen to helium, then helium to carbon, and on to heavier elements. The encounter of hydrogen nuclei makes helium simply (though indirectly) because they hold together: arrangements are briefly formed which in time form the more complex configuration that is helium. Each helium nucleus is a new unit which is stable, and can therefore be used as a new raw material to build up still higher elements.

The most telling example is the creation of carbon from helium. Two helium nuclei which collide do not make a stable element, and fly apart again in less than a millionth of a millionth of a second. But if in that splinter of time a third helium nucleus runs into the pair, it binds them together and makes a stable triad which is a nucleus of carbon. Every carbon atom in every organic molecule in every cell in every living creature has been formed by such a wildly improbable triple collision in a star.

Here then is a physical model which shows how simple units come together to make more complex configurations; how these configurations, if they are stable, serve as units to make higher configurations; and how these higher configurations again, provided they are stable, serve as units to build still more complex ones, and so on. Ultimately a heavy atom such as iron, and perhaps even a complex molecule containing iron (such as hemoglobin), simply fixes and expresses the potential of stability which lay hidden in the primitive building blocks of cosmic hydrogen.

The sequence of building up stratified stability is also clear in living forms. Atoms build the four base molecules, thymine and adenine, cytosine and guanine, which are very stable configurations. The bases are built into the nucleic acids, which are remarkably stable in their turn. And the genes are stable structures formed from the nucleic acids, and so on to the sub-units of a protein, to the proteins themselves, to the enzymes, and step by step to the complete cell. The cell is so stable as a topological structure in space and time that it can live as a self-contained unit. Still the cells in their turn build up the different organs which appear as stable structures in the higher organisms, arranged in different and more and more complex forms.

Two special conditions have assisted this mode of climbing from simple to complex. First, of course, there is the energy which comes to us from the sun, which increases the number of encounters between simple units and helps to lift them over the next energy barrier above them. (In the same way, simple atomic nuclei encounter one another reasonably often, and are lifted over the next energy barrier above them, by the energy in hot stars.) And second, natural selection speeds up the establishment of each new stratum of stability in the forms of life.

The stratification of stability is fundamental in living systems, and it explains why evolution has a consistent direction in time. Single mutations are errors at random, and have no fixed direction in time, as we know from experiments. And natural selection does not carry or impose a direction in time either. But the building up of stable configurations does have a direction, the more complex stratum built on the next lower, which cannot be reversed in general (though there can be particular lines of regression, such as the viruses and other parasites which exploit the more complex biological machinery of their hosts). Here is the barb which evolution gives to time: it does not make it go forward, but it prevents it from running backward. The back mutations which occur cannot reverse it in general, because they do not fit into the level of stability which the system has reached: even though they might offer an individual advantage to natural selection, they damage the organization of the system as a whole and make it unstable. Because stability is stratified, evolution is open, and necessarily creates more and more complex forms.[3]

There is therefore a peculiar irony in the vitalist claim that *the progress of evolution from simple to complex* cannot be the work of chance. On the contrary, as we see, exactly this *is how chance works, and is constrained to work by its nature.* The total potential of stability that is hidden in matter can only be evoked in steps, each higher layer resting on the layer below it. The stable units that compose one layer are the raw material for random encounters which will produce higher configurations, some of which will chance to be stable. So long as there remains a potential of stability which has not become actual, there is no other way for chance to go. It is as if nature were shuffling a sticky pack of cards, and it is not surprising that they hold together in longer and longer runs.

VII. THE SECOND LAW OF THERMODYNAMICS

It is often said that the progression from simple to complex runs counter to the normal statistics of chance that are formalized in the Second Law of Thermodynamics. Strictly speaking, we could avoid this criticism simply by insisting that the Second Law does not apply to living systems in the environment in which we find them. For the Second Law applies only when there is no overall flow of energy into or out of a system: whereas all living systems are sustained by a net inflow of energy.

But though this reply has a formal finality, in my view it evades the underlying question that is being asked. True, life could not have evolved in the absence of a steady stream of energy from the sun – a kind of energy wind on the earth. But if there were no more to the mechanism of molecular evolution than this, we should still be at a loss to understand how *more and more complex* molecules came to establish themselves. All that the energy wind can do, in itself, is to increase the range and frequency of variation around the average state: that is, to stimulate the *formation* of more complex molecular arrangements. But most of these variant arrangements fall back to the norm almost at once, by the usual thermodynamic processes of degradation; so that it remains to be explained why they do not all do so, and how instead some complex arrangements *establish* themselves, and become the base for further complexity in their turn.

It is therefore relevant to discuss the Second Law, which is usually interpreted to mean that all constituent parts of a system must fall progressively to their simplest states. But this interpretation quite misunderstands the character of statistical laws in general in non-equilibrium states. The Second Law describes the final equilibrium state of a system; if we are to apply it, as here, to stable states which are far from equilibrium, we must interpret and formulate it differently. *In these conditions, the Second Law of Thermodynamics becomes a physical law only if there is added to it the condition that there are no preferred states or configurations.* In itself, the Second Law merely enumerates all the configurations which a system could take up, and it remarks that the largest number in this count are average or featureless. Therefore if there are no preferred configurations (that is, no hidden stabilities in the system on the way to equilibrium) we must expect that any special feature that we find is exceptional and temporary, and will revert to the average in the long run. This is *a true theorem in combinatorial arithmetic*, and (like other statistical laws) *a fair guess at the behavior of long runs.* But it tells us little about the natural world which, in the years since the Second Law seemed exciting, has turned out to be full of preferred configurations and hidden stabilities, even at the most basic and inanimate level of atomic structure.

The Second Law describes the statistics of a system around equilibrium whose configurations are all equal, and it makes the obvious remark that chance can only make such a system fluctuate around its average.[4] There are no stable states in such a system, and there is therefore no stratum that can establish itself; the system stays around its average only by a principle of indifference, because numerically the most configurations are bunched around the average.

But if there are hidden relations in the system on the way to equilibrium which cause some configurations to be stable, the statistics are changed. The preferred configurations may be unimaginably rare; nevertheless, they present another level around which the system can bunch, and there is now a counter-current or tug-of-war within the system between this level and the average. Since the average has no inherent stability, the preferred stable configuration will capture members of the system often enough to change the distribution; and in the end, the system will be established at this level as a new average. In this way, local systems of a fair size can climb up from one level of stability to the next, even though the configuration at the higher level is rare. When the higher level becomes the new average, the climb is repeated to the next higher level of stability; and so on up the ladder of strata.

So, contrary to what is usually said, the Second Law of Thermodynamics does not fix an arrow in time by its statistics alone. Some empirical condition must be added to it before it can describe time (or anything else) in the real world, where our view is finite.

When there are hidden strata of stability, one above another, as there are in our universe, it follows that the direction of time is given by the evolutionary process that climbs them one by one. Indeed, if this were not so, it would be impossible to conceive how the features that we remark could have arisen. We should have to posit a miraculous beginning to time at which the features (and we among them) were created ready-made, and left to fall apart ever since into a tohubohu of individual particles.

Time in the large, open time, takes its direction from the evolutionary processes which mark and scale it. So it is pointless to ask why evolution has a fixed direction in time, and to draw conclusions from the speculation. It is evolution, physical and biological, that gives time its direction; and no mystical explanation is required where there is nothing to explain. The progression from simple to complex, the building up of stratified stability, is the necessary character of evolution from which time takes its direction. And it is not a forward direction in the sense of a thrust towards the future, a headed arrow. What evolution does is to give the arrow of time a barb which stops it from running backward; and once it has this barb, the chance play of errors will take it forward of itself.

NOTES

[1] For example, it is a familiar paradox in projective algebraic geometry that an argument like Wigner's can be used to prove that every conic section is a pair of straight lines.
[2] For example, if Wigner's reasoning here were applied to nuclear fission, it would persuade us that a chain reaction is impossible – because we can never stipulate that the entry of a neutron into a specified nucleus will *certainly* release two neutrons.
[3] For a fuller account, see my Condon lectures for 1967 at the University of Oregon, published under the title *Nature and Knowledge*. Recently David Bohm has put forward a similar scheme for an inherent hierarchy of complexity, in which he calls *levels of order* what I call *strata of stability*.
[4] It should be remarked that von Neumann's quantum theoretical proof of the Second Law, like Wigner's argument above, assumes that the behavior of a system to which it is applied can be represented by a randon symmetric Hamiltonian matrix – that is, contains no hidden inner relations.

BIBLIOGRAPHY

Bernal, J. D.: 1965, 'Molecular Structure, Biochemical Function, and Evolution', in *Theoretical and Mathematical Biology* (ed. by T. H. Waterman and H. J. Morowitz), pp. 96–135.

Bethe, H. A.: 1939, 'Energy Production in Stars', *Physical Review* **55**, 434.

Bohm, D.: 1969, 'Some Remarks on the Notion of Order', in *Towards a Theoretical Biology* (ed. by C. H. Waddington), pp. 18–40.

Bronowski, J.: 1965, *The Identity of Man.*

Bronowski, J.: 1969, *Nature and Knowledge.*

Delbrück, M.: 1949, 'A Physicist looks at Biology', *Transactions of the Connecticut Academy of Arts and Sciences* **38**, 173; and 1966, in *Phage and the Origins of Molecular Biology* (ed. by J. Cairns, G. S. Stent, and J. D. Watson) pp. 9–22.

Elsasser, W. M.: 1958, *The Physical Foundation of Biology.*

Fisher, R. A.: 1930, *The Genetical Theory of Natural Selection.*

Harrison, B. J. and Holliday, R.: 1967, 'Senescence and the Fidelity of Protein Synthesis in Drosophila', *Nature* **213**, 990.

Hayflick, L.: 1966, 'Cell Culture and the Aging Phenomenon', in *Topics in the Biology of Aging* (ed. by P. L. Krohn), pp. 83–100.

Holliday, R.: 1969, 'Errors in Protein Synthesis and Clonal Senescence in Fungi', *Nature* **221**, 1224.

Neumann, J. von: 1932, *Mathematische Grundlagen der Quantenmechanik.*

Orgel, L.: 1963, 'The Maintenance of the Accuracy of Protein Synthesis and its Relevance to Ageing', *Proceedings of the National Academy of Science* **49**, 517.

Paley, W.: 1794, *Evidences of Christianity.*

Polanyi, M.: 1967, 'Life Transcending Physics and Chemistry', *Chemical and Engineering News* **45**, No. 35, 54.

Polanyi, M.: 1968, 'Life's Irreducible Structure', *Science* **160**, 1308.

Schrödinger, E.: 1944, *What Is Life?.*

Watson, J. D. and Crick, F. H. C.: 1953, 'A Structure for Deoxyribose Nucleic Acid', *Nature* **171**, 737.

Wigner, E. P.: 1961, 'The Probability of the Existence of a Self-Reproducing Unit', in *The Logic of Personal Knowledge. Essays Presented to Michael Polanyi on his Seventieth Birthday.*

PART III

THE GEORGE SARTON MEMORIAL LECTURE

1969

MARTIN J. KLEIN

BOLTZMANN, MONOCYCLES AND MECHANICAL EXPLANATION

I

For more than two hundred years – from Isaac Newton in the seventeenth century to Ludwig Boltzmann at the close of the nineteenth – physical theory had a fixed, clear goal. Christiaan Huygens expressed it in his *Treatise on Light* in 1690, when he referred to "the true Philosophy, in which one conceives the causes of all natural effects in terms of mechanical motions."[1] Huygens was convinced that either one sought for mechanical explanations or else one had to "renounce all hopes of ever comprehending anything in Physics." Newton's *Principia* became the great model that scientific theory was to follow, and Newton was very definite in his views on this subject. "The whole burden of philosophy," he wrote, "seems to consist in this – from the phenomena of motions to investigate the forces of nature, and then from these forces to demonstrate the other phenomena." To this end he had developed the rational mechanics of the first two Books of the *Principia*, and his 'explication of the System of the World' in the third Book was merely 'an example' of his method – a superb example, but an example nevertheless. Newton expressed the wish that "we could derive the rest of the phenomena of Nature by the same kind of reasoning from mechanical principles."[2] This wish described the program of generations of Newton's successors who tried to do for the other forces of nature what he had done for gravitation.

By the latter part of the nineteenth century the program of mechanical explanation had made enormous progress. Mechanics itself had grown into an extremely flexible and powerful mathematical theory. Its concepts and methods had been expanded so that it could deal with rigid and elastic solids, and fluids of various kinds, as well as with systems of particles. There was now a mechanical theory of heat: the new science of thermodynamics incorporated the old idea that heat is motion into a conceptual system whose great scope was exhibited in Josiah Willard Gibbs's memoirs of the 1870's. Thermodynamics was accompanied by a

R. J. Seeger and R. S. Cohen (eds.), Philosophical Foundations of Science, 155–175.

more detailed mechanical theory of heat, the kinetic theory of gases, which promised a complete mechanical explanation of "the nature of the motion we call heat."[3] This theory also offered new possibilities for exploring the nature of matter, as it combined the atomic hypothesis with Newtonian mechanics and the theory of probability to yield results that could be brought to experimental test.

The Newtonian program had also had much success in dealing with electric and magnetic phenomena. The laws of force for both electric charges and magnetic poles at rest were determined before the end of the eighteenth century. These laws were quickly developed into a mathematical theory of electrostatics. When the phenomena of electromagnetism were discovered, more elaborate laws of force for current elements and moving charges were proposed by Ampère, Weber and others. But the most fruitful electromagnetic theory was James Clerk Maxwell's dynamical theory of the electromagnetic field. Maxwell gave up forces acting at a distance in favor of Michael Faraday's local action in a medium, but he viewed his theory as being essentially dynamical, as its name implied. Heinrich Hertz's experiments in the 1880's gave solid support to the most striking result of Maxwell's theory – its identification of light as an electromagnetic wave.

It was Hertz who restated the classic goal of physical theory in almost the same words that had been used in the seventeenth century: "All physicists agree that the problem of physics consists in tracing the phenomena of nature back to the simple laws of mechanics."[4] But by the time Hertz's words appeared in print in 1894, they were no longer strictly true. Physicists did not *all* agree then about the nature of mechanical explanation, nor did they all believe that mechanical explanation was the goal towards which to strive. There were even some who doubted that mechanics was the most fundamental science, and other candidates for this honor were being seriously considered.

The mechanical theories had not been uniformly successful. The kinetic theory of gases, for example, had been beset by complications and difficulties from its very beginnings. Doubts and controversies accompanied its positive results. The most general deduction from the kinetic theory was the equipartition theorem, and in 1875 Maxwell, himself one of the creators of the theory, referred to its failure as "the greatest difficulty the theory has yet encountered."[5] Almost twenty years later proponents of the kinetic theory were still hoping that this failure was "not necessarily a fatal objection to the theory."[6]

Maxwell's theory of the electromagnetic field provided other difficulties for the mechanical world picture. Maxwell had arrived at the view that light is an electromagnetic wave with the help of a detailed and rather complicated mechanical model, but he never claimed that this model faithfully represented reality. In the later and more complete version of his theory the model disappeared completely, although Maxwell continued to treat the electromagnetic field by the methods of analytical dynamics. His rather subtle views on the nature of mechanical explanation puzzled most of his contemporaries. Some (like Lord Kelvin) thought he had not gone far enough in constructing mechanical models; others (like Pierre Duhem) drew back in dismay from the model Maxwell had used, and tried to manage without any mechanical assistance at all.[7] All agreed on one point: the relationship between Maxwell's theory and the old goal of mechanical explanation was not clear.

The generally received idea of what scientific thought was like in the last quarter of the nineteenth century was expressed by Alfred North Whitehead, who is not known as an exponent of received ideas. He described it as "one of the dullest stages of thought since the time of the First Crusade," and called it "an age of successful scientific orthodoxy, undisturbed by much thought beyond the conventions."[8] This description will not do. The long tradition of mechanical explanation was coming to an end in those years, and its difficulties were reflected in a wide variety of specific problems. The debates during the 80's and 90's over the validity of energetics, the need for atomism, the status of the second law of thermodynamics, the nature of Maxwell's theory, the possibility of a purely electromagnetic explanation of mass – all of these were related to this central issue. Questions of fundamental principle drew an unusual amount of attention in this supposedly stodgy era.[9]

If we are to go beyond an empty discussion of generalities we have to look at the ways in which individual physicists constructed, used and argued about mechanical explanations. I have chosen Ludwig Boltzmann as the central figure in my analysis since he was one of the most ardent practitioners of mechanical physics and a most articulate and vigorous spokesman for his point of view.[10] Boltzmann worked and wrote on these problems for forty years, but I shall concentrate on the part that Helmholtz's monocycles played in Boltzmann's thought for a decade or so. It sounds like a very particular question, but then it is only in William Blake's 'minute particulars' that one can find historical truth.

II

Hermann von Helmholtz published his studies on monocyclic systems in 1884.[11] He had found a class of purely mechanical systems whose behavior paralleled in a remarkable way the behavior required by the second law of thermodynamics. Helmholtz had at least one avid reader for his work – Ludwig Boltzmann.

Boltzmann began his scientific career in 1866 with an attempt to find the mechanical explanation of the second law of thermodynamics.[12] Only the previous year Rudolf Clausius had given the second law its final thermodynamic formulation: every system is characterized by its entropy, a well-defined function of the variables of state, and this entropy must increase when a closed system carries out an irreversible process.[13] Boltzmann, then only twenty-two, claimed that he had found 'a purely analytical, completely general proof' of this law which reduced it to a theorem in mechanics. Boltzmann's claims were too strong. His mechanical theorem allowed one to define the entropy in terms of mechanical concepts, but only when the system was strictly periodic, an unreasonably severe limitation for anything as complicated as a sample of gas. Even with this limitation Boltzmann had not done anything to explain the irreversibility required by the second law.

In the course of the next few years Boltzmann studied Maxwell's papers on the kinetic theory of gases. He learned from Maxwell that it was essential to describe a gas by statistical methods, that the statistical distribution of molecular velocities was basic to any analysis of the properties of a macroscopic system. Boltzmann made this approach his own, generalizing it and applying it to new problems.[14] Foremost among these was the problem he had started with, and in 1871 Boltzmann gave a new 'analytic proof' of the second law.[15] This time the second law, or rather those aspects of the second law that deal with equilibrium and reversible processes, appeared as theorems in a new and as yet unnamed discipline – statistical mechanics. The laws of probability had as essential a part to play as the laws of mechanics in the new explanation of the second law. A year later Boltzmann showed that this general approach could also account for the nature of irreversibility, when he proved the result he later called the H-theorem.[16] But this H-theorem had a paradoxical aspect, since it derived *irreversibility* from the *reversible* laws of mechanics. The paradox was pointed out first by Josef Loschmidt and repeatedly thereafter by a variety of critics. Boltzmann answered this objection by insisting upon the statistical character of his result and, indeed, of the second law.[17] The observed irreversible behavior of nature, as exemplified by the flow of heat from hot bodies to colder ones and not the other way, was only a probable and not a certain occurrence. It was overwhelmingly more probable than the reverse behavior, to be sure, but it was still only a statistical result. Boltzmann reinforced this interpretation of the second law by showing that the entropy, whose increase signified irreversibility, was itself a measure of the probability of finding the system in the state in question.[18] Irreversibility and increasing entropy then meant simply that a system, when left alone, will evolve into the state of maximum probability, the state that can be achieved in the greatest number of molecular configurations.

Boltzmann reached this fully statistical interpretation of the second law of thermodynamics in 1877. He evidently believed that the problem was settled, that he had explained the essential features of the second law, and he turned his attention to other matters. His later discussions of this problem, in the 90's, were undertaken only in response to new criticisms, and always consisted of elaborations and more careful restatements of his statistical point of view.[19] There is an exception to this general statement, however, and it brings us back to Helmholtz and his monocycles. When Boltzmann read Helmholtz's first papers on monocyclic systems in the spring of 1884, he immediately set to work to explore this apparently very different approach to the meaning of the second law. To understand Boltzmann's response we must now look at what Helmholtz had done.[20]

Helmholtz considered a mechanical system, described by Lagrange's equations of motion, whose generalized coordinates could be separated into two distinct classes. Coordinates of the first type were cyclic: these coordinates did not themselves appear in the Lagrangian function of the system. Their time derivatives, the generalized velocities, did appear in the kinetic energy and so in the equations of motion. These velocities were considered to be large, so large in fact that the velocities of the other class of coordinates could be neglected by comparison. The system was called monocyclic if it had only one independent cyclic coordinate. But why consider such systems at all?

Helmholtz had a definite analogy in mind. In thermodynamics one has to treat complicated systems, such as a container full of a gas, composed of many molecules in rapid motion. The locations of the molecules in the container at any particular time, the molecular coordinates, have no effect on the thermodynamic properties of the gas (to a first approximation). The molecular velocities, however, determine the molecular kinetic energy, which is proportional to the temperature. The volume of the gas, on the other hand, is a coordinate or parameter of the system that is varied at a rate negligibly slow compared to molecular speeds. The two quantities whose interplay constitutes the essence of thermodynamics, heat and work, are the energy changes of the system produced by changes in the molecular velocities and in the slowly varying parameters (like the volume), respectively.

Now, if one has a mechanical device like a wheel rotating rapidly about a fixed axis, its energy depends only on the angular velocity of the wheel and not on its instantaneous orientation. In this respect, the rotating wheel is analogous to the molecular motions in a gas. One can imagine a device, like a governor, attached to the axle in such a way that the energy of the system also depends on the coordinates of certain masses, coordinates that can be varied at a rate which is slow compared to the speed of rotation. Helmholtz's idea was that such a purely mechanical system, monocyclic in the sense described before, might provide an analogy for the complicated systems of thermodynamics, not only with respect to the classification of its coordinates but also with respect to the relationships between heat and work. The essential point was to show that energy provided to the system as heat (that is, as a change in the kinetic energy of the cyclic coordinate) could not be completely converted into work (that is, into a slow change in the auxiliary parameters, the second class of coordinates referred to above). In other words Helmholtz had to show that the analogue of the differential heat had an integrating factor proportional to the kinetic energy of the system.

To see how this goes in the simplest possible case let L be the Lagrangian of the system. Lagrange's equations have the form,

$$\frac{\mathrm{d}}{\mathrm{d}t}\frac{\partial L}{\partial \dot{q}} - \frac{\partial L}{\partial q} = F \tag{1}$$

where q is a generalized coordinate, $\dot{q}$ is the corresponding generalized velocity, and F is the external generalized force acting on this coordinate. For the single cyclic coordinate q_a of our monocyclic system one has, by definition,

$$\frac{\partial L}{\partial q_a} = 0 \tag{2}$$

so that

$$F_a = \dot{p}_a \tag{3}$$

where $\dot{p}_a$ is the corresponding generalized momentum,

$$p_a = \frac{\partial L}{\partial \dot{q}_a}. \tag{4}$$

For the remaining coordinates, q_b, which are slowly varying, one can drop the terms in $(\partial L/\partial \dot{q}_b)$ and write simply

$$F_b = -\frac{\partial L}{\partial q_b} \tag{5}$$

for these coordinates. The kinetic energy T of the system can be written simply as

$$T = \tfrac{1}{2}\dot{q}_a p_a \tag{6}$$

and the differential of the total energy U can easily be put in the form

$$\mathrm{d}U = \dot{q}_a \, \mathrm{d}p_a + \sum_b F_b \, \mathrm{d}q_b. \tag{7}$$

The first term is the energy change produced by a change in the velocity of the cyclic coordinate, and is identified with the heat $\mathrm{d}Q$ added to the system. The second term is the work done on the system by a change in the slowly varying parameters, q_b. One can now write $\mathrm{d}Q$ in the form

$$\mathrm{d}Q = \dot{q}_a \, \mathrm{d}p_a = T \, \mathrm{d} \ln (p_a)^2 \tag{8}$$

with the help of equation (6). We see that $\mathrm{d}Q$ is the product of the kinetic energy and the differential of a state function, $\ln (p_a)^2$. If kinetic energy is identified with temperature, then the state function $\ln(p_a)^2$ will be the analogue of the entropy of the system.

Helmholtz carried the analysis a great deal further, treating more general monocyclic systems and showing how two such systems can be coupled in much the same way that thermodynamic systems are coupled isothermally. In other words, he developed a rather complete mechanical analogy for the laws of thermodynamics. Helmholtz was quite aware of the fact that real thermodynamic systems are not monocyclic, and that he was giving a mechanical *analogy* for thermodynamics and not a mechanical *explanation* of thermodynamics.

He made this point emphatically in response to a criticism from Clausius. The actual thermal motions in a system like a gas or a solid were unknown but certain to be complicated. Wrote Helmholtz:

In these circumstances it seems completely rational to me to look for the most general conditions under which the most general physical characteristics of thermal motion can appear in other well-known classes of motion. It is in this sense, of course, that I have called special attention to the analogies between the behavior of thermal motion and of the monocyclic motions I have investigated, but I have declared from the outset that thermal motion is not strictly monocyclic. I have, accordingly, never claimed to have given 'an explanation' of the second law of thermodynamics.[21]

III

Boltzmann, who *had* tried to give an explanation of the second law, and who was convinced that he had succeeded in giving one in terms of statistical mechanics, was, nevertheless, fascinated by Helmholtz's work. He was quite ready to accept Helmholtz's approach to the second law in parallel with his own and even to develop it further himself. Boltzmann already knew and appreciated the idea of a mechanical analogy: he had learned it from the writings of the master of this concept – James Clerk Maxwell. Maxwell had made effective use of dynamical analogies in developing his electromagnetic theory, to the bewilderment of his continental critics. We must have a look at Maxwell's analogies.

In his first paper on electromagnetism Maxwell carefully distinguished between the approach to the subject he had chosen and those he had rejected.[22] "The present state of electrical science," he wrote, "seems peculiarly unfavourable to speculation." A large number of experimental results and "a considerable body of most intricate mathematics" had to be mastered, if one were to make any progress. Maxwell rejected two obvious ways of trying to reduce these results to a form one could readily grasp. He would not begin by constructing "a purely mathematical formula," because then one would "entirely lose sight of the phenomena to be explained." Nor would he prematurely adopt "a physical hypothesis," since that would make him "see the phenomena only through a medium," and therefore make him "liable to that blindness to facts and rashness in assumption which a partial explanation encourages." What was wanted was "some method of investigation which allows the mind at every step to lay hold of a clear physical conception without being committed to any theory..., so that it is neither drawn aside from the subject in pursuit of analytical subtleties, nor carried beyond the truth by a favorite hypothesis."

Maxwell proposed the use of analogies, and meant by an analogy "that partial similarity between the laws of one science and those of another which makes each of them illustrate the other." The laws of electrostatic potential theory, for example, apply to a physical situation essentially different from that described by the laws of heat conduction. Not only that: one theory starts with discrete charges and the other with a continuous medium. But because the mathematical equations of the two theories are identical in form, one can solve a problem in the theory of attractions by thinking physically about the analogous problem in the theory of heat conduction, and vice versa. Maxwell's paper exploited just such an analogy between the flow of an incompressible fluid and Faraday's lines of magnetic force. He emphasized that however suggestive such an analogy might be, it was no substitute for "a mature theory, in which physical facts will be physically explained." The subject was not ready for such a theory, however, in 1855. Maxwell was arguing against a premature commitment to the physical hypothesis of a force acting at a distance between moving charges, like the force proposed by Weber. He wanted to keep open the possibility of a theory based on Faraday's local action of a field of force, even though he could not yet construct such a theory.

A few years later Maxwell elaborated his analogy in the series of papers entitled, 'On Physical Lines of Force.'[23] He could now represent the relationships of the electromagnetic field variables with the help of an elaborate mechanical construction involving a latticework of vortex rings in an ideal fluid, with adjacent vortices coupled by means of electrical particles acting like idle wheels. Maxwell carefully pointed out once again that all this was only an analogy. The use of particles to couple the vortex rings was not intended to be "a mode of connexion existing in nature,"[24] but rather "a mode of connexion which is mechanically conceivable and easily investigated," one that would help rather than hinder the "search after the true interpretation of the phenomena." The analogy did its work: it made it possible for Maxwell to deduce that "light consists in the transverse undulations of the same medium which is the cause of electrical and magnetic phenomena."[25]

In his full-fledged 'Dynamical Theory of the Electromagnetic Field'[26] of 1867 the detailed mechanical analogy almost completely disappeared. It was mentioned only in a passing reference halfway through the long memoir. Maxwell had now dropped the specific mechanical analogy, but he was still constructing a mechanical theory. He viewed the field as "a complicated mechanism capable of a vast variety of motion," and asserted that "such a mechanism must be subject to the general laws of Dynamics."[27] Maxwell compared the old and new versions of his theory in a letter to his friend Peter Guthrie Tait, writing:

> The former is built up to show that the phenomena [of electromagnetism] are such as can be explained by mechanism. The nature of the mechanism is to the true mechanism what an orrery is to the Solar System. The latter is built on Lagrange's Dynamical Equations and is not wise about vortices.[28]

Maxwell's intentions have often been misunderstood. The new approach was not the complete mechanical explanation of electromagnetism that he would have wanted to achieve. His use of Lagrange's methods allowed him to proceed without a detailed knowledge of "the nature of the connexions of the parts of the system,"[29] and this was an advantage, but it did not mean that his search for a detailed mechanical explanation was abandoned. That search simply had to be postponed, but only until the true nature of the electric current became known. Then, Maxwell wrote in his *Treatise*, we would have

> the beginnings of a complete dynamical theory of electricity in which we should regard electrical action, not, as in this treatise, as a phenomenon due to an unknown cause, subject only to the general laws of dynamics, but as the results of known motions of known portions of matter, in which not only the total effects and final results, but the whole intermediate mechanism and details of the motion, are taken as the objects of study.[30]

Maxwell did use both particular mechanical analogies and schematic theories based on general dynamics, but this fact should not obscure his goal. That goal was still mechanical *explanation*, just as it had been since the time of Newton.

IV

Boltzmann had learned to appreciate the subtlety and variety of the ways in which Maxwell used mechanics in trying to understand the physical world. His sympathetic study of Maxwell prepared him to accept Helmholtz's monocycle analogy for the second law in the proper spirit. He set to work on it as soon as it appeared, extending and developing Helmholtz's ideas in a number of directions.[31]

Although Helmholtz proposed a mechanical analogy, one must not imagine that he had described a particular mechanical model, like Maxwell's vortex rings and idle wheels. His analogy was of the general, schematic sort, comparable rather to Maxwell's dynamical theory of the *Treatise*. It is typical of Boltzmann, however, that he began by illustrating Helmholtz's ideas with the help of several very specific mechanical examples. These examples were more than just illustrations. They allowed Boltzmann to show that the monocycle analogy was less general than Helmholtz had claimed it to be. The kinetic energy was not always an integrating factor for the differential heat in monocyclic systems, and there were even situations when no such integrating factor existed at all, cases without thermodynamic parallels.

Boltzmann explored the relationships between the monocycle analogy and his own previous work in several different ways. Although the individual systems used in the statistical approach to the second law were not monocycles, Boltzmann showed how an ensemble of such systems could exhibit monocyclic properties. It was in the course of this investigation that he introduced the notion of an ergodic system, as Stephen Brush has recently pointed out.[32] Boltzmann also discussed the connection between Helmholtz's work and his own early attempt to find a purely mechanical explanation of the second law. He emphasized the value of Helmholtz's clear distinction between two types of coordinates – the slowly varying parameters of the system and the rapidly varying cyclic coordinates, analogues of the molecular coordinates. This distinction made it possible to define the differences between heat and work within a mechanical theory, a crucial point in developing such a theory.[33]

The three papers Boltzmann wrote[31] on monocyclic systems between 1884 and 1886 did not exhaust his interest in the subject. He had already noted in the first of these papers that Maxwell used a monocyclic system when he discussed electromagnetic induction in the *Treatise*. Boltzmann developed this use of monocycles at great length in his Munich lectures on Maxwell's theory a few years later.[34] In these lectures the cyclic system becomes the central mechanical concept for the construction of a theory of electric circuits and their interactions. The reasoning is simple: when a current flows in a closed circuit, some kind of motion is taking place. When the current is a steady one, the motion must be such that whenever one particle leaves a certain point, another identical one moving with the same velocity must immediately arrive at that point, just as in the uniform rotation of a disk. In other words, the coordinates of the electricity must be cyclic, and the energy of the current can only depend on the corresponding velocities.

Boltzmann's *Lectures on Maxwell's Theory* demonstrates the art of constructing mechanical analogies in a very highly developed form. Paul Ehrenfest, who studied with Boltzmann, has described this aspect of his master's work, an aspect rather neglected by Boltzmann's biographers.

Boltzmann's presentation of the Maxwell theory starts with just these mechanical analogies – in contrast to all other presentations of this theory. In Boltzmann's lectures one had ample time at the beginning just to learn the amazing reactions that relatively very primitive, purely mechanical systems can carry out – reactions of a type that one would never expect from a mechanical system so long as one instinctively thinks only of the planetary system. The imagination had to be trained by studying these systems so that it could push on further and further to the construction of a mechanism, free of contradictions in all its details, which could explain the complicated properties of the luminiferous aether.[35]

In another passage Ehrenfest wrote:

Mechanical representations were the material from which Boltzmann preferred to fashion his creations. ... He obviously derived intense aesthetic pleasure from letting his imagination play over a confusion of interrelated motions, forces and reactions until the point was reached where they could actually be grasped. This can be recognized at many points in his lectures on mechanics, on the theory of gases, and especially on electromagnetism. In lectures and seminars Boltzmann was never satisfied with just a purely schematic or analytical characterization of a mechanical model. Its structure and its motion were always pursued to the last detail. If, for example, several strings were used to illustrate certain kinematical relations, then the conceptual arrangement had to be devised in such a way that the strings would not become entangled. In those great works of his whose results always encompass an immense domain, the simple example is also given an exhaustive and loving treatment.[36]

Boltzmann's models for inductively coupled circuits show exactly what Ehrenfest was describing. He began by treating a simple device, a monocycle, which nevertheless could serve as a mechanical model for a Carnot cycle.[37] An appropriately coupled combination of three of these satisfied the equations of two coupled circuits, with self and mutual induction.[38] This did not satisfy Boltzmann, however, because the coefficients of self and mutual induction could not be varied independently in the model, and he proceeded to redesign the system to allow for this possibility.[39] This perfected design not only could be constructed – it actually was constructed to Boltzmann's specifications at a workshop in Graz.[40] An unwary reader glancing through the pages of Boltzmann's book on electromagnetic theory might easily imagine that he had picked up a treatise on the design of engineering mechanisms by mistake.

V

Helmholtz's monocyclic analogy for the second law made a powerful impression on at least one other reader in addition to Boltzmann. That was Heinrich Hertz, Helmholtz's own best-beloved pupil. Hertz worked at a book on mechanics during the last three years of his tragically short life. This book, *The Principles of Mechanics Presented in a New Form*, was practically complete at his death and it appeared posthumously in 1894.

Hertz re-examined the fundamental ideas of mechanics and arrived at a new way of organizing, presenting and thinking about the subject. He wrote the whole book in a strictly deductive style, following the Euclidean pattern except for a long philosophical introduction. Philosophers of science have found Hertz's work to be rich in suggestions, and have admired its logical structure.[41] Almost seventy years ago Bertrand Russell referred to its principles as being "so simple and so admirable."[42] Physicists have generally considered Hertz's system to be beautiful but valueless; treatises on dynamics rarely do much more than mention Hertz's principle of the straightest path or least curvature.

The best known feature of Hertz's system of mechanics is his elimination of force as a fundamental concept. He was dissatisfied with what he called the "logical obscurity" that surrounded the idea of force. Hertz described one result of this logical obscurity in a sentence that most teachers of mechanics would be happy to endorse warmly:

I would mention the experience that it is exceedingly difficult to expound to thoughtful hearers that very introduction to mechanics without being occasionally embarrassed, without feeling tempted now and again to apologize, without wishing to get as quickly as possible over the rudiments, and on to examples which speak for themselves. I fancy that Newton himself must have felt this embarrassment....[43]

Hertz argued that force was introduced into physics because we cannot account for motion, even for such a simple motion as that of a falling stone, by using only "what can be directly observed." "If we wish to obtain an image of the universe which shall be well-rounded, complete, and conformable to law," he wrote, "we have to presuppose, behind the things which we see, other, invisible things...." Force is such an invisible thing, as is energy, but Hertz proposed another choice. He argued:

We may admit that there is a hidden something at work, and yet deny that this something belongs to a special category. We are free to assume that this hidden something is nought else than motion and mass again – motion and mass which differ from the visible ones not in themselves but in relation to us and to our usual means of perception.[44]

This was to be Hertz's basic assumption – that the motions of visible masses are to be accounted for by their couplings to other concealed masses which are themselves in motion. Hertz was proposing an explanation in the Cartesian tradition, a mechanics from which dynamics would be eliminated and which would consist exclusively of kinematics.[45]

Hertz was attempting to explain force as the effect of kinematical constraints linking the observed motion of the visible masses to the hidden motions of hidden masses. He took this to be a natural next step in the process of dynamical explanation, referring to the success of the kinetic theory of heat and of Maxwell's electromagnetic theory, both of which were based on hidden motions of unseen masses. And Hertz emphasized the significance of Helmholtz's studies on cyclical systems, in which he had "treated the most important form of concealed motion fully, and in a manner that admits of general application." The hypothesis of hidden motions had explained "the forces connected with heat" and at least some aspects of electromagnetic forces; Helmholtz's analysis suggested to Hertz that one could explain all forces this way.[46] Hertz might well have described his work as a mechanical explanation of mechanics!

The reason for wanting to construct a mechanics without forces, a purely kinematical mechanics, was to provide physicists with the proper instrument for the next great task facing them – the full mechanical description of the aether. Hertz thought that the success of the field theory of electromagnetism was an argument against the old view of forces. One could attain a better approximation to the truth "by tracing back the supposed actions at a distance to motions in an all-pervading medium whose smallest parts are subjected to rigid connections." It was over the ground of the aether that "the decisive battle between these different fundamental assumptions of mechanics must be fought out."[47] Hertz's system was to furnish the mechanics of that electromagnetic aether which his own experiments did so much to establish.

In one central passage of his book Hertz discussed the general concept of dynamical analogies or models. He took mechanical explanation to be only the construction of such dynamical analogies, and explicitly stated that "it is impossible to carry our knowledge of the connections of natural systems further than is involved in specifying models of the actual systems." Hertz saw "the relation of a dynamical model to the system of which it is regarded as the model" to be "precisely the same as the relation of the images which our mind forms of things to the things themselves."[48]

There was one crucial trouble with Hertz's system. He gave no examples to show how one could account for particular mechanical phenomena by means of a particular arrangement of hidden masses in motion. Worse than that – it was apparently not possible to construct such explanations for even very simple mechanical problems. Boltzmann, whose mechanical ingenuity has already been amply demonstrated, was unable to find a mechanism, of the general type allowed in Hertzian mechanics, which would give a kinematic reduction for such a simple and basic problem as the elastic collision of two spheres. He questioned, therefore, whether the theory would be of much value for physics, "in spite of all its philosophical beauty and completeness,"[49] and he did not follow Hertz's approach in his own lectures on mechanics.[50]

VI

The physicists I have discussed – Maxwell and Boltzmann, Helmholtz and Hertz – certainly differed about how mechanical explanations were to be achieved, but they all worked toward that goal. One can, to be sure, cite passages that seem to contradict this statement. Helmholtz made such a remark in his preface to Hertz's book – a preface summing up Hertz's career and expressing grief over Hertz's death, to be followed in only a few months by Helmholtz's own death. Helmholtz wrote that in contrast to Hertz and the English physicists he preferred "the simple representation of physical facts and laws in the most general form, as given in systems of differential equations." He felt this to be "safer" than attempts at mechanical explanation.[51] There is also an often-quoted remark by Hertz himself which is taken as an expression of the same position: "To the question, 'What is Maxwell's theory?'," Hertz wrote, "I know of no shorter or more definite answer than the following – Maxwell's theory is Maxwell's system of equations."[52]

The answer to criticism based on such remarks is simply to follow Einstein's advice:

> If you want to find out anything from the theoretical physicists about the methods they use, I advise you to stick closely to one principle: don't listen to their words, fix your attention on their deeds.[53]

That is what I have been trying to do, and among the deeds of Helmholtz and Hertz are their serious and repeated acts of mechanical explanation. Helmholtz, for example, may have said he preferred noncommittal theories that merely describe the facts, but he wrote a series of papers on the monocyclic analogy for the second law and tried hard to make the principle of least action into the basis for all of physics.[54]

There were physicists who really did reject the goal of mechanical explanation, in any of its forms. Ernst Mach was the leading thinker among these anti-mechanists, whose ranks included Pierre Duhem, Wilhelm Ostwald and Georg Helm. Although they attacked the primacy of mechanics from different positions, and their arguments had various degrees of cogency, they would all have accepted Mach's statement:

> The view that makes mechanics the basis of the remaining branches of physics, and explains all physical phenomena by mechanical ideas, is in our judgment a prejudice.

The historical priority of mechanics was not an adequate reason for giving it logical priority. Mach went on:

We have no means of knowing, as yet, which of the physical phenomena go *deepest*, whether the mechanical phenomena are perhaps not the most superficial of all, or whether all do not go *equally deep*.

Mach favored giving up this "artificial conception" and restricting physics to "the expression of actual facts,"[55] without superfluous hypotheses and so with the maximum economy of thought. This attack was aimed as much at atomism as at mechanism, and the men I have named are the most notorious scientific disbelievers in atoms to be found at the end of the nineteenth century.[56]

Ludwig Boltzmann took it upon himself to defend the mechanical world view against these critics. He argued long, hard and often for both mechanism and atomism, claiming that they were, at the very least, extremely fruitful analogies, in Maxwell's sense. Boltzmann's polemics on the indispensability of atomism, the importance of mechanics and the fundamental errors in the position of the energeticists (as Ostwald and Helm, in particular, called themselves), constitute a significant part of his writings in the 90's.[57]

It would appear from Boltzmann's lectures and papers that he was fighting a last ditch battle against the victorious forces of the energeticists. He liked to portray himself as the last survivor of the classical approach to physics, or at least the only one left who would struggle to keep its values alive despite the success of the new schools of thought. "I take my stand before you," he said in a lecture at Munich in 1899, "as a reactionary, a survivor, who is still an enthusiast for the old and the classical as opposed to the modern."[58] The most famous example of this is Boltzmann's preface to the second volume of his *Theory of Gases*, written a year earlier, where he described himself as "only an individual struggling weakly against the stream of time," trying to record the essential results so that "when the theory of gases is again revived not too much will have to be rediscovered."[59]

Boltzmann's cry from the depths expressed his own view of his position, and that position had certainly been attacked from many directions, but I do not think it can be taken completely at face value. In the great debate between Boltzmann and the energeticists at Lübeck in 1895, it was Boltzmann who carried the day against Helm and Ostwald, and, even more important, it was Boltzmann who had the support of such leaders of the younger generation as Walther Nernst and Arnold Sommerfeld.[60]

The threat offered by energetics to the mechanical physics did not bring the old tradition to an end.

That tradition did not come to a sudden end. In 1892, H. A. Lorentz started a new stage in the development of electromagnetic theory with his paper on the applications of Maxwell's theory to moving bodies. This paper contained the first version of his theory of electrons, but Lorentz began it by treating the electromagnetic field itself as a dynamical system. The success of the theory of electrons did not depend on this dynamical theory of the field, however, and Lorentz quietly dropped it. In the later versions of his theory he made no attempt to arrive at the field equations from a mechanical starting point.[61] This is typical of the way mechanical explanation ceased to be a program for the development of physics. One could try to treat the electromagnetic field as a dynamical system, but it was awkward and profitless to do so. If the supposed mechanical explanation provided no new insight, and if it led to no further progress, in what sense did it provide an explanation? As Einstein put it:

> One got used to operating with these fields as independent substances without finding it necessary to give oneself an account of their mechanical nature; thus mechanics as the basis of physics was being abandoned, almost unnoticeably, because its adaptability to the facts presented itself finally as hopeless.[62]

The second volume of Boltzmann's *Lectures on Mechanics*, which contained an extended discussion of monocyclic systems was published in 1904.[63] By the time it appeared the search for a new and nonmechanical foundation for physics had begun. Monocycles would soon seem to be as far from the current concerns of physicists as epicycles.[64]

Yale University

NOTES

[1] Christiaan Huygens, *Treatise on Light* (transl. by S. P. Thompson), London 1912, p. 3.

[2] Isaac Newton, *Mathematical Principles of Natural Philosophy* (A. Motte's translation revised by F. Cajori), Berkeley 1934, pp. xvii–xviii.

[3] Rudolf Clausius, 'The Nature of the Motion Which We Call Heat,' *Phil. Mag*. **14** (1857) 108.

[4] Heinrich Hertz, *The Principles of Mechanics Presented in a New Form* (transl. by D. E. Jones and J. T. Walley), London 1899, Author's Preface.

[5] James Clerk Maxwell, 'On the Dynamical Evidence of the Molecular Constitution of Bodies', *Nature* **11** (1875) 357, 374. Reprinted in *The Scientific Papers of James Clerk Maxwell* (ed. by W. D. Niven), Cambridge 1890, **2**, p. 433.

[6] Henry William Watson, *A Treatise on the Kinetic Theory of Gases*, 2nd ed., Oxford 1893, p. 87.
[7] Pierre Duhem, *The Aim and Structure of Physical Theory* (transl. by P. P. Wiener), Princeton 1954, pp. 55–104, especially pp. 71–72.
[8] Alfred North Whitehead, *Science and the Modern World*, New York 1925, p. 148.
[9] See, for example, John Theodore Merz, *A History of European Thought in the Nineteenth Century*, Edinburgh, **1** (1896), **2** (1903), Chapters 4–7.
[10] (a) Engelbert Broda, *Ludwig Boltzmann*, Vienna, 1955. (b) René Dugas, *La théorie physique au sens de Boltzmann*, Neuchâtel 1959.
[11] Hermann von Helmholtz, (a) 'Studien zur Statik monocyklischer Systeme', *Berliner Berichte* (1884), pp. 159, 311; (b) 'Principien der Statik monocyklischer Systeme', *Journal für die reine und angewandte Mathematik* **97** (1884) 111, 317.
[12] L. Boltzmann, 'Über die mechanische Bedeutung des zweiten Hauptsatzes der Wärmetheorie', *Wiener Berichte* **53** (1866) 195. Reprinted in L. Boltzmann, *Wissenschaftliche Abhandlungen* (ed. by F. Hasenöhrl), Leipzig, 1909, **1**, 9–33. This collection will be referred to as Boltzmann, *Wiss. Abh.* and page references to Boltzmann's work are made to this reprint.
[13] R. Clausius, Über verschiedene für die Anwendung bequeme Formen der Hauptgleichungen der mechanischen Wärmetheorie', *Pogg. Ann.* **125** (1865) 353.
[14] L. Boltzmann, 'Studien über das Gleichgewicht der lebendigen Kraft zwischen bewegten materiellen Punkten', *Wiener Berichte* **58** (1868) 517. *Wiss. Abh.* **1**, 49–96.
[15] L. Boltzmann, 'Analytischer Beweis des zweiten Hauptsatzes der mechanischen Wärmetheorie aus den Sätzen über das Gleichgewicht der lebendigen Kraft', *Wiener Berichte* **63** (1871) 712. *Wiss. Abh.* **1**, 288–308.
[16] L. Boltzmann, 'Weitere Studien über das Wärmegleichgewicht unter Gasmolekülen', *Wiener Berichte* **66** (1872) 275. *Wiss. Abh.* **1**, 316–402.
[17] L. Boltzmann, 'Bemerkungen über einige Probleme der mechanischen Wärmetheorie', *Wiener Berichte* **75** (1877) 62. *Wiss. Abh.* **2**, 116–122.
[18] L. Boltzmann, 'Über die Beziehung zwischen dem zweiten Hauptsatze der mechanischen Wärmetheorie und der Wahrscheinlichkeitsrechnung respektive den Sätzen über das Wärmegleichgewicht', *Wiener Berichte* **76** (1877) 373. *Wiss. Abh.* **2**, 164–223.
[19] I have analyzed the development of Boltzmann's ideas in Chapter 6 of my book, *Paul Ehrenfest*, Vol. 1, *The Making of a Theoretical Physicist*, Amsterdam 1970.
[20] In addition to the references of note 11, see also (a) J. Larmor and G. H. Bryan, 'On Our Knowledge of Thermodynamics, Specially with Regard to the Second Law', *B.A.A.S. Reports* **61** (1891) 96; (b) H. von Helmholtz, *Vorlesungen über Theoretische Physik, VI. Vorlesungen über Theorie der Wärme* (ed. by F. Richarz), Leipzig 1903, p. 338.
[21] H. von Helmholtz, 'Studien zur Statik monocyklischer Systeme III', *Berliner Berichte* (1884) 757.
[22] J. C. Maxwell, 'On Faraday's Lines of Force', *Trans. Cambr. Phil. Soc.* **10** (1855) 27. *Scientific Papers* **1**, 155.
[23] J. C. Maxwell, 'On Physical Lines of Force', *Phil. Mag.* **21** (1861) 161, 281, 338; **23** (1862) 12, 85. *Scientific Papers* **1**, 451.
[24] J. C. Maxwell, *Scientific Papers* **1**, 486.
[25] *Ibid.*, p. 500.
[26] J. C. Maxwell, 'A Dynamical Theory of the Electromagnetic Field', *Trans. Roy. Soc.* **155** (1865) 459. *Scientific Papers* **1**, 526.
[27] J. C. Maxwell, *Scientific Papers* **1**, 533.
[28] J. C. Maxwell to P. G. Tait, 23 December 1867. Quoted in C. G. Knott, *Life and Scientific Work of Peter Guthrie Tait*, Cambridge 1911, p. 215.

[29] J. C. Maxwell, *A Treatise on Electricity and Magnetism*, 3rd ed., 1891 (Reprinted New York 1954), **2**, p. 213.
[30] *Ibid.*, p. 218.
[31] L. Boltzmann, (a) 'Über die Eigenschaften monozyklischer und anderer damit verwandter Systeme', *Journal für die reine und angewandte Mathematik* **98** (1884) 68; (b) 'Über einige Fälle, wo die lebendige Kraft nicht integrierender Nenner des Differentials der zugeführten Energie ist', *Wiener Berichte* **92** (1885) 853; (c) 'Neuer Beweis eines von Helmholtz aufgestellten Theorems betreffend die Eigenschaften monozyklischer Systeme', *Göttinger Nachrichten* (1886) 209. *Wiss. Abh.* **3**, 122, 153, 176.
[32] Stephen G. Brush, 'Foundations of Statistical Mechanics 1845–1915', *Archive for History of Exact Sciences* **4** (1967) 169.
[33] This point had been strongly emphasized by Maxwell. Helmholtz's sharp distinction between the two classes of coordinates, and a corresponding restriction of the possible ways in which energy could be added to the system, made it possible to avoid Maxwell's criticism of mechanical theories of the second law. For further discussion and references to Maxwell see my paper, 'Maxwell, His Demon, and the Second Law of Thermodynamics', *American Scientist* **58** (1970) 84.
[34] L. Boltzmann, *Vorlesungen über Maxwells Theorie der Elektricität und des Lichtes* (Leipzig, **1**, 1891; **2**, 1893).
[35] Paul Ehrenfest, 'Ludwig Boltzmann', *Mathematisch-Naturwissenschaftliche Blätter* **3** (1906). Reprinted in Paul Ehrenfest, *Collected Scientific Papers* (ed. by M. J. Klein), Amsterdam 1959, p. 134.
[36] *Ibid.*, p. 135.
[37] L. Boltzmann, *op. cit.*, note 34, **1**, p. 8.
[38] *Ibid.*, p. 26.
[39] *Ibid.*, p. 42.
[40] *Ibid.*, pp. 48–49.
[41] H. Hertz, *op. cit.*, note 4. See the Introductory Essay and Bibliography by Robert S. Cohen in the reprinted edition, Dover Publications, New York, 1956.
[42] Bertrand Russell, *The Principles of Mathematics*, 2nd ed., New York, 1943, p. 494. (This edition was reprinted unchanged from the original edition of 1903.)
[43] H. Hertz, *op. cit.*, note 4, p. 6.
[44] *Ibid.*, p. 25.
[45] See P. Duhem, *L'évolution de la mécanique*, Paris 1903, pp. 157–168.
[46] H. Hertz, *op. cit.*, note 4, p. 26.
[47] *Ibid.*, p. 41.
[48] *Ibid.*, p. 177.
[49] L. Boltzmann, 'Anfrage, die Hertzsche Mechanik betreffend', *Wiss. Abh.* **3**, p. 641.
[50] L. Boltzmann, *Vorlesungen über die Principe der Mechanik* **1**, Leipzig 1897, pp. 1–6, 37–42.
[51] H. von Helmholtz, 'Preface' in H. Hertz, *op. cit.*, note 4.
[52] H. Hertz, *Electric Waves* (transl. by D. E. Jones), London 1893, p. 21.
[53] Albert Einstein, *The World As I See It*, New York 1934, p. 30.
[54] H. von Helmholtz, 'Über die physikalische Bedeutung des Prinzips der kleinsten Wirkung', *Journal für die reine und angewandte Mathematik* **100** (1887) 137, 213.
[55] Ernst Mach, *The Science of Mechanics* (transl. by T. J. McCormack), 6th ed., Lasalle, Ill., 1960, pp. 596–597.
[56] P. Duhem, *op. cit.*, note 7.
[57] L. Boltzmann, *Populäre Schriften*, Leipzig 1905, pp. 104, 137, 141, 158, 198.

[58] *Ibid.*, p. 205.
[59] L. Boltzmann, *Lectures on Gas Theory* (transl. by S. G. Brush), Berkeley 1964, pp. 215–216. This preface is dated August, 1898.
[60] See E. Broda, *op. cit.*, note 10, p. 13; R. Dugas, *op. cit.*, note 10, p. 95. Also see the letters from G. Helm to his wife, 17 September 1895 and 19 September 1895, reprinted in *Aus dem wissenschaftlichen Briefwechsel Wilhelm Ostwalds* **1** (ed. by H.-G. Körber), Berlin 1961, pp. 118–120.
[61] See Tetu Hirosige, 'Origins of Lorentz' Theory of Electrons and the Concept of the Electromagnetic Field', *Historical Studies in the Physical Sciences* **1** (1969) 191–209.
[62] Albert Einstein, 'Autobiographical Notes', in *Albert Einstein: Philosopher-Scientist* (ed. by P. A. Schilpp), New York 1949, pp. 25, 27.
[63] L. Boltzmann, *Vorlesungen über die Principe der Mechanik* **2**, Leipzig 1904, pp. 162–212.
[64] I gratefully acknowledge the support for this work provided by a grant from the National Science Foundation.

PART IV

CURRENT PROBLEMS OF COSMOLOGY

(*Chairman:* John Stachel)

INTRODUCTION TO THE SYMPOSIUM ON COSMOLOGY

This section of the AAAS Proceedings contains the papers delivered at a session on Current Problems in Cosmology, with the exception of Prof. E. R. Harrison's paper, 'The Grin Cosmologies' (*Q. J. Roy. Astron. Soc.* **11** (1970) 214–217).

In addition to the papers delivered, Professors de Vaucouleurs and Hoyle kindly agreed to my request to reprint papers of theirs, so that some important points of view not represented in the talks could appear in the printed survey: de Vaucouleur's challenge to the existence of a homogenous matter distributed in the universe, which forms the basis of most relativistic cosmology; and an appraisal by Hoyle of the steady state theory.

Like so many other fields of active research, cosmology has not stood still in the years since 1969. Thus, it might be helpful to the reader interested in following up on this survey to consult a quite recent survey, *Modern Cosmology*, by Dennis Sciama (Cambridge University Press, 1972) which also contains further references.

The 1969 Boston meeting of the AAAS saw the first interventions at its meetings by groups loosely described by the phrase 'science for the people'. Although our session saw no such action (whether this is to be taken as praise or slight of cosmologists I leave an open question), I took advantage of my chairmanship to open the session with some remarks suggesting that one not be put off by the form of such interventions, but rather concentrate on the content of their challenge to scientists to situate their work in a context which made clear its relevance (oh, much abused word!) to wider human issues than were traditionally considered proper to raise in discussions among scientists. I suggested that modern cosmology was the heir to an ancient tradition in which the image of the cosmos influenced and was influenced by the image of man in this cosmos. Another speaker commented that the reason he did cosmology was because it was fun – which I took as a confirmation of my thesis, since it certainly isn't the cosmos which is having the fun.

R. J. Seeger and R. S. Cohen (eds.), Philosophical Foundations of Science, 179–180.

In any case, cosmology is alive and flourishing, as this symposium demonstrated, and hopefully will continue to live and flourish as long as humanity does.

Boston University

PETER G. BERGMANN

COSMOLOGY AS A SCIENCE

ABSTRACT. In recent years observational techniques at cosmological distances have been improved so that cosmology has become an empirical science, rather than a field for unchecked speculation. There remains the fact that its object, the whole universe, exists only once; hence we are unable to separate 'general' features from particular aspects of 'our' universe. This might not be a serious drawback if we were justified in the belief that presently accepted laws of nature remain valid on the cosmological scale. In the author's opinion, however, there are grounds for doubting that belief. The three arguments presented are (1) the possibility that apparent constants of nature may during cosmological times turn out to vary (Dirac); (2) the effective break-down of the principle of relativity caused by the effects of the cosmological environment on local experiments; and (3) the fact that present theory leads to field singularities at the early stages of the expanding universe, which might be a signal that currently accepted theoretical concepts are inadequate for an understanding of highly condensed matter.

Looking at my distinguished colleagues at this session, I am afraid that I am the one who has contributed least in the way of original research to the present state of cosmology. Throughout much of my life as a scientist I retained the feeling that cosmology as a science was not quite 'nice' and that it lacked an empirical foundation more than almost any other discipline among the so-called exact sciences. Whether or not my feelings were ever justified, the situation in cosmology has changed so rapidly, and so profoundly, within the last few years that my original reserve has lost any basis in fact that it might have had. Nevertheless, cosmology remains a peculiar field, with characteristics that set it apart from the other exact sciences. With your permission I shall attempt a review of these distinctive features.

As a starter, let me call your attention to two outstanding discussions of cosmology, one being Appendix I of *The Meaning of Relativity* by A. Einstein (1955) which was first published in 1945, the other a Cambridge Monograph by H. Bondi, entitled *Cosmology* (Bondi, 1960), which appeared in a second edition in 1961. If you compare these two discussions with the status of our observational evidence today, you discover that the basic evidence considered by Einstein was Olbers' paradox (Olbers, 1826), the Hubble effect (Hubble and Tolman, 1935), and estimates of the mean density of mass in the universe. In Bondi's book the

R. J. Seeger and R. S. Cohen (eds.), Philosophical Foundations of Science, 181–188.

value of Hubble's constant is corrected (and hence the age of the universe no longer embarrassingly close to that of the earth's crust). Besides, Bondi already discusses extragalactic radio sources and number counts, the chemical composition of stars, galaxies, and meteorites, and mentions cosmic rays as a source of cosmological information.

To this array of information and techniques of observation the sixties have added the discovery and detailed examination of the primordial fireball, of the quasars, and most recently of the pulsars. Besides, novel techniques have made possible the detection of forms of radiation that cannot penetrate the atmosphere, and have brought us to the threshold of neutrino astronomy and to the investigation of gravitational waves. In the very near future we may anticipate a 'cosmological information explosion' beyond anything we have as yet experienced.

It looks very much as if we shall be able to look for specific evidence that will help to rule out whole classes of cosmological models. Conversely, when we have made a successful guess, confirmation will come from a diversity of modes of observation. Theories of the universe, hypotheses as to its past, will be subject to the same kind of critical checking as theories of the composition and history of the earth, or of the solar system.

This development is of course most welcome. Clearly, cosmology will not hereafter fade back into the limbo of idle speculation, inaccessible to observational procedures. Moreover, with the value of the Hubble constant now fairly well fixed, it is clear that investigations assume cosmological significance if they deal with linear dimensions exceeding 10^9 parsecs. In other words, there is a rough dividing line between 'ordinary' astronomical domains and cosmological dimensions.

The importance of the existence of such a dividing line appears to me to lie in the special character of cosmology as a discipline, the fact that cosmology deals with a class of objects of which there is but one, the universe. By contrast, the astronomer who examines our solar system and who endeavors to understand its development and its dynamics, has reason for hoping that ours is not the only system consisting of one or more fixed stars surrounded by satellites. True, we have not yet discovered incontrovertible evidence that other solar systems do indeed exist. But we proceed in confidence that the set of circumstances that led to the development of ours will have been duplicated elsewhere, with similar results. Whether this conjecture is indeed correct remains for future generations to determine.

But we know with certainty that there is only one universe accessible to our investigations, no matter how much the astronomer's techniques may develop. If we can observe something at all, then, by definition, it forms part of our universe. If we contend that the laws of gravity contain the potentialities for both bounded and unbounded universes, we know that within our grasp, now and forever, only one particular universe exists, and that other possibilities cannot be subjected to empirical confirmation or rejection.

Bondi (1960) has discussed this peculiarity of cosmology in his book. He suggests that there might be two fundamentally different approaches to deal with it. He calls one the *deductive* attitude. It starts out from certain *a priori* postulates about space and time, and the universe as a whole, and deduces from these postulates, with minimal help from observational information, a model, or models, of our universe. The other attitude, called *extrapolating*, starts with the laws of nature that have been obtained on the terrestrial scale, and constructs model universes primarily on the basis of extrapolation, which accords to these laws of nature validity beyond the scale on which they have been confirmed experimentally. I believe that it is fair to say that actual model construction has proceeded on an eclectic basis, by and large. The (terrestrial) laws of nature are assumed to be valid, but a selection from conceivable model universes is made by the adoption of more or less stringent *cosmological principles*, that is to say principles concerning the large-scale homogeneity and isotropy of the universe. Though cosmological principles are subject to observational confirmation, to be sure – I remind you of the experiments regarding the degree of isotropy of the background radiation – these principles are not terrestrial-scale laws of nature. They have no application to physical structures on any lesser than cosmological scale.

One can, of course, take the attitude that the construction of model universes on this twin foundation is perfectly all right. There is no sense in doubting the validity of the previously discovered laws of nature on the cosmological scene. Likewise, assumptions with regard to the homogeneity and isotropy of the universe are simpler to make, and to apply, than assumptions concerning ultimate inhomogeneity or anisotropy; it would appear reasonable to try simpler models first, and to abandon them in favor of more complex constructions only if forced to do so by observational evidence. The uniqueness of the universe would then be no worse an impediment to our scientific inquiry than is the uniqueness of 'life', as long as life has been observed only on earth, where presumably it had a common origin and hence failed to exhibit the full diversity of life that may be realized on other worlds.

Nevertheless, in the remainder of this paper I should like to present arguments to the effect that one might well entertain doubts as to the eventual success of this enterprise, and that one might suspect that the laws of nature will require substantial modification on the cosmological scale. These doubts are of three kinds.

(1) *The Dirac conjecture*. Dirac (1937, 1938) has suggested that the very large dimensionless constants that are met with in nature might not be constants at all, but variables that change their values noticeably only during cosmological periods. Not all of these constants, incidentally, refer directly to cosmological data, such as the number of particles in the universe. One of them is the ratio of electric to gravitational forces acting between elementary particles.

Dirac's conjecture has been taken seriously by P. Jordan and, more recently, R. H. Dicke, who have constructed field theories in which a parameter, interpreted by them as the constant of gravitation, obeys a wave equation. The Brans-Dicke theory, being one particular response to Dirac's suggestion, may be subjected to observational tests, which would permit one to make a decision as between the classical general theory of relativity and the Brans-Dicke modification. One can well imagine other responses, some of which might be subject to observational tests only on the cosmological scale.

(2) *The Break-Down of the Principle of Relativity*. The foundation of the (special) theory of relativity is the so-called principle of relativity: among the inertial frames of reference (which are, relative to each other, in states of irrotational uniform linear motion) there is no privileged frame representing absolute rest; no conceivable experiment will reveal any difference in quality between any two inertial frames of reference. This principle is incompatible with the notion that an experiment may be affected by the large-scale environment, more particularly by the state of motion of the surrounding large masses.

Today there is very obviously one type of experiment that is affected by the cosmological environment, and that is any experiment involving interaction with the pervasive radio-frequency background radiation, the primordial fireball. As far as is known today, this radiation represents black-body radiation at a temperature of perhaps 2.7 K. Any black-body radiation changes under a Lorentz transformation. In all but one frame it is anisotropic, being in fact canonical with respect to a linear combination of energy and linear momentum. In one frame the coefficient of the linear momentum vanishes; it is the frame of reference in which this radiation is isotropic, and in that sense that frame represents 'rest'. There is no conceivable distribution of frequencies that is unchanged (invariant) under Lorentz transformations, except a (gray) distribution corresponding to an infinite temperature; such a distribution leads to an ultraviolet catastrophe, that is to say, the associated energy density is infinite.

There is, of course, nothing strange about the fact that locally there are frames that represent the prevailing state of motion of the surrounding matter. The interior of a star is best studied in terms of a frame of reference centered on the star's center of mass. The same is true of a galaxy. But the background radiation is not a phenomenon of local origin. As far as we understand it today, this radiation represents a sampling of radiation that originated all over the universe and during the very early stages of its history. There is every reason to believe that if we could examine this background radiation over a very large domain, much larger than our galaxy, and use it to fix a local rest frame at every point of that domain, then the resulting field of time-like directions would be very smooth, and it would represent indeed a cosmological background, against which the motion of local structures, such as fixed stars, galaxies, or even clusters of galaxies, could be determined with considerable accuracy.

In other words, the principle of relativity would hold only for certain types of experiments (those excluding interaction with the background radiation, for instance), or provided experiments are not refined beyond a certain degree of accuracy or sensitivity.

In the foregoing, I have pinned the break-down of the principle of relativity to the background radiation; but this is only by way of emphasis. One can construct local frames of rest also by averaging over the observed proper motions of the surrounding galaxies; the field of directions obtained by this procedure will not deviate grossly from the one gained from observing the background radiation. Either way, permitting large-scale samplings to enter, one is led inexorably to the break-down of the principle of relativity.

Here, perhaps, is a specific illustration of the difficulties engendered by the uniqueness of the universe. One is tempted to postulate that the principle of relativity is a very general law of nature, valid without limit or equivocation, but that in our particular universe the cosmological structure leads to a smooth flow pattern of matter-at-large that leaves its imprint on the outcome of appropriate experiments. We can conceive of universes without such flow patterns. If neighboring galaxies had randomly distributed relativistic velocities relative to each other, attempts to calculate mean velocities by averaging over ever-increasing space and time domains could lead to diverging, and hence inconclusive results. Having available only one universe we cannot design experiments that will help us separate 'general' structural characteristics from those 'peculiar' to our particular universe.

(3) *Degenerate Conditions.* If our universe at one time in the distant past was characterized by an exceedingly high density of matter and radiation, it is possible that there were forces at work which today play but a negligible role. It is well known that with fairly reasonable assumptions (of which one is that the mass density of matter can never be negative in any frame of reference) a universe that is expanding now must have passed through a stage in which its metric (according to the general theory of relativity) was singular, in a sense independent of the choice of coordinate system. From the point of view of a classical field theory, a singularity represents a break-down of that theory. Whether quantization would help any is not entirely clear, but at least to me appears unlikely.

Einstein (1955) has expressed the hope that modifying the classical general theory of relativity along the lines of a unitary field theory might lead to a modification of the field equations for high densities and, incidentally, avoid the primordial singularity. It appears now fairly clear that the appearance of that singularity cannot be circumvented by a minor adjustment, such as the postulation of some additional force field presently unknown, but only by a major recasting even of the geometric ideas peculiar to the general theory of relativity. Einstein undoubtedly envisaged such a fundamental overhauling of the laws discovered primarily by himself. Today we are no further in this respect than Einstein himself at the time of his death, almost fifteen years ago.

In one respect I believe we have gained a new possible approach. Extreme conditions might not only have prevailed at the cradle of the universe, but might also be present at the graves of individual stars. I am referring to the theory of stellar collapse. Under certain conditions collapse that has progressed beyond a certain stage results not only in enormous condensation of matter, first into white dwarfs and then into neutron stars, but eventually in the formation of 'black holes'. Actually, from a purely theoretical point of view the collapse is not nearly as inevitable as it is frequently pictured; Einstein has constructed an admittedly artificial model of a star of arbitrarily large mass that will not collapse (Einstein, 1939). On the other hand, the discovery of pulsars last year has, in the opinion of most of us, demonstrated the existence of neutron stars in our very own galaxy, and in fairly large numbers.

It would be rash to assume that high-density matter in partly collapsed stars resembles in every respect matter as it may have existed everywhere in the early stages of our expanding universe. For one, the average amounts of energy per baryon may have been very different; for another, neutron stars have surfaces on which matter must be in a state of quasi-equilibrium with the surrounding space, whereas the universe presumably had no such surface (or any surface at all). Nevertheless, one might hope to learn something about highly condensed matter from a study of neutron stars as they become accessible to the astronomer.

By now, cosmology has developed far beyond the stage of a fascinating parlor game, in which the rules are fixed by common agreement among the participants, inaccessible to empirical verification. Information bearing on the structure of the universe is rapidly being developed, and may by now be adequate to remove some models previously enjoying a measure of popularity from further consideration. Because of the uniqueness of the universe, cosmology has some of the aspects of historical research; in other respects it resembles the earth and space sciences. It differs from both in that its results may contribute in a very basic way to the exact sciences, while it draws on the exact sciences for its own sustenance. I do not believe that as yet we understand fully these relationships; but we must give serious attention to the foundations of cosmology if we are to appreciate fully its potentialities, as well as its pitfalls.

ACKNOWLEDGMENT

The author's work has been supported by the Air Force Office of Scientific Research under Grant AF AFOSR 789-66 and by Aerospace Research Laboratories, U.S. Air Force, under Contract F 33615-70-C-1110.

Syracuse University

BIBLIOGRAPHY

Bondi, H., *Cosmology*, Cambridge University Press, Cambridge, Mass., 1960, second edition.

Dirac, P. A. M., *Nature* **139** (1937) 323; *Proc. Roy. Soc. London, Ser. A* **165** (1938) 199.

Einstein, A., *Ann. Math.* **40** (1939) 922.

Einstein, A., *The Meaning of Relativity*, Princeton University Press, Princeton, N.J., 1955, fifth edition.

Hubble, E. P. and Tolman, R. C., *Astrophys. J.* **82** (1935) 426.

Olbers, H. W. M., *Bode's Jahrbuch* **110** (1826).

PHILIP MORRISON

OPEN OR CLOSED ?

A great philosophical tradition is at work here. You saw Professor Harrison admit to a rude barbarism compared with the rational structures, I take it, of the preceding talk by Professor Bergmann. I didn't hear Professor Bergmann's talk, but I'm certain it was rational, even elegant. I want rather to sound the note of crude empiricism even more evidently, to use numbers, and notions like that which up to now were somehow foreign to the cosmologist's trade. Fortunately, they have been imported into the whole enterprise, mainly in the last decade, as the result of a set of surprises, of which the most important was surely the by now properly hallowed discovery of the microwave background.

The topic that I'm speaking about is a somewhat trivial feature, allowing discrimination only among those classical kinematic models which arise from the postulates of symmetry which are surely not permanent foundations of the theory. As Professor Harrison is strongly implying, it is clear that the investigations of the early universe will sooner or later come to grips with the lumpy reality beneath the symmetrical grin. But for the moment, since I don't want to engage in investigations that plumb so deeply into the past as those we've heard described, I will stick pretty much to a simple view of the universe.

What we are studying, to put it indeed very simply, is not the structure of space and time, which are only ways of describing physical phenomena, but the behavior of gravitation in the large, with a small admixture of microwaves. That seems to be the picture that we now have; and of course that is indeed the classical picture. The observations, both of microwaves, of discrete optical and radio sources, and of other diffuse backgrounds of various kinds, are interpretable as physical tracers, so to speak, of the large-scale motions, dominated, as far as we know, (but of course we do not know for sure) by gravitation. If you ask why this is the case, it is merely a scale problem; it is the absence of saturation of the gravitational forces with their very large range, perhaps infinite.

R. J. Seeger and R. S. Cohen (eds.), Philosophical Foundations of Science, 189–202.

The relationship of these facts to the whole deep question of the foundations of Newtonian mechanics was, I think, adequately reviewed. I shall assume that we can argue successfully on the surface, as we so often do in physics, and simply discuss the behavior of matter in motion in a way already familiar in the 19th century. It turns out that the conclusions of this argument are precisely those of the more sophisticated arguments that arise from a better understanding of the nature of gravitation, that is, from the Einsteinian understanding. Since we come to the same conclusions it is probably not wrong to discuss it in a more intuitive language which will lead to the same result.

The whole point is to consider a uniform, infinitely extended (or bounded, for that doesn't make any practical difference provided the boundary is very far away) mixture of Newtonian gravitating particles. We impose one very heavy constraint, which we can verify in some large sense today, and which we can imagine to have failed only in some distant time in the past: on a sufficiently large scale of averaging, things are uniform, and indeed isotropic. The evidence for this is today much better than ever before – as I say, not as an ultimate hypothesis, but as the present state of affairs in an evolving situation. If we do that, it's quite natural to ask what will happen to that system. It was Newton's great idea, apparently, that one had such a system infinitely extended in absolute space. Out of this there appeared, by means which he is not very clear about, the 'plasma instability' characteristic of the imaginary charge which is part of gravitation of course, in Newtonian language. So we reached to condensation of the particles – planets, stars, galaxies – which make up the world we see.

But we do not see the world as static but extended; instead, we see a headlong homogeneous expansion. Let us treat the uniform world in a Newtonian way. There is an inertial frame, and everybody measures with respect to his own coordinate system. If you worry about light propagation – and of course you must in fact do that – you can't remain in this happy Newtonian world forever, because light propagation, strictly speaking, is not Newtonian; if you make it so then you will introduce an ether. That's a very plausible thing to do: an expanding ether will then carry the background and produce light signals that move in the right way. It is only the conflict of this with our understanding of electrodynamics that set physics into the Einstein direction.

I'm trying to argue that one does not make much of a mistake, once he is willing to say this is only a first approximation, if he is also willing to overlook the detailed study of the propagation of light. Those are, of course, two heavy prices to pay. I don't mean to say this is any substitute for the proper relativistic point of view – not at all; it is simply a heuristic path towards understanding that point of view. It is then quite straightforward to show that you preserve all the properties of uniformity and lack of rotation, assuming neither preferred position, nor direction in space. You do however single out a position in time: and that is intrinsic even to the relativistic theory. If we are right in our view of the present evolution of the universe, this position in time is physically real; this is so because there is a great clock whose hand points to the temperature of the isotropic black-body radiation. We are at 3K o'clock, and there is no question we have a pretty good time signal, which everybody can share.

Now having said this, you have next to ask what is the gravitational force that a test particle sees? It sees only the force from a sphere within a radius R from any observer – any observer looking out at the world from his inertial frame sees that force; as usual, the stuff that is outside of the sphere plays no role; it is only what is internal to the sphere that counts. The internal-sphere attraction goes like the mass divided by the square of the distance. Thus we can write down the ordinary Newtonian laws on a test particle, which must now be regarded as representing any one of the test particles that occupy this whole fluid. If I ask for a solution for the motion of the fluid under this force, then the Newtonian equations lead to a solution in the spherically symmetric case; of course it is exactly the same solution exemplified in throwing a coin into the air (or nowadays throwing a satellite above the atmosphere). Namely, we have a critical velocity – a velocity which might depend on our position, but the position is not important in the cosmological case; so it can depend upon density only, and thus gives the rate of change of density, which we observe by watching the galaxies expand.

Two numbers must come in: the famous Hubble constant, and the density of matter in the universe at any one time. Now if the expansion velocity is great enough, then the distant galaxy we view at R, being dragged back only by the gravitational force due to the sphere within R, is moving fast enough and can escape from those gravitational forces, so the motion will continue indefinitely. If, on the other hand, the density is great enough and the expansion velocity we observe is not too high, then the particle at R, while continuing to move for some time in the outward direction – that is, the continued expansion – is destined to come to rest and fall back again, exactly because it has not reached the velocity of escape. The escape velocity, then, is a critical velocity, so that on one side of this, the system will forever dissipate into infinite space; and on the other side of it, the system will eventually reverse the motion we now see: the universal expansion becomes a universal contraction, and all comes back towards the center.

That is the story of Newtonian motion. It turns out – as I think everyone knows who has looked into the topic a little – that the famous Friedmann equation which is obtained by applying the kinematical postulates of the early cosmologists, the grin cosmologists (and we don't do better yet, at least for the latter stages of the universe), to the equations of general relativity, gives rise to one and the same differential equation. The velocity of escape, and the characteristic numbers which we would interpret in Newtonian language as allowing either infinitely distant motion in the future, diluting the universe infinitely far, or expansion only to some finite time and then return, are also found in general relativity. There it can be interpreted if you like as a curvature of space-time, and this is a language which one often uses. But I think it is in no case misleading simply to say that this is a test of the dynamical behavior of gravitating particles: are they moving so fast in this overall expansion that they will fly apart forever, or will they go outward only to some distance, different for each test particle, but all slowing down as they move out, eventually to fall back down again? That is the issue. In the language of differential geometry, the question becomes: is the universe open or closed? We import into the theory the notion of a single parameter, namely the local curvature, to describe the nature of the space-time manifold within which we are operating. As geometers, we then speak in the metaphor of curves or surfaces, open or closed, whereas the physicist, I think, would simply have wanted to say, in this naive way, are we beyond or within the velocity of escape? And that's the whole situation.

I will mention one more interesting point. We know the fundamental law of gravitation, which is the fundamental Newtonian guarantee of all that I've said, tolerably well. We have excellent accuracy for its measurement. Perhaps *the* principal subject of theoretical physics so far, integrated over time, has been the validity of the law of gravitation. It is only since the end of the nineteenth century that electrodynamics, atomic physics and particle physics entered the domain of theoretical physics. Most of the great minds in theoretical physics – or at least most of the heavy books from Euler to Riemann and Kelvin – were really studying the details and the accuracy of the gravitational description. We are pretty satisfied with it. And yet it has to be admitted that it is certainly not perfect: it is possible in a very natural way to add various modifications to the gravitational equations. One such modification is a finite range. Another such modification is given by an assumption that in the absence of matter we *still* would have the equivalent of a finite density – which might be set positive or negative – that is, to introduce essentially a new zero point for counting gravitational source density. And that is an interpretation I'd like to give to what is called the cosmological constant in the Friedmann equation. If you add that in, then by so regauging the density, so to speak, you get another set of cosmological solutions; you can then use some negative density, otherwise unheard of, and this gives you a chance to seek more complicated solutions.

The entire family of solutions then shows either the particles moving so that they are beyond the velocity of escape from any observer's point of view, who will see that the sphere of mass attracting a particle is not adequate to hold it in. Or else they are moving with a speed within the velocity of escape, in which case, since we see them definitely moving away from us now, we know that at some future time they must start falling back in. That is really the Newtonian result. If you want to import the somewhat slippery notion of a negative density, which does enter the equations in a reasonably natural way if one looks at it without prejudice, it admittedly is a major import into a theory which has no real need of it. In addition, you can have a few other jazzy possibilities. For example, you can have one model that contracts, but doesn't come in very dense before it starts out expanding again, which is rather interesting; and you can have one that hits a plateau – it expands brilliantly as though it really were going to be an oscillating universe in the old Newtonian way, and then suddenly remembers that it has this fixed pseudo-density which is trying to blow it up – a density that does not scale with distance in the way a real, material density does. This failure of scaling means that once you come to a certain density the repulsion just matches the gravitational attraction, so the universe can stand still on a plateau for a long time; and if you're bold enough to adjust the parameters delicately, it will rest for an arbitrarily long time, and nevertheless can start to expand again. There has been no dearth of claims in the past two or three years by several learned workers that they even see some experimental evidence for that. Personally, I think they are wrong, but it's fair to keep the matter in the back of one's mind.

So there we have the situation. The real question is – escape or not? That sounds like a straight-forward physical argument, and so it is. It's a simple enough thing, having to do with a parameter which we might call the intrinsic energy of the motions of the universe; but like all things in cosmology it *can* be regarded in a very grand way. So I *can* say it touches the question of whether the universe is infinite in any physical sense, or finite in the same physical sense. Because openness, that is infinite escape, clearly implies an infinite domain of space and time: it may be very slow and dull at the end, like the jokes about Philadelphia, but still infinite. A universe that closes, that reverses, presents quite different problems. It's simply gravitation we're talking about, nothing much more mysterious.

Now gravitation itself is extraordinarily mysterious, I'd like to emphasize that point; gravitation is something grand and profound; and the simple division of Newtonian mechanics into inertial frame and laws of motion is quite an inadequate account of what is really contained there. But I'm trying to stick to that level of description, a sufficient level of description. On that level we can say simply that gravitation not only keeps in the matter, that is, makes the matter fall back again, but it is the very essence of the whole business that all forms of radiation also must behave just the same way. Therefore there is simply no signaling that will get away – you cannot acquire the velocity of escape, not matter what you do, if that's the way the universe goes. Whatever you can possibly learn about the universe by anything that does not travel locally at greater than the speed of light (and nothing does), you'll never be able to learn anything more than what is contained in a finite domain – albeit it is a very large domain. Therefore this is the great question, infinity or finiteness – and of course touches on the great eschatological questions of theology and philosophy and those vast domains of comment. I shall not say more about these than to remark that this is a straightforward physical question: given two galaxies or clusters of galaxies which are now moving apart, will they at some future time begin to come together again? That's the physical way of putting it.

And another question is: do we live in a universe which is knowably infinite, that is, knowably without bound; or knowably with a bound? And that of course is a nice question. One must admit that the importation of a fictitious negative density, which is admissible in the equations but which is otherwise not known, will change the results. That may be a very interesting point, but it is still a little early to add such a new concept. Absent that, the whole question is: what density of matter is adequate, given the known motion of things, to make things just come back together?

The answer is quite straightforward. We know the motions fairly well – not perfectly, we might be off a factor of 2, but reasonably well – we know the motions for a given distance. We know the constants that come into the equation, essentially G and c^2, arbitrarily well for this purpose. We can calculate the critical density with an error which is not as great as an order of magnitude, I feel. That density, which is the critical density that limits these things, is only complicated, as far as we know now, by the possibility of the negative, fictitious density that I'm going to put aside – (any lovers of the cosmological constant had better seek equal time), comes out to be about one proton per cubic foot, which is not much, but it is something. In fact, since there are lots of cubic feet, it is everything. I'm going to translate that into electron volts of energy per cc, so that I can compare it more easily with radiation, which we find easier than matter to measure directly; and say that the critical density is 20000 eV cc^{-1}, give or take a factor of a four. More density than that, the universe is bound to come back (to repeat: galaxies go to a maximum distance and then come back again); less density than that, the galaxies will drift apart infinitely far and will eventually go to a genuine age of darkness, from our present age of galaxies which at least are neighbors, if not in intimate connection. We look back, as Professor Harrison implied, to an age of light; and before that, to an age of contact. These are the ages of the universe as I see it; it is so, whether we go through it only once; or whether we come back again, with I believe many profound, and in my opinion not yet sufficiently understood, influences on the physics of the local behavior in the contracting phase.

On the other hand, if we look around the galaxies, using our measure of the mass in the galaxy obtained by gravitational means, as well as by direct study of stars and the luminosity-mass relationship, we can give limits for the actual density of the universe. Applying this to the distant galaxies and making appropriate corrections for luminosity of other kinds of galaxies – that is, they have fainter stars with more mass per watt of light – one cannot get a very close answer, clearly, because we need a lot of data which we really don't have. Even in our own galaxy, I think the total mass is not known with reliability to a factor of two (probably we could say the error is about $1\frac{1}{2}$, but I think you could not exclude a factor of two error). Within the same error, the amount of matter we see smeared over to produce that grin effect is about 200 to 500 eV per cc, so we are short by a factor of from 50 to 100 – or say 30 to 100 – of having enough mass for closure, for universal repetition. That is really the question of the day. *Prima facie*, we appear to be the observers, the participants, in a once only performance; we had better watch it while we can.

However, of course, nature, like the oracle of Apollo, neither reveals nor conceals, but gives signs: we have to look more carefully at what the signs mean. What else could there be that we have left out of account in calculating our density? Could it be radiation of some sort that is closing the universe? Well, the most powerful electromagnetic radiation which is familiar to everyone is the starlight. It is certainly very ubiquitous, but it amounts to only about $\frac{1}{10}$ of an electron volt per cc at the present time – neither a few hundred like the known matter nor 20 kV like the matter requires for closure, but only $\frac{1}{10}$ of one volt. Well, the 'big bang' decayed radiation, which is the big flash, the most radiation we know about, is 0,4 V per cc, and that's very far short. Maybe if the gravitational wave pulses which have been seen by Weber are given a spectral shape, for which there is no evidence but which again could not be excluded, you can inflate the doubtful gravitational radiation enormously enough to do the job. I myself feel that doing this would be most premature and speculative; I dont't think gravitational radiation is yet really known to exist, but it has to be admitted as a possibility. It's one of the interesting possibilities, and one of those about which we are least competent to judge. But let us ask what other means are there for finding our what's going on?

The first possibility, of course, is a direct kinematic study – let's just watch and see if the galaxies are going to slow down – by watching a particular galaxy we might see its red shift really change. But we don't have patience for that. Our own individual generation's part in this show is so nearly instantaneous that there's no hope of doing that; even the entire history of science is not sufficient time to do that. We do it by looking, not at a single galaxy over a long time, but at the whole collection of galaxies at a single time to see whether marshalling the history of many galaxies at a single time, by looking further and further away so that we gain information about older and older times, gives you at least a sense of the direction of change. That corresponds to looking for a curvature in the famous Hubble law, i.e. looking for a deceleration, or an acceleration, in the past as we look out at distant galaxies. In principle that could be done, and it was for that reason that Professor Hale pressed to make the Palomar 200″ telescope. When it was finished, the telescope worked beautifully, but it could not solve this problem.

The problem is still unsolved, and is unsolved in spite of the repeated claims to have solved it by the observers at Palomar, by the radio astronomers, even by the interferometer operators: I suspect that we simply know too little about the physics of galaxies and of galaxy clusters, of their luminosity distribution and its evolution in time; that it is at least quite premature to expect that a highly accurate kinematical measurement of this sort can be done. You hardly know what the objects are that you're looking at, and very little about their history. That has been the history of this subject for twenty years. Even in very recent years, within the last three or four, there has been a claim of solution of this problem followed by a quick denial. I don't think that we're going to solve it by kinematical means, though if it it were so solved everybody would be very happy to see that. It doesn't look easy. Certainly it could come only as a result of a very deep understanding of the evolution of the typical kinds of radio and optical sources.

Well, if this is the case, we must look around and see whether we can find any kind of argument for the accuracy of the count of particles that has been made in the number that I gave – 200 to 500 V per cc, measured indirectly by looking at starlight. Of course starlight does not reveal everything: there are even a few signs that there does exist hidden matter which does not shine in visible light.

Point one: the clusters of galaxies, or at least some of them, seem to show internal motions too rapid for galaxies in the field of each other's influence. We can't watch them move, of course; we measure only instantaneous velocities by Doppler shift; so we can't tell what the history has been. But you can tell what the average velocity of the galaxy is, compared with its neighbors. Those velocities are a bit higher than one would like – as though in the ancient and distant gravitational collisions of those galaxies, they had gained recoil velocities rather higher than you would expect. This gives you a mass greater than the best guess for the visible mass of the same galaxies; but it does not look to be 100 times greater. In a few cases it might be for a few special clusters. But by and large we don't see enough to give it that factor of 50 or 100 times. It's certainly quite important; it's one place where we will continue to work, but I think it is not going to give us closure.

Second, could there be matter which is hidden in the sense that it doesn't shine at all? Or can we look for the kind of concealed shining that it does by some means other than classical optical astronomy? Of course, that is the second interesting contact between this work and high-energy particle physics in the laboratory. In the early epoch of the universe, particle physics was essential because the whole universe was made of colliding unstable particles. At the present time we can get information in electromagnetic spectral ranges not available in the classical domain of astronomy. For example, X-rays, infrared radiation, ultraviolet radiation, etc. I shall not give all the possibilities, but I want to mention two that are interesting – very much on the agenda for study at the present time, and on which results are flowing in every few months. It turns out, first of all, that matter which is compact shines by some surface phenomenon, which does not reveal its full mass. We can measure the mass behind visible matter only if it is not compacted any more than are the stars with we are familiar, so we know its internal nature.

There are two kinds of compact matter, one of which I think is known to exist on a modest scale, and one of which is surmisable, and may be very important. The first I would like to call by the word *spinar* – 'spinar' is a word of my own which stands for a motion like a pulsar – a rapidly spinning, condensed object which may occur on any scale of spatial dimension; such that its energy evolution is not controlled by thermal radiation at all; but, on the contrary, controlled by the loss of rotational work to the surroundings, usually through the manufacture of relativistic particles. Spinars thus give rise to synchrotron radiation, cosmic rays, quasars, and all these strange non-thermal things which we now see. We know some such objects – quite a few of them – in our near galactic neighborhood, which we call *pulsars*; they have a special quality, because they have the density of nuclear matter as well, so they represent a kind of equilibrium end-point in the evolution of some stars. It is not too much to conjecture that similar objects may be found; and Ambartsumian has for 15 years and Arp for at least five, pointed to the centers of galaxies, the nuclei of galaxies, as most unusual places, where something strange is going on. By now this is undoubted; there are recent data, for example, showing the presence of a point radio source in the galaxy M87, both in the center of the nucleus and way out along the unusual jet of that object; these two radio sources contribute, each one of them, of the order of some percent of the total radio flux; and yet their diameter is measured in light years or smaller, whereas the diameter of the galaxy is 100000 light years. That's an enormous concentration of energy, even though we see with certainty only the radio emission from these objects. I think it's very likely that M87 is the Crab Nebula of the galaxies; and just as the Crab Nebula reveals to us the nature of a pulsar, so M87 may be revealing the nature of those objects which may be active centers of galaxies, and so account for much of the galactic pyrotechnics for which we have some vague evidence. If so, this will allow us to find more mass in this condensed and therefore hidden form, not at all obeying the rules of the evolution of stellar matter. Even here it does not seem anything like enough. We don't see how we could find 100 times more mass there in the center, so I don't think this is going to work as a mechanism of closure.

A sub-case of this, of course, is to let the matter disappear entirely from the universe, to be represented only by the memory of its gravitational effects. That is the famous 'black hole': sooner or later the thing will stop spinning, sooner or later the thing will have no means of support against gravitational collapse, sooner or later it generally will fall into a singularity. Its presence in our universe will be manifest only by the gravitational effects it produces, tending to be an obstacle to light coming from behind it, or a sink for radar waves sent at it. Maybe these black holes will, on some scale, make up the matter we are missing. Again, it could be. If the gravitational pulses now claimed to have been seen are indeed as indicated, they may represent the presence of such black holes, snaffling bits and pieces of matter to enrich themselves at the expense of the surrounding universe. I think that is pretty speculative; and I wouldn't care to bet that it is true.

The other possibility, apart from condensed matter, is diffuse matter. Cold gas is out. If cold gas were present in anything like adequate quantities, its atomic structure would imply individual spectral lines; no sign of enough of it is shown across distant galaxy spectra in optical or radio work. There cannot be enough cold gas. There might be hot gas. It must be ionized; yet it cannot be too hot gas, for if it is too hot the X-rays from it would be detectable. And while some argue that the X-rays are seen in the $\frac{1}{2}$ kV to 1 kV range, they certainly aren't visible beyond that. Thus, it seems that we might find the missing matter in the form of hot plasma at $\frac{1}{2}$ to 1 kV per particle occupying the universe. I don't think it is at all demonstrated; there are many difficulties with these early experiments, but it is certainly one possibility.

Another fascinating possibility, which has attracted attention only in the last six months, is that the gas is there; it is hot in the sense that it is ionized – it contains almost no neutral atoms – but it is not hot in the sense that it is at very high temperature. That is to say, it was ionized by non-thermal processes, by ultraviolet radiation and X-radiation from some catastrophe in the past. It was ionized, it remains ionized; but it is very slow to recombine. Its slow recombination from a luke-warm temperature is not fast enough to make detectable X-rays. It may yield ultraviolet radiation, and perhaps even occasional lines in the visible spectrum, and therefore it is accessible to experiment, though it has not yet been demonstrated. It's hidden in the worst possible place for our measurements – in the ultraviolet, both the near ultraviolet and the far ultraviolet, where we have a very hard time indeed making measurements, whether from the earth or even from anywhere in the galaxy.

One feature of this is very nice, because if this gas does now exist in sufficient density in the universe, then at a red shift of a factor of 3 or 4 or more into the past, the gas was proportionately denser. That corresponds to a time in the past before we see any galaxies. All we see of structure that old (we see the microwaves, which, if our theory is right, are very old indeed), the only structured objects we see, are quasars.

Curiously enough, quasars become few and far between beyond a red-shift factor about equal to 3. The highest red-shift ratio one has is 3.8 or thereabouts, and between 2.95 and 3.3 there are a dozen of the 100 or so known; almost 10% of them are in there. This is very striking, a pile-up in the past. Now a pile-up in the past is in itself not unlikely, because the evolution of such objects may be rapid, and therefore if there were more of them in the past and none very near us, that would be quite sensible. If there were a lot of ionized gas back there, it could hide such early objects by the light scattering it induces, without itself being very conspicuous in any of our present observations. On the other hand, the gas itself is now hidden, if you like, because the quasars have excited it so much that it is slow to recombine; so all one can see is some weak ultraviolet background, very hard to measure. It looks possible that a mutual conspiracy exists in the early universe to conceal luke-warm plasma from us and at the same time to conceal the formative stage of the quasars, while no normal galaxies were bright enough to be seen at such enormous distances. That is really a possibility; this early luke-warm plasma is a very interesting suggestion which has been made in recent months by McCrea, by Rees, and by Axford in different forms, and it may be that we should look for this very intensely.

We have two choices: condensed and diffuse. I think both are present. That is, I do believe there is condensed matter of a kind we have not previously seen, which is important, significant in the evolution of the structure of the universe. I do believe also that there will be found a considerable pool of ionized gas hiding the early quasars and cutting off the red shift distribution. But it does not seem to me easy to put enough matter to close the universe into either of these two bins. Possibly it is there. That will be the genuine test, one no longer subject only to kinematical studies of solutions of Friedmann's equation; but now attacked with the whole battery of modern astronomical equipment.

The next few years will certainly see a large-scale growth of information on these two bins. That will give us an answer, unless we have severely misread evolutionary history (I don't think we know the universe as far back as those condensed-particle times very well; but I am fairly confident we know it as far back as the microwave radiation implies, back to a mean temperature of some volts). Apart from the possible errors and deficiencies in our general theory, it seems to me we will be able to answer confidently if not finally (for the absence of a mass will never be quite probative) the question whether we are present at a unique show, or one which is going to reverse. Maybe there is some reason why we do not have a single show, but we have an everlasting scalloped, cycloidal repetition of the history.

It is fine, and in the nature of the subject, that such grand and somewhat metaphysical conclusions can be based on matter-of-fact observations of spectral lines and X-ray intensities and the rest. That is the place to look in the near future for that experimental evidence that will help us to answer the question: once only? or again and again?

Massachusetts Institute of Technology

DAVID LAYZER

COSMIC EVOLUTION

I should like to try to explain the basic premises and results of a theory of cosmic evolution that addresses a number of related problems. Let me begin by describing one of these which has special relevance to the subject of this colloquium.

We are often told that the universe is running down. This is a consequence of the second law of thermodynamics, which requires the entropy of every closed system to be a non-decreasing function of the time, and of plausible assumptions about the large-scale structure of the universe. A wide range of physical processes – including molecular transport, turbulent mixing, and macroscopic electromagnetic processes – implement the approach to cosmic heat death.

But there is another side to the coin. We are also familiar with many processes that work in the opposite direction. These occur in open systems or subsystems and increase their store of information. Life and all other evolutionary processes, physical as well as biological, have this character. They accumulate information and thus produce growing records of the past. So the universe is winding up as well as running down.

Thus two distinct arrows of time coexist (Figure 1). I shall refer to

	Type of system	Entropy variation	Examples of processes
Thermo-dynamic arrow	Closed	$S\uparrow$	Molecular transport Turbulence Radiation processes
Historical arrow	Open	$S\downarrow$	Evolution of planets, stars, galaxies, etc. Biological evolution Life processes Consciousness

Fig. 1. Two arrows of time.

R. J. Seeger and R. S. Cohen (eds.), Philosophical Foundations of Science, 203–213.

them as the thermodynamic arrow and the historical arrow. Because closed systems are easier to study than open systems, we know more about the thermodynamic arrow than about the historical arrow. But the two are closely related. This becomes apparent when we consider the structure of current theories of the approach to equilibrium.

In all such theories the first step is to replace an exact reversible – or nearly reversible – description of a many-body system by a statistical description – which in the first instance is also symmetric with respect to the two directions of time.

The next step is coarse-graining. One combines the microstates into aggregates within which the macroscopic variables of the description, such as the energy, do not vary appreciably. From the original probability distribution or density matrix one constructs the corresponding coarse-grained quantities. Obviously, this step does not disturb the symmetry of the description with respect to time-reversal.

It is the final step that introduces a preferred time direction. Fifteen years ago Van Hove (1955–1957) proved that if the Hamiltonian of a closed system satisfies a certain rather general criterion and if the off-diagonal elements of the coarse-grained density matrix vanish *at some initial instant*, then the system will exhibit irreversible behavior at all subsequent times. In particular, the coarse-grained entropy will increase until it assumes the maximum value compatible with the macroscopic constraints. Since the total entropy remains constant, by Liouville's theorem, the residual entropy, $S-\bar{S}$, associated with the 'microscopic' degrees of freedom must decrease in the process. Similar theorems have been estabished in different formal contexts by a number of other authors.

The content of all these theorems can be concisely expressed in terms of information, defined (see Figure 2) as minus the difference between the

$$S = -\sum p_k \ln p_k \quad \text{or} \quad S = -Tr\{\rho \ln \rho\}$$

p_k = probability of microstate k, ρ = density matrix

Coarse-graining: $\bar{S} = S(\overline{p_\alpha})$, $\overline{p_\alpha} = \sum_{k \text{ in } \alpha} p_k$

$$\text{Information:} \quad \begin{array}{l} I = S_{\max} - S \\ I = \bar{I} + \tilde{I} \end{array}$$

Fig. 2. Entropy and information.

entropy and its maximum possible value subject to given constraints. The macroscopic information $\bar{I}$ and microscopic information $\tilde{I}$ correspond to the coarse-grained entropy $\bar{S}$ and the residual entropy $\tilde{S}=S-\bar{S}$. Van Hove's and similar results can be stated in the following terms:

If microscopic information (suitably defined) is initially absent in a closed system, then the macroscopic information will subsequently decrease monotonically with time. The microscopic information increase at an equal rate, so that the total information of the system remains constant; information flows from the macroscopic into the microscopic degrees of freedom.

Thus if current statistical theories of the approach to equilibrium are essentially correct, the existence of a universal thermodynamic arrow has an obvious cosmogonic implication: in newly formed physical systems macroscopic information is regularly present and microscopic information is regularly absent.

This statement has a paradoxical quality. It purports to express an objective property of the universe. Yet according to currently held views, the absence of microscopic information in a theoretical description does not imply that such information is unattainable. It merely reflects certain practical or aesthetic considerations. Moreover, the boundary between the macroscopic and microscopic domains and the choice of macroscopic variables are to a large extent arbitrary. How then can the absence of microscopic information be regarded as an objective property of newly formed systems? Within the framework of current physical theories, no satisfactory way of resolving this paradox has so far been found. I suggest that existing theories of the approach to equilibrium need to be supplemented by considerations that (a) enable us to attach precise and objective significance to the concept of microscopic information, (b) explain why microscopic information is regularly absent in newly formed systems, and (c) explain the genesis of macroscopic information.

The required considerations seem to flow naturally from a strict interpretation of Einstein's cosmological principle, which states that the spatial structure of the universe, as described in some coordinate system, is statistically homogeneous and isotropic: no average property of the spatial distribution of matter and motion serves to define a preferred position or direction in space.

The cosmological principle has two well-known consequences (see Figure 3). In the first place, it leads to a unique resolution of space-time

(1) Unique resolution of space-time into space + time,

(2) Cosmic expansion (contraction),
---Define scale factor a through $\mu a^3 = \text{const.}$,
$\mu =$ Baryon number density. Then as $t \to 0$

$$a \sim \begin{cases} t^{1/2} \text{ (radiation-dominated)} \\ t^{2/3} \text{ (matter-dominated)} \end{cases}$$

Fig. 3. Two consequences of the cosmological principle.

into space and time, and thus reestablishes absolute simultaneity at the cosmological level of description. For it is clear that only spatial symmetry operations – and not, for example, Lorentz transformations – preserve spatial homogeneity and isotropy. In the second place, together with Einstein's theory of gravitation, it implies that the universe is expanding from or contracting toward a singular state of infinite density. So cosmic time is both absolute and directed.

The cosmological principle has a third important consequence, which does not seem to have been noted previously. By way of illustration, consider a statistically uniform (Poisson) distribution of points on a straight line divided into equal cells of width h (Figure 4). This realization, characterized by a set of occupation numbers infinite in both directions, has two remarkable properties that, taken together, distinguish it from realizations of a Poisson distribution on a finite or semi-infinite segment.

First, the law of large numbers ensures that from a single realization one can estimate with arbitrary precision the mean occupation number of a cell. This statistical parameter, or macroscopic variable, completely characterizes the distribution. Second, two infinite realizations of a Poisson distribution characterized by the same mean occupation number are operationally indistinguishable, for we can always match them up – in infinitely many ways – over any finite domain. This is evidently *not* true

OCCUPATION NUMBERS: ...10210031112130102111...

$p_k = (\lambda^k / k!)\, e^{-\lambda}, \quad \langle k \rangle = \langle k^2 \rangle = \lambda$

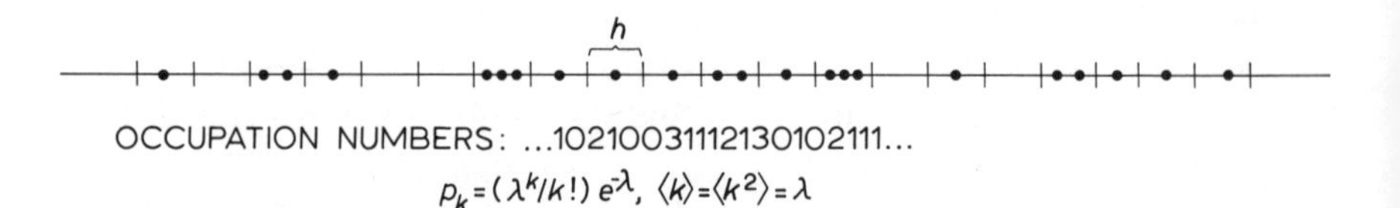

Fig. 4. Poisson distribution in one dimension.

for semi-infinite or finite realizations, because the existence of an end-point makes it possible to construct demonstrably different realizations having the same statistical properties. To sum up, a single infinite realization of a Poisson distribution contains all the macroscopic information needed to define it, and contains no additional microscopic information. For microscopic information, by definition, is what distinguishes different realizations characterized by the same statistical parameters.

These conclusions can be generalized to spatially infinite models of the universe that satisfy the cosmological principle and in which all correlation distances are less than some finite value. (The last property ensures that the distributions are ergodic.)

In short, the cosmological principle, strictly interpreted, automatically creates a distinction between macroscopic and microscopic information. Macroscopic information is that supplied by a complete statistical description of the universe. Microscopic information, according to this definition and to what has been said, is entirely absent at the cosmological level of description. So if a theory based on the cosmological principle can account for the emergence of effectively isolated systems, it will automatically provide a link with statistical theories of the approach to equilibrium. For any given system the definitions of macroscopic variables will be determined by the system's prenatal history.

Next let us consider how macroscopic information comes into being. Again this question can be elucidated through a simple example. Consider a universe uniformly filled with a mixture of nonrelativistic gas and electromagnetic radiation, initially at the same temperature. The cosmic expansion is adiabatic, but the specific entropy of the mixture increases monotonically at a rate depending on how strongly the two components interact. Suppose first that the gas and the radiation do not interact at all. In this limit – as in the opposite limit of instantaneous thermalization – the specific entropy does not change. Each component behaves as if the other was not present. But the laws for adiabatic temperature variation are different for the gas and the radiation, so a temperature difference must develop between the two components (Figure 5). Although the specific entropy of the mixture does not change in this process its maximum possible value, for given specific energy, increases monotonically as the model expands or contracts away from the initial configuration of local thermodynamic equilibrium. So the specific informa-

$$e = e^{\text{gas}} + e^{\text{rad}}, \qquad p = p^{\text{gas}} + p^{\text{rad}}$$

$$p^{\text{gas}} = \tfrac{2}{3} e^{\text{gas}}, \qquad p^{\text{rad}} = \tfrac{1}{3} e^{\text{rad}}$$

Adiabatic law: $\dfrac{\mathrm{d}(\mathrm{eV})}{\mathrm{d}t} + p \dfrac{\mathrm{d}V}{\mathrm{d}t} = 0$

Zero interaction: $VT_{\text{rad}}^{3} = \text{const.}, \qquad VT_{\text{gas}}^{3/2} = \text{const.}$

Fig. 5. A mixture of nonrelativistic gas and radiation.

tion increases monotonically.

The same qualitative result follows if the radiation and the gas are allowed to interact. Only in the limit of a quasi-static expansion or contraction, when the relaxation time is much shorter than the characteristic expansion or contraction time, does the mixture remain in a state of equilibrium and the specific information remain constant.

The relaxation process involves a transfer of energy between gas an radiation, which affects the pressure and hence the rate of change of the specific energy (see Figure 6). So the variation of the specific energy depends on the relaxation rate. It is for this reason that the rate of growth of specific information depends on the relaxation rate, being greatest when this rate is least.

This example illustrates a general result: *Expansion or contraction from a state of thermodynamic equilibrium generates macroscopic information.*

Under what conditions can thermodynamic equilibrium be expected to prevail in an expanding universe? A necessary and sufficient condition for *local* thermodynamic equilibrium is that the rates of the reactions

If $T_{\text{rad}} > T_{\text{gas}}$, $\quad \Delta e^{\text{gas}} = -\Delta e^{\text{rad}} > 0$, Where Δe^{gas} denotes a change due to thermalization.

$$\therefore\ \Delta p = \tfrac{2}{3}\Delta e^{\text{gas}} + \tfrac{1}{3}\Delta e^{\text{rad}} = \tfrac{1}{3}\Delta e^{\text{gas}} > 0$$

$$\therefore\ \Delta(\mathrm{d}E/\mathrm{d}t) = -\Delta p\, \mathrm{d}V/\mathrm{d}t < 0 \quad (E \equiv \mathrm{eV})$$

Increasing the thermalization rate during expansion from an initial state with $T_{\text{rad}} = T_{\text{gas}}$ *increases* the rate of decrease of the specific energy and thus *decreases* the rate of increase of specific information.

Fig. 6. Dependence of E on thermalization rate.

that dominate the relaxation process greatly exceed the expansion rate. Figure 7 compares two-body reaction rates with the expansion rate in a Friedmann universe. Whether matter or radiation dominates, the ratio of the expansion rate to any given two-body reaction rate approaches zero as $t \to 0$. So in spite of the fact that the expansion rate becomes infinite in the limit $t \to 0$, the 'big-bang' is the exact opposite of an explosive process. The universe expands from a limiting state of local thermodynamic equilibrium.

This conclusion suggests the assumption that the limiting initial state was one of global, and not merely local, thermodynamic equilibrium. The question then arises whether structure – that is to say, large macroscopic density fluctuations – can be generated in an initially structureless universe. If it can be, we will have answered – at least on a qualitative level – the questions I raised at the beginning of this talk. First, physical systems will separate out with precisely the qualitative properties required by current statistical theories of irreversible processes. Secondly, the theory will supply, at least in principle, the needed prescriptions for defining macroscopic variables. Finally, the theory will predict a continual growth of macroscopic information in the universe and hence account for manifestations of the historical arrow in systems capable of absorbing or extracting information from their environment.

In seeking to formulate a theory for the development of structure in an initially structureless Friedmann universe, one must begin by choosing between a hot and a cold initial state. The main features of the two hypotheses are compared in Figure 8. The chief advantage of the hot initial state is that it affords a simple interpretation of the microwave background, as radiation left over from the primordial fireball. The disadvantage of a hot universe is its well-documented stability against macro-

$$\text{Two-body reaction rate } \alpha \propto \mu \propto a^{-3} \propto \begin{cases} t^{-2} & \text{(matter-dominated)} \\ t^{-3/2} & \text{(radiation-dominated)} \end{cases}$$

$$\text{Expansion rate } H \propto t^{-1}$$

$$\frac{H}{\alpha} \propto \begin{cases} t \\ t^{1/2} \end{cases} \to 0 \quad \text{as} \quad t \to 0$$

Fig. 7. Two-body reaction rates compared with expansion rate in a friedmann universe.

	Hot	Cold
Microwave background	Remnant of early radiation-dominated phase	Radiation emitted around $t \simeq 3 \times 10^7$ yr by supernovae in the mass range 5–10 $M_\odot$ – subsequently thermalized by interaction with dust formed from heavy elements ejected in the supernova outbursts
Primordial helium abundance	$\simeq 28\%$	Arbitrary: determined by early degenerate neutrino pressure
Large-scale density fluctuations	Must have been present at supernuclear densities	Initially uniform cosmic medium solidifies ($t \simeq 10^3$ s), then shatters into fragments of mass $M \simeq 10^{27}$ gm; large-scale fluctuations form subsequently by clustering process

Fig. 8. Hot and cold universes.

scopic density fluctuations. If one accepts the hot initial state, one is driven to postulate initial structure.

The interpretation of the microwave background is less direct in a cold universe. According to a recent discussion (Layzer and Hively, 1972), most of the radiation was emitted around $t \simeq 3 \times 10^7$ yr by an early generation of supernovae containing a large fraction of the mass of the universe. The emitted radiation was subsequently thermalized by dust grains formed from heavy elements synthesized in the interiors of these stars and ejected in the supernova outbursts. We suggest that the spent cores account for the so-called 'missing mass' whose presence has been inferred from dynamical studies of galaxy clusters.

Hively (1971), in a detailed quantum-mechanical study of the early phases of the cold universe, found that the medium solidifies in the metallic state around $t \simeq 10^3$ s. Hively and I have argued that the subsequent expansion causes the medium to shatter into planet-sized fragments of mass $M \simeq 10^{27}$ gm (Layzer and Hively, 1972).

According to a still speculative theory of the subsequent evolution (Layzer, 1969), this event triggers a form of turbulence in which density fluctuations and the accompanying gravitational fields play a central role. A wide spectrum of density fluctuations is excited, and ultimately self-gravitating systems begin to separate out. The theory predicts that the least massive systems separate out first, followed by progressively more massive systems. In this way a hierarchy of self-gravitating systems, still in process of formation at the present time, comes into being.

The theory predicts a definite relation between the average energy per unit mass and the mass of newly formed self-gravitating systems. Although still in a rather primitive state, the theory contains no adjustable parameters. So the predicted binding energies, masses, times of formation, and so on, are all expressed in terms of atomic constants and the gravitational constant. Some examples of such formulae are shown in Figure 9. Figure 10 shows the predicted relation between binding energy

Dimensionless constants:

$N = e^2/Gm_e m_p \simeq 2.3 \times 10^{39}$ (Eddington's number)

$\alpha = e^2/\hbar c \simeq 1/137$ (Fine-structure constant)

$r = m_p/m_e \simeq 1836$ (Proton-electron mass ratio)

Mass of most tightly bound protosystems

$$M^\dagger \simeq 23 N^{3/2} \alpha^{-15/4} r^{3/8} m_p \simeq 4.4 \times 10^{69} m_p \simeq 3.5 \times 10^{12} M_\odot$$

Energy of most tightly bound protosystems

$$E^\dagger \simeq -0.35 \alpha^{-1/2} r^{1/4} (e^2/m_p a_0) \simeq -7 \times 10^{14} \text{ erg gm}^{-1}$$

Mass of smallest self-gravitating protosystems

$$M^* \simeq 10^{-3} N^{3/2} r^{-3/2} \simeq 10^{27} \text{ gm}.$$

Initial energy of a system of one solar mass

$$E_\odot \simeq (M/M^\dagger)^{1/3} E^\dagger \simeq -8 \times 10^{10} \text{ erg gm}^{-1}$$

Mass and initial energy of globular star clusters

$$M^\circ \simeq (M^* M^\dagger)^{1/2} \simeq 5 \times 10^5 M_\odot$$

$$E^\circ \simeq (M^*/M^\dagger)^{1/6} E^\dagger \simeq -6 \times 10^{12} \text{ erg gm}^{-1}$$

Mass of largest protosystems

$$M_{\max}(t) \simeq 5[\rho^\dagger/\rho(t)]^{1/5} M^\dagger \simeq 10^{15} M_\odot$$

Fig. 9. Some predictions of an approximate cosmogonic theory.

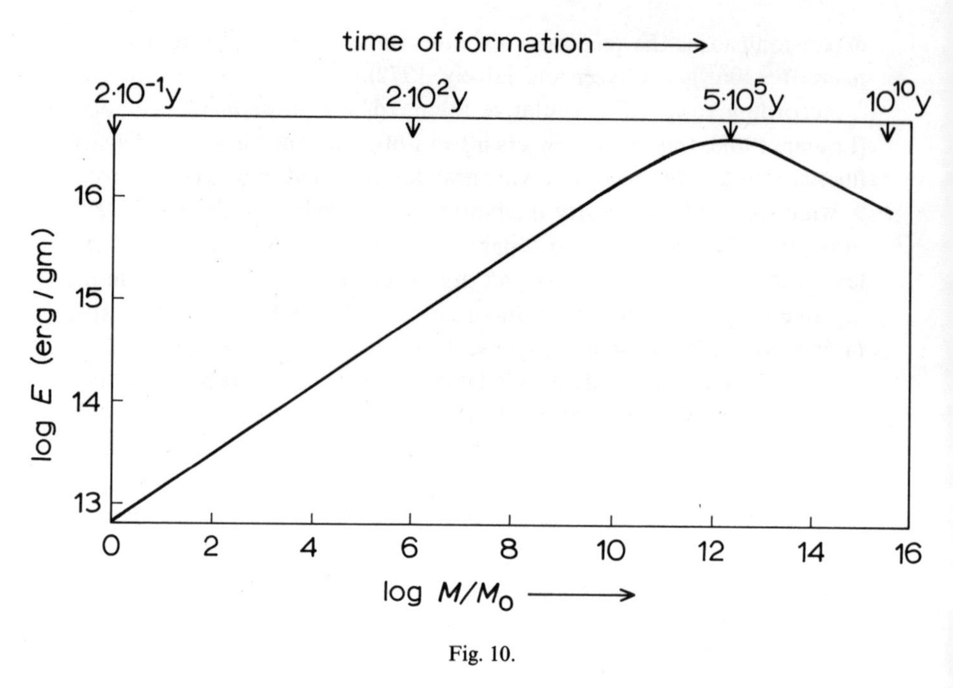

Fig. 10.

per unit mass and mass. Considering the semi-quantitative nature of the theory, the agreement with observation is encouraging.

Finally, as this talk is sponsored by the Boston Colloquium for the Philosophy of Science, it may be appropriate to end on a philosophical note. We have seen that the specific information of the universe increases monotonically with time (see Figure 11). But this implies that no future

	Type of system	Entropy and information
Thermodynamic arrow	Closed	$S\uparrow$
Historical arrow	Open	$I\uparrow$
Cosmological arrow	Infinite and unbounded	$S\uparrow$ and $I\uparrow$

Fig. 11. Three arrows of time.

state is wholly predictable: the future always contains more information than the present. In short, the future grows from the past, as a plant from a seed, but is not contained in the past. This view of the physical universe and its evolution differs radically from the one currently held by most physicists, but it harmonizes with our intuitive ideas about the nature of time. And I have tried to show that it emerges as a straightforward consequence of a few simple and natural cosmological postulates.

Harvard College Observatory

BIBLIOGRAPHY

Hively, R. M., Ph. D. thesis, Harvard University, 1971.
Layzer, D., *Astrophysics and General Relativity (New York)* **2** (1969), 155–233.
Layzer, D. and Hively, R. M., *Astrophys. J.* **179** (1973), 361–369.
Van Hove, L., *Physica (The Hague)* **21** (1955), 512; *ibid.*, **22** (1956), 343; *ibid.*, **23** (1957), 441.

FRED HOYLE

HIGHLY CONDENSED OBJECTS*

The first highly condensed object to become known to astronomers was the white dwarf. The observed white dwarfs lie on a band stretching from moderately high left to low right in the HR diagram. They follow what is known as a cooling line, along which they evolve when no further nuclear fuel is being burned inside them. High on the left of the cooling line the ground-based astronomer loses these objects because the main radiation from them lies too far in the ultra-violet. Indeed, high enough to the left the main radiation comes into the X-ray region, and it is likely that such stars constitute X-ray sources which are now beginning to be observed with the aid of rockets and satellites.

How do stars arrive at this cooling band? We have no evidence of a continuous evolutionary sequence coming across from high right to left in the HR diagram. This agrees with theoretical expectation since there are no appropriate models which provide for such a smooth evolution. If you surround a white dwarf by even a small amount of material containing hydrogen which burns as nuclear fuel by H → He, the star appears to the observer as a giant, far to the right in the HR diagram. Indeed this is just what a giant is – a white dwarf in the middle surrounded by an envelope in which nuclear fuel is being burned. But, of course, once all the fuel has been burned and the whole envelope becomes added to the inner white dwarf then we must be far over to the left, at the beginning of our cooling line. So we expect a quick transition from far right – the giant phase – to far left – the white dwarf cooling line. Because the transition must be fast by astronomical evolutionary standards we cannot expect to see many stars in the transition phase. Indeed, some 10 years ago Shklovskii suggested that the transition from right to left might be essentially discontinuous. The idea was that the star manages to throw off the last of its envelope into space. Only a small velocity of ejection is necessary because gravity is so weak at the surface of a giant. The ejected envelope would expand rather slowly and be observed as a planetary nebula. Indeed, Shklovskii arrived at

R. J. Seeger and R. S. Cohen (eds.), Philosophical Foundations of Science, 215–226.

this point of view by considering the low expansion velocities of planetary nebulae. He wondered what kind of star would eject material at such small speeds and decided that only giants would do so.

This argument has recently been discussed by Roxburgh and his colleagues. I have always thought it was probably correct, because a plausible mechanism exists for promoting the ejection. The long-period variables are giant stars far to the right in the diagram. Such stars are presumably unstable to oscillation for reasons similar to the cepheids and RR Lyrae stars. But in cepheids and RR Lyrae stars gravity is stronger and the oscillations do not build up to large enough amplitude to promote ejection of anything like a planetary nebula (although some steady mass loss probably does occur in these stars). Certainly the amplitudes of long-period variables are known to be very large.

If the transition from far right to far left is essentially discontinuous we would expect the hot surface of the central white dwarf component of the star to be suddenly exposed. The hot surface would have a temperature of $\sim 10^7$ K and would emit X-rays. Hence X-rays from compact stars far to the left and high in the HR diagram fit naturally into this evolutionary picture.

White dwarfs have central densities that range from $\sim 10^5$ g cm^{-3} to $\sim 10^{10}$ g cm^{-3} according to the mass of the star, the largest masses having the highest densities. The pressure inside these stars is due to electrons packed in a Fermi sea, the so-called degeneracy pressure. Because of relativistic effects there is a limit to the mass that can be supported by this kind of pressure – i.e. by an electron pressure that will still continue to operate as the temperature falls towards zero, which it does as the star evolves along the cooling line. The mass upper limit lies between about 1.3 and 1.4 $\mathfrak{M}_\odot$ according to the chemical composition.

What happens to stars with masses greater than this? Such stars would have to be non-degenerate if a substantial fraction of their mass evolves to densities of white dwarf order – i.e. 10^6 g cm^{-3} or more. Pressure support for non-degenerate material must then come from high temperature. Temperatures up to, and in excess of, 10^9 K are needed, and it is at such temperatures that complex nuclear reactions occur, reactions which have been intensively studied during the past decade and which have been shown to lead to the synthesis of all elements from carbon upwards. I do not wish to enter this fascinating study today, but rather

I wish to ask whether at these very high temperatures explosive reactions can occur which lead to sufficient material being ejected for the remaining residue to have a low enough mass for it to become a white dwarf.

In attempting to answer this question one encounters a curious situation. The potentially most explosive nuclear reactions are those which involve carbon-burning and oxygen-burning. But such reactions deliver less energy than the non-explosive $H \rightarrow He$, by an order of magnitude. In fact, the energy yield is barely sufficient to shatter the star, since by the stage at which the inner density has risen above 10^6 g cm^{-3} the main part of the star has become compact and its gravitational binding energy is large. Extensive calculations on this question have been performed and even so the question is undecided. The explosive reactions occur at precisely the evolutionary stage where it has just become difficult to disrupt the star. If they had occurred earlier the star would have been easily shattered. If they occurred later there would be no possibility of disruption.

It is useful to think of such problems in terms of the gravitational parameter $GMc^{-2}R$, where c is the velocity of light and R is the radius inside which the main part of the mass is located. This parameter is a dimensionless number. For the Sun and other main-sequence stars it is about 10^{-6}, for the average white dwarf it is about 10^{-4}, for stars in which we are considering nuclear explosions it is about 10^{-3}. Stellar evolution is characterized by a steady trend in which this number increases in the main inner regions of the star. The whole subject of my lecture today is to consider how large this number can become in astronomical objects. In beginning with white dwarfs and with stars in general I am starting with objects that we understand fairly well, but my main attention will eventually be devoted to objects of a different kind which we understand less well, in particular with quasars and with the centres of galaxies.

Let me go back to nuclear explosions for a moment and consider the problem of supernovae. We might expect perhaps one star per year to reach the critical evolutionary phase I am speaking about. Yet the observed frequency of supernovae is about two orders of magnitude less than this, a circumstance that would seem to fit the theoretical difficulty we have in deciding whether the nuclear reactions are sufficient to give explosive disruption or not. It is attractive to suppose that explosive disruption does not occur except when some special condition is satisfied. An appropriate degree of rotation would help disruption and this may well be the right special condition for a supernova to occur. This would suggest that the outburst of a supernova should be axially symmetric rather than spherically symmetric.

We have reached the stage of considering stellar masses that are too large for a white dwarf structure, and where nuclear explosions are thought insufficient to reduce the mass appreciably. What happens to such stars? With continuing evolution the density and the gravitational parameter both increase. With rising density, electrons plus protons turn more and more into neutrons. Eventually at densities $> \sim 10^{14}$ g cm^{-3} the neutrons themselves form a Fermi sea, and there is a pressure from the neutrons analogous to the degeneracy pressure in white dwarfs. Consequently it is possible to have stable configurations supported by the degeneracy pressure of the neutrons. Once again there is a maximum mass for such a neutron star. How does the maximum mass for a neutron star compare with that for a white dwarf? Unless one assumes a special, and perhaps rather unlikely, form of interaction between neutrons at close range, the white dwarf maximum mass exceeds that of neutron stars. Hence if a star continued to collapse because its mass exceeded the limit for a white dwarf the star could not evolve into a stable neutron configuration. At best it would have to form a cluster of neutron stars. On the other hand it may be that the neutron interaction does take such a form as to give a possibly greater mass for a neutron star than for a white dwarf. In this case one could have a neutron star form as a final stable configuration. To settle this matter it would be necessary to examine the short-range neutron interaction in the laboratory.

However, the maximum mass of a neutron star cannot exceed 2–3 $\mathfrak{M}_{\odot}$. What happens if the stellar mass exceeds this limit? The answer given by present-day theoretical physics is that contraction proceeds until the gravitational parameter becomes of order unity, at which stage the star collapses inside its Schwarzschild radius. Signals from the star continue to reach an external observer but after a short while any such radiation is subject to a strong red-shift and the object becomes essentially unobservable. To summarize we have the situation shown in Table I.

This brings me to the subject of pulsars. I have been so much concerned with white dwarfs and neutron stars because essentially all theories of

TABLE I

Star	$2\,GMRc^{-2}$
Main-sequence	$\sim 10^{-6}$
White dwarf (maximum mass $\sim 1.4\,\mathfrak{M}_{\odot}$)	10^{-4}
Supernovae	10^{-3}
Neutron stars (maximum mass probably $\sim 1\,\mathfrak{M}_{\odot}$ but possibly 2–3 $\mathfrak{M}_{\odot}$)	10^{-1}
Stars of mass $> \sim 3\,\mathfrak{M}_{\odot}$ (ultimate evolution)	1

pulsars have worked in terms of models involving white dwarfs or neutron stars. The evidence we have recently heard from De Jager, Lyne, Pointon, and Ponsonby, showing that CP 0328 is beyond the 4 kpc Perseus arm, has profound implications for these remarkable objects. In the first place it makes it seem likely that PSR 1749–28 lies close to the galactic centre. The latter pulsar has galactic coordinates $l = 1.6°$, $b = -1.0°$, and is therefore either projected against the centre or is in the centre. So long as there seemed to be reasons why pulsars lay at distances no greater than a few hundred parsecs one had to accept a coincidence of direction, but now that we have observational confirmation of a pulsar distance comparable to the distance of the galactic centre it seems preferable to suppose that PSR 1749–28 really does lie close to the centre.

This conclusion has a number of interesting implications. It suggests ~ 10 kpc as the typical pulsar distance. The typical pulsar period is 1s, and this requires the characteristic dimensions of pulsars to be less than 3×10^{10} cm, which means that these objects have a radio surface brightness comparable with, or even greater than, quasars or radiogalaxies. It is therefore wrong to think of our Galaxy as a feeble manifestation of the radio emission phenomenon. Our Galaxy possesses intensely bright sources but it happens that they are small in size and there are probably not very many of them. However, one would expect the physical conditions in the few that there are to be quite as extreme as the conditions in the larger sources. The crucially important characteristic of the latter, as we shall see later, is that the gravitational parameter, $GMc^{-2}R$, is probably near unity.

I wish to come next to the objects which are present in the great radio-sources, quasars and the nuclei of radiogalaxies. Recent interferometric measurements have shown that the main emission at high radio frequencies, 20 cm to 6 cm, comes from a region with a radius of only $\sim\frac{1}{3}$ light year, $R \simeq 10^{12}$ km. This has been established explicitly for the sources 3C 84 (NGC 1275) and 3C 120. (The situation is less certain for quasars because of uncertainties over distance, so that angular measurements are not immediately translatable into radii.)

We also know that we must associate large masses with radio-sources. This follows from the very large energy output of the sources – only a very large mass could yield such enormous quantities of energy. The largest energy estimates fall in the range 10^{62}–10^{63} erg. These are obtained from the observed infra-red output of the nuclei of galaxies like 3C 120 – more than 10^{46} erg s^{-1}. The length of time for which the emission takes place must exceed 10^8 years, otherwise galaxies with such very large emission rates would be rarer objects than they seem to be. This leads to the energy range just quoted. Again there is uncertainty in the case of quasars, partly because of the uncertainty of distance, partly because the lifetimes of quasars are unknown.

If the energy source were nuclear, H $\rightarrow$ He, it would need 10^{10}–10^{11} $\mathfrak{M}_\odot$ to supply 10^{62}–10^{63} erg. This is too large for the centres of Seyfert galaxies, some of which also show these high infra-red values. Hence it seems unlikely that the source of energy is nuclear, particularly as H $\rightarrow$ He is not an explosive reaction. I remarked earlier that carbon and oxygen are the most explosive nuclear fuels and for these the energy yield is appreciably less, and the necessary mass range is further increased to 10^{11}–10^{12} $\mathfrak{M}_\odot$. This seems to place nuclear energy quite outside the range of what is possible. Moreover, nuclear energy yields particles with energies that are typically ~ 1 MeV, whereas the radio-sources demand cosmic-ray energies – three orders of magnitude greater than this. Hence the efficiency of conversion of nuclear energy into cosmic-ray energy would also have to be taken into account, and this would increase the mass estimates still further.

The most reasonable model for radio-sources and quasars is of a mass $\sim 10^9$–10^{11} $\mathfrak{M}_\odot$ situated inside a sphere with a radius $\sim 3 \times 10^{17}$ cm, and for the energy to be derived from gravitational rather than from nuclear sources. We cannot at present exclude the possibility that there may be a number of such objects in the source and that the mass in each individual object is reduced appropriately – so that the total mass of the whole sources lies in the range 10^9–10^{11} $\mathfrak{M}_\odot$. But since the high-frequency radio emission is found to come from just one region a model with a large number of sub-units seems rather unlikely.

In Figure 1 we have a plot of red-shifts versus magnitude for quasars, compact radio-sources, N-galaxies, the nuclei of Seyfert galaxies, all ob-

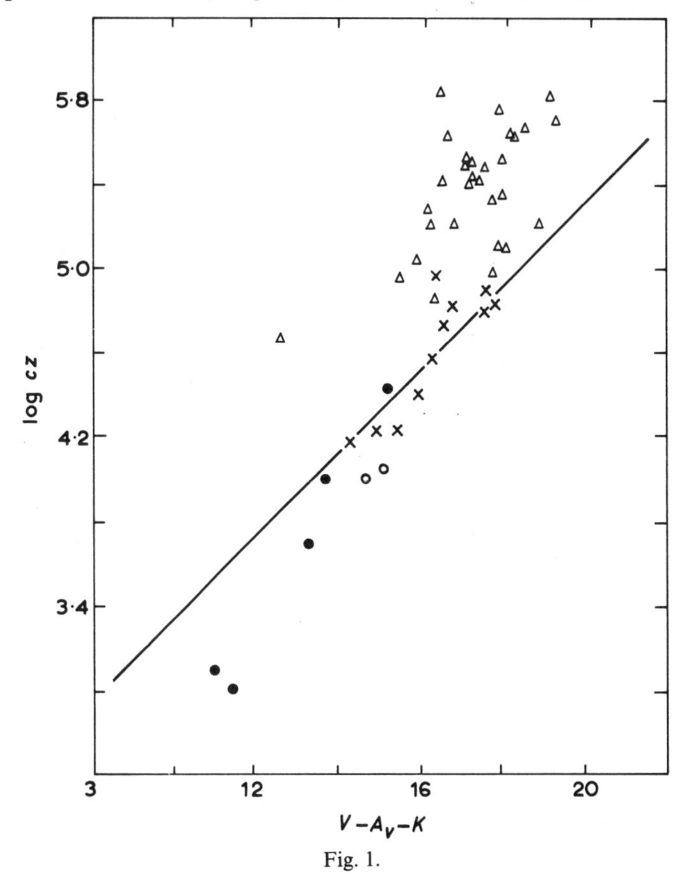

Fig. 1.

jects with spectroscopic similarities. Indeed, the compact structures of these objects, together with spectroscopic resemblances suggest – quite apart from the results of Figure 1 – that we are dealing with phenomena belonging to the same family. The objects in question form a sequence in Figure 1 very different from the characteristic relationship due to the expansion of the Universe. This property was first noted by Burbidge and Hoyle (1966), whereas Figure 1 shows Arp's more recent plot (1968) using a greater volume of data than was available two years ago.

The implication of Figure 1 is that there are large red-shifts which do not arise from the expansion of the Universe. Other evidence pointing to the same conclusion is also accumulating. Burbidge has found regular patterns in the emission line red-shifts of a fairly large sample of recently measured galaxies and radio-sources, while Sargent has found a grossly discrepant red-shift in one galaxy among five in the chain VV 172, again curiously the one galaxy in the chain to show emission lines.

Most astronomers are puzzled and worried by these results. We have seen that strong gravitational fields almost certainly exist in the class of objects we are now considering, and strong gravitational fields also produce red-shifts. The puzzle is how strong fields can produce changes of red-shift not just in the small volumes of the highly compact sources themselves, but over volumes with dimensions of several kpc. My own point of view is that these worries are exaggerated. The apparent difficulty comes from the refusal of most astronomers to face the fact that the behaviour of sources cannot in any case be explained in terms of conventional physical theory. According to conventional theory large masses in small volumes plunge into singularities. I believe it is impossible to imagine such a structure persisting for upwards of 10^8 years. I would expect the sources to fade away (as seen by an external observer) in a time scale of the order of a year. Instead we observe violent outbursts in a time scale of the order of a year. The observed properties of the sources are exactly opposite to what we would expect according to conventional physical theory. The implication is that physical theory is wrong, and that it is wrong at precisely the point where the theoretician should most be on his guard, namely the collapse of matter into singularities. Observation shows that the opposite is true, matter emerges from strong gravitational fields, it does not collapse into singularities.

In the light of these modern discoveries it is as well to remember a most remarkable paragraph in Jeans' *Astronomy and Cosmology*:

> ... until the spiral arms have been satisfactorily explained, it is impossible to feel confidence in any conjectures or hypotheses in connection with other features of the nebulae which seem more amenable to treatment. Each failure to explain the spiral arms makes it more

and more difficult to resist a suspicion that the spiral nebulae are the seat of types of forces entirely unknown to us, forces which may possibly express novel and unsuspected metric properties of space. The type of conjecture which presents itself, somewhat insistently, is that the centres of the nebulae are of the nature of 'singular points', at which matter is poured into our Universe from some other, and entirely extraneous, spatial dimension, so that, to a denizen of our Universe, they appear as points at which matter is being continually created.

The controversies between Eddington and Jeans, which were a source of entertainment and wonderment to members of the Royal Astronomical Society for many years, have recently assumed a new significance for me. To my generation, Eddington was the man who had said all the right things and Jeans the man who had said all the wrong things. I agreed with Rosseland's criticism of *Astronomy and Cosmology* as 'more a work of art than of science'. But this was to do Jeans an injustice.

In the first quarter of the present century astronomers were faced by a mass of data concerning stars and stellar evolution. We know today that the explanation of these data lay in the science of nuclear physics, which did not emerge until the 1930s. Men like Eddington and Russell, who allowed themselves to be guided by the observations, were led to suspect many of the results that emerged later from a knowledge of nuclear physics. In doing so they were derided by contemporary physicists. It is strangely ironic that the main reason why British physicists cannot bring themselves to regard astronomy as a serious branch of physics is because of the extent to which Eddington was derided by the Cavendish Laboratory, in particular over his suggestion that the energy of the stars was due to H → He. You remember how they told him the stars were not hot enough, and how in something of a temper he told them to go and find a hotter place.

Jeans on the other hand was a physicist himself. He was the most experienced theoretical physicist to engage on the problems of stellar evolution. He restricted himself to conventional physics and attempted to force the astronomy to fit the pattern of observation – on the basis that the physics was better known than the astronomy.

I believe we have exactly the same situation today in relation to highly condensed objects. We can stick to conventional physics and force the astronomy, as Jeans did. Or we can follow Eddington by sticking to the astronomy and forcing the physics. Most present-day theoretical astronomers follow Jeans, mainly for the reason that they do not like to risk the derision of the contemporary physicist. For myself, I believe the right course is to follow Eddington, or rather to follow where the evidence points. The extraordinary development of radio-astronomy over the past 20 years parallels the first 20 years of the century. Previously it was stellar evolution pointing to nuclear physics. Now it is highly condensed objects pointing to strong gravitational fields and to physics that at present we cannot describe.

When Jeans abandoned his conventional physics and permitted himself to be guided by the empirical situation he was capable of arriving at the remarkable speculation which I just quoted. The remark about 'pouring' matter into our Universe is particularly significant.

If we grant that contemporary physics is wrong in its attitude to singularities we must also accept that the big-bang cosmologies are wrong, because the singularity which represents the origin of the Universe in these cosmologies has the same mathematical structure as the local singularities. The two go together. If one is wrong so almost certainly must the other be.

This leads me to say a word about the steady-state cosmology. In two decades I have never been able to make my fellow astronomers understand what is being said in this theory, partly no doubt from a lack of clarity in my exposition, but partly also from the emotional atmosphere which unfortunately has always interfered with a rational discussion of the theory. I will now state the essential result. *If* matter is poured into the Universe at some assigned average rate per unit proper volume then the Universe will take up an expansion rate that can be calculated. If you imagine yourself to be controlling the pouring-in process – you could turn up the rate or turn it down like a gain control – you would find that the Universe would expand faster as you turned up the control, and the Universe would slow down as you turned the rate down. As long as you turned the control slowly enough, so that the Universe had time to settle down to the changes, you would find a one-to-one correspondence between the expansion rate and the pouring-in rate, and you would find the one-to-one correspondence obeyed an important rule, that the average density of matter always attained a steady value. For example, if you turned up the gain control – you poured in matter more rapidly – you would find that the average density of matter in space would rise, but it would not rise indefinitely because the increasing rate of expansion of the Universe would eventually hold it to a steady value.

This is the result that was proved mathematically. The result is independent of how the matter comes to be poured in. It depends only on the pouring-in process taking place in some way. At first, the pouring-in was assumed to occur uniformly throughout space, essentially because this was the easiest example to investigate. Now it seems clear, however, that the pouring-in process, *if* it takes place, must come from localized sources. This more difficult case was considered by Narlikar and myself about three years ago. Logically the problem is to consider the feedback loop

$$\text{Sources} \rightleftarrows \text{Matter}$$

and to decide whether the 'gain control' for this loop becomes set at a fixed value – in which case there is a completely steady-state situation – or whether the gain control is subject to slow drift, as if you were indeed moving it gently up and down. In the latter case we should have a Universe that was 'steady' over moderate lengths of time but which was subject to slow drifts. Our work tentatively suggested that slow drift would take place but our answer ignored the possible cosmological influence on the local properties of matter.

Everyone is familiar with the apparent coincidence of very large dimensionless numbers that can be calculated in apparently distinct ways. This leads one to suspect a connection between cosmology and local physics. The ratio of the electrical and gravitational forces between an electron and a proton, $e^2/G\, m_e\, m_p$, is 2.3×10^{39}, close to the ratio between c/H, the cosmological distance scale, and the characteristic length $4\pi e^2/m_e c^2$ associated with the electron. Indeed, if we write

$$\frac{c}{H} = 2.3 \times 10^{39}\, \frac{4\pi e^2}{m_e c^2} = 8 \times 10^{27} \text{ cm}$$

we obtain $H \simeq 100$ km per Mpc, in agreement with the observational determination of the Hubble constant. This suggests that the above equation may possibly be satisfied *exactly*.

We have made some progress in understanding why the force ratio for an electron and proton must be taken, rather than the ratio for two electrons or two protons. We also think there are reasons connected with the action formulation of the gravitational and electromagnetic theories to explain why $4\pi e^2/m_e$ should be used rather than e^2/m_e – note the combination $4\pi e^2/m_e$ in important physical formulae such as the critical frequency of an ionized plasma.

The volume $(c/H)^3 \simeq 5 \times 10^{83}$ cm^3. The steady-state average density corresponding to this value of H is $\sim 3.10^{-5}$ atoms cm^{-3}, the atoms being taken as mainly hydrogen. Thus in a typical cosmological volume there are $\sim 10^{79}$ atoms $= N$ say, so that the dimensionless ratios are close to $N^{1/2}$. This strongly suggests that the connection between cosmology and local properties involves statistical fluctuations. If this is so we must suppose that the cosmological boundary conditions that influence the local properties of matter are subject to statistical variation, and hence that the behaviour of matter in other places and at other times may be different from what is now found in the terrestrial laboratory. It is possible that in the highly condensed objects I have been discussing in this lecture the local fields are strong enough to compete with the cosmological boundary conditions, with the consequence that local physical properties are modified. This is the kind of idea that comes to my mind in considering the strange red-shift data which I described earlier in this lecture. Admittedly the situation is still very vague, but it is to this manner of concept that I believe the evidence is systematically pointing. We may not be able at present to guess the explicit nature of the new physics, but just as some astronomers in the 1920s were convinced that the evolution of the stars pointed to new physical concepts, so the highly condensed objects point today to a situation that may well be wholly new.

Institute of Theoretical Astronomy, Cambridge, U.K.

NOTE

* George Darwin Lecture delivered in the Lecture Theatre of the Royal Society on October 11, 1968.

Reprinted with kind permission from Quart. J. Roy. Astron. Soc. ***10*** *(1969) 10–20.*

G. DE VAUCOULEURS

THE CASE FOR A HIERARCHICAL COSMOLOGY*

Recent Observations Indicate that Hierarchical Clustering is a Basic Factor in Cosmology

In questions of science the authority of a thousand is not worth the humble reasoning of a single individual.

GALILEO GALILEI

True knowledge can only be acquired piecemeal, by the patient interrogation of nature.

SIR EDMUND WHITTAKER

Once upon a time philosophers and cosmographers insisted that the motions of the planets must be circular and uniform. An irrelevant aesthetic concept of 'perfection' and a more valid mathematical need for simplicity were at the root of this long-held error. Nowadays, theoretical cosmologists insist that the large-scale distribution of galaxies most be homogeneous and isotropic, and most astonomers believe that the expansion of the universe is linear and isotropic and that it proceeds at a uniform rate measured by the Hubble 'constant' H.[1,2]

1. SOME HISTORICAL PERSPECTIVE

Modern theoretical cosmology was born 50 years ago with General Relativity; its subject matter was determined a few years later with Hubble's final proof of the old concept that nebulae are 'island universes'. Its first observational test was Hubble's discovery just 40 years ago of the universal red shift in the spectra of distant galaxies. Its second potential test by galaxy counts, which has been available for more than 30 years, has so far miserably failed for a variety of technical reasons, although the principle remains valid and further observational progress may reestablish its value. The third and presently most exciting test came with the discovery just a few years ago of the so-called 3 °K background radiation (this discovery is so new, in fact, that its relevance to cosmology still awaits the test of time). Other data, of course, may have a bearing on the problem –

R. J. Seeger and R. S. Cohen (eds.), Philosophical Foundations of Science, 227–256.

if, in fact, the universe is 'one and indivisible', almost every fundamental law or principle of physics, chemistry, and astronomy must have some cosmological connotation. Familiar and perhaps overworked examples are Mach's principle, the abundance of the elements, and the (extra-atmospheric) brightness of the night sky.[3,4] Nevertheless, the few facts and figures which in the past 40 years have been given prominence as particularly relevant to cosmology are still too little understood and often too poorly established or too recently discovered to form a solid basis for a 'final' solution. Also we may well still lack some fundamental knowledge of physical laws on the very large (cosmic) scale or on the very small (particle) scale, or both, to even hope for a realistic solution at the present time. Is it not possible, indeed probable, that our present cosmological ideas on the structure and evolution of the universe as a whole (whatever that may mean) will appear hopelessly premature and primitive to astronomers of the 21st century? Less than 50 years after the birth of what we are pleased to call 'modern cosmology', when so few empirical facts are passably well established, when so many different oversimplified models of the universe are still competing for attention, is it, may we ask, really credible to claim, or even reasonable to hope, that we are presently close to a definitive solution of the cosmological problem?

Those who are so optimistic as to answer affirmatively have in effect already made a choice, primarily for philosophical, aesthetic, or other extraneous reasons, from among the vast array of possible homogeneous isotropic universes of general relativity; thus solving the cosmological problem reduces to the almost trivial matter of fitting a few empirical constants, which, some suggest, may take only a matter of a few years.

I cannot subscribe to this view, first, because promotors of other cosmologies will not meekly 'abjure, renounce, and detest' their errors. Quite recently, for instance, opponents of the steady-state theory confidently announced the demise of this concept which was said to be inconsistent with some counts of radio sources and the existence of the 3 °K background radiation. But the latest pronouncements of at least one defender of this particular theory show that he still maintains his original views.[5]

A second reason is that even within the framework of the orthodox 'primeval atom' or 'big bang' theory we have witnessed in the past 40 years frequent and drastic changes in the fundamental 'constants'. For example, estimates of the Hubble constant decreased from $H=560$ km per sec per megaparsec in 1931 to present values in the range $50<H<110$ km per sec per megaparsec (1 megaparsec $=3.25\times10^6$ light-years $=$ $=3.18\times10^{24}$ cm). The situation is worse still for values of the so-called deceleration parameter which have fluctuated wildly from year to year, without any clear trend. And so it seems unbelievable that we are now in A.D. 1969 on the threshold of reaching the promised land of the true and only cosmology.

Let us look, for example, at a graph (Figure 1) of successive estimates of

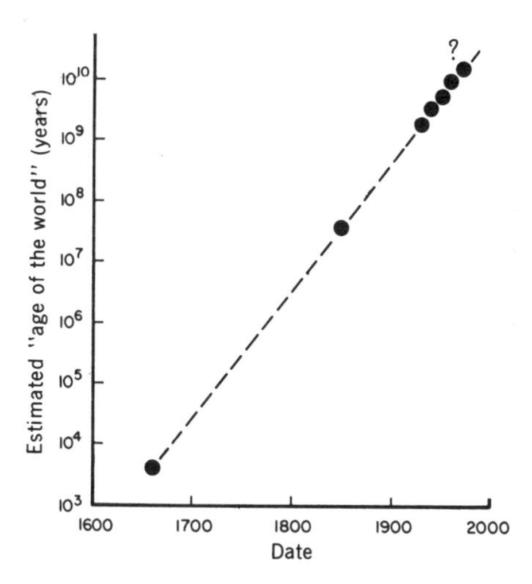

Fig. 1. Estimates of the 'age of the world' have grown exponentially during the past three centuries. What is the probability that a limit has finally been reached?

'the age of the world' made during the past three centuries, from Bishop Ussher's specific, if unfortunate, 17th century assertions to the Helmoltz-Kelvin gravitational contraction ages in the mid-19th century, to the ages based on early 20th century radioactive dating, to the expansion ages of the mid-20th century, and to present estimates based on the evolution of stars in globular clusters. During this entire three-century span, estimates of the age of the universe have increased exponentially at the surprisingly uniform logarithmic rate of 1.9 per century (the doubling time of 16 years just about matching the growth rate of astronomical progress in general).

It is true that the curve may well suddenly stop rising and level off beyond A.D. 1969. If so, we live at a truly remarkable time, the time when the age of the universe is finally fixed, this despite the fact that the expression 'age of the universe' cannot be defined without a prior defintion of 'universe' and of a universal time scale, or without a scheme of universal evolution, and, indeed, a solution of the cosmological problem. The least that can be said is that the historical record is a warning against excessive optimism.

2. Questionable Assumptions of Orthodox Cosmology

With few exceptions modern theories of cosmology have come to be variations on the homogeneous, isotropic models of general relativity. Other theories are usually referred to as 'unorthodox', probably as a warning to students against heresy. When inhomogeneities are considered (if at all), they are treated as unimportant fluctuations amenable to first-order variational treatment. Mathematical complexity is certainly an understandable justification, and economy or simplicity of hypotheses is a valid principle of scientific methodology; but submission of all assumptions to the test of empirical evidence is an even more compelling law of science. Facts of observation cannot be ignored indefinitely or dismissed as unimportant. A stubborn discrepancy of 8′ between Tycho's observations of the longitude of Mars and the most elaborate pyramiding of circular epicycles led in Kepler's exuberant but correct view "to a complete revolution of astronomy." The history of science, of course, is full of examples of stubborn, 'ugly' little facts that destroy 'beautiful' theories, but most scientists have learned not only to live with this reality but also to hunt zealously for such discrepancies, which usually lead to further progress, to improved theories, and sometimes to completely new ideas.

Unfortunately, a study of the history of modern cosmology [2] reveals disturbing parallelisms between modern cosmology and medieval scholasticism; often the borderline between sophistication and sophistry, between numeration and numerology, seems very precarious indeed. Above all I am concerned by an apparent loss of contact with empirical evidence and observational facts, and, worse, by a deliberate refusal on the part of some theorists to accept such results when they appear to be in conflict with some of the present oversimplified and therefore intellectually appealing theories of the universe.[6] It is not merely that, as Otto Struve once remarked, "In a sense the observer knows too many facts to be satisfied with any theory"; it is due to a more basic distrust of doctrines that frequently seem to be more concerned with the fictious properties of ideal (and therefore nonexistent) universes than with the actual world revealed by observations.

If this sounds too harsh a judgment, let us consider some of the questions that are routinely raised and even 'answered' in our modern cosmological symposia and congresses: what is the mean density ρ of the universe? Granting that we agree on a definition of the universe (countable? observable? horizon limited?), seldom, if ever, do we hear raised the following *a priori* questions: what precisely do we mean by the average density? What is the evidence to support the notion that a mean density can be defined? In short, how do we *know* that the universe is homogeneous and isotropic? In fact, since ρ is so evidently *not* a constant independent of space coordinates in our neighborhood, how large a volume of space do we need to consider before the average density in this volume may be accepted as a valid estimate of ρ? And what proof do we have that the same value of ρ would obtain in another equal, disjoint volume of space or in a still larger volume? Or again, for another central question of modern cosmology: What is the value of the Hubble constant H? Granting again that we agree on the interpretation of galactic red shifts as classical Doppler shifts (for which there *is* good evidence), seldom is there a discussion of the *a priori* question: What is the concrete evidence to support the assumption that the expansion parameter is a universal constant? Why must it be a constant independent of place and direction? In short, how do we *know* that expansion is linear and isotropic? And since ρ is not in fact a constant in our neighborhood, how can H be a constant? Or is it possible that it is a stochastic variable that fluctuates with ρ and, if so, again, how large a distance must we consider before a stable statistical average value emerges for H, irrespective of direction in space? Or again for a third example of a standard topic: what precisely is the age of the universe? This question requires the adoption of a very specific class of cosmological models before it can be given any sense. For an irreverent comment on this question, see Figure 1.

These and similar questions cannot be answered by aesthetic prejudices or considerations of mathematical simplicity; correct answers can only be discovered by a searching, critical study of the empirical evidence. Clearly, simplifying assumptions and first-order (or even zero-order) approximations are legitimate tools of the theoretical trade; their value is not in question here, and occasionally nature will cooperate. Not infrequently, the simplest assumptions will give a fair – even a good – approximation of observations. Newton's law is a shining example.

But if nature refuses to cooperate, or for a time remains silent, there is a serious danger that the constant repetition of what is in truth merely a set of *a priori* assumptions (however rational, plausible, or otherwise commendable) will in time become accepted dogma that the unwary may uncritically accept as established fact or as an unescapable logical requirement. There is also the danger inherent in all established dogmas that the surfacing of contrary opinion and evidence will be resisted in every way.

3. Clustering and Superclustering over 10 to 100 Million Light-Years

Let us now turn to specific facts and figures that warn against premature confidence in current 'orthodox' models, in particular, against their fundamental assumptions of homogeneity, isotropy, and the existence of a definite mean density.[7] First, let us recall some of the drastic changes in the observational evidence on the large-scale distribution of matter in the universe since the early surveys of faint galaxies of the 1930's. From sampling surveys with small-field reflectors (primarily the Mount Wilson 60-inch and 100-inch reflectors),[8] a picture emerged, about 1935, of a so-called 'general field' of more or less 'randomly' distributed galaxies, broken only in rare places by an occasional large globular cluster of galaxies, a few megaparsecs across, of which the nearest and best known example is in Coma. These great clusters, being easy to recognize at large distance, were (and still are) used as convenient markers for a study of the velocity-distance relation, the proof of an expanding universe. Except for effects attributed to local absorption in our galaxy, the number density N of galaxies at the magnitude limit of the Mount Wilson counts (about $m = 19.4$ on the current scale) seemed to be roughly independent of direction over at least one hemisphere (there are serious experimental difficulties in maintaining a constant limiting magnitude in the celestial southern hemisphere from an observatory in the northern hemisphere). Such results certainly encouraged theoretical cosmologists to adopt homogeneous, isotropic models as realistic approximations of the physical universe. Thus galaxies played the role of molecules in a gas, and from galaxy counts, estimates of distances, and average masses derived from the rotation of a few nearby galaxies, it was in principle a simple matter to derive the mean density of the corresponding volume of space, and this, evidently, was the required value of ρ.

Astronomers soon realized, however, that the concept of a randomly distributed general field of galaxies was the result of poor statistics: first, random sampling with a field of view much smaller than the angular scale of density fluctuations will tend to mask large-scale clustering; second, the Gaussian distribution of the logarithm of N observed by Hubble really means that N is subject to contagion, that is, clustering, as several authors quickly pointed out.[9] At about the same time, shortly before World War II, surveys with wide-field astrographs at Harvard Observatory and especially with the 18-inch Schmidt camera at Mount Palomar, proved that clusters and groups are the rule rather than the exception[10, 11]; apparently most, if not all, galaxies are members of some group or cluster, with typical populations of, perhaps, 10 to 100 in the first three magnitudes. The exhaustive galaxy counts with the Lick 20-inch astrograph[12] over two-thirds of the sky to $m = 19.0$, and the searching surveys of galaxy clusters with the 48-inch Schmidt camera at Mount Palomar[13, 14] to $m = 20.5$ provided in the 1950's overwhelming evidence of the universal prevalence of galaxy clustering on a wide range of linear scales from a few megaparsecs up to at least 50 megaparsecs; this scale of galaxy clustering corresponds to the so-called 'superclusters'.

Evidence has accumulated in favor of the hypothesis, first advance by Zwicky[11, 14] in 1938, that clusters or more precisely 'cluster cells' are 'space fillers' that occupy all space available as 'suds in a volume of suds'.[15] Indeed, in a recent systematic, exhaustive survey of the 55 nearest groups of galaxies (those within 16 megaparsecs), there were very few galaxies that could not be assigned to a definite group or cluster.[16] Isolated, intercluster galaxies (or, if you will, clusters of $N = 1$ member in statistical terminology) are apparently very rare, a fact that has obviously important, if still hidden, physical as well as cosmological implications. Diameters of typical groups are generally between 1 and 3 megaparsecs[16]; clusters have diameters between 2 and 5 megaparsecs; and there is now good evidence for clustering on a much larger scale of, say, 30 to 60 megaparsecs, that is, for 'superclusters'.

Some controversy has arisen concerning the concept and reality of 'superclusters', that is, of condensations of galaxies on a scale much larger than conventional groups or clusters which typically do not exceed a few megaparsecs in diameter. On the one hand Zwicky and his collaborators[17, 18] have repeatedly asserted that clusters of (globular) clusters of galaxies do not exist and their evidence is not denied. On the other hand, Abell – working from his catalog of clusters based on plates from the same Mount Palomar 48-inch camera – has given definite statistical evidence that at least some of these large clusters have a (nonrandom) clumpy distribution on a typical clustering scale of 50 megaparsecs; he has offered specific examples of such associations or loose groups of clusters, all having about the same red shift.[19, 20] Statistical analyses by Neyman, Scott, and Shane[21] of the counts made at the Lick Observatory have shown that galaxy distribution models based on the assumption of single clustering (that is, of a random distribution of independent cluster centers) do not account in detail for the parameters of the observed galaxy distribution and that the hypothesis of multiple clustering, that is, clusters of clusters, is, therefore, probably necessary. Recently, Karachentsev[22] has analyzed the distribution of 'very distant' and 'extremely distant' clusters at high northern galactic latitudes in Zwicky's catalog; these clusters are at distances of several hundred megaparsecs. Again, as in the Lick Observatory counts, positive correlation, indicative of clustering of cluster centers, exists over areas several degrees in diameter, corresponding to a linear diameter of some 40 megaparsecs for the average supercluster; some five to ten major clusters are included in each supercluster (and, of course, many more small groups). The mean volume of space ('supercluster cell') occupied by a supercluster has a diameter on the order of 60 megaparsecs. Finally, I have discussed on several occasions since 1953 the growing evidence for a Local Supercluster,[23–25] encompassing the majority of the nearby galaxies and groups with a center approximately in or near the Virgo cluster. The influence of this supercluster on galaxy counts can be detected at least down to $m = 16$ in the northern galactic hemisphere.[26] Our Galaxy is in an outlying location, in our Local Group, near the southern edge of the system. Here, too, the major diameter is on the order of 30 to 60 megaparsecs, depending on the distance scale adopted.

Furthermore, studies of radial velocities have indicated that in our neighborhood, say, within 100 million light-years, the velocity field is neither isotropic nor linear; this is apparently the result of differential expansion and rotation of the flattened supersystem.[25, 27–29] A total mass of the order of 10^{15} solar masses was derived from the rotation of the supercluster.[27, 28] The total solar motion due to galactic and supergalactic rotation[25] causes a slight asymmetry of the order of 0.1% in the intensity of the 3 K background radiation; this asymmetry may have been detected recently by refined observations.[30]

Most astronomers who have studied the problem closely have been convinced by this evidence and now accept the reality of superclustering on a scale of the order of 50 megaparsecs. Some of the controversy is more a matter of nomenclature than of fact; Zwicky apparently designates condensations as large as this as 'clusters' and admits that such clusters exhibit much 'subclustering'[13, 18]; Kiang[31] and Kiang and Saslaw[32] have concluded from a statistical analysis of Abell's catalog[19] that clustering exists on all possible scales from small groups to the largest superclusters, and they contend that in this respect the concept of 'cluster' or 'supercluster' is not significant. I presume that a statistical study of human agglomerations would disclose a continuous spectrum of city sizes from isolated farms, hamlets, and townships, to major towns, capital cities, and perhaps megalopolises; I would *not*, however, conclude from this argument that, because a distinction between, say, Johnson City, Texas, and Washington, D.C., has no clear-cut statistical basis, it is therefore not physically significant. This argument remains valid even if clusters occasionally overlap (as cities do too), a fact that can be taken into account in the statistical theory of 'interlocking' clusters.

There is a further danger to a purely statistical or 'fluctuations' approach to the theory of clustering or superclustering; it neglects the fundamental effects of collective gravitation and its possible counterbalancing of the general expansion. In other words, the physics of the formation, evolution, and possible dissolution of clusters and superclusters in the framework of the expanding background or frame of reference must be considered, not merely the instantaneous aspect of a fluctuating field of massless particles. Unfortunately, concepts and theories of cluster formation and dynamic evolution are still in a very primitive state.[33] In a sense, a self-consistent theory of clustering can be developed only within the framework of a specific cosmological model; such a model is necessary to fix initial conditions and the surrounding cluster field.

4. Higher-order Clustering over 100 to 1000 Million Light-Years

The question naturally arises whether still larger organizations, that is, higher-order clustering, exist on a scale much greater than the typical supercluster. At present there is no definitive quantitative evidence, principally because no effort has been made to detect fluctuations on this enormous scale, but also in part because one runs out of data. I believe, nevertheless, that there is some indication of nonrandom density fluctuations on a scale of the order of 60° of arc in the smoothed isopleths of Hubble's galaxy counts [8] with the 100-inch Mount Wilson reflector (Figure 2) reaching the faint limiting magnitude ($m \simeq 19.4$). At the estimated limit of the survey, about 300 megaparsecs, this size corresponds to a diameter of about 300 megaparsecs, which is large enough to accommodate many typical superclusters. Similarly, in the galaxy counts to $m \simeq 18.6$ in the southern galactic polar cap, made long ago at Harvard Observatory, Shapley [34] noted the presence of a strong density gradient along a 90° arc as one moves away in the northwest direction from a giant galaxy cloud in Fornax. Here, too, the linear scale is in the hundreds of megaparsecs. [35]

If, then, the deepest, most encompassing surveys of faint galaxies do not begin to approach the required statistical uniformity, if clustering forces still operate strongly on a scale of hundreds of megaparsecs, what is the evidence for large-scale homogeneity and isotropy? How far do we need to go before, at last, we begin to encompass a volume of space big enough to be a fair sample of the universe, with its presumed characteristic mean density ρ? We have been talking about this 'mean' density for so long that we almost believe it exists, and many authors have attempted to estimate it from counts of galaxies to $m = 18$ (or even $m = 13$!) in blissful ignorance of the overwhelming, pervasive influence of this hierarchical clustering on ever larger scales – demonstrated by the present, concrete evidence of existing galaxy counts. We can only conclude that, if indeed there is a definite mean density (of normal galaxies) in the universe, it

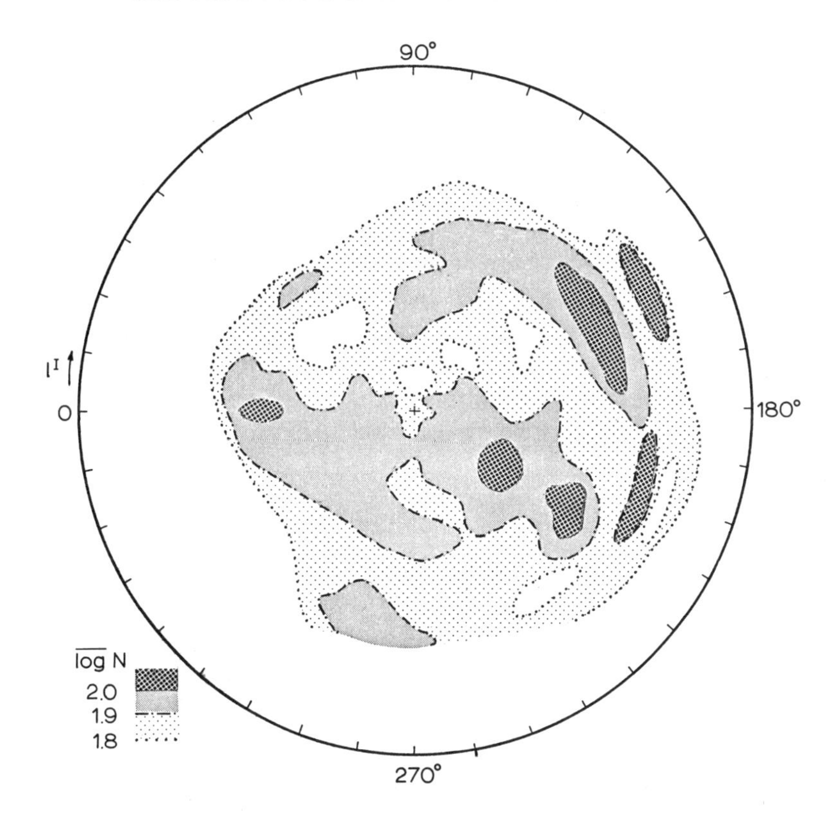

Fig. 2. Smoothed isopleths of the Mount Wilson counts of faint galaxies in the northern galactic hemisphere show large areas of above-average density separated by lanes of lower density. Three or four possible supersuperclusters, or third-order clusters, on a scale of 200 to 300 megaparsecs, are in evidence. North galactic pole is at center of map, equator at circumference.

can be estimated only for the whole counted region, unless a still larger volume is required to constitute a genuine 'fair sample'. There is no proof that even the whole of the presently counted region is a fair sample, because to test this hypothesis would require that one count a roughly equal volume, entirely outside the present one, to check whether the two disjoint volumes lead to approximately equal densities. But if we could do that, that is, reach out to at least twice the present range, we could just as well (in principle) count a volume of space almost an order of magnitude greater than the presently counted volume. Then, however, we would probably run into serious difficulties of interpretation because these deeper counts, to an estimated limiting magnitude $m \simeq 21$ or fainter, involve such questions as the effects of red shifts on apparent magnitudes (the so-called K-correction) [7], possible effects of intergalactic absorption over very long paths, and problems of galaxy evolution over the correspondingly long time intervals ($> 2 \times 10^9$ yr).

To get a feel for this problem of the mean density ρ of the universe, we can replace the hypothetical test out to, say, twice the present distance by actual tests over the range of smaller distances for which data do now exist. This range for galaxy counts is of the order of 300 megaparsecs, thus defining a volume of space of the order of 10^8 cubic megaparsecs, (10^{82} cm^3), not a negligible test domain.

5. Definition of a Mean Space Density

For clarity let us restate the definition of ρ and the object of the test: if space is homogeneous and isotropic, the average density ρ of the universe is defined as the mean density of a 'big enough' volume of space such that the same mean density obtains for any arbitrary increase in the radius of the sample region or for any other, disjoint region of volume at least equal to that of the original test region.

Since matter is evidently clustered on a small scale, this definition implies that, except perhaps for statistical fluctuations, the average density is that of a volume of space large enough to contain at least several clusters of the largest order of clustering (say, on current ideas, simple cluster centers) and that there are no larger clusters of a higher order of clustering, that is, the cluster centers of the highest order of clustering actually realized have a statistically uniform (Poisson) spatial distribution.

In the 1930's astronomers stated, and cosmologists believed, that, except perhaps for a few clusters, galaxies were randomly distributed throughout space; in the 1950's the same property was assigned to cluster centers; now the hope is that, if superclusters are here to stay (and apparently they are), at least they represent the last scale of clustering we need to worry about, and that *their* centers may be denizens of an isotropic homogeneous expanding universe [see, however, [35]]. Ignoring for a moment the evidence of superclustering on the scale of 200 to 300 megaparsecs foreshadowed by the galaxy counts, we can at least check whether there is empirical evidence for a leveling off of the average space density of galaxies as sampling regions of increasing radii are considered out to the range of the Lick Observatory survey, the deepest with a well-established limiting magnitude.

In order to place the problem in proper perspective, we will consider the average density of astronomical bodies from the highest value, that of neutron stars, to the lowest, that of the whole counted region. We may always assume that the observer happens to be located at the center of each volume of space in which we estimate the mean density, just as we are, perforce, in the center of the observable region. The problem, then, is that of the relation between the radius R and the mean density ρ of various domains of space. Further, to express all results in conventional physical units, say, centimeters and grams per cm^3, we will need to adopt masses derived either from rotation or from the virial theorem. The well-known discrepancy of one and one-half orders of magnitude between rotational masses of individual galaxies and statistical masses of galaxies in pairs, groups, and clusters is not at issue here. The old debate on the respective merits or flaws of various methods of estimating galaxy masses is not closed [36–38], although it is quiescent at the moment. Whether the 'missing mass' is located in an extensive corona of low-mass stars with large mass-luminosity ratios around each galaxy or is thinly spread out in the form of ionized gas or other optically invisible forms of matter scattered in intergalactic space between cluster galaxies is of little import here; in all cases, a given mean mass can be statistically attached to each counted galaxy and the actual mass must be bracketed by the lower and upper limits set by the rotational and statistical methods.

6. Density-radius relation and Carpenter's density restriction

Now we believe that, to be optically observable, no stationary material sphere can have a radius R less than the Schwarzschild limit

$$R_M = 2GM/c^2$$

corresponding to its mass M (G is the gravitation constant and c is the velocity of light). In a plot of the correlation between mean density ρ and characteristic radius R of cosmical systems of various sizes (Figure 3), the line

$$\text{(1)} \qquad \rho_M = 3c^2/8\pi G R_M^2$$

or, in cgs units,

$$\text{(1a)} \qquad \log \rho_M \simeq 27.2 - 2 \log R_M$$

defines an extreme upper limit or envelope. The ratio $\phi = \rho/\rho_M$ of the actual density to the limiting value for a system of observed radius R may be called the Schwarzschild filling factor. For most common astronomical bodies (stars) or systems (galaxies), the filling factor is very small, on the order of 10^{-4} to 10^{-6}.

A second, lower natural limit to the density of a nonrotating system of free particles is that which is fixed by the virial theorem condition for statistical equilibrium between the total kinetic energy T and the gravitational potential energy Ω,

$$2T + \Omega = 0.$$

If ρ^* is the equilibrium density, this condition may be written in the form

$$\text{(2)} \qquad \rho^* \simeq 3\sigma_v^2/4\pi R^2 G$$

where σ_v is the velocity dispersion, or

$$\text{(2a)} \qquad \log \rho^* \simeq 6.5 + 2 \log \sigma_v - 2 \log R.$$

If ρ is less than ρ^*, the system is unstable and will evaporate in a relatively short time; if ρ is greater than ρ^*, the system is dynamically stable and will tend to shrink toward the equilibrium condition (Equation (2)). Real systems with nonzero values of net angular momentum depart from this relation, but the departure is relatively minor. In large systems of stars and galaxies, σ_v is often in the range of 100 to 1000 km per sec, so that

$$\text{(2b)} \qquad \log \rho^* \simeq (21.5 \pm 1) - \log R$$

which corresponds for the filling factor to

$$\log \phi = \log(\rho^*/\rho_M) \simeq -5.7 \pm 1.$$

There is naturally no definite *lower* limit to the density of matter in a given volume of space, if σ_v is negligible and forces other than gravitation are important; for example, $\phi \simeq 10^{-8}$ for the planets and $\phi \simeq 10^{-12}$ for the solar system as a whole.

We can now compare observational data on stars and stellar systems with these theoretical limits. The no-longer so hypothetical neutron stars may come close to the Schwarzschild limit as shown in Figure 3, where the dashed segment in the upper left corner illustrates a range of theoretical models for which $-2.5 < \log\phi < -0.6$. The next group of very dense stars, the white dwarfs, is represented (Figure 3 and Table I) by some well-observed white dwarfs for which $-5.0 < \log\phi < -2.7$. The sequence of ordinary stars is illustrated by the sun and a few representative points for main sequence stars from M8 dwarfs to O5-B0 supergiants and on down to M2 supergiants. Giants with distended atmospheres fall three to six orders of magnitude lower. Infrared stars would extend the density-radius relation downward by several orders of magnitude (dashed line) as illustrated by a hypothetical star of 100 solar masses at the stage where it just begins to radiate ($R \simeq 1000$ AU). Here the filling factor is in the range $10^{-7} < \phi < 10^{-5}$ for stars and $10^{-9} < \phi < 10^{-7}$ for protostars. It is rather remarkable that on the grand view of Figure 3, where details of stellar models matter little, all families of stars follow closely the same density-radius relation, although each is governed by very different basic physical laws (perfect gases as opposed to degenerate and nuclear matter) and do not form a continuous evolutionary sequence (for example, no stable objects populate the gap between white dwarfs and neutron stars).

An order-of-magnitude relation is thus defined for stellar bodies

$$\log\rho \simeq -2.7\,(\log R - 11.0) \tag{3}$$

which applies at least in the range $-14 < \log\rho < +14$, or $6 < \log R < 16$, and in this range the filling factor ϕ decreases from about 10^{-1} to about 10^{-9} (Figure 3).

An even more intriguing and as yet unexplained situation develops as we move from stars to star clusters, galaxies, clusters of galaxies, and eventually to the whole countable extragalactic space depicted in the lower half of Figure 3. As one moves downward in the figure, the symbols refer to (i) compact dwarf elliptical galaxies, (ii) normal giant elliptical galaxies, (iii) normal giant spiral galaxies, (iv) small, compact groups of galaxies, (v) larger groups and clouds of galaxies, (vi) small clusters, of the Virgo or Fornax I type, (vii) large clusters of the Coma type, (viii) the

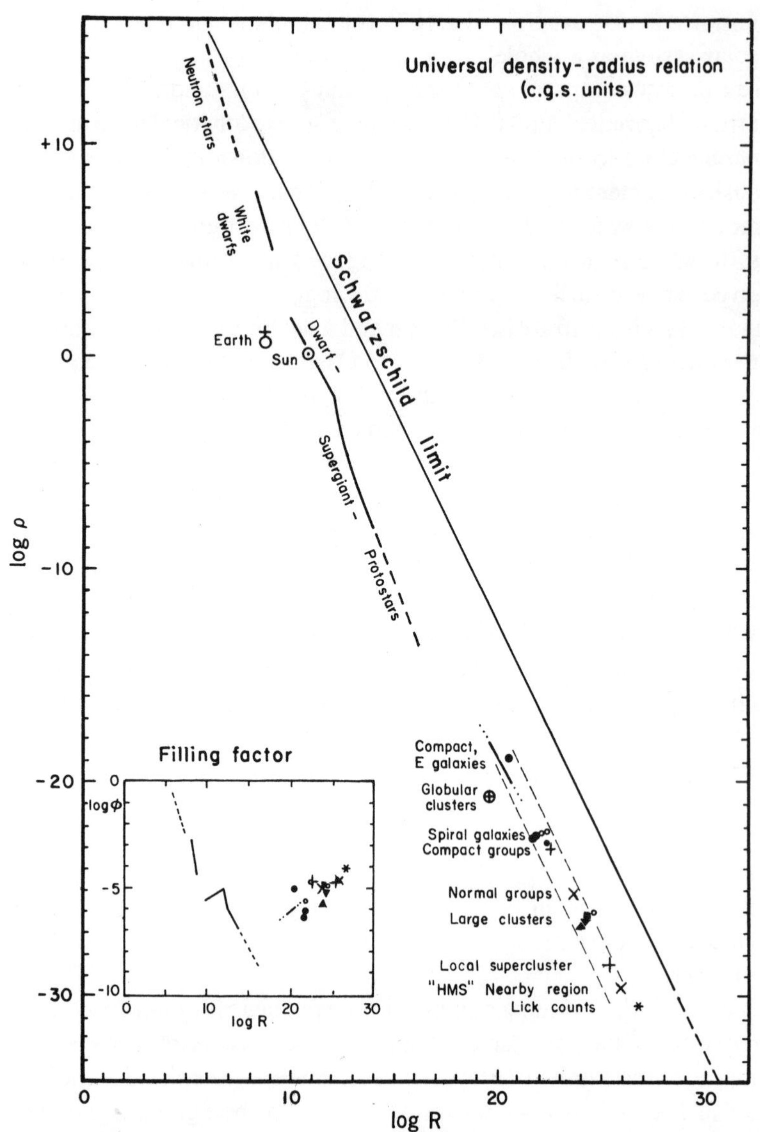

Fig. 3. Universal density-radius relation gives the maximum average density of matter (in g per cm^3) in spherical volumes of radius R (in cm) from neutron stars (dashed line at top) to the largest domain in which galaxies have been counted (asterisk at bottom). The Schwarzschild limit (thin line) and filling factor ϕ (inset) are shown. The range of densities by the virial theorem for stellar and galactic clusters is shown (thin dashes).

TABLE I

Mass-radius-density data

Class of objects	Examples	$\log M$ (g)	$\log R$ (cm)	$\log \rho$ (g cm^{-3})	$\log \phi$	Ref.
Neutron stars		33.16	5.93	14.75	−0.6?	53
		32.54	7.44	9.60	−2.5	53
White dwarfs	L930-80	33.45	8.3:	7.93	−2.7	54
	αCMaB	33.30	8.77	6.37	−3.2	54
	vM2	32.90	9.05	4.13	−5.0	54
Main sequence stars	dM8	32.2	9.95	1.76	−5.6	55
	Sun	33.30	10.84	0.15	−5.5	55
	A0	33.85	11.25	− 0.55	−4.7	55
	O5	34.9	12.1:	− 2.0	−5.0:	55
Supergiants stars	F0	34.4	12.65	− 4.2	−6.1	55
	K0	34.4	13.15	− 5.7	−6.6	55
	M2	34.7	13.75	− 7.2	−6.9	55
Protostars	IR	35.3?	16.2?	−13.9?	−8.7?	55
Compact dwarf elliptical galaxies	M32, core	41.0	19.5?	−18.1	−6.3	56, 57
	M32, effective	42.5	20.65	−20.0	−5.9	58
	N4486-B	43.4	20.5	−18.75	−5.0	56, 59
Spiral galaxies	LMC	43.2	21.75	−22.65	−6.3	60
	M33	43.5	21.8	−22.5	−6.1	61
	M31	44.6	22.3	−22.9	−5.5	62
Giant elliptical galaxies	N3379	44.3	22.0	−22.35	−5.6	38, 63
	N4486	45.5	22.4	−22.3	−4.7	56, 64
Compact groups of galaxies	Stephan	45.5	22.6:	−23.1:	−4.7	38, 65
Small groups of spirals	Sculptor	46.2	24.1	−26.7	−5.7	38, 66
Dense groups of ellipticals	Virgo E, core Fornax I	46.5	23.7	−52.2	−5.0	38, 67
Small clouds of galaxies	Virgo S Ursa major	47.0	24.3	−26.5	−5.1	38, 67
Small clusters of galaxies	Virgo E	47.2	24.3	−26.3	−4.9	38, 67
Large clusters of ellipticals	Coma	48.3	24.6	−26.1	−4.9	36, 38
Superclusters	Local	48.7:	25.5:	−28.4:	−4.7	24
			26.0:	−29.6	−4.6	68
			26.8	−30.5	−4.1	([12], p. 55)

HMS sample to $m \simeq 12.5$
Lick Observatory counts to $m \simeq 19.0$

Local Supercluster, (ix) the nearby region ($R<30$ megaparsecs) in which detailed studies, for example, of red shifts and luminosity function, are possible, and (x) the largest volume of space ($R<250$ megaparsecs), for which reliable galaxy counts are available. Numerical details and references are given in Table I.

This lower half of Figure 3 is a revised, updated version of a graph first published 10 years ago,[24, 36] and which was itself inspired by Carpenter's discovery of a 'density restriction' governing the maximum population of groups and clusters of galaxies.[39] If we consider all clusters with a characteristic radius $R \pm \mathrm{d}R$, there is apparently an upper limit to the space density of galaxies ν_R or to the mass density ρ_R that can exist within the corresponding volume; this limit defines a linear envelope in the scatter diagram of $\log \rho$ as a function of $\log R$ above which no system is observed. This envelope is well below the Schwarzschild limit ($10^{-6} < \phi < 10^{-5}$), and it is clearly not caused by observational selection, since a system having the same diameter as, say, the Coma cluster, but with 10 to 100 times its galaxy population would have been among the first to be discovered. In order to define the envelope as shown in Figure 3, we need therefore to consider only a sample of the densest systems corresponding to a given range of typical radii.

It is remarkable that these points also define a linear relation, but it is not a direct extension of the line for the stellar bodies. For systems of stars and galaxies the filling factor ϕ apparently increases with R from about 10^{-6} to 10^{-4} when R increases from 10^{18} to 10^{27} cm. A linear fit gives the relation

$$\log \rho = -21.7 - 1.7 (\log R - 21.7). \tag{4}$$

The slope lies within the range (-1.5 to -1.9) of earlier preliminary estimates.[24, 36, 39] Comparing Equations (3) and (4), we see that the slope

$$\partial (\log \rho)/\partial (\log R) = S$$

of the density-radius relation is significantly different for stellar bodies for which $S = -2.7$ and for star and galaxy systems for which $S = -1.7$. Nevertheless, most of the data points are within a strip parallel to the Schwarzschild relation (Equation (1a) for which $S = -2$; this strip corresponds to a mean filling factor $\phi \simeq 10^{-4}$ to 10^{-6}. There are few reliable data for individual objects in the range $13 < \log R < 18$. This may result from observational selection if most such objects are dark; for example, infrared stars, protostars, and the small dark clouds of interstellar matter known as globules have diameters in this range. Luminous objects in this range, which would appear as highly compact globular clusters with diameters of a fraction of a parsec, should be easily detected if they were present in the galactic neighborhood of the sun, but no such object is known. Quasars are other possible candidates that might fill this gap; but we know too little of their masses and radii to place them in Figure 3 with any degree of confidence at present.

7. Cosmological Implications and Charlier's Hierarchical Models

The density-radius relation for stellar and galactic systems has interesting cosmological implications. There is no indication out to the largest observed value of R defined by the Lick Observatory counts to $m = 19.0$ that a limiting constant value of ρ is reached which could be *the* average density ρ_0 of the universe (or at least of optically visible condensed matter). If a constant nonzero value of ρ_0 exists, it is not reached within the range of distances sampled by galaxy counts. In the range $18 < \log R < 27$ the larger the volume of space sampled, the lower the mean density of countable galaxies; that is, the volume of space in which galaxies have been counted is not the 'fair sample' of space postulated by the isotropic homogeneous models of cosmology, and the value of ρ inserted in such models on the basis of existing galaxy counts may well be irrelevant. *A fortiori* values of the space density (and also of the expansion parameter H) derived from studies of nearby galaxies within the Local Supercluster are even less likely to fulfill the theoretical assumptions.

There is a little-discussed class of cosmological models in which the average density ρ can converge to an arbitrarily small value as the radius of the test volume increases indefinitely in a manner which could be consistent with Equation (3).

The concept of a hierarchy of systems, that is, galaxies, clusters of galaxies, superclusters, or clusters of the second order and so on, was first introduced by Charlier[40] in 1908 and refined in 1922 as a possible classical solution to Olbers' paradox.[41] Charlier showed how with this hierarchical concept a Euclidean-Newtonian universe could be built up

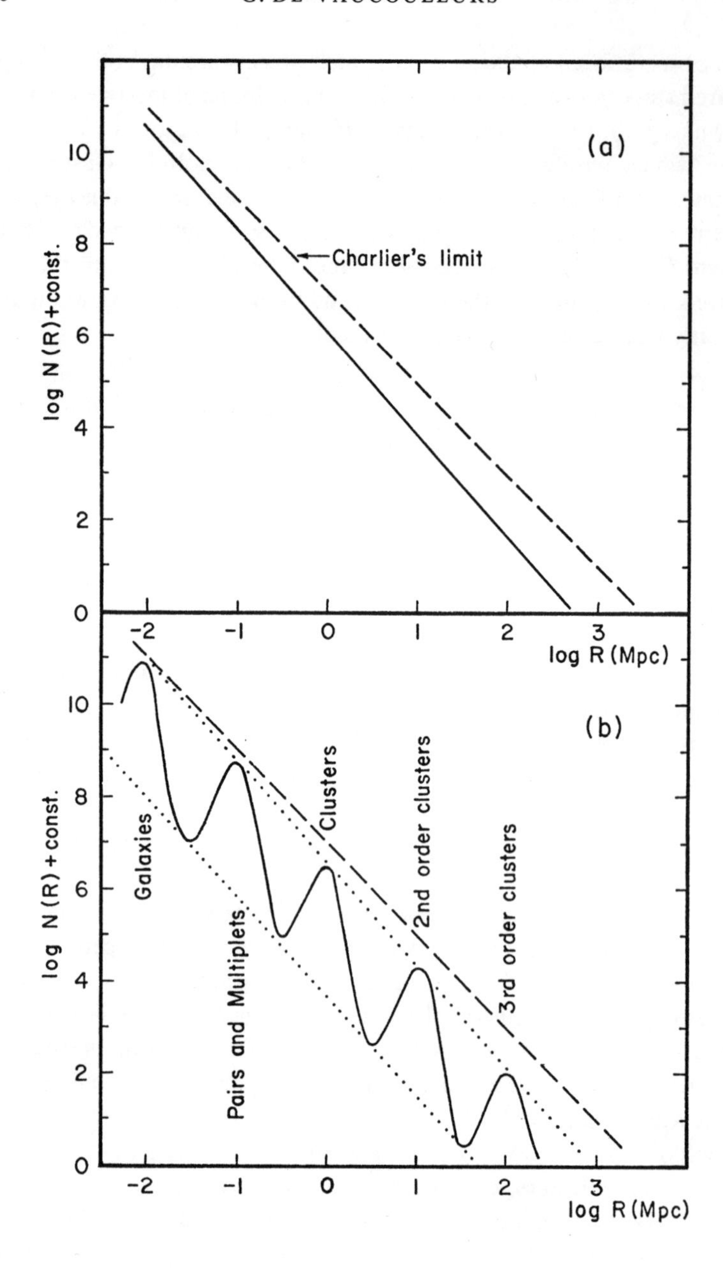
(a)
Charlier's limit
log N (R) + const.
log R (Mpc)
(b)
Galaxies
Pairs and Multiplets
Clusters
2nd order clusters
3rd order clusters

which would avoid the Olbers' disvergence as long as the radii R and population N of systems of order i and $i+1$ satisfy the inequality

$$R_{i+1}/R_i > \sqrt{N_{i+1}} \tag{5}$$

where N_{i+1} is the number of systems of order i in the system $i+1$. It is easy to demonstrate that, if this condition is met, the infinite series which gives the total light flux is finite and the contribution of the distant systems is arbitrarily small. The discovery of the universal red shift, which decreases the energy of photons from distant sources, and the successful development of finite world models neatly solved the Olbers' paradox and removed the necessity of invoking the Charlier solution. Nevertheless, the concept of a hierarchical structure was not disproved by the convenient emergence of different types of solutions to the Olbers' paradox. Nor is there any need to adhere to an oversimplified geometric description of a hierarchical structure, which is perhaps a convenient model to demonstrate the theorem but is probably not an essential one. In reality, the scale of density fluctuations may very well form a continuous spectrum; that is, clustering could probably occur on all possible scales – for example, as in a theory of hydrodynamic turbulence on a cosmic scale, in which we might compare the clusters of various orders to a hierarchy of eddies – and still satisfy the rather weak condition expressed by Charlier's inequalities, as long as enough relative 'void' is left between the eddies. Here two cases must be distinguished, depending on the shape of the frequency function of cluster radii $N(R)$. In the first case $N(R)$ is a monotonically decreasing function of R (Figure 4a), as implied by Kiang[31] and Kiang and Saslaw.[32] In the second case $N(R)$ has a series of relative maxima at some preferred values of R (Figure 4b), say, $R_{-2} \simeq 0.01$ megaparsec (galaxies), $R_{-1} \simeq 0.1$ megaparsec (pairs and multiplets), $R_0 \simeq 1$ megaparsec (groups and clusters), $R_1 \simeq 10$ megaparsecs (superclusters or second-order clusters), $R_2 \simeq 100$ megaparsecs

Fig. 4. Idealized frequency functions of star and galaxy clusters illustrate two possible pictures of a hierarchical universe. (a) Clustering of galaxies occurs on all possible scales with no preferred sizes; the number density of clumps per unit volume decreases smoothly as their radius R increases; (b) clustering of galaxies occurs on all scales but with greater numbers of clumps with characteristic radii near 10 kiloparsecs (galaxies), 100 kiloparsecs (pairs and multiplets), 1 megaparsec (groups and clusters), 10 megaparsecs (superclusters), 100 megaparsecs (third-order clusters), and so forth. In both (a) and (b) the average slope of $\log N$ as a function of $\log R$ must be steeper than -2, the Charlier limit for convergence.

(third-order clusters), and so forth, as suggested by the data presented here. The empirical evidence and the statistical methods of analysis are not yet good enough to clearly distinguish between these two cases.[42,43] It is clear that there is a need for an extension of Charlier's work to quasi-continuous models of density fluctuations that would replace the original, oversimplified discrete hierarchical model. It is equally clear that this fluctuating density field should be considered within the framework of the world models of general relativity.

It would seem, in principle, that nothing prohibits the introduction of the hierarchical concept of an indefinitely clustered density distribution in the more promising relativistic or steady-state models. In practice, of course, major mathematical difficulties may be encountered. A first-order treatment of small fluctuations (say, $\sim 10\%$) has been developed successfully,[44] but it is not sufficient for a discussion of the actual situation in the real universe where 'fluctuations' are of the order of 1000% of the local average density and, as indicated by Figure 4 and Equation (4), the average density decreases by nearly two orders of magnitude when the radii of the systems considered differ by one order of magnitude.

In a sense a hierarchical model is homogeneous because, except possibly for statistical fluctuations, two equally large disjoint volumes of radius R_n, each encompassing at most one supersystem of order n, can have the same average density $\rho(R_n)$. This type of homogeneity satisfies the restricted cosmological principle (no privileged position in space) but, because of Charlier's inequality, it does not imply that ρ should be a constant independent of n. All observers, wherever located (but within the hierarchy), will find that the average density decreases as the range R of their counts increases. In a Newtonian-Euclidean universe devoid of intergalactic matter, R could presumably increase indefinitely, and, if so, $\rho(R)$ would asymptotically approach zero as in Charlier's original concept. In a relativistic world model, all the basic concepts of the homogeneous models would presumably still remain, including the possibility of closed or open universes. In the case of a closed static universe, the radius of curvature (R) would be related to the density ρ of the largest-scale supersystem defined by $R_M \simeq (R)$, and $\rho(R) = \rho_M$ is the Schwarzschild radius defined by Equation (1). As a purely arithmetic curiosity, if the value of ρ given by Equation (4) is substituted into Equation (1), then $\rho = \rho_M$ for $\log R \simeq 40$, where $\log \rho \simeq -52.8$(!) Obviously this result cannot be taken seriously because of the enormous extrapolation and observational uncertainties in the basic data.

Nevertheless, the point remains that if $\log \rho = -30 \pm 1$ throughout (independent of R for all observers), as is currently assumed in most orthodox theories, we must postulate that intergalactic space is filled by a uniform and relatively homogeneous invisible gaseous medium comprising 90 to 99% of the smoothed-out mean density. If so, we are forced to conclude that the mass condensed in galaxies, clusters, and superclusters makes an almost trivial contribution to ρ and that it is possible, indeed probable, that the values of ρ currently derived from galaxy counts are of little significance with respect to the problem of the mean density of homogeneous models.

Even if we grant, as a working hypothesis, that galaxies may not be the major contributors to the mass density, we may at least use them as tracers, which show where matter is condensed; it would be very strange indeed if condensations of visible and invisible matter were mutually exclusive. If so, it seems difficult to believe that, whereas visible matter is conspicuously clumpy and clustered on all scales, the invisible intergalactic gas is uniform and homogeneous. This is perhaps conceivable for radiation, but not for matter, whether it be diffused or condensed. Certainly interstellar matter is extremely clumpy and irregularly distributed; so are the high-latitude, high-velocity neutral hydrogen clouds in the galactic corona; why then should intergalactic gas be smoothly spread out throughout the universe?

If indeed there is something to the latest version of the steady-state hypothesis,[5] where matter is injected (by some as yet unknown law of physics) into the visible universe at the centers of galaxies,[45] then, evidently, the distribution of invisible matter must be closely related to and just as clumpy as the distribution of galaxies, and a hierarchical structure *à la* Charlier must be included in any realistic cosmological model.

8. Cosmology in a Clumpy Universe

If, therefore, we grant that clumpiness in the distribution of matter in the universe is a basic property of fundamental importance for cosmology and not merely a local nuisance that can be ignored in the grand smoothed-out view, we must pay much more attention than we have thus far to the possible consequences of this situation.

Although, clearly, much detailed work will have to be done before we can assess the consequences of this hierarchical concept for different cosmological models, we may perhaps consider the existing homogeneous, isotropic models, as analogous to the osculating parabolic elements of a cometary orbit. The curvature, expansion rate, and deceleration of a given homogeneous model may approximate conditions over some distance range (volume) within which a local average density ρ_R obtains, but it cannot be extrapolated to the whole universe past and present, any more than the parabolic orbit can be extrapolated to infinity in space and time. Within the volume considered, density fluctuations will wrinkle the geodesics and alter the expansion parameters; there is already clear evidence that the Hubble expansion rate is reduced by gravitation within the Local Supercluster,[27, 29] perhaps almost cancelled within the Local Group,[46] and, of course, completely overwhelmed by it within individual galaxies.

This leads one to view the Hubble parameter as a stochastic variable, subject in the hierarchical scheme to effects of local density fluctuations on all scales. A simple analogy is a light ray weaving to and fro as it traverses successive domains of different sizes, densities, and inhomogeneous gravitation fields. The path of the light ray will experience more deflection, but over a shorter length, as it crosses small domains of relatively high density (for example, a cluster), or less deflection, but over larger paths, as it propagates through larger domains of lower density (for example, a supercluster). Detailed calculations for specific models of the hierarchical structure will be needed to evaluate the net effect of this mechanism over increasing path lengths.[47, 48]

Such calculations may require a kind of statistical approach to relativity in which the model parameters will depend in some complicated manner on characteristic scale lengths, all satisfying Charlier's inequality. On any scale an osculating homogeneous model may be defined, but it should not be extrapolated to much larger (or smaller) scales. In this sense if the current homogeneous-isotropic models seem to converge or point to a definite solution of the cosmological problem, it may well be merely a reflection of the limited range of present data on red shifts, galaxy counts, and source counts. And so, perhaps, once again we are mistaking the horizon for the end of the world.

9. Summary

We have to choose between two sets of possibilities. First set: (i) estimates of the age of the world which have grown exponentially for the past 3 centuries (Figure 1) will suddenly reach a final constant level within a few years of A.D. 1969 – obviously possible, but a little surprising; (ii) the obvious non-random clustering which dominates the galaxy distribution on all scales out to the limit of the deepest survey (Figure 2) will suddenly vanish and be followed by the emergence of statistical uniformity as soon as we consider volumes equal to or greater than the totality of the presently counted volume – again possible, but, on present evidence, a not too likely event; (iii) as ever larger volumes of space are considered, the correlation between the maximum density of matter and radius, demonstrated over a range of more than 20 orders of magnitude in radius and 45 orders of magnitude in density, suddenly stops operating beyond the last observed point (Figure 3) to level off at the presumed value of the mean density postulated by current homogeneous models – again not completely impossible, but certainly a highly artificial hypothesis on the basis of the evidence at hand (what other law of physics or astronomy has been checked over a greater range?).

Now in order to accept the present orthodox cosmologies we must assume that all three suppositions above will be verified. The combined probability of all three being simultaneously resolved in the affirmative appears to be very small indeed, since each requires a sudden break just beyond the last datum, of which there is no hint in the observed range.

The second set of possibilities involves merely accepting the empirical evidence as it stands: (i) the concept of 'age of the world' has a long historical record of rapid change and, anyway, lacks definiteness, except in specialized world models which are far from established; (ii) clustering of galaxies, and presumably of all forms of matter, is the dominant characteristic of the structure of the universe on all observable scales with no indication of an approach to uniformity; (iii) the average density of matter decreases steadily as ever larger volumes of space are considered out to the limit of the counted region, and there is no observational basis for the assumption that this trend does not continue out to much greater distances and lower densities.

It would seem that the time has come to give serious consideration to hierarchical models either within the framework of otherwise conventional relativistic cosmology, if that is possible, or within the context of the steady-state theory which is eminently compatible with the kinds of large-scale properties discussed here.[4]

A beginning has already been made in this direction; thus Hoyle and Narlikar[47] have considered the effects of a creation rate which varies with pre-existing local density in a kind of hierarchical steady-state model, and Rees and Sciama[48] have discussed possible effects of large-scale inhomogeneities in evolutionary relativistic models.[49] Wertz[50] made a preliminary survey of possible variations on the hierarchical scheme, starting with the simplest homogeneous isotropic hierarchical (or 'polka dot') cosmological models and branching out to include hierarchical-homogeneous as well as inhomogeneous anisotropic models, both finite and infinite, and even hierarchical schemes with homogeneous background and inhomogeneous models with varying elementary (that is, galaxian) masses. It is evident that a systematic morphological analysis of all possible hierarchical models will reveal the great richness of the concept compared with the already large range of choices offered by homogeneous models. The task of developing self-consistent theoretical models within the framework of this vast array of hierarchical systems and of rigorously testing each of them against the observed properties of the universe appears rather formidable. It seems safe to conclude that a unique solution of the cosmological problem may still elude us for quite some time!

University of Texas at Austin

NOTES AND REFERENCES

* Reprinted with kind permission from *Science* **167** (1970) 1203–1213.

[1] See, for example, W. H. McCrea, *Science* **160** (1968) 1295; H. Y. Chiu, *Sci. J.* **4** (1968) 33.
[2] Among the better introductions to modern cosmology the following are especially recommended: P. Couderc, *The Expansion of the Universe*, Faber & Faber, London, 1952; G. C. McVittie, *Fact and Theory in Cosmology*. Macmillan, New York, 1961; H. Bondi, *Cosmology*, Cambridge Univ. Press, New York, 1960. For philosophical perspectives on the history of modern cosmology I recommend the following: J. Singh, *Great Ideas and Theories of Modern Cosmology*, Dover, New York, 1961; J. D. North, *The Measure of the Universe*, Clarendon Press, Oxford, 1965. The last reference is exceptionally thorough and comprehensive as well as more technical.
[3] H. Bondi, *Rep. Progr. Phys.* **22** (1959) 97.

[4] W. Davidson and J. V. Narlikar, *ibid.*, **29** (1966) 541.
[5] F. Hoyle, *Quart. J. Roy. Astron. Soc.* **10** (1969) 10.
[6] For example, in a recent review article on 'The Cosmology of Our Universe' [1], we read that "In cosmological theory it is convenient to regard galaxies as the basic elements of the Universe" (even though Zwicky showed more than 30 years ago that clusters, not galaxies, are the basic blocks); more amazingly, as a summary of the observational evidence, we read: "The distribution of galaxies is generally uniform up to a distance of 2×10^9 light-years, with a variation of a few percent. In general, the distribution is isotropic and homogeneous.... The matter density due to galaxies... is 3×10^{-31} g per cm^3." This is little more than sheer fiction as I will show in this paper.
[7] Readers unfamiliar with some astronomical terms or concepts may find useful the following explanations: Galaxies are counted on photographic plates by visual inspection through low-power magnifiers: anything fuzzy (unless it is obviously a galactic nebulosity) is counted as a galaxy down to the faintest detectable smudge that can be just differentiated from the sharper star images; the limiting brightness at the detection threshold evaluated in the traditional stellar magnitude scale (which increases logarithmically with decreasing intensity) is the magnitude limit m of the counts. The results, corrected for various instrumental factors, are expressed as the areal number density of galaxies on the celestial sphere, that is, the number of galaxies per square degree $N(m)$ brighter than the limiting magnitude m of the survey. The first systematic survey by Hubble at Mount Wilson Observatory in the early 1930's was limited to small telescopic fields, each covering less than the area of the moon, at 5° and 10° intervals regularly spaced in galactic latitude and longitude. The most recent and thorough survey by Shane and his collaborators at Lick Observatory completely covered the sky north of declination −23° (including about two-thirds of the area of the sphere) to a slightly brighter limit of magnitude. At the large distances of the faintest galaxies counted in these surveys, the apparent luminosity is reduced not only by distance itself but also by geometric and spectral effects of the expansion of the universe which must be corrected (the so-called K-correction) before the results can be used in cosmology. The existence of clustering among galaxies far in excess of the accidental clumping that arises by chance in a random (Poisson) distribution of independent particles is quantitatively demonstrated by a comparison of the frequency distribution of $N(m)$ determined in many small different areas with that expected for a Poisson distribution. The average size of the clumps is derived by correlation techniques which make possible the evaluation of the degree of coherence or incoherence of counts in separate nearby areas of variable separation S as a function of S. For example, positive correlation exists in the Lick Observatory counts over areas up to 8° in diameter, corresponding to an average clump size of 30 to 40 megaparsecs at the estimated limiting distance of the survey (this size is typical of the second-order clustering).
[8] E. P. Hubble, *Astrophys. J.* **79** (1934) 8.
[9] B. J. Bok, *Bull. Harvard Coll. Observ.* No. **895** (1934); P. Bourgeois and J. Cox, *C. R. Hebd. Seances Acad. Sci. Paris* **204** (1937) 1622; *Ciel Terre* **54** (1938) 287; A. G. Mowbray, *Publ. Astron. Soc. Pac.* **50** (1938) 275.
[10] H. Shapley, *Bull. Harvard Coll. Observ.* No. **880** (1932) 1; F. Zwicky, *Astrophys. J.* **86** (1937) 217.
[11] F. Zwicky, *Publ. Astron. Soc. Pac.* **50** (1938) 218.
[12] C. D. Shane and C. A. Wirtanen, *Publ. Lick Observ.* **22** (1967) Part 1.
[13] F. Zwicky, *Morphological Astronomy*, Springer-Verlag, Berlin, 1957.
[14] F. Zwicky, *Publ. Astron. Soc. Pac.* **64** (1952) 247.
[15] Zwicky and his collaborators call 'cluster' any grouping of galaxies larger than small

groups, irrespective of population and diameter, at least up to 50 megaparsecs. Also a 'cluster cell' need not be completely filled by a cluster but may be thought of as delimited by the surface of minimum space density between the 'spheres of influence' of adjacent clusters, which, of course, may be of unequal sizes and populations.

[16] G. de Vaucouleurs, in *Stars and Stellar Systems*, vol. 9 (ed. by A. R. Sandage) Univ. of Chicago Press, Chicago; *Astron. J.* **72** (1967) 325 (abstract).

[17] F. Zwicky *et al.*, *Astrophys. J.* **137** (1963) 707; *ibid.* **141** (1965) 34; *ibid.* **142** (1965) 625; *ibid.* **146** (1966) 43.

[18] F. Zwicky and K. Rudnicki, *Z. Astrophys.* **64** (1966) 246.

[19] G. O. Abell, *Astrophys. J. Suppl.* **3** (1957) 211.

[20] G. D. Abell, *Astron. J.* **66** (1961) 607; *ibid.* **72** (1967) 288.

[21] J. Neyman, E. L. Scott, and C. D. Shane, *Astrophys. J.* **117** (1953) 92; *Astrophys. J. Suppl. 8* (1954); *3rd Berkeley Symposium on Statistics*, (Univ. of California Press, Berkeley, 1955, p. 75.

[22] I. D. Karachentsev, *Astrofizika* **2** (1966) 307; [English translation in *Astrophysics* **2** (1966) 159].

[23] G. de Vaucouleurs, *Astron. J.* **58** (1953) 31; *Vistas in Astronomy*, vol. 2, Pergamon Press, London, 1956, p. 1584; *Sov. Astron.* **3** (1960) 897.

[24] G. de Vaucouleurs, *Astron. J.* **66** (1961) 629.

[25] G. de Vaucouleurs and W. L. Peters, *Nature* **220** (1968) 868.

[26] E. Holmberg, *Ann. Lund Observ.* **6** (1937) 52; A. Reiz, *ibid.* **9** (1941) 65; R. L. Carpenter, *Publ. Astron. Soc. Pac.* **73** (1961) 324.

[27] G. de Vaucouleurs, *Astron. J.* **63** (1958) 253.

[28] G. de Vaucouleurs, *Nature* **182** (1958) 1478.

[29] G. de Vaucouleurs, *Publ. Dep. Astron. Univ. Tex. Ser. I* 1 No. 10 (1966).

[30] E. K. Conklin, *Nature* **222** (1969) 971.

[31] T. Kiang, *Monthly Notices Roy. Astron. Soc.* **135** (1967) 1.

[32] T. Kiang and W. C. Saslaw, *ibid.*, **143** (1969) 129.

[33] G. B. van Albada, *Bull. Astron. Inst. Neth.* **15** (1960) 165; *Int. Astron. Union Symp. 15, Santa Barbara, Calif., 1961* (1962) p. 411; S. J. Aarseth, *Monthly Notices Roy. Astron. Soc.* **126** (1963) 223; *ibid.* **132** (1966) 35.

[34] H. Shapley, *The Inner Metagalaxy*, Yale Univ. Press, New Haven, 1957.

[35] In the conclusion of his study of superclustering of cluster centers in the Zwicky catalog, Karachentsev[22] is led "to question the correctness of the initial premise of a random distribution of the supercluster centers" and concludes that "some associative property is conspicuous in their distribution," in agreement with the evidence of Figure 2 and other data discussed in this article.

[36] G. de Vaucouleurs, *Astrophys. J.* **131** (1960) 585.

[37] J. Neyman, T. Page, and E. Scott (eds.), *Proceedings of the Conference on Instability of Systems of Galaxies* [*Astron. J.* **66**, No. 1295, (1961) 533]; D. N. Limber, *Int. Astron. Union Symp. 15, Santa Barbara, Calif., 1961* (1962), 239.

[38] T. L. Page, *Smithsonian Astrophys. Observ. Spec. Rep.* **195** (1965).

[39] E. F. Carpenter, *Publ. Astron. Soc. Pac.* **43** (1931) 247; *Astrophys. J.* **88** (1938) 344.

[40] C. V. L. Charlier, *Ark. Math. Astron. Fys.* **4** No. 24 (1908); *ibid.* **16** No. 22 (1922); *Medd. Lund Observ.* No. **98** (1922).

[41] The Olbers' paradox (*Bode's Jahrbuch*, Berlin, 1826, p. 110) concerns the question "Why is the sky dark at night?" In an infinite, transparent Euclidean-Newtonian universe uniformly populated with stars, any line of sight would eventually encounter the surface of a star and everywhere the sky should be about as bright as the sun. The problem was first discussed

in the 18th century by de Chéseaux, of Lausanne, and the related question of infinite gravitational potential was further analyzed by Seeliger in 1895 (*Astron. Nachr.* No. 3273). One possible solution which survived as late as 1918 in Shapley's views of the galactic system and 1922 in Kapteytn's heliocentric stellar universe was the hypothesis that the world of stars is not infinite but rather a single island universe in an empty cosmos. Charlier's concept was a neat geometric solution (in nonrelativistic terms) to the problem of "how an infinite world may be built up." A hierarchical structure of the cosmos was first evisioned by the Alsatian J. Lambert, in his somewhat fanciful 'Kosmologischen Briefen' (Augsburg, 1761). An exellent historical summary (in French) on the Olbers' paradox and its sequels was published in 1966 by R. Chameaux [*Bull. Soc. Astron. Toulouse* **57**, No. 485 (1966)]. Another very interesting discussion of the paradox and its cosmological implications was presented in June 1965 by A. G. Wilson in an unpublished lecture to the Los Angeles Astronomical Society (Rand Corporation reprint P-3256) [see also [49]].

[42] For example, statistical analyses of Abell's cluster catalog have led to the following conflicting conclusions: definite superclustering on a 50-megaparsec scale for at least a fraction (but not the totality) of the cluster population, according to Abell[20]; some superclustering on scales of 50 to 200 megaparsecs, according to Kiang and Saslaw[32]; no significant superclustering, according to Yü and Peebles[43]. [However, after reading a preliminary version of this paper, Dr. Abell informed me (personal communication) that the results of Yü and Peebles "followed only by not counting those clusters in distance group 5 in the southern hemisphere, where the superclustering appears most obvious. In fact, the superclustering was so pronounced in that part of the catalog that they felt it was not representative and so discounted it."] Similarly, analyses of the Zwicky catalogs have led to the widely diverging conclusions of Zwicky and his collaborators[13–18] and of Karachentsev[22]. One additional remark should be made here; in the study of superclustering it may not be strictly equivalent to study the distribution of galaxies in general (as was done in the analysis of the Local Supercluster and of the Lick Observatory counts), on the one hand, or the distribution of large or rich clusters (as listed in the Mount Palomar surveys) on the other. It is entirely possible that superclustering of large clusters is much less pronounced and prevalent than superclustering of groups and small clusters which are automatically excluded in the Abell and Zwicky catalogs by the very definition and method of selection of 'rich clusters'. For example, only one cluster – and not a particularly rich one – the Virgo cluster, is known in the Local Supercluster, and if this supercluster were seen from a great distance, it would not be recognized as such from cluster counts because it would be represented by only one cluster (and perhaps none). To use again the human population analogy, it is doubtful that the obvious 'superclustering' of the general population indicated by statistics of agglomerations of all sizes (that is, complete 'counts') would be readily detected by an analysis restricted to the worldwide distribution of the great capital cities only (the 'rich clusters'). I suspect that the apparent disagreement in the conclusion reached by various investigators arises, at least in part, from a failure to recognize that only a small fraction of the total galaxy population is concentrated in the relatively rare 'rich clusters'.

[43] J. T. Yü and P. J. E. Peebles, *Cal. Inst. Tech. Orange Preprint Ser.* No. 168 (1969); *Astrophys. J.* **158** (1969) 103.

[44] W. M. Irvine, *Ann. Phys.* **32** (1965) 322; J. Kristian and R. K. Sachs, *Astrophys. J.* **143** (1966) 379; R. K. Sachs and A. M. Wolfe, *ibid.* **147** (1967) 73.

[45] This concept was foreshadowed as long ago as 1928 when Jeans wrote in *Astronomy and Cosmogony* (Cambridge Univ. Press, Cambridge, p. 352) his well-known, possibly prophetic speculation: "The type of conjecture which present itself, somewhat insistently, is that the centers of the nebulae are of the nature of 'singular points', at which matter is

poured into our universe from some other, and entirely extraneous, spatial dimension, so that to a denizen of our universe, they appear as points at which matter is being continually created."

[46] M. L. Humason and H. D. Wahlquist, *Astron. J.* **60** (1955) 254.

[47] F. Hoyle and J. V. Narlikar, *Monthly Notices Roy. Astron. Soc.* **123** (1961) 133.

[48] M. J. Rees and D. W. Sciama, *Nature* **217** (1968) 511.

[49] In recent years A. G. Wilson has given much thought to the concept of a hierarchical cosmos and has searched for signs of a possible 'discretization' in 'modular' structures [51] that might perhaps relate cosmic and atomic constants through quantized relations. Although the numerological aspects of this approach – reminiscent of Eddington's brilliant but futile *Fundamental Theory* – are admittedly highly speculative, the basic concepts and evidence of a hierarchical world structure discussed by Wilson are very much the same as those presented here. I am indebted to Dr. A. Wilson for calling my attention to his own extensive work in this area and for a preprint of his chapter *Hierarchical Structure in the Cosmos.* [52]

[50] J. R. Wertz, thesis, University of Texas.

[51] A. G. Wilson, *Proc. Nat. Acad. Sci U.S.* **52** (1964) 847. *Astron. J.* **70** (1965) 150; *ibid.* **71** (1966) 402; *ibid.* **72** (1967) 326.

[52] A. G. Wilson, in *Hierarchical Structures* (ed. by L. L. Whyte, A. Wilson, and D. Wilson), American Elsevier, New York, 1969, p. 113 [see also T. Page, *Science* **163** (1969) 1228; A. G. Wilson, *ibid.* **165** (1969) 202].

[53] J. A. Wheeler, *Annu. Rev. Astron. Astrophys.* **4** (1969) 393.

[54] E. Schatzman, *White Dwarfs*, Interscience, New York, 1958.

[55] C. W. Allen, *Astrophysical Quantities*, Athlone Press, London, 1961, p. 203.

[56] A. Poveda, *Bol. Tonantzintla* No. **17** (1958) 3; *Bol. Tonantzintla* No. **20** (1960) 3.

[57] I. King, *Astrophys. J.* **134** (1961) 272.

[58] G. de Vaucouleurs, *Monthly Notices Roy. Astron. Soc.* **113** (1953) 134.

[59] R. Minkowski, *Int. Astron. Union Symp. 15, Santa Barbara, Calif., 1961* (1962) p. 112.

[60] G. de Vaucouleurs, *Astrophys. J.* **131** (1960) 265; *ibid.* **137** (1963) 373.

[61] K. J. Gordon, *Astrophys. J.* **169** (1971) 235–270.

[62] S. T. Gottesman, R. D. Davies, and V. C. Reddish, *Monthly Notices Roy. Astron. Soc.* **133** (1966) 359.

[63] E. M. Burbidge, G. R. Burbidge, and R. A. Fish, *Astrophys. J.* **133** (1961) 393, 1092; *ibid.* **134** (1969) 251.

[64] J. C. Brandt and R. G. Roosen, *Astrophys. Lett.* **156** (1969) L59.

[65] G. R. Burbidge and E. M. Burbidge, *Astrophys. J.* **130** (1959) 15; *ibid.* **134** (1961) 244.

[66] G. de Vaucouleurs, *ibid.* **130** (1959) 718.

[67] G. de Vaucouleurs, *Astrophys. J. Suppl. Ser.* **6**, No. 56 (1961) 213.

[68] T. Kiang, *Monthly Notices Roy. Astron. Soc.* **122** (1961) 263.

[69] I am indebted to G. O. Abell, P. Couderc, T. Page, and A. G. Wilson for their helpful criticisms of a preliminary version of this paper.

JOHN ARCHIBALD WHEELER

FROM MENDELÉEV'S ATOM TO THE COLLAPSING STAR*

ABSTRACT. A decade before Planck's 1900 formulation of the quantum principle, Mendeléev out of his researches on chemistry and the periodic system of the elements recognized the general character that the new principle must have, dealing with no "deathlike inactivity", establishing "individuality amid continuity", and destined to "hasten the advent of true chemical mechanics". Implicit in the Bohr-Rutherford 1911 model of the atom was the paradox of atomic collapse. No cheap way out offered itself. Only application of the quantum principle resolved the paradox. This development led (1927) to the 'chemical mechanics' envisaged in outline by Mendeléev. It may be symbolized today by an electron orbit with the shape of a double necklace, the 'chemical orbit' (Powers). A new crisis confronts physics today in the predicted phenomenon of gravitational collapse, both at the level of a star ('black hole physics') and at the level of the universe itself. Again no way out is evident except to call on the quantum principle. It leads to the conclusion that the dynamics of the universe goes on in superspace. In superspace alternative dynamical histories of the universe (cycles of expansion and recontraction) not only dynamically couple to each other in the era of collapse, as well as one can judge, but also 'coexist'. If the vision of Clifford and Einstein in updated form is taken as guide and particles are regarded as quantum states of excitation of a dynamic geometry, then it is natural to believe that each period of collapse sees the universe 'reprocessed', with the previous spectrum of particle masses extinguished, and a new pattern of masses established. On this view particle masses are as far removed from any possibility of being calculated from first principles as the 'initial conditions of dynamics' themselves. For an early test of this framework of ideas nothing looks so promising as the prediction that a black hole, formed by whatever combination of baryons, photons, leptons, and other entities, is characterized by nothing but mass, charge, and angular momentum ('transcendence of the law of conservation of baryons and leptons').

1. MENDELÉEV'S GOAL: MUCH OUT OF LITTLE

This centenary of the birth of a great law is also by a coincidence the year of the death of a great architect. Mies van der Rohe expressed the theme of his life work, striving for simplicity and unity, in his famous motto, 'less is more'. What shall be our motto for Dimitri Ivanovich Mendeléev? He sought to the end to see the fantastic wealth of facts of chemistry as consequences of a few central principles. How more briefly can we state this theme than, 'much out of little'? And if the two worlds of physics and chemistry ever float a flag over their long allied forces,

R. J. Seeger and R. S. Cohen (eds.), Philosophical Foundations of Science, 257–301.

what happier motto could they find for it than *multum ex parvo*? No words would honor more the cause for which Mendeléev stood.

2. GRAVITATIONAL COLLAPSE: MUCH INTO LITTLE

If 'much out of little' epitomizes the triumphs of the physical sciences, then, also its direct opposite, 'much *into* little', or *multum in parvum*, summarizes the greatest crisis that one can easily name in the theoretical physics of our day: to understand what happens in complete gravitational collapse.[1]

Figure 1 illustrates schematically the geometry around a mass, comparable to the sun or greater, that has undergone complete collapse. The details have vanished. The Cheshire Cat in *Alice in Wonderland* also vanished. Only its smile remained behind. What remains behind here?

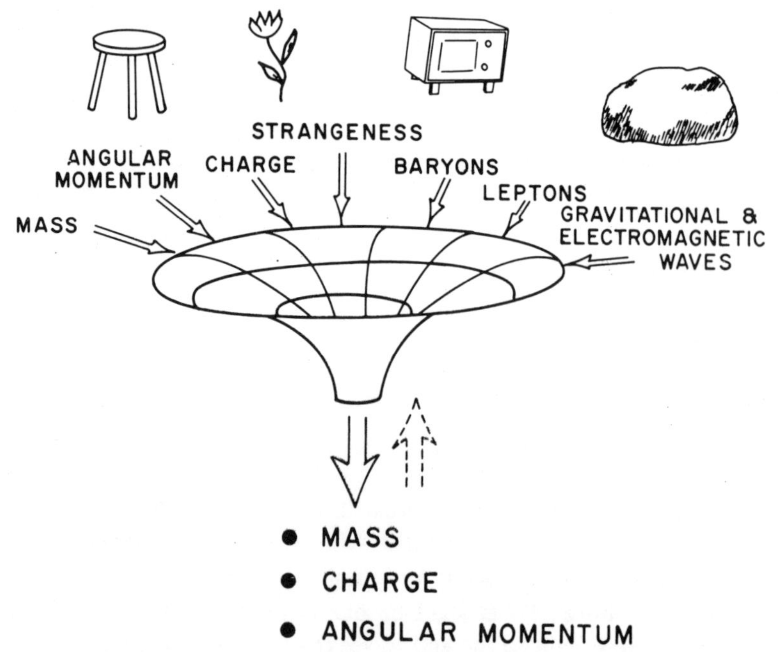

Fig. 1. "A black hole has no hair." All particularities of objects dropped in fade away in a characteristic time of less than a millisecond for a black hole of solar mass. No known means whatsoever will distinguish between black holes of the most different provenance if only they have the same mass, charge and angular momentum.

Mass, first of all. This mass binds a planet in orbit as firmly as ever. In addition to mass the collapsed object possesses electric charge and angular momentum.[2]

3. "A BLACK HOLE HAS NO PARTICULARITIES"

The collapse takes place on a characteristic time scale. For an object with a mass comparable to the mass of the sun this time is less than a millisecond. Let the original object have a hill on it. Then the effective height of this hill decreases to half its value in a characteristic relaxation time also less than a millisecond. Dropping to half value, then to quarter value, then to eighth value, and so on, with each stage lasting less then a millisecond, every geographical feature of the system by the end of a second is erased away to the utmost perfection.

Drop in a meteorite. It makes a momentary disturbance in the geometry. It perturbs briefly the centrality of the gravitational attraction exerted on a planet. That perturbation, that 'protruding particularity' also shrinks to half height in each successive relaxation time. Drop in familiar objects of the greatest variety of sizes and shapes. All details quickly disappear. We end up with an object characterized, so far as we can tell, by mass, charge and angular momentum, and by nothing more. If we call the resulting entity a 'black hole', then we can summarize the perfection of its final state by saying, "A black hole has no particularities."

4. PARTICLE CONSERVATION LAWS TRANSCENDED

Fire in neutrons, protons, antiprotons, particles and radiations of whatever kind one chooses. In the final object not one of these particularities remains. Make meticulous count of the leptons that one drops in. Check in the baryons with equal care. Compare the resulting black hole with another hole built from a very different number of baryons and leptons. Only require that the two objects have the same mass, charge and angular momentum. Then the one black hole can be distinguished from the other by no known means whatsoever. No measurable meaning of any kind do we know how to give to the baryon number and lepton number of a black hole of unknown provenance.[3] Gravitational collapse deprives baryon number and lepton number of all significance.

'Baryon number is conserved'.[4] 'Lepton number is conserved'.[5] Of all principles of physics these familiar conservation laws belong among the most firmly established. Yet with gravitational collapse the content of these conservation laws also collapse. The established is disestablished.

Baryon and lepton number have long served as indices to the discreteness of nature. Today their status changes. Their conservation laws, though often useful, cannot be absolute. The distinction between particles and geometry, though often useful, cannot be absolute. Particles that melt into geometry under one set of circumstances surely cannot be resolved from geometry under other extreme conditions. But what? And when? And how? And what is the higher dynamic law that links particles and geometry together into a larger unity? No theory of particles can correctly describe particles that deals only with particles!

Gravitational collapse, *multum in parvum*, by linking particles with geometry, raises an old issue in a new form. Is nature continuous or is it discrete? Is Plato to be at last the victor, who tells us[6] that all is geometry? Or is Pythagoras again to prove the better prophet, who teaches us[7] that all is number?

5. A WALK WITH MENDELÉEV

Troubling about the mystery of gravitational collapse, great issue of our day, turn for a time on this happy anniversary to Mendeléev's life and work. Find ourselves unexpectedly transported to a bright new world, rich with the glitter of the chemistry of hundreds of compounds high and low and thousands of reactions. Walk in the company of this modest and wise man. Partake of insights as relevant to today as yesterday. And come back refreshed at the end for a new look at gravitational collapse.

Of all the writings of Mendeléev none is more famous than his *Principles of Chemistry*,[8] and of all treatises on chemistry none is a greater pleasure to read, nor better suited to attract the uncommitted reader to chemistry. Fresh and new is one's first impression; and fresh and new it is to the end. Never before had the elements of the same group of the periodic table found themselves together in the same chapter: zinc, cadmium and mercury in one; silicon and the other elements of group IV in another; and so on. Never was there a guide that one would follow to clay pit or tannery, mine or mill, with more interest in his observations, more profit in his predictions, or more assurance of discoveries along the way.

If Mendeléev shows us all the world and even something of the moon and stars, he nevertheless is more and does more than any guide. He stands for a cause, "unity [amid]... individuality and... apparent diversity."[9] His enthusiasm communicates itself for the mystery hidden in everything around. He is as exciting as the nature he tells us about. He carries in his hand his magic wand, the periodic law:

> It was in March 1869 that I ventured to lay before the then youthful Russian Chemical Society the ideas upon the same subject which I had expressed in my just written *Principles of Chemistry*...: (1) The elements, if arranged according to... (2) ... (3) ... (8) ...[10]

We see him touch it to this frozen fact and that, only to make each in turn spring into life and meaning. As we walk on, see new substances and new reactions, and hear him discourse thoughtfully about one and another, we are the more in suspense about where he will next use his magic wand because we sense that he too is all the time pondering this very question. The periodic law: how much of nature does it subsume? Every day to pose this point was every day to find a fresh new face on all creation.

6. TEN YEARS FROM ANOTHER UNCOVERING OF THE UNITY HIDDEN IN DIVERSITY

As we read on in Mendeléev, we suddenly sense that there is something strangely familiar about the man and the message. Whom else have we met equally in love with the richness of the world around? Equally master of it? Equally modest? What other overarching principle, emergent midway on a voyage of discovery, yielded equal thrill to its discoverer? Equal power to find the unity hidden in diversity? Equal radiance so to light up nature that it never looked the same again? And then we remember: Charles Darwin[11] and Dimitri Mendeléev; the *Origin of Species*[12] and the periodic law. Only twenty-two years separated the two men in age; and only ten years, their great publications. Each was anticipated by a lesser formulation: Darwin by Alfred Russell Wallace[13]; Mendeléev by Lothar Mayer.[14] In each case it was the greater man who understood the new principle the more deeply and made the greater lifetime commitment to testing and applying it. Whether we accompany Darwin on his sand path or join Mendeléev in his lecture hall, we walk with an inspiring companion through a nature suffused with new light and color.

7. Homely Facts and Industrial Forecasts

Mendeléev tells us about the chemistry and mechanics of a running mountain stream and the constitution of the cloudy water.

The coarser particles are first deposited ... whilst the clay, owing to its finely divided state, is carried on further, and is only deposited in the still parts of ... the estuaries of rivers, lakes, seas, and oceans ... A detached account of the theory of falling bodies in liquid, and of the experiments bearing on this subject, may be found in my work, *Concerning the Resistance of Liquids and Aeronautics*, 1880 ... Thus gold and other heavy ores are washed free from sand and clay ...[15]

These are homely facts, but they are vital to one who sees the central role of atomic volumes in the establishment of the periodic system, and who has been concerned with volumes and densities since he was twenty:

[I] occupied myself since the fifties (my dissertation for the degree of M.A. concerned the specific volumes, and is partially printed in the Russian Mining Journal for 1856) with the problems concerning the relations between the specific gravities and volumes, and the chemical composition of substances ...[16]

From clay he goes on to describe the different kinds of soil and the reasons for their differing suitability for cultivation, and ends by giving detailed analyses of four soils from very different parts of Russia.[17]

Mendeléev does not finish the chemistry of lead without telling us how white lead is manufactured for paint and how it acquires its covering power.[18] He explains the inflammability of phosphorous, and the operation of safety methods.[19] He describes the precautions to be taken in testing for poisoning by arsenic.[20] In his eyes not one element lacks interesting features, often puzzling features. Mendeléev communicates his omnivorous interest to us, and often gives us his simple explanation for a puzzle.

If a prismatic crystal of sulphur be thrown into one branch of the U tube containing the liquid sulphur at 100°, and an octahedral crystal be thrown into the other branch, then, as Gernez showed, the sulphur in each branch will crystallize in corresponding forms, and both forms are obtained at the same temperature; therefore it is not the influence of the temperature only which causes the molecules of suphur to distribute themselves in one or another form, but also the influence of the crystalline parts already formed. This phenomenon is essentially analogous to the phenomena of supersaturated solutions.[21]

Not only a great investigator, Mendeléev was also a great teacher. To the immense throngs of students who came to his St. Petersburg lectures he gave a new vision of chemistry and at the same time a new view of the future.

There is no doubt that in time the alloys of aluminium will be very widely used... [Aluminium] is distinguished for its capacity to fill up the most minute impressions of the mold into which it may be cast.[22]

... if rich deposits of nickel are discovered a wide field of application lies before them.[23]

... in England alone the annual output of cast iron is above 8 million tons, and that of the whole of Europe and America about 15 million tons. Russia contributes but a small part proportionately – namely, about one-fortieth of the whole production – although the Ural, the Don district, and other parts of Russia represent the combination of all advantageous conditions for the future successful development of a vast iron industry.[24]

The fuel of blast furnaces consists of wood charcoal..., anthracite (for instance, in Pennsylvania and in Russia at Pastouhoff's works in the Don district), coke, coal, and even wood and peat. It must be borne in mind that the utilization of naphtha and naphtha refuse would probably give very profitable results in metallurgical processes. Here experiments are absolutely necessary, and particularly important for Russia, because the Caucasus is capable of yielding vast quantities of naptha.[25]

8. Colleagueship

If Mendeléev gives, he also receives. He not only listens to the quiet voice of nature, he also seeks out and responds warmly to the best thinking of his colleagues. He cites more than 500 authors in a single volume of his book. For example, referring to the numerical value of atomic weights, he notes,

Ten years earlier such knowledge did not exist, as may be gathered from the fact that in 1860 chemist from all parts of the world met at Karlsruhe in order to come to some agreement, if not with respect to views relating to atoms, at any rate as regards their definite representation. Many of those present probably remember how vain were the hopes of coming to an understanding, and how much ground was gained at that Congress by the followers of the unitary theory so brilliantly represented by Cannizzaro. I vividly remember the impression produced by his speeches, which admitted of no compromise, and seemed to advocate truth itself, based on the conceptions of Avogadro, Gerhardt, and Regnault, which at that time were far from being generally recognized. And though no understanding could be arrived at, yet the objects of the meeting were attained, for the ideas of Cannizzaro proved, after a few years, to be the only ones which could stand criticism, and which represented an atom as – 'the smallest portion of an element which enters into a molecule of its compound'.[26]

9. The Periodic Law

Besides judgment Mendeléev had courage. His bold and clear formulation of the periodic law, as he himself puts it, "by insisting on the necessity for the revision of supposed facts, exposed itself at once to destruction."[27] He notes for example,

> In my first memoirs... I particularly insisted on the necessity of altering the then accepted atomic weights of cerium, lanthanum and didymium. Cleve, Höglund, Hillebrand, and Norton, and more especially Brauner, and others accepted the proposed alteration, confirmed my determination of the specific heat of cerium, and gave fresh proofs in favour of the proposed alterations of the atomic weights.[28]

Such examples, numerous though they are, are far from measuring Mendeléev's boldness. Lothar Mayer had made a classification of the known elements, but he had not ventured to predict new elements. Mendeléev did. Eighteen years later (1889) he could say,

> We now know three cases of elements whose existence and properties were foreseen by the instrumentality of the periodic law. I need but mention the brilliant discovery of *gallium* (1875), which proved to correspond to ekaaluminium of the periodic law, by Lecog de Boisvaudran; of *scandium* (1879), corresponding to ekaboron, by Nilson; and of germanium (1886), which proved to correspond in all respects to ekasilicon, by Winkler. When, in 1871, I described to the Russian Chemical Society the properties (density, valences, chemical activity, etc.), clearly defined by the periodic law, which such elements ought to possess, I never hoped that I should live to mention their discovery... as a confirmation of the exactitude and generality of the periodic law.[29]

Our guide stresses that "even now the periodic law needs further improvements in order that it may become a trustworthy instrument in further discoveries."[30]

How prophetic! In taking atomic weight as guide to the construction of his table, Mendeléev was operating, we realize today, not with one periodic system, but two: one for the atoms, the other for the nuclei. The one was being played off against the other, a remarkable balancing act. That out of this double puzzle Mendeléev could come to the correct order of the elements and reach within an ace of the concept of atomic number is a tribute in part to the relatively simple course of the valley of stability[31] in the chart of the nuclei. It is more; it is a tribute to Mendeléev's insight and judgment.

10. THE QUANTUM PRINCIPLE BEFORE THE QUANTUM PRINCIPLE

The periodic law to Dimitri Ivanovitch was not the end of chemical science, but the beginning; it was a clue to the central mystery: the machinery of matter. Ask him to spare a few steps more with us to tell us of his reflections on this mystery. He replies that he cannot accept a static model of a molecule, nor follow chemists who "represent... the interior of molecules... as being in a condition of death-like inactivity."[32] It may or may not be true that "our atoms may... be compared to... solar systems;"[33] but to him, "the conviction that motion pervaded all things... has now extended to the unseen world of atoms."[34] We must "harmonize... chemical theories with the immortal principles of Newtonian natural philosophy, and so hasten the advent of true chemical mechanics."[35]

"Hasten the advent of true chemical mechanics." The words come just as our guide is saying farewell. We hold him a minute more.

A new mechanics inside an atom! What an inspiring thought! How you must have pondered this vision! Can you spare a last word on the central feature of this newness?

Mendeléev turns his eyes to the distance, stands a time silently, then slowly frames his great reply:

... while we admit... throughout the universe a unity of plan, a unity of forces, and a unity of matter... we none the less must explain the individuality and the apparent diversity which we cannot fail to trace everywhere. ... After a long and painstaking research, natural science has discovered the individualities of the chemical elements... Unity and the general, like time and space, like force and motion, vary uniformly. The uniform admit of interpellations, revealing every intermediate phase; but the multitudinous, the individualized – ..., like the chemical elements, ..., like Dalton's multiple proportions – (are) characterized in another way. We see... – side by side with general connecting principles – leaps, breaks of continuity, points which escape from the analysis of the infinitely small – a complete absence of intermediate links. However, the place for individuality is... limited by the all-grasping, all-powerful universal....[36]

And then he vanishes. This is Mendeléev, in 1889 revealing to us the nature of the quantum principle, eleven years before the discovery of the quantum principle! No one came close to being so good a prophet.

Our edition of Mendeléev's book appeared in 1891, at the beginning of the 'great decade': 1895, X-rays; 1896, radioactivity; 1897, the electron; and 1900, the quantum. Turn back in the volume that we left behind to this passage:

The alteration or doubling of the atomic weight of uranium – *i.e.*, the recognition of U=240 – was made for the first time in the first (Russian) edition of this work (1871), ..., because with an atomic weight 120 (the value previously generally accepted by chemists), uranium could not be placed in the periodic system.[37]

Read later in the book that now "its high atomic weight is received without objection, and it endows that element with special interest".[38] And as if Mendeléev himself were staging the surprises to come, read his note, "Becquerel, Bolton and Morton have made some most interesting researches on the phosphorescent spectra of the uranium compounds..."[39]

11. Nuclear physics opening up atomic physics

We know the story[40]: Henri Poincaré and A. H. Becquerel sitting together at a session of the Paris Academy, discussing like everyone else at the time the mystery of the rays just discovered by Konrad Roentgen; the remark of Poincaré to the effect, you have seen and told us in the past about strange radiations from uranium; his query, would it not be worthwhile to see if uranium produces anything like Roentgen's rays; and the sequel: Becquerel's wonderful discovery.[41]

From 1896 and radioactivity to 1926 and the 'true chemical mechanics' envisaged by Mendeléev was an unplanned voyage of discovery, a zig-zag journey from paradox to paradox.

Radioactivity itself was a double paradox. Belying any small vision of 'deathlike inactivity' in the atom, it inaugurated a 'nuclear chemistry' that burst the bounds of molecular chemistry and outdid it in energy by six orders of magnitude. Paradox No. 1: Products of nuclear radioactivity had to be scattered by the nucleus to reveal the planetary structure of the atom. Paradox No. 2: The completely probabilistic character of elementary quantum jumps revealed itself in nuclear transformations a decade sooner than in atomic transitions. The planetary structure plus Earnshaw's theorem, no stable equilibrium for any static system of point charges, led to paradox No. 3: Why do not solids and liquids collapse in 10^{-17} sec? Once confronted by the prediction that all matter should collapse, one had no cheap way out, any more than one has today any evident escape from the issue of gravitational collapse. Collapse was the crisis of 1910 as collapse is again the crisis of 1970.

Don't give up Coulomb's law of force between charged particles.

Don't deny that an accelerated charge radiates. Accept instead the consequences of the inescapable quantum of Planck for the motion of an electron in an atom. With this daring conservatism Bohr resolved the crisis of 'atomic collapse' and opened the door to the understanding of atomic structure. However, molecular structure remained a mystery pending a still deeper elucidation of the quantum principle.

12. Electron orbits and chemical bonds

Many there were in the early 1920's who could not believe that the tetrahedral valences of carbon had the slightest connection with the circular and elliptic orbits of Bohr; and many were deeply convinced that chemistry is chemistry and physics is physics: that chemical forces are chemical forces and electrical forces are electrical forces.

Superpose s- and p-state wave functions to build quantum states of directed valence! That was the key of Heitler and London[42] to unlock the central mystery of chemistry. This idea in its rich unfolding and consequences[43] gave the world at last Mendeléev's 'true chemical mechanics'. In one way this development was an anticlimax. There was no new force hidden in chemistry, nor any new principle. Something so 'accidental' as the near identity in energy of two quantum states allowed the building of a directed probability wave. At issue in producing this degeneracy was only the modicum of energy of binding of the last electrons. The appetite of the atomic field for negative charge had been almost satiated before this point was reached. 'Tiny residuals' was the final diagnosis of 'chemical forces' as compared to atomic binding. Little surprise there was at the end that homopolar bonds should differ in strength by orders of magnitude from Van der Waals couplings.

In another sense 'directed valence' was the climax of four decades of elucidation of the quantum principle. If Darwin reduced the richness of life to the principle of natural selection, the quantum principle reduced the richness of solids, liquids, and gases, and of all chemistry, to the dynamics of a collection of moving point charges.[44] In no way was it required or right to meet each complication of physics or chemistry with a corresponding complication of principle. One had emerged into a world of light, where nothing but simplicity and unity was to be seen. If sovereign light extended its rule so far by capitalizing on the crisis of atomic collapse, how can it fail to seize the crisis of gravitational collapse as opportunity to add elementary particles and the universe itself to its destined empire?

One cannot leave the atom without taking a moment to contrast it with the nucleus. In the nucleus the energy of an individual particle is approximately $E=(2n_r+l+\frac{3}{2})\hbar\omega$. Moreover the effective potential approaches more closely the assumed ideal harmonic potential, $V=\frac{1}{2}m\omega^2r^2$, as the number of particles increases. In contrast, in the same many-particle limit the atomic potential in the bulk of the atom deviates more and more widely from the simple Coulomb law, $V=-Ze^2/r$. The energy levels do not follow the elementary formula $E=-(Z^2e^4/2\hbar^2)(n_r+l+1)^{-2}$. Thus in $_{55}$Cs, $_{56}$Ba, $_{57}$La and the lanthanide series (ends at $Z=71$), the $5d$ and $4f$ levels have nearly the same energy (Table I):

TABLE I

d- and f-levels of the same value $k=n+l=$ $=n_r+2l+1=7$ have nearly the same energy for atomic numbers of the order of $Z=60$

Orbit	n	n_r	l	$k=n+l$
$5d$	5	2	2	7
$4f$	4	0	3	7

13. The 'Chemical Orbit'

The energy of an outer electron is governed neither by $2n_r+l$ nor by n_r+l but in a certain approximation and over a certain range by $k=$ $=n+l=n_r+2l+1$; and not only in this region of Mendeléev's table, but also, as has long been known,[45] in other stages of the level-filling process (from $_{48}$Cd through $_{54}$Xe, comparable energies for $4d$ and $5p$ with $k=n_r+2l+1=6$; and comparable energies for $6d$ and $5f$ with $k=$ $=n_r+2l+1=8$ in a significant part of the range $_{87}$Fr, $_{88}$Ra, $_{89}$Ac and the actinide series; similarly earlier in Mendeléev's Table). It is known that no other local potential admits a level degeneracy so extreme as one finds for the Coulomb and oscillator potentials.[46] Therefore the energy of the electron cannot depend exclusively upon $k=n_r+2l+1$ over the full range of orbits.[47] What kind of power law potential will make the

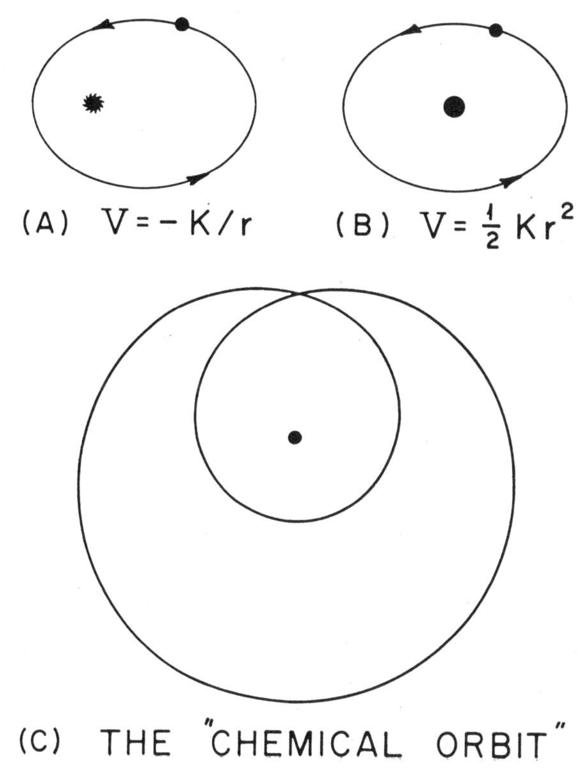

Fig. 2. (a) Kepler orbit. The periods of revolution and of excursion of the radial distance are identical, regardless of the ellipticity of the orbit ('degeneracy of states of the same $n = n_r + l + 1$'). (b) Harmonic oscillator orbit. Two complete radial oscillations in the course of one revolution ('degeneracy of states of the same $2n_r + l$'). (c) Double necklace or 'chemical orbit'. One complete radial oscillation in the course of two revolutions (energy dependent upon $n_r + 2l$).

energy depend upon this particular combination of n_r and l for a limited range of orbits has been analyzed in detail by Robert Powers. He also determines the classical orbits themselves. They are very different for the condition under discussion here from what they are either for the harmonic oscillator potential or the Coulomb potential (Figure 2 and Table II).

It follows from the equation

$$E = E(k) = E(n_r + 2l + 1)$$

that the circular frequency of revolution in the classical approximation,

$$\omega_\theta = \frac{\partial E}{\hbar\, \partial l} = \frac{2E'}{\hbar}$$

is twice the frequency of excursions in the radial direction,

$$\omega_r = \frac{\partial E}{\hbar\, \partial n_r} = \frac{E'}{\hbar}.$$

In other words, the electron, as treated in the semiclassical approximation, must make two complete revolutions to go from perihelion to perihelion: one revolution to go from perihelion to aphelion, and another to go back from aphelion to perihelion (Figure 2). The orbit has the character of a double necklace. Robert Powers and the writer join in calling this the 'chemical orbit'.

TABLE II

'Chemical orbit' compared and contrasted with orbits in Coulomb potential and simple harmonic oscillator potential

Potential	Orbit	Features of orbit preserved over entire semi-classical range of n_r and l?	Reaches r_{max} how many times per circuit of the orbit?	Passes $\theta = 0°$ how many times per circuit of the orbit?
Oscillator	Centered ellipse	Yes	2	1
Coulomb	Kepler ellipse	Yes	1	1
Atom	'Chemical orbit'	Restricted range	1	2

Naturally, no single quantum state ever gives a distribution of probability amplitude like that predicted in the classical chemical orbit. For a semiclassical description of motion in an orbit in the spirit of the correspondence principle, one requires a wave packet. To form the required wave packet we superpose quantum states which have very nearly the same energy, E, by virtue of having the same k value, but values of l

differing by one unit:

$$\psi(r, \theta, g, t) \simeq [u_{L, n_r}(r) (\sin\theta)^L e^{iL\phi} + \\ + u_{L-1, n_r+2}(r) (\sin\theta)^{L-1} e^{i(1-1)\phi}] \, e^{-iEt/\hbar}.$$

What requirements must a point of observation fulfill if it is to maximize this probability amplitude? First, $\sin\theta$ must be as large as possible ($\theta = \pi/2$; equatorial plane; this special orientation for orbit chosen for simplicity of analysis). Second, the remaining factors inside the square bracket should have the same phase (constructive interference!). Specifically each radial wave function can be viewed as a standing wave, the linear superposition of an outgoing wave and an incoming wave. The standard JWKB approximation gives for the phase of the outgoing wave

$$\int [(2m/\hbar^2)(E - V(r)) - L^2/r^2]^{1/2} \, \mathrm{d}r + L\phi,$$

apart from an additive phase constant, irrelevant in the present analysis. Constructive interference between two such outgoing waves with angular momentum quantum numbers L and L-1 does not demand that the phase of either wave individually should be zero. It demands instead that the difference in the phases should vanish at the point of observation. This requirement becomes

$$\int \frac{L \, \mathrm{d}r/r^2}{[(2m/\hbar^2)(E - V(r)) - L^2/r^2]^{1/2}} = \phi \tag{1}$$

where we have replaced the first difference by the derivative evaluated for an angular momentum halfway between L and L-1. We have here the location of the point of constructive interference. The formula just derived is identical with that contained in classical mechanics for the track of the particle in its double necklace orbit. The special feature that distinguishes the chemical orbit from other orbits is the fact that ϕ increases by 2π as r runs from perihelion to aphelion. Thus we have to deal with the left hand side of (1) considered as a definite integral between the two turning points (defined by the vanishing of the denominator). However, this definite integral, divided by π, gives according to classical theory exactly the ratio of the fundamental frequencies for revolution and for excursions of the radius. In summary, the conditions encountered in segments of the periodic table, where orbits of the same k value are filled nearly simultaneously, besides being conditions favorable for formation of states of directed valence, are also conditions favorable for the formation of a wave packet that realizes approximately the chemical orbit, in the sense of Figure 2. A closer analysis shows that there can be two ranges of r values where such motions can occur, separated by a barrier. Moreover, 'electron leakage' can take place through this barrier[48]. So much can one easily do (more in the report of Robert Powers[49]) to tie the particularities and periodicities of chemistry to the continuity of dynamics!

14. Six lessons from the history of the atom for gravitational collapse

In the history of 'atomic collapse' and its epilogue of 'true chemical mechanics' six lessons stand out of relevance to the crisis of gravitational collapse ('much into little'). First, much emerges from little. The physics of solids, gases, liquids and light, as well as chemistry with all its richness, spring out of something so simple as the quantum mechanics of a system of point charges. Second, such astonishing features of nature as chemical valence and bond directivity give little hint of the simplicity of the underlying machinery. Third, 'interactions' as different in strength as ionic forces, homopolar forces, and Van der Waals forces all take their origin in elementary electrostatics. Fourth, to bring to light even the gross anatomy of an atom took probing at the scale of the nucleus (alpha particle scattering). Fifth, most of the dynamics of nature is frozen out. Only so does the individuality of the lowest quantum state so often manage to assert itself. This apparent annihilation of degrees of freedom, seen in a nucleus, in Dirac's sea of negative energy states, in an atom, a molecule, or a crystal lattice, is nowhere more clearly illustrated than in the famous formula for the freezing out of freedom for an oscillator:

$$E = \hbar\omega/(e^{\hbar\omega/kT} - 1) \begin{cases} \nearrow kT \text{ for high } T \text{ (unfrozen)} \\ \searrow \hbar\omega e^{-\hbar\omega/kT} \text{ for low } T \text{ (frozen!)} \end{cases}$$

(Einstein[50]). Sixth, like Merlin the magician, the quantum principle kept changing its shape as it was pursued.

To Mendeléev the Merlin principle was 'individuality limited by the universal'. To Planck the principle appeared as the element of discreteness in oscillator energies. To Rutherford and Soddy the everchanging mystery showed itself in a strictly probabilistic law for radioactive transformations. To Bohr the mutating magic first manifested itself as a principle for the quantization of angular momentum. Then to Bohr and Einstein it showed itself in the guise of transition probabilities. To Sommerfeld it became a prescription for the quantization of action variables. To Heisenberg the quantum principle showed in the non-commutative algebra of physical quantities, and then in the uncertainty principle. To de Broglie and Schroedinger it manifested itself in the laws of standing and running waves. Pursued further by Bohr, Merlin's mystery grew in stature, becoming in turn the 'principle of complementarity' and the 'principle of wholeness'. To Feynman it became the principle of 'sum of probability amplitudes over all histories', with a democratically equal weight for every history. Today all these forms are recognized as belonging as separate attributes to one logically self-consistent but still not fully apprehended grandeur. The greatness of an advance in physics may be measured by the enlargement it makes in our vision of the quantum principle!

15. Much out of little

No firing line where a new advance may be sought calls more clearly today than the crisis of gravitational collapse, nor holds out greater challenges, nor promises more lessons. (1) Much out of little? The collapse of a star is of a kind with, and a 'laboratory' model for, the collapse envisaged by Einstein for the universe itself.[51] It is therefore also a model, at the other end of time, for the original burst that brought into being the present universe and everything within it, 'processed' through fantastically small dimensions and fantastically high densities. The stakes in the analysis are hard to match: the dynamics of the largest object, space, and the smallest object, an elementary particle – and how both began.

(A) There is no 'spontaneous generation'.
(B) But life began!

(A) There is no 'creation or destruction of baryons'.
(B) But baryons began!

16. Astonishing Features from Simple Machinery

(2) Astonishing features from simple machinery? The central doctrine is Einstein's standard battle-tested 1915 geometrodynamics.[52] It is the first description of nature to make geometry a part of physics. No purported discrepancy with its predictions has ever stood the test of time. No logical inconsistency in its foundations has ever been detected. No acceptable alternative has ever been put forward of comparable simplicity and scope. It predicted something so preposterous as the expansion of the universe[53] before that expansion had been observed[54]. It predicted that the time linearly extrapolated back to the start of the expansion ('Hubble time', $t_H = H^{-1}$) should be greater than $1.5t_0$, where t_0 is the actual time back to the start of the expansion (effect of slowing down of expansion) at a date when the measurements indicated $t_H < 0.3t_0$ (era of theories of 'continuous creation of matter' and 'steady state universe') and years before the more than five-fold error in the Hubble scale of distances had been discovered. In general relativity the equation for the motion of a particle in a known field has been shown not to be needed as a special postulate, as in every other formulation of physics; it follows as a consequence of Einstein's geometrodynamic field equations themselves. Einstein's theory counts the topology as well as the geometry of space as information to be derived out of nature itself, not as decrees of Euclid. The existence of electric charge no longer requires one to assume that Maxwell's equations fail at certain points, or to accept the hypothesis that there exists in nature a magic electric jelly beyond further explanation. Electricity admits of interpretation as lines of force trapped in the topology of a multiply connected space[55].

Whether matter also can be understood as a manifestation of curved empty space, as proposed by W. K. Clifford of the Clifford algebras at the Cambridge Philosophical Society, 21 February 1870, is one of the greatest of unanswered questions. It is of no help in answering this question to know that a geon[56] in principle can be constructed out of a collection of electromagnetic radiation or gravitational radiation. Such an object, to admit of classical analysis, must be more massive than the sun!

At small distances the natural object of attention is not the deterministic evolution of the geometry of space with time as predicted by Einstein's equations, but the quantum fluctuations, $\delta g_{\mu\nu}$, in this geometry in a region of observation of extension L,

$$\delta g_{\mu\nu} \sim L^*/L,$$

where

$$L^* = (\hbar G/c^3)^{1/2} = 1.6 \times 10^{-33} \text{ cm}$$

is the Planck length. These fluctuations occur everywhere throughout all space. At distances of the order of the Planck length they are so large ($\delta g \sim 1$) that no natural escape is evident from concluding that fluctuations occur as well in the topology of space as in its geometry. On this view those geometries of space which occur with appreciable probability amplitude are characterized everywhere by the most varied submicroscopic structure and varied topology ('foamlike structure of space in the small'). For the effective energy density of these fluctuations order-of-magnitude estimates give

$$\begin{aligned}\rho &= (\text{energy density}/c^2) \sim \hbar c/L^{*4}c^2 \\ &= c^5/\hbar G^2 \sim 10^{95} \text{ g cm}^{-3}.\end{aligned}$$

A bit of nuclear matter with its density of $\sim 10^{14}$g cm^{-3} is completely unimportant by comparison. A particle means less to the physics of the vacuum than a cloud (10^{-6} g cm^{-3}) means to the physics of the sky (10^{-3} g cm^{-3}). No single fact points more powerfully than this to the conclusion that a 'particle' is not the right starting point for the description of nature.

If a particle is of geometrodynamical origin, it is not a 'wormhole' in the geometry of space ($\sim 10^{-33}$ cm), for its extension is enormous by comparison ($\sim 10^{-13}$ cm). Moreover its energy ($\sim 10^{-27}$ g to 10^{-24} g) is negligible by comparison with the energy associated with a single 'wormhole' (the Planck mass-energy, $(\hbar c/G)^{1/2} = 2.2 \times 10^{-5}$ g). Neither can it be be a classically describable deformation in the geometry of space (energy density negligible compared to the effective energy density of the quantum fluctuations!) No possibility has ever presented itself for a quantum geometrodynamic interpretation of a particle but this: the particle is not any individual 10^{-33} cm fluctuation in the geometry of space; instead, it is a fantastically weak alteration in the pattern of these fluctuations, extending over a region containing very many such 10^{-33} cm regions. In brief, a particle is a quantum state of excitation of the geometry; it is a *geometrodynamical exciton* [55]. If this updating of the vision of Clifford and Einstein makes sense, then one can summarize its content in one sentence: The physics of particles is the chemistry of the continuum.

17. Interactions of Varied Strength from a Single Origin

(3) Interactions of the most varied strength derive from a single elementary origin? Bring together two regions, each of extension 10^{20} times as great as the size of a single wormhole, and let the pattern of fluctuation be modified in each as would be suggested by the concept of 'exciton'. Then in the region of overlap the pattern will be modified still further. However small the resulting fractional alteration in effective energy density of fluctuations may be in the region of overlap, that basic density itself is so enormous ($\sim 10^{95}$ g cm^{-3}) that resulting net energy of interaction can as well have one magnitude as another. Thus the possibility would not seem to be excluded to conceive of strong forces, weak forces, and intermediate forces as no more distinct in their character than Van der Waals forces, ionic force, and valence forces. On this interpretation elementary particle interactions are not primordials; they are residuals.

18. Work up from Small Distances

(4) Work up from small distances rather than work down from large distances? If one had first to know of the existence of the nucleus before he could make progress on the constitution of the atom, it would not seem unnatural to recognize the fluctuation in the geometry of space at the Planck scale of distances as the first step in understanding the constitution of a particle. However, between the one layer of structure at 10^{-33} cm and the other at 10^{-13} cm it is not clear in advance how many intermediate layers of structure interpose themselves to complicate the situation. It is conceivable that the situation is much simpler than one might fear. In any case three remarks are in order about the significance of the Planck length. First, it measures the unavoidable indeterminism imposed by the uncertainty principle on the geometry of space (known precisely

at one instant, it cannot be predicted precisely at the next!) In effect the phenomenon of gravitational collapse is going on all the time at this submicroscopic scale of distances, and all the time is being undone again. Second, as collapse proceeds the effective radius of a star, or of the universe itself, drops faster and faster, according to classical geometrodynamics. However, it is difficult to imagine that any such deterministic dynamics continues when the dimensions fall below the Planck value. What sets in is more reasonably believed to be much the same aimless doing and undoing of collapse that characterizes the fluctuation regime, followed eventually (in the case of the universe) by a reexpansion stripped of all memory of the phase of contraction. Third, any truly quantitative treatment of either the fluctuations or the probabilistic coupling between collapse and reexpansion would seem to be excluded until the analysis can take into account fluctuations as well in topology as in geometry. Physics burst out of the straightjacket of Euclidean geometry in coming to an understanding of gravitation. That it must also unlock itself from the fixed connectivity postulated by classical differential geometry would seem an inescapable consequence of the quantum principle. If in coming to terms with 'indeterminism' one also finds it unavoidable to abandon the concept of the dimensionality of space at the smallest distance, and retreat to the position that 'three' has a meaning only on the average, that development will come as no surprise ('pregeometry'[57]). Thus, when a handle in the topology breaks, the handle becomes thinner and thinner until at last only a single point remains to connect two fingers of space. Then even that contact ends. Two points that were neighbors have parted company. However, no such transition can be discontinuous, according to quantum theory. The two points must retain some residual connection after they have separated. If these two points have a residual connection, then every two points must have some connection, one with another – a view of the structure of space very different from that supplied by classical differential geometry!

19. The Freezing Out of Dynamics

(5) Most of the dynamics 'frozen out'? What goes on at the scale of the Planck distance, and distances larger by many orders of magnitude, has associated with it characteristic quantum energies of the order of 10^{28} eV

and many powers smaller but, even so, far too high to be unfrozen by any devices at the command of physics today. The vacuum is going to go on looking as innocent of structure to the explorer of the atom and the nucleus as a sheet of glass emerging from the rolling mill looks innocent of structure to the operator!

20. An Enlarged Vision of the Quantum Principle

(6) An enlargement in our vision of the quantum principle? None has come out of general relativity – yet! History has gone rather the other way around. The quantum principle has been more effective than any single force in the last two decades in driving one into a deeper understanding of the content of Einstein's standard geometrodynamics – even at the classical level. That understanding, won through the labors of many investigators,[58] is summarized by no phrase so well as the single word 'superspace'.[59]

Superspace is the arena in which the dynamics of space unfolds. One point in superspace represents one 'geometry' or one special distribution of curvature over a closed 3-dimensional manifold. Another point in superspace represents another such '3-geometry'. Figure 3 illustrates a simplified version of superspace. It is endowed with 98 dimensions, whereas superspace proper has infinite dimensionality. A single point in this 98-dimensional space, via its projections on the 98 coordinate axes, determines in one stroke 98 'edge lengths', $L_1, L_2, \ldots, L_{98}$. Ninety-eight 'bones' with precisely these lengths build up (Regge[60]) a skeleton 3-geometry (simplified to a 2-geometry in Figure 3 for ease of visualization). Make a small displacement of the representative point in the abbreviated 98-dimensional superspace. Then the 98 edge lengths all change by a small amount. The skeleton 3-geometry undergoes a small deformation. No better illustration can one easily supply as to what it means to speak of the 'dynamics of *space*'!

In all the difficult investigations that led at length to the understanding of the dynamics of geometry, both classical and quantum, the most difficult point was also the simplest: the dynamic object is not spacetime. It is space. The geometrical configuration of space changes with time. But it is space, three-dimensional space, that does the changing (compare with particle dynamics, Table III).

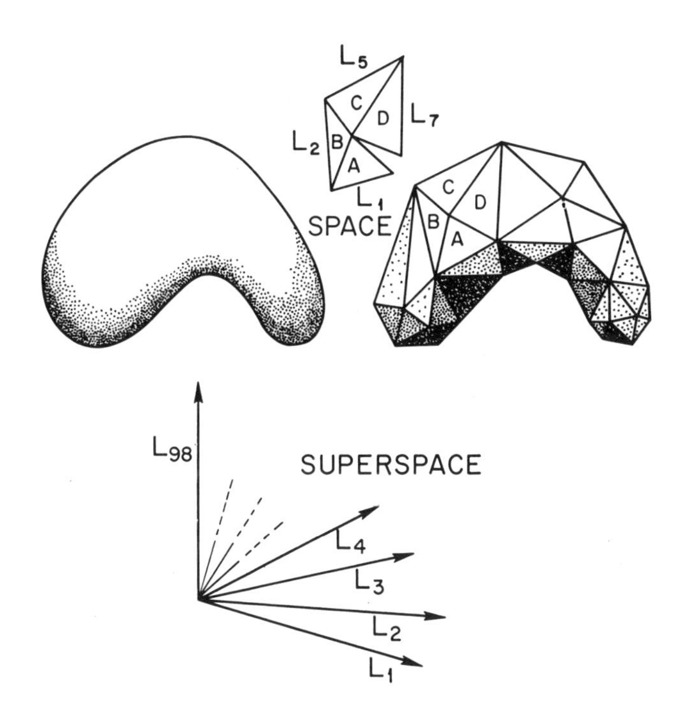

Fig. 3. Simplified version of superspace. Upper left: a 2-geometry (a stand-in for the 3-geometry of the Einstein universe). Upper right: approximation of this 2-geometry by decomposition into triangles (these 2-simplexes replaced by 3-simplexes (tetrahedrons) in the actual analysis!). The curvature at any vertex (cf. insert) is governed by the nearest neighbor lengths, and by nothing more. The geometry in the simplical approximation is completely fixed by the specification of the 98 edge lengths. Equivalently (lower diagram) the geometry is completely specified by a single point in a space of 98 dimensions. To go to superspace proper (replacement of simplicial decomposition by actual geometry) one has to go from 98-dimensional space to ∞-dimensional space.

TABLE III

Geometrodynamics compared with particle dynamics

Concept	Particle Dynamics	Geometrodynamics
Dynamic entity	Particle	Space
Descriptors of momentary configuration	x, t('event')	${}^{(3)}\mathcal{G}$('3-geometry')
History	$x=x(t)$	${}^{(4)}\mathcal{G}$('4-geometry')
History is a stockpile of configurations?	Yes. Every point on world line gives a momentary configuration of particle	Yes. Every spacelike slice through ${}^{(4)}\mathcal{G}$ gives a momentary configuration of space
Dynamic arena	Spacetime (totality of all points x, t)	Superspace (totality of all ${}^{(3)}\mathcal{G}$'s)

Space, the dynamic entity, and '3-geometry' or 'metric', the precise measure for the momentary configuration of space ('distribution of curvature over space'), have sharply defined meanings as well in quantum geometrodynamics as in classical geometrodynamics. However, given the precise 3-geometry at one instant, one cannot predict ahead to the precise configuration of space at the next instant in quantum theory. The principle of indeterminism forbids. Therefore the concept of spacetime is deprived of any well-defined meaning in quantum physics. In contrast, 'spacetime' does have a sharp content in classical physics. It records the deterministic classical history of space changing with time. Only in a classical approximation does it make sense to say that there exists any such object as 'spacetime', standing like a rigid framework or a marble tablet, ready to record every event, past, present and future, with the Einstein interval from each event to its neighbor eternally established.

Go from classical spacetime as one way of specifying a classical history, to a 'leaf cutting through superspace' (Figure 4) as another way of specifying the same classical history. Then turn attention to the quantum description and see why one cannot return from superspace to spacetime! In the spacetime depicted schematically at the right in Figure 4, each spacelike slice, such as A, through the 4-geometry represents an entire 3-geometry, a momentary configuration for space, a single point in superspace (lower part of Figure 4). The given spacetime, the given 'deterministic classical history of space evolving with time', admits many such slices, A, B, B′,...... etc. For clarity one may speak of these as 'YES 3-geometries'. This term distinguishes them from the enormously more numerous 'NO 3-geometries' which can never be obtained, try as one will, by making spacelike slices through the *given* 4-geometry, the *given* classical history. The YES 3-geometries, depicted as points in superspace, do not fall on a single line.

No single 'time' parameter has the power to order the YES 3-geometries into a one-parameter family. The reason is simple. 'Time' in general relativity is not a one-parameter concept. It is a 'many-fingered time'. From a given spacelike slice one can move attention forward a little way into spacetime in an infinitude of ways – 'pushing the hypersurface ahead in time' at one rate here, at another rate there. Therefore the given classical history of space, the given spacetime, the given 4-geometry, is mapped, not into a line, but into a leaf cutting through superspace.[61]

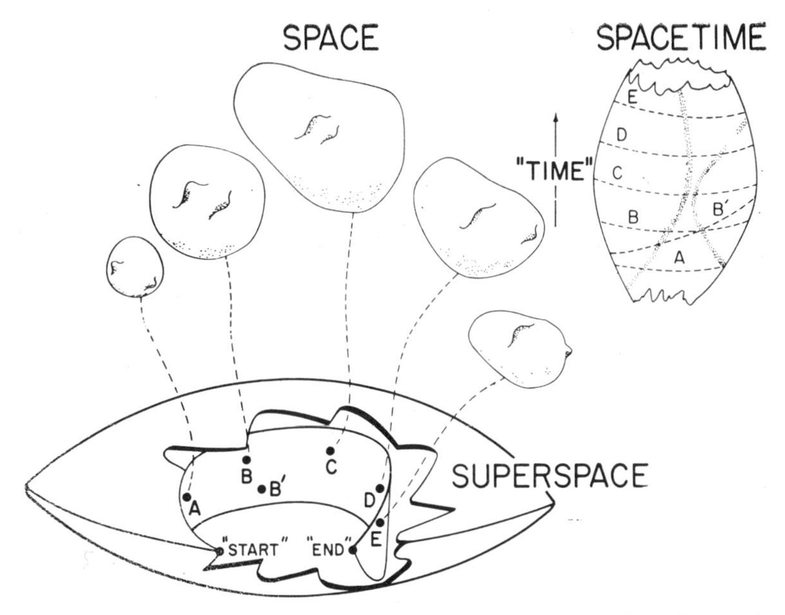

Fig. 4. Space, spacetime and superspace. Upper left: five sample configurations attained by space in the course of its expansion and recontraction. Upper right: spacetime. A spacelike cut, like A, through spacetime, gives a momentary configuration of space (upper left: the early 3-geometry A). Below: superspace. The one bent 'leaf' in superspace (running from 'START' to 'END') comprises all those 3-geometries which are obtainable as spacelike slices through the given spacetime.

Given the leaf of history cutting through superspace, one can go back and reconstruct the original classical spacetime in three steps: first, from leaf to the points on the leaf; second, from each of these representative points to the corresponding 3-geometry; finally, assemble these 3-geometries into the 4-geometry, as a child assembles and internests a multitude of boxes into a rigidified construction.

Quantum physics does away with the sharp distinction between 'YES 3-geometries' and 'NO 3-geometries'. It speaks of a probability amplitude $\psi = \psi({}^{(3)}\mathscr{G})$ for this, that and the other 3-geometry. It contrasts strikingly with classical geometrodynamics, with its sharp confinement to a single leaf in superspace. Quantum geometrodynamics gives an equation (last three books cited in [58]) for the propagation of the probability amplitude $\psi({}^{(3)}\mathscr{G})$ throughout superspace. Almost every 3-geometry has a non-zero probability amplitude. The ${}^{(3)}\mathcal{G}$'s that occur with significant probability amplitude do not fit and cannot be fitted into any spacetime, any 4-geometry, any classical history of space evolving dynamically with many-fingered time. That ${}^{(4)}\mathcal{G}$, that 'magic structure' of classical physics, simply does not exist. Without that building plan to organize the ${}^{(3(}\mathcal{G}$'s of significance into a definite relationship, one to another, even such a familiar notion as the 'time ordering of events' is devoid of all immediate significance. Only under circumstances where one can neglect the quantum-mechanical spread of the probability wave-packet in superspace can one treat the dynamical evolution of geometry in the deterministic context of classical general relativity[62].

Nowhere do these consequences of the quantum principle have more immediate application than in the phenomenon of gravitational collapse. Nowhere are the differences greater between the classical and the quantum pictures of the dynamics. A computer following the collapse according to the sharply deterministic prescription of classical geometrodynamics comes to the point where it cannot go on (heavy arrow in Figure 5 leading directly to the singular collapse domain of superspace). This prediction can only be compared with a prediction of classical mechanics. A particle headed directly towards a point center of Coulomb attraction will arrive in a finite time at a condition of infinite kinetic according to Newtonian theory. In actuality the particle is described by a wave packet. It experiences, not deterministic collapse, but probabilistic scattering. No more for the universe than for the particle can one believe that physics comes to the limit of its predicative power! Little one may know about some central details of quantum geometrodynamics, among them especially how to describe quantum mechanical transformations of the connectivity of space. This circumstance in no way impairs one's perception of the broad outline of quantum geometrodynamics and the central lesson of superspace: the universe begins a new cycle of expansion from the same singular region of superspace where collapse lost its way. Moreover, there is not a unique history that leads out of this singular region. There is a probability distribution of histories. Each of these alternative cycles of the universe has a different set of determinants: a different volume at the phase of maximum expansion and a different lapse of time from the start of expansion to the end of recontraction. No escape is evident from this picture of what goes on: the universe transforms, or transmutes, or 'transmigrates' probabilistically from one cycle of history to another in the era of collapse.

If to describe this process requires one to burst the bounds of classical differential geometry, this will not be the first time that physics has given the lead to mathematics, nor the quantum principle the lead to physics.

However straightforwardly and inescapably this picture of the transformations of the universe would seem to follow from general relativity and the quantum principle, the two overarching principles of 20th century physics, it is nevertheless fantastic to contemplate. How can the dynamics of a system so incredibly gigantic be switched, and switched at the whim of probability, from one cycle that has lasted 10^{11} years to another that

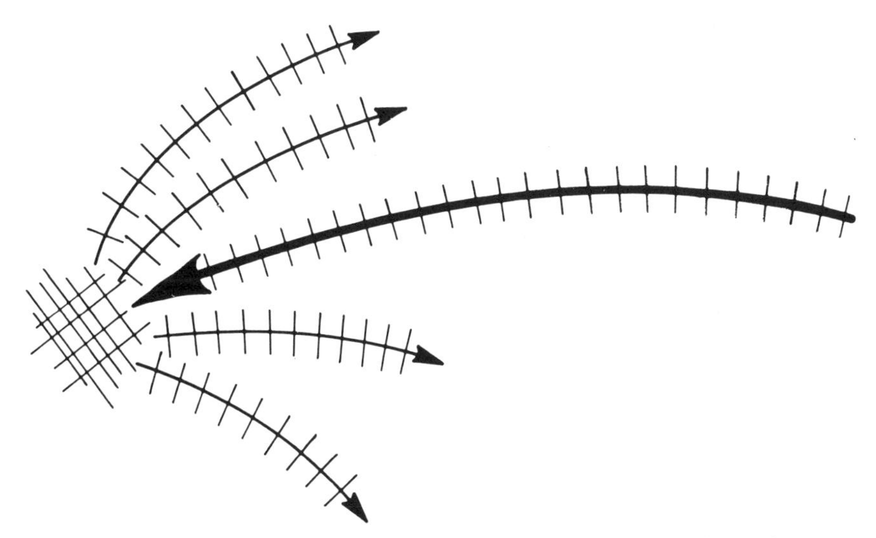

Fig. 5. Collapse depicted symbolically as coupling one cycle of expansion and recontraction of the universe (heavy 'line' in superspace) in region of collapse (crosshatched part of superspace) with other cycles of the universe (alternative 'histories of the geometry of space' depicted as 'lines' emergent from region of collapse). A fuller picture, still in the context of classical geometrodynamics, would show each one dimensional 'line' through superspace expanded to a multidimensional 'leaf of history' as shown in the lower part of Figure 4. This deterministic track through superspace is replaced in quantum general relativity by a propagating wave packet (symbolized by wave crest in diagram). Any deterministic classical analysis breaks down in the crosshatched 'region of complete collapse' (dimensions of universe of order of the Planck length $(hG/c^3)^{1/2} = 1.6 \times 10^{-33}$ cm). What goes on there is envisaged as analogous (a) to the scattering of the de Broglie wave of an electron by the Coulomb field of a point charge or (b) to the coupling of waves in separate wave guides at a wave guide junction.

will last only 10^6 years? At first only the circumstance that the system gets squeezed down in the course of its dynamics to incredibly small distances reconciles one to a transformation otherwise so unbelievable. Then one looks at the upended strata of a mountain slope, or a bird not seen before, and marvels that the whole universe is incredible:

mutation of a species
metamorphosis of rock
chemical transformation
spontaneous transmutation of a nucleus
'transmigration' of the universe.

If it cast a new light on geology to know that rocks can be raised and lowered thousands of meters and hundreds of degrees, what does it mean for physics to think of the universe as from time to time 'squeezed through a knot-hole', drastically 'reprocessed', and started out on a fresh dynamic cycle? Four considerations press themselves to one's attention. (a) No known quantity is conserved in the dynamic evolution of a closed model universe endowed with no special symmetry. Thus, there is no platform on which to stand 'outside the universe' to measure its mass-energy or angular momentum, via effects on Kepler orbits or otherwise. Therefore it is not surprising that the energy and angular momentum of the universe are not even defined, let alone conserved. Consequently it is difficult to see what determinant could continue unchanged in value from cycle to cycle of the universe, even any determiner of the spectrum and masses of the elementary particles.

21. Reprocessing: matter: molecules, nuclei, particles

(b) Molecules are reprocessed in a flame to new molecules. Nuclei are reprocessed in a star to new nuclei.[63] In the overwhelming rage of gravitational collapse what escape is there from believing that the spectrum of elementary particles is reprocessed (Table IV) to a new spectrum? Must not the particles of today be viewed as fossils from that early turmoil as the molecules and nuclei of today are also viewed as fossils from events of lesser violence? With no determinant left alive from cycle to cycle of the dynamics to perpetuate the spectrum of the particles, no alternative is evident but to conclude that that spectrum must be extin-

TABLE IV

Lesson, as envisaged here, from reexamining collapse of universe predicted by classical theory, compared and contrasted with lesson learned in past from reexamining classically predicted collapse of atom

System	Atom	Universe
Dynamic entity	System of electrons	Geometry of space
Nature of classically predicted collapse	Electron headed towards point center of attraction is driven in a finite time to infinite energy	Not only matter but space itself arrives in a finite proper time at a condition of infinite compaction
One rejected 'way out'	Give up Coulomb law of force	Give up Einstein's field equations
Another rejected proposal	'Accelerated charge need not radiate'	'Matter cannot be compressed beyond a certain density by any pressure, however high'
How this proposal violates principle of causality	Coulomb field of point charge cannot readjust itself with infinite speed out to indefinitely great distances to sudden changes in volocity of charge	Speed of sound cannot exceed speed of light; pressure cannot exceed density of mass-energy
A major new consideration introduced by recognizing quantum principle as overarching organizing principle of physics	Uncertainty principle; binding too close to center of attraction makes zero-point kinetic outbalance potential energy; consequent existence of a lowest quantum state; can't radiate because no lower state available to drop to	Uncertainty principle; propagation of representative wave packet in superspace does not lead deterministically to a singular configuration for the geometry of space; leads rather to a probability distribution of outcomes, each outcome describing a universe with a different size, a different set of particle masses, a different number of particles, and a different length of time required for its expansion and recontraction

guished and start with a new pattern after each collapse. One is reminded of the fixity of geological strata between one period of the metamorphosis and the next!

On this view the present spectrum of particle masses, and even the net numbers of baryons and leptons in the universe today, cannot be regarded as fundamental constants of nature. They are rather to be understood as in the nature of initial conditions relevant to this cycle of the universe alone. The so-called 'big numbers' ($\sim 10^{80}$ particles in the universe; $\sim 10^{40}$ ratio of radius of universe at maximum expansion to effective size of an elementary particle; $\sim 10^{20}$ ratio between elementary particle dimensions and the Planck length)[64] have never received a physical explanation. They never will, if they are initial value data! Physics can unravel dynamic law, but it has never found what principle fixes the other half of dynamics, the 'initial conditions' of nature.

Physical explanation for the large numbers, no; biological explanation, perhaps. Brandon Carter has given reasons to believe that determinants substantially different from those that characterize the present cycle of the universe would have made the development of life difficult if not impossible.[65]

(c) No debate of old seems more fruitless on rereading than this: can there exist a multiplicity of universes?[66] Warned by the inconclusive nature of the arguments advanced on both sides, how can a thinking man return today to that old quagmire of reason? Today it is not a matter of choice. The superspace of Einstein's standard 1915 geometrodynamics, one now realizes, compels one to ask how one cycle of the dynamics of the universe is coupled to another, and even supplies the beginning of a framework for the answer. Is there no way, however, to tie down such cosmic considerations to something more concrete? Happily there is.

The collapse of the white dwarf core of a star to a black hole provides a kind of 'laboratory model' in which one can hope to see and test some of the effects predicted in the collapse of the universe itself. Of these effects none is more striking than the wiping out of every possibility to tell how many baryons and how many leptons went in (Figure 1). To start with particles, and end with something indistinguishable from pure geometry, will be encouragement to believe that nature can start with pure geometry and end with particles. To start with something that resists collapse and to end with everything crushed will allow one to continue to believe that the universe itself must collapse.

No one can accept the geometrodynamic interpretation of particles if no one can verify predictions about black holes. Happily three ways offer themselves to test for the existence of these objects and to study their properties: pulses of gravitational radiation given out at the time of formation; X-rays given out by gas heated by compression as it converges onto a black hole; and activity associated with black holes in galactic nuclei.[67]

(d) The black hole does not model in every respect the dynamics of the universe. It has an outside, a far away asymptotically flat 'platform', where the observer, as payment for not being caught up in the collapse himself, also cannot see the final stages of the collapse, much less any reexpansion. Closely associated with this circumstance is the onesidedness in time of the dynamics of the black hole: it absorbs particles and quanta of radiation, but never emits them.

This onesidedness in time is to be read not out of Einstein's field equations, but out of experience.[68] One does not need equations to know that heat flows from hot to cold, or that radiation diverges outward from an accelerated charge, but does not converge inward from infinity upon it. What is true of faraway infinity must be true also of 'captured infinity', if one may be permitted to give that name to the horizon of a black hole. It can receive. It cannot give (except slowly: S. W. Hawking, 1974).

Nowhere in physics can one name a process where friction ('onesidedness in time') more dominates dynamics than in the fusion of two black holes: fusion takes place, but fission does not. Is some small modicum of this onesidedness in time inherited by particles, that they should violate CP-invariance?[69] Is the process of CP-violation a small premonitory signal of the deeper world of geometric physics, as Becquerel's radioactivity betrayed the then hidden world of nuclear physics?

Does the universe today contain equal amounts of matter and antimatter? Or does matter everywhere have the same composition which it does in this vicinity? This question, always interesting, also seemed at one time a deep issue of principle. Not so now! Given equal amounts of the two kinds of matter, in principle one can arrange for a black hole to swallow all the antimatter, leaving only matter behind. No amount of inspection of the black hole will ever reveal the secret of what it did with a once symmetric past.

22. FROM EINSTEIN'S GEOMETRODYNAMICS TO AN ENLARGED VISION OF THE QUANTUM PRINCIPLE

All of these considerations came up in asking, what does the quantum principle have to contribute to geometrodynamics? Turn back now to the original formulation of question 6. What can geometrodynamics do to unveil more of the content of the quantum principle?

23. Everett's 'Relative State' Formulation of Quantum Mechanics

(a) To ask about the 'quantum state of a closed universe' is to displace the 'observer' from his usual position outside the system under study. He is inside. No one has found a way to deal rationally with this kind of situation except through Everett's 'relative state formulation' of quantum mechanics.[70] It abandons the postulate that an observation starts the system off in a fresh quantum state. Instead, it envisages the wave function after an 'observation' as the sum of terms ('branches of history'), each the product of (i) a factor describing the non-observer part of the system in one quantum state, and (ii) a factor describing the observer in a corresponding 'relative state'. All branches of the history are viewed as 'coexisting' in an ethereal sense, and only in an ethereal sense, a sense perhaps best described by William James.[71] "Actualities", he wrote, "seem to float in a wider sea of possibilities from out of which they were chosen; and *somewhere*, indeterminism says, such possibilities exist, and form part of the truth."

How can these considerations be brought to a sharp focus and definitive resolution? For this purpose no issue would seem more timely or more compelling than the quantum mechanics of the universe. Here, in the context of superspace, one finds oneself confronted not only with the coupling between alternative histories of the universe, but even with the 'coexistence' of these histories.[72]

24. Monads and Fluctuations

(b) Of all the developments that Leibniz gave the world, from the calculus in its modern form to the principle of least action, and from the principle of sufficient reason to the concept of the best of all possible worlds, there was none on which he laid greater weight, and none more puzzling to his contemporaries, than the idea of a world made of monads,[73] each of which however subsumes inside itself the whole world, as a microcosm.[74] A fluctuation in the geometry of space at the Planck length is a small scale version of the expansion and recontraction of the universe itself. Is the similarity between the two levels of dynamics more than superficial? Are the two really one and the same? Is this the message of the conformal invariance that stares out from so much of physics,[75] as from a friend without vocal chords whose intelligence one vainly tries to hear? Are particles identical because each carries within itself the same image of the universe in travail giving birth to them all?[76] One will have a larger view of the quantum principle when one knows the answers to these questions!

25. MUCH INTO LITTLE AS THE KEY TO WINNING MUCH OUT OF LITTLE

From Mendeléev's atom to the collapsing star, from the dynamics of a system of particles to superspace and the dynamics of geometry, a central theme stands out: much out of little, richness from simplicity, individuality out of continuity, Pythagoras and Plato reconciled in the principle of Planck; and the emblem of this theme? A continuous potential binding a continuous wave with a discrete number of nodes and a discrete energy value![77]

Much into little is the other theme: *collapse*; the collapse of the atom that was not collapse; the collapse of the star and the universe itself, rated as inescapable by Einstein's battle-tested theory; the collapse satisfying the guiding principle of Bohr's life work: no progress without a paradox!

The paradox of atomic collapse, attacked with the weapon of the quantum principle, led to 'chemical mechanics'. No cheap way offered itself out of that paradox. None offers itself now out of the new paradox, crisis in the theoretical physics of our day. Not long can one believe it will be until the advance of physics and astrophysics brings direct evidence on black holes to contradict or confirm a central prediction and, either way, sharpen the crisis.

Black holes exist, if standard 1915 general relativity and the simplest elements of astrophysics are right. They are formed from time to time. There is no disposable parameter for a black hole except its mass, charge, and angular momentum. In a black hole all memory is erased of the particularities of the particles that went to build it. In effect a collection of $\sim 10^{57}$ particles is ground down into geometry. Einstein's theory says more: that the universe itself with its $\sim 10^{80}$ particles is also ground down.

Collapse at both levels gives incentive to consider afresh the vision of Clifford and Einstein, that every particle is constructed out of geometry; or, in updated language, a particle is a 'geometrodynamic exciton'. The atom is preposterous, with its unbelievable wealth of chemistry built on the dynamics of a few electrons. The nucleus is preposterous, with its many-radiant myriad of energy levels built on the motions of a few nucleons. Can the elementary particle be less preposterous? A Sibyl seems to say, "Choose: paradox or nothing!" Nothing offers for building a particle except the dynamics of space itself. Nothing offers for giving discreteness to this dynamics except the quantum, Mendeléev's 'individuality amid continuity'.

The arena for the quantum dynamics of geometry is superspace. Superspace is forced on physics by the union of the quantum principle and standard Einstein geometrodynamics. The mathematical framework of superspace, though understood in broad outline today, still requires some augmentation to accommodate the quantum fluctuations in the topology of space that seem so inescapable at the Planck scale of distances.

As viewed in the context of superspace, gravitational collapse appears as a probabilistic process of scattering (Figure 5). In each such act of 'collapse scattering', space is reprocessed. It is transformed. It 'transmigrates'.

Each new cycle of the universe, according to these views, begins with a new spectrum of particle masses. To specify the determiners of the new spectrum is to give initial conditions for the dynamics of the new expansion, a prediction beyond the power of physics to make. The probabilistic character of the 'collapse scattering' forbids. On this view it is unrealistic to believe that one will ever be able to calculate the masses of the elementary particles from the principles of physics. To ask physics to give these masses would seem a mistaken question. A better question has rather another turn. Among all the cycles of the universe that run through their histories in superspace, long cycles and big universes being less probable than short cycles and little universes, in which do the determinants have values which will permit life as we know it?[78] And what is the spectrum of masses that goes with this set of determinants?

Preposterous it surely is to imagine that the dynamics of the universe can have anything to do with the mass of the proton. Nothing could be more preposterous – except to suppose that the mass of the proton has nothing to do with the dynamics of the universe! The crisis of gravitational collapse has launched the physics of our time on an incredible journey, surely a new zig-zag course from paradox to paradox.

If we want a guide on this odyssey of exploration, one who had a serene and happy approach to mystery, an eye for what is central, the energy to work around and around the vital paradox wherever work was possible, and the sense of harmony to recognize simplicity in the midst of complexity, who better can we choose for our patron saint than Dimitri Ivanovitch Mendeléev?

ACKNOWLEDGMENT

I wish to thank my colleages, especially V. Bargmann, Ugo Fano, Robert Geroch, V. I. Gol'danski, Gertrude and Maurice Goldhaber, S. Gorodetsky, Robert Powers, Emilio Segré, and Claudio Teitelboim for many discussions. I also wish to express my appreciation to S. A. Goudsmit for emphasizing in a conversation some years ago the then already long known approximate dependence of energy upon the combination $n_r + 2l$.

Preparation of this report was assisted in part by National Science Foundation Grant GP 7669 to Princeton University.

Princeton University

* Reprinted from *Atti del Convegno Mendeleeviagno, Accademia delle Scienze di Torino – Accademia Nazionale dei Lincei – Torino-Roma, 15–21 Settembre 1969*, Torino, Vincenzo Bona, 1971.

[1] For an account of gravitational collapse, including a summary of the pioneering contributions of L. D. Landau, S. Chandrasekhar, J. R. Oppenheimer, R. Serber, G. Volkoff, H. Snyder and others, see for example Chapter 6 in B. K. Harrison, K. S. Thorne, M. Wakano and J. A. Wheeler, *Gravitation Theory and Gravitational Collapse*, University of Chicago Press, Chicago, 1965; or K. S. Thorne, 'Relativistic Stellar Structure and Dynamics', a section in *High Energy Astrophysics*, Proceedings of Course XXXV of the International School of Physics 'Enrico Fermi', ed. by L. Gratton, Academic Press, New York, 1967, pp. 167–280. For more on many physical and astrophysical issues connected with the topic see Ya. B. Zel'dovich and I. D. Novikov, 'Relativistic Astrophysics', *Usp. Fiz. Nauk* **84** (1964) 377 and **86** (1965) 447 (English translation in *Sov. Phys. Usp.* ~**6**

(1965) 763 and **8** (1966) 522) and their book on relativistic astrophysics, the English version of which, *Stars and Relativity*, translated by Eli Erlock and edited by K. S. Thorne and W. D. Arnett, was published in 1971 by the University of Chicago Press.

[2] The geometry associated with a black hole endowed with angular momentum, an exact solution, like the Schwarzschild geometry, of Einstein's field equations for a region of space ('outside the horizon') free of all matter, was first reported by R. P. Kerr, *Phys. Rev. Lett.* **11** (1963) 237 and generalized to include charge by E. T. Newman *et al.*, *J. Math. Phys. N.Y.* **6** (1965) 918. A detailed analysis of this geometry is given by B. Carter, *Phys. Rev.* **174** (1968) 1559.

[3] The mass of a black hole follows from the period of a test particle in Keplerian orbit around it. The angular momentum follows from the difference in rates of precession of two two orbits, one corevolving, the other counterrevolving. The charge follows from an application of the theorem of Gauss. Why should one not then be able to determine still other features of the black hole by still other measurements on test particles outside; and, most immediately, the number of leptons which have collapsed into the black hole by way of the standard (l/r^5)-interaction between lepton and lepton (cf. especially G. Feinberg and J. Sucher, *Phys. Rev.* **166** (1968) 1638 and J. B. Hartle, *Phys. Rev.* **D1** (1970) 394; (note also that the sign reverses when one of the leptons is replaced by its antiparticle!) intermediated by the exchange of two neutrinos? This important question of principle was raised by James B. Hartle in a discussion with the writer at Santa Barbara. Further discussion made it appear reasonable that no influence would remain behind to reveal to the faraway test lepton the lepton number of the particle falling into the black hole. First, there is no Gauss theorem or analogous conservation law that applies to the (l/r^5)-interaction. Second, that potential describes an interaction between essentially static leptons, whereas the lepton being captured is in effect fleeing with a speed ever closer to the speed of light. Summer 1970 calculations kindly communicated by Professor Hartle in a preprint of a paper to be submitted by him to *Phys. Rev.* now give a field theoretical analysis showing that the lepton-lepton interaction is extinguished as one of the leptons approaches the horizon of the black hole, and giving a formula for the rate of this extinction.

[4] For evidence on the conservation of baryons in elementary particle processes, see F. Reines, C. L. Cowan, Jr., and M. Goldhaber, *Phys. Rev.* **96** (1954) 1157; F. Reines, C. L. Cowan, Jr. and H. W. Kruse, *Phys. Rev.* **109** (1957) 609; G. N. Flerov, D. S. Klochov, V. S. Skobkin and V. V. Terentiev, *Sov. Phys. Dokl.* **3** (1958) 78; the review of G. Feinberg and M. Goldhaber, 'Microscopic Tests of Symmetry Principles', *Proc. U.S. Nat. Acad. Sci.* **45** (1959) 1301 and 'Experimental Tests of Symmetry Principles', *Science* **129** (1959) 1285; and H. S. Gurr, W. R. Kropp, F. Reines and B. Meyer, 'Experimental Test of Baryon Conservation', *Phys. Rev.* **158** (1967) 1321. Appreciation is expressed to Gertrude and Maurice Goldhaber for a 16 June 1970 phone conservation in which they note the new figure of 10^{21} yr for the spontaneous fission of Th^{232}. Hollander and Kalman (*Table of Isotopes*, 6th ed.) enable one to deduce from the 1954 results of Reines, Cowan, and Goldhaber a limit of 10^{23} yr for the life of a baryon against an 'inconspicuous' mode of disappearance, not very far from the limit set on the life against any conspicuous mode of disappearance. The author is also indebted to Frederick Reines for the following remarks in a 19 June 1970 letter from him regarding the limits that one can place on a 'silent process' of baryon disappearance:

"Let us consider two possibilities which together, I believe, give the least restrictive and hence most believable test of baryon conservation to date under terrestrial conditions. Assume that a proton (or a neutron) disappears from a nucleus and the universe in some undefined manner leaving behind a nucleus with a hole in it; i.e. missing a proton (or a neutron). No assumption as to the validity of the other conservation laws is made in this view. It is, however, assumed that what is left behind behaves normally.

In this circumstance, if the original nucleus is a deuteron, the disappearance of a proton would leave a neutron behind. Therefore, detection of a neutron capture pulse in a suitably designed deuterated system could signal the deuteron breakup – or at least set a lower limit for the occurrence of such a process. An experiment of this kind was actually performed at my suggestion by F. Dix at Case Western Reserve University. He obtained a limit $>3\times10^{23}$ yr (90% confidence level) for the process

$$D \to n + ?$$

The case of neutron disappearance can be signaled by the radioactivity of the residual nucleus such as

$$C^{12} \to C^{11} + ?$$
$$C^{11} \to B^{11} + e^{+} + \nu_e \text{ (1.98 MeV total)}$$
20 min.

Two pieces of data can be used to establish a limit for neutron disappearance $>10^{20}$ yr. One is the residual rate seen (>1.5 MeV) in a 500 gallon liquid scintillation detector located in a salt mine. The other is the positron annihilation rate seen in the old Savannah River neutrino experiment where the target was water

$$O^{16} \to O^{15} + ?$$
$$O^{15} \to N^{15} + e^{+} + \nu_e$$
2 min.

I believe it possible to significantly improve these limits if one wishes to do so."

[5] For evidence of the law of conservation of leptons in elementary particle transformations, see M. Goldhaber, 'Weak Interactions: Leptonic Modes – Experimental', a report in *1958 Annual International Conference on High Energy Physics at CERN*, Proceedings, pp. 233–250; V. R. Lazarenko, *Usp. Fiz. Nauk* **90** (1961) 601 (English trans. in *Sov. Phys. Usp.* **9** (1967) 860); B. Pontecorvo on neutrino experiments and the problem of conservation of leptonic charge in *Sov. Phys. JETP* **26** (1968) 984; K. Bohrer *et al.* on the conservation of μ-lepton number, *Helv. Phys. Acta* **43** (1970) 111; and C. Franzinetti, 'Experimental Limits on the Validity of Conservation Laws in Elementary Particle Physics', in *Atti del Convegno Mendeleeviano, Accademia delle Scienze di Torino – Accademia Nazionale dei Lincei, Torino – Roma, 15–21 Settembre, 1969*, Torino, Vincenzo Bona, 1971.

[6] Plato (428 B.C.–348 B.C.), *Collected Dialogs* (ed. by Edith Hamilton and Huntingdon Cairns), Bollingen Series LXXI, Bollingen Foundation, New York, p. 1177: "For if the matter were like any of the supervening forms, then whenever any opposite or entirely different nature was stamped upon its surface, it would take the impression badly, because it would intrude its own shape. Wherefore that which is to receive all forms should have no form."

[7] Pythagoras (~530 B.C.), "all things are numbers".

[8] Dimitri Ivanovitch Mendeléev (Tobolsk, 7 Feb. 1834 – St. Petersburg, 20 Jan. 1907), *Principles of Chemistry* (transl. from the Russian 5th ed. by George Kamensky, ed. by A. J. Greenaway, in 2 Vols), Longmans, Green and Co., London and New York, 1891; cited hereafter as PC I or PC II. The first Russian edition dates from 1869.

[9] PC II, p. 407.

[10] D. I. Mendeléev, *The Periodic Law of the Chemical Elements:* Faraday lecture delivered before the Fellows of the Chemical Society in the theatre of the Royal Institution on Tues-

day, June 4, 1889; reprinted in PC II, pp. 435ff., and referred to hereafter as FL. The cited passage occurs on p. 436.

[11] Charles Darwin (Shrewsbury, 12 Feb. 1809–Down, 19 April 1882).

[12] C. Darwin, *Origin of Species*, 24 Nov. 1859.

[13] Alfred Russel Wallace (1823–1913), 'On the Tendencies of Varieties to Depart Indefinitely from the Original Type', Feb. 1858.

[14] J. Lothar Meyer (1830–1895), *Die modernen Theorien der Chemie*, Breslau, 1864.

[15] PC II, pp. 67–68.

[16] PC II, p. 31.

[17] PC II, p. 69.

[18] PC II, p. 135.

[19] PC II, p. 149.

[20] PC II, p. 176.

[21] PC II, p. 198.

[22] PC II, p. 82.

[23] PC II, p. 345.

[24] PC II, p. 316.

[25] PC II, p. 310.

[26] PC II, p. 327.

[27] PC II, p. 439.

[28] PC II, p. 89.

[29] PC II, p. 447.

[30] PC II, p. 447.

[31] 'Valley of stability', in G. Gamow, *Structure of Atomic Nuclei and Nuclear Transformations*, Oxford University Press, 1937.

[32] PC II, p. 418.

[33] PC II, p. 418.

[34] PC II, p. 417.

[35] PC II, p. 419.

[36] PC II, pp. 407–408.

[37] PC II, p. 287.

[38] PC II, p. 449.

[39] PC II, p. 288.

[40] I owe the story to the kindness of Professor Gorodetsky of the University of Strassbourg at Warsaw 17 October 1967 who tells me that he heard it in turn from the son of Becquerel. Another part of the story recounts the admiration that the elder Becquerel expressed on his return from London to Paris for the then very young Rutherford, "Il y a en Angleterre un jeune homme qui devine tout."

[41] H. A. Becquerel, *C.R. Acad. Sci. Paris* (1896) pp. 122, 420, 501, 559, 689, 762, 1086.

[42] W. Heitler and F. London, *Z. Phys.* **44** (1927) 455 and later papers.

[43] See for example L. Pauling, *The Nature of the Chemical Bond*, Cornell University Press, Ithaca, New York, 1960, 3rd ed.

[44] P. A. M. Dirac in 'The Quantum Mechanics of Many-Electron Systems', *Proc. Roy. Soc. London Ser. A* **123** (1929) 714 writes, "The underlying physical laws necessary for the mathematical theory of a large part of physics and the whole of chemistry are thus completely known, and the difficulty is only that the exact application of these laws leads to equations much too complicated to be soluble."

[45] For a quick survey, see for example L. D. Landau and E. M. Lifshitz, *Quantum Mechanics*, Oxford 1958, p. 235, 245 or E. U. Condon, chapter on 'Quantum Mechanics and

Atomic Structure', *Handbook of Physics* (ed. by E. U. Condon and H. Odishaw) McGraw-Hill, New York, 2nd ed., 1967, pp. 7–22. For his pioneering 1926 correlation between the order of shell-filling and $k=n+l$, see the early reference to Madelung on p. 670 in S. A. Goudsmit and P. I. Richards, *Proc. U.S. Nat. Acad. Sci.* **51** (1964) 664, who also stress this correlation; see also E. Madelung (himself), *Die mathematischen Hilfsmittel der Physikers*, Springer, Berlin, 1950, p. 611. (These references from E. Neubert, kind personal communication and *Z. Naturforsch.* **25a** (1970) 210.) For fuller references to the literature on the theory of shell-filling see A. Sommerfeld, *Wave Mechanics*, Vol. II, 1933; Yeou Ta, *Ann. Phys. Leipzig* **1** (1946) 88; L. Simmons, *J. Chem. Educ.* **24** (1947) 588; R. Hakala, *J. Phys. Chem.* **56** (1952) 178; V. M. Klechkovski, *Dokl. Akad. Nauk SSSR* **80** (1951) 603; **86** (1952) 691; **92** (1953) 923; and **135** (1960) 655; also *Zh. Fiz. Khim.* **27** (1953) 1251 and *Zh. Eksp. Teor. Fiz.* **25** (1953) 179; V. I. Goldanskii, *J. Chem. Educ.* **47** (1970) 406. For the order of filling of levels as given in the Fermi-Thomas statistical atom model (cf. for example the treatise by P. Gombas, *Die statistische Theorie des Atoms und ihre Anwendungen*, Springer, Vienna, 1949) see A. Sommerfeld, *Atombau und Spektrallinien, Wellenmechanische Erganzungsband*, Braunschweig, Vieweg, 1921; M. G. Mayer, *Phys. Rev.* **60** (1941) 184; D. Ivanenko and S. Larin, *Dokl. Akad. Nauk SSSR* **88** (1953) 45; V. M. Klechkovskii, *Dokl. Akad. Nauk SSSR* **92** (1953) 923; *Zh. Eksp. Teor. Fiz.* **26** (1954) 760; **30** (1956) 199 (English translation in *Sov. Phys. JETP* **3** (1956) 125; R. Latter, *Phys. Rev.* **99** (1955) 510; T. Tietz, *Ann. Phys.* **15** (1955) 186 and **5** (1960) 237; and V. M. Klechkovskii, *Zh. Eksp. Teor. Fiz.* **41** (1961) 465 (English translation in *Sov. Phys. JETP* **14** (1962) 334).

[46] For some of the analysis of potentials in which a particle has energy levels of high degeneracy, see W. Lenz, *Z. Phys.* **24** (1924) 197; V. A. Fock, *Z. Phys.* **98** (1935) 145; D. I. Fivel, 'Solutions of Scattering and Bound State Problems by Construction of Approximate Dynamical Symmetries', University of Maryland Technical Report No. 460, May 1965; H. Bacry, H. Ruegg and J. M. Souriau, *Commun. Math. Phys.* **3** (1966) 323; R. Hermann, *Lie Groups for Physicists*, W. A. Benjamin, New York, 1966; P. B. Guest and A. Børs, *Proc. Phys. Soc. London* **92** (1967) 525; and J. S. Alpher and O. Sinanoğlu, *Phys. Rev.* **177** (1969) 77.

[47] Note on quantum defect and chemical orbit. At the conclusion of this report Emilio Segré raised with the writer the question whether out of the formula often employed in spectroscopy for the correlation of atomic energy levels, $E=-R/(n-\Delta)^2$, one can read out under appropriate circumstances a constancy of E with l for fixed $n+l=n_r+2l+1$. The answer turns out to be yes for even one of the simplest and oldest of the models of how the effective atomic field increases in strength as the electron reaches deeper into the atom: $V_{\text{eff}}(r)=-e^2/r-B\hbar^2/2mr^2$. Here for a certain range of quantum states one takes B to be an appropriately chosen constant. The added term is equivalent in its consequences to a decrease of the coefficient in the repulsive potential of the centrifugal force from $l(l+1)$ to $l^*(l^*+1)=l(l+1)-B$. Add $\frac{1}{4}$ to both sides of this equation and take the root. In this way find that the Rydberg formula undergoes the transformation $E=-R/(n_r+\frac{1}{2}+l+\frac{1}{2})^2\rightarrow$

$$\rightarrow E=-R/(n_r+\tfrac{1}{2}+l^*+\tfrac{1}{2})^2=-R/(n_r+\tfrac{1}{2}+l+\tfrac{1}{2}-\Delta_l)^2.$$

In this model the familiar calculated quantum defect is $\Delta_l=(l+\frac{1}{2})-[(l+\frac{1}{2})^2-B]^{1/2}$. Identical energy for two states means identical corrected Rydberg denominator. If in a certain range of atomic number the $5d$ and $4f$ states are to have nearly the same energy, then the difference between the values of $n=n_r+l+1$ for those two states, $n_d-n_f=5-4=1$, must be compensated by the difference in quantum defects, $\Delta_d-\Delta_f=1$. This requirement leads to the equation $[\frac{5}{2}-(\frac{25}{4}-B)^{1/2}]-[\frac{7}{2}-(\frac{49}{4}-B)^{1/2}]=1$. From it one finds $B=6$ for the states in question and for the range of atomic numbers in question (formula meaningless for s and

p states for this B value!). The calculated individual quantum defects in the example are $\Delta_d = 2.0$, $\Delta_f = 1.0$. For a brief account of empirical data on quantum defects and screening constants, see for example *Handbook of Physics* (ed. by E. U. Condon and H. Odishaw) McGraw-Hill, New York, 2nd ed., 1967, Table 2.2 on pp. 7–40 in the chapter by J. R. McNally, Jr., 'Atomic Spectra' and Table 8.2 on pp. 7–129 and Fig. 8.1 on pp. 7–130 in the chapter of E. U. Condon, 'X-Rays'; and for more see especially C. E. Moore, *Atomic Energy Levels*, National Bureau of Standards Circular No. 467, U.S. Government Printing Office, Washington, D.C., 1949; and S. T. Manson, 'Dependence of the Phase Shift on Energy and Atomic Numbers for Electron Scattering by Atomic Fields', *Phys. Rev.* **182** (1969) 97, Table I. Appreciation is expressed to Ugo Fano for reference to this last work.

[48] Note on 'electron leakage' as a periodic phenomenon yet to be observed: periodic variation with atomic number as it shows itself in chemical binding and directions of valence bonds is not the only type of periodicity. Ionization potentials also vary periodically. So does the calculated cross-section (1) for scattering of low energy electrons (data of Ramsauer and Kollath and others, and quantum mechanical interpretation thereof, as summarized in N. F. Mott and H. S. W. Massey, *The Theory of Atomic Collisions*, Oxford University Press, London, 1965, 3rd ed.; see also L. B. Robinson on Ramsauer resonances in the Fermi-Thomas atom model, *Phys. Rev.* **117** (1960) 128 and S. T. Manson, *Phys. Rev.* **182** (1969) 97) and (2) for absorption of light by an inner electron of an atom (U. Fano and J. W. Cooper, cited in the following reference). In this connection A. R. P. Rau and Ugo Fano (*Phys. Rev.* **167** (1968) 7) emphasize, and give data to support the conclusions that, "properties that depend primarily on *different layers* of the atomic structure are seen to attain extreme values at *different columns* of the periodic table." One of the most interesting of effects is the double minimum in the effective potential experienced by an electron of angular momentum $l = 3$ (and other l values). At the outer minimum the Coulomb attraction of the almost fully screened nucleus dominates. At the barrier separating the two minima the centrifugal potential dominates. At the inner minimum the attraction of the almost unscreened nucleus dominates. The barrier is predicted by the elementary Fermi-Thomas atom model (E. Fermi in *Quantentheorie und Chemie*, Leipziger Vorträge (ed. by H. Falkenhagen) S. Hirzel Verlag, Leipzig, 1928, p. 95, reprinted in *Collected Papers of Enrico Fermi*, University of Chicago Press, Chicago, 1962, p. 291; M. Goeppert-Mayer, *Phys. Rev.* **60** (1941) 184; R. Latter, *Phys. Rev.* **99** (1955) 510; and further works cited in reference 1 of the paper of R. T. Powers, *Mendeléev Proceedings*, 1970) to vary smoothly with atomic number, Z. In contrast, Rau and Fano point out, "the height and the very existence of the barrier depend upon Z nonmonotonically... (and) maximum barrier heights occur for Cu, Ag (or Pd) and Au." This barrier powerfully influences the scattering of slow electrons and the absorption of light by inner electrons, as emphasized by Rau and Fano, and considered in more detail by Fano and J. W. Cooper (cited by Rau and Fano). In addition another effect ('electron leakage') has to be anticipated when the barrier summit exceeds the ionization limit, a condition that is met for f-electrons over a significant range of Z-values. An electron of positive energy can be trapped inside the potential barrier, as an alpha particle of positive energy is trapped inside the nuclear potential. As the leakage of the alpha particle out through the barrier manifests itself in alpha radioactivity, so the quantum mechanical tunneling of the atomic barrier will lead to 'electron leakage'. For more than one excited state of a nucleus one sees evidence of the competition between two modes of decay: alpha emission vs. emission of a gamma-ray. In the kind of excited state of an atom under discussion here there will be a similar competition between electron and photon emission (electron dropping from f-state to an unoccupied d-state). In consequence the spectral line will have a breadth $\Gamma = \Gamma_{\text{rad}} + \Gamma_{\text{leak}}$ over and above what would

be expected from the rate of radiation alone. It is not clear that this effect has ever been observed in an atom. It is to be distinguished from two well-known effects, predissociation in a molecule (atoms 'leaking apart' by penetration through molecular potential barrier) and autoionization or Auger effect in an atom (one electron dropping to a tightly bound state, giving enough energy to another electron to eject it from the atom, a two-electron process in contrast to the one-electron process of 'electron leakage'). For some of the literature on autoionization see the original paper of Russell and Saunders on Russell-Saunders coupling, where they speak of evidence for states beyond the ionization limit; the work of A. G. Shenstone on the spectrum of copper; the section on autoionization on pp. 7–52 in the chapter of J. R. McNally, Jr. *Handbook of Physics* (ed. by E. U. Condon and H. Odishaw) McGraw-Hill, New York, 2nd ed., 1967 and comments on barrier penetration in the 1935 book of Condon and Shortley cited there; the 'Introductory Remarks' of U. Fano in *Radiation Research 1966*, North-Holland, Amsterdam, 1967; and the recent review article by U. Fano, 'Doubly Excited States of Atoms' in *Atomic Physics*, Plenum Press, 1969). It would seem possible to make approximate estimates of the mean life $\tau_{\text{leak}} = \hbar/\Gamma_{\text{leak}}$ with respect to this process by straightforward application of the Gamow theory of decay.

$$I = (1/\hbar) \int_{\text{barrier}} (2m)^{1/2} (V_{\text{eff}} - E)^{1/2} \, \mathrm{d}r$$

with the Hill-Wheeler correction,

$$\tau_{\text{leak}}^{-1} = \Gamma_{\text{leak}}/\hbar = v_{\text{class}}/(e^{2I} + 1)$$

making use of the atomic potentials listed by F. Herman and S. Skillman (*Atomic Structure Calculations*, Prentice-Hall, Englewood Cliffs, New Jersey, 1963 as employed by S. T. Manson and J. W. Cooper, *Phys. Rev.* **165** (1968) 126, by A. R. P. Rau and U. Fano, *Phys. Rev.* **167** (1968) 7, and by S. T. Manson, *Phys. Rev.* **182** (1969) 97). Compared to this leakage rate it is difficult to name any quantity characteristic of an atom which will vary periodically through the atomic table over a greater range of orders of magnitude. The very smallness of some of the expected line widths would seem to pose difficulties for the observation of the resonances expected in (a) the inverse process of electron scattering, and to make more appropriate as means to detect and observe 'electron leakage' in its 'direct outgoing form': (b) measurements of line broadening and (c) direct measurements of lifetimes of the 'leakage states'.

[49] Robert Powers, 'Frequencies of Radial Oscillation and Revolution as Affected by Features of a Central Potential', in *Atti del Convegno Mendeleeviagno, Accademia delle Scienze di Torino – Accademia Nazionale dei Lincei, Torino – Roma, 15–21 Settembre 1969*, Torino, Vincenzo Bona, 1971.

[50] A. Einstein, 'Die Plancksche Theorie der Strahlung und die Theorie der spezifischen Wärme', *Ann. Phys. Leipzig* **22** (1907) 180–190'

[51] Einstein counted relativity theory as consisting not only of the field equations, but also of something in the nature of a boundary condition, based upon Mach's principle, requiring that the universe be closed, and leading to the consequence that the universe expands, reaches a maximum dimension, and recontracts. In this connection see the remarks of Einstein at the end of the chapter on Mach's principle in his book, *The Meaning of Relativity*, Princeton University Press, Princeton, New Jersey, 3rd ed., 1950, p. 107. For more on Mach's principle envisaged as a boundary condition for the solution of Einstein's field equations, see the relevant parts in the chapters by J. A. Wheeler, and the references cited there, in the books *Relativity, Groups and Topology* (ed. by C. and B. DeWitt), Gordon and Breach, New York, 1964 and *Gravitation and Relativity* (ed. by H.-Y. Chiu and W. F. Hoffmann), W. A. Benjamin, New York, 1964.

[52] For an account of Einstein's theory oriented towards modern physical applications see for example C. W. Misner, K. S. Thorne and J. A. Wheeler, *Gravitation*, Freeman, San Francisco, 1973.
[53] A. Friedmann, "Über die Krümmung des Raumes", *Z. Phys.* **10** (1922) 377.
[54] E. P. Hubble, 'A Relation between Distance and Radial Velocity Among Extra-Galactic Nebulae', *Proc. Nat. Acad. Sci. U.S.* **15** (1929) 168–173.
[55] J. A. Wheeler, 'Superspace and the Nature of Quantum Geometrodynamics', *Battelle Rencontres: 1967 Lectures in Mathematics and Physics* (ed. by C. DeWitt and J. A. Wheeler), Benjamin, New York, 1968. This paper, augmented, appears in German as the book *Einsteins Vision*, Springer, Berlin, 1968. For literature on the topological interpretation of electric charge see reference 36 in either the paper or the book.
[56] Geons: J. A. Wheeler, *Geometrodynamics*, Academic Press, New York, 1962.
[57] For tentative attempts to assess the nature of 'pregeometry', see J. A. Wheeler, 'Geometrodynamics and the Issue of the Final State', in *Relativity, Groups and Topology* (ed. by C. DeWitt and B. DeWitt), Gordon and Breach, New York, 1964, pp. 335, 495–500; 'Particles and Geometry', in *Relativity* (ed. by Carmeli, Frick and Witten) (1969 Cincinnati conference), Plenum Press, New York, 1970, p. 40; and report, 'Pregeometry', presented at Gwatt, Bern, 15 May 1970 (unpublished). The author is indebted to Robert Geroch for pointing out that topology as studied in many modern contexts does not deal with points individually, but with collections or sets, *called* collections of points, even though one never narrows down the focus to the sharpness where one deals with any individual point as such. This noted, he points out, one recognizes that certain relations that might otherwise seem obvious really require proving along such lines as those spelled out for example by J. W. Kelley, *General Topology*, Van Nostrand, Princeton, New Jersey, 1955. In connection with this thought that the concept of set lies deeper than the concept of point, Valentine Bargmann kindly stresses the interest that attaches to Alexander lattices (J. W. Alexander, 'The combinatorial theory of complexes', *Ann. Math.* **31** (1930) 294 and later publications).
[58] Quantization of general relativity in a context not specially emphasizing that it deals with the dynamics of *three*-dimensional geometry: L. Rosenfeld, *Ann. Phys. Leipzig* **5** (1930) 113; *Z. Phys.* **65** (1930) 589; *Ann. Inst. Henri Poincaré* **2** (1932) 25; P. G. Bergmann, *Phys. Rev.* **75** (1949) 680; P. G. Bergmann and J. H. M. Brunings, *Rev. Mod. Phys.* **21** (1949) 480; Bergmann, Penfield, Schiller, and Zatzkis, *Phys. Rev.* **78** (1950) 329; P. A. M. Dirac, *Can. J. Math.* **2** (1950) 129; F. A. E. Pirani and A. Schild, *Phys. Rev.* **79** (1950) 986; P. Bergmann, *Helv. Phys. Acta*, Suppl. IV, (1956) 79; *Nuovo Cimento* **3** (1956) 1177; *Rev. Mod. Phys.* **29** (1957) 352; C. W. Misner, *Rev. Mod. Phys.* **29** (1957) 497; B. S. Dewitt, *Rev. Mod. Phys.* **29** (1957) 377; P. A. M. Dirac, *Proc. Roy. Soc. London Ser. A*, **246** (1958) 326, 333; *Phys. Rev.* **114** (1959) 924; B. S. De Witt, 'The Quantization of Geometry', in *Gravitation: An Introduction to Current Research* (ed. by L. Witten), Wiley, New York, 1962; J. Schwinger, *Phys. Rev.* **130** (1963) 1253; **132** (1963) 1317: R. P. Feynman, mimeographed letter to V. F. Weisskopf dated 4 January 18–11, February 1961; *Acta Physica Polonica* **24** (1963) 697; *Lectures on Gravitation* (notes mimeographed by F. B. Morinigo and W. G. Wagner, California Institute of Technology, 1963); report in *Proceedings of the 1962 Warsaw Conference on the Theory of Gravitation* (PWN-Editions Scientifiques de Pologne, Warszawa, 1964); S. N. Gupta, report in *Recent Developments in General Relativity*, Pergamon, New York, 1962; S. Mandelstam, *Proc. Roy. Soc. London Ser. A* **270** (1962) 346; *Ann. Phys.* **19** (1962) 25; J. L. Anderson in *Proceedings of the 1962 Eastern Theoretical Conference* (ed. by M. E. Rose), Gordon and Breach, New York, 1963, p. 387; I. B. Khriplovich 'Gravitation and Finite Renormalization in Quantum Electrodynamics' (mimeographed report, Siberian Section Academy of Science, U.S.S.R., Novosibirsk, 1965); H. Leutwyler,

Phys. Rev. **134** (1964) B1155; B. S. De Witt, 'Dynamical Theory of Groups and Fields', in *Relativity Groups and Topology* (ed. by C. De Witt and B. De Witt) Gordon and Breach, New York, 1964; S. Weinberg, *Phys. Rev.* **135** (1964) B1049, **138** (1965) B988, and **140** (1965) B516; M. A. Markov, *Progr. Theor. Phys. Yukawa Suppl.* (1965) 85.

Quantization recognizing the *three*-dimensional character of the dynamic entity: P. W. Higgs, *Phys. Rev. Lett.* **1** (1958) 373, and **3** (1959) 66; R. Arnowitt, S. Deser and C. W. Misner, a series of papers summarized in 'The Dynamics of General Relativity', in *Gravitation: An Introduction to Current Research* (ed. by L. Witten), Wiley, New York, 1962; A. Peres, *Nuovo Cimento* **26** (1962) 53; R. F. Baierlein, D. H. Sharp, and J. A. Wheeler, *Phys. Rev.* **126** (1962) 1864; cf. also the Princeton A. B. Senior Thesis of D. H. Sharp, May 1960 (unpublished); J. A. Wheeler, *Geometrodynamics*, Academic Press, New York, 1962; 'Geometrodynamics and the Issue of the Final State', in *Relativity, Groups and Topology* (ed. by C. De Witt and B. De Witt), Gordon and Breach, New York, 1964; B. De Witt, *Phys. Rev.* **160** (1967) 1113, **162** (1967) 195, and **162** (1967) 1239; J. A. Wheeler, 'Superspace and the Nature of Quantum Geometrodynamics', in *Battelle Rencontres: 1967 Lectures in Mathematics and Physics* (ed. by C. De Witt and J. A. Wheeler), W. A. Benjamin, New York, 1968 (appears in expanded form, in German, as J. A. Wheeler, *Einsteins Vision*, Springer, Berlin, 1968); and chapters by J. R. Klauder, A. Komar, J. A. Wheeler, C. W. Misner, and B. De Witt, and especially the chapter by A. Fischer, 'The Theory of Superspace', in (ed. by M. Carmeli, S. I. Fickler and L. Witten), *Relativity* Plenum Press, New York, 1970.

[59] 'Superspace': phrase inspired not least by listening to V. Bargmann, B. De Witt and C. W. Misner expound at various times the content of the Arnowitt-Deser-Misner canonical formulation of general relativity; first mention found in print, pp. 453, 459, 463, and 495 of 'Geometrodynamics and the Issue of the Final State', cited in [58].

[60] For a complete formulation of classical geometrodynamics in the language of a skeleton 4-geometry, see T. Regge, *Nuovo Cimento* **17** (1961) 558. For more on this Regge calculus, see pp. 463–494 in 'Geometrodynamics and the Issue of the Final State', cited in [58]; and for an actual application on a computer, see C.-Y. Wong, 'Application of Regge Calculus to the Schwarzschild and Reissner-Nordstrøm Geometries at the Moment of Time Symmetry', accepted fall 1970 for publication in *J. Math. Phys.* For the earlier background of 'skeletonization' as relevant in topology and differential geometry, see especially S. Lefschetz, *Algebraic Topology*, Amer. Math. Soc., New York, 1942. In quantum geometrodynamics, where one recognizes that the dynamic object is 3-geometry, the skeletonized manifold differs from that considered by Regge only in this, that it has three dimensions, rather than four. In either three or four dimensions one has options as to the density of skeletonization – denser here, thinner there – for a specified number of bones. Thus some of the displacements in the 98-dimensional version of superspace illustrated in Figure 3 represent, not true deformations, but mere readjustments as between one place and another in the density of coverage. By adopting a prescription as to how density of vertices will be adjusted to magnitude of curvature, one will wring out these surplus degrees of freedom from the analysis and end up with an abbreviated superspace with a much reduced number of dimensions. In Regge calculus, which is entirely free of coordinates, the variety of ways to reskeletonize one and the same geometry bears some similarities to the freedom that one has as to choice of coordinates in the Ricci calculus of conventional differential geometry.

[61] There is an infinitude of ways to make a small deformation in a 3-geometry and yet leave its representative point on the leaf. There is a higher order of infinity for the number of ways to deform the 3-geometry so that its representative point moves off the leaf into the surrounding NO-portion of superspace (infinite dimensional 'leaf of history' imbedded

in a superspace with an overridingly infinite dimensionality).

[62] These considerations reveal that the concepts of spacetime and time itself are not primary but secondary ideas in the structure of physical theory. These concepts are valid in the classical approximation. However, they have neither meaning nor application under circumstances when quantum-geometrodynamical effects become important. Then one has to forgo that view of nature in which every event, past, present, or future, occupies its preordained position in a grand catalog called 'spacetime'. There is no spacetime, there is no time, there is no before, there is no after. The question what happens 'next' is without meaning.

[63] Long before there was any direct evidence for the thermonuclear reactions, it will be recalled, Eddington gave reasons why it was inescapable that nuclear reactions deep in the interior of a star must be responsible for its output of energy, and remarked that if a critic did not believe the temperature there to be high enough, "let him go and find a hotter place."

[64] A. S. Eddington, *Relativity Theory of Protons and Electrons*, Cambridge Univ. Press, 1936, and *Fundamental Theory*, Cambridge Univ. Press, 1946; also *Proc. Cambr. Phil Soc.* **27** (1931) 15; P. A. M. Dirac, *Nature* **139** (1937) 323; *Proc. Roy. Soc. London Ser. A* **165** (1938) 199; P. Jordan, *Schwerkraft und Weltall*, Vieweg und Sohn, Braunschweig, 1955, and *Z. Physik* **157** (1959) 112; R. H. Dicke, *Science* **129** (1959) 3349; *Nature* (1961) 440; *The Theoretical Significance of Experimental Relativity*, Gordon and Breach, New York, 1964, p. 72; S. Hayakawa, *Progr. Theor. Phys.* **33** (1965) 538 and *Progr. Theor. Phys. Suppl* (1965) 532.

[65] That entropy will increase, radiation will be retarded, life will evolve forward, and black holes will absorb after the universe moves from the phase of expansion to the phase of recontraction is guaranteed by no experience whatsoever and may even be questioned: L. Boltzmann and N. Wiener, as referred to in *The Nature of Time* (ed. by T. Gold), Cornell University Press, Ithaca, New York, in which see also J. A. Wheeler, pp. 28–29, 46–47, 60–61, 73–74, 78–79, 90–107, 116–117, 142, 186, 233–240; also G. W. Leibniz, *Essais de Théodicée sur la bonté de Dieu, la liberté de l'homme, et l'origine du mal*, Amsterdam,, 1747; S. Clarke, *A Collection of Papers which passed between the late Learned M. Leibniz, and Dr. Clarke, In the years 1715 and 1716 – Relating to the Principles of Natural Philosophy and Religion*, London, 1717; L. D. Landau and E. M. Lifshitz, *Statistical Physics* (transl. by E. Peierls and R. F. Reierls), Addison-Wesley, Reading, Mass., and Pergamon Press, London, 1958; R. H. Dicke, 'Dirac's Cosmology and Mach's Principle', *Nature*, November 1961; B. Carter, 'Large Numbers in Astrophysics and Cosmology', preprint, Institute of Theoretical Astonomy, Cambridge, England, September 1968; J. A. Wheeler, 'Man's Place in Cosmology', Léon Lecture, University of Pennsylvania, 11 November 1969 (unpublished); pp. 71–72 in J. A. Wheeler, 'The Universe in the Light of General Relativity', *The Monist* **47** (1962) 40; W. J. Cocke, *Phys. Rev.* **160** (1967) 1165; F. Zerilli, unpublished calculations, Princeton, 1967, on effect of approach to moment of statistical symmetry on relative rates of radioactive transformations of long and short half lives.

[66] Ruling of Bishop of Paris in 1228 that it is wrong to deny God's power to create as many worlds as He pleases; cf. also writings of William of Ockham and a modern recapitulation by Gordon Leff.

[67] For a review, see sections on black holes and gravitational radiation in R. Ruffini and J. A. Wheeler 'Relativistic Cosmology and Space Platforms', in *The Significance of Space Research for Fundamental Physics*, European Space Research Organization Book SP–52, Paris, 1970.

[68] See however [65].

[69] J. H. Christenson, J. W. Cronin, V. L. Fitch and R. Turlay, *Phys. Rev. Lett.* **13** (1964)

138; T. T. Wu and C. N. Yang, *Phys. Rev. Lett.* **13** (1964) 380.
[70] H. Everett, III, *Rev. Mod. Phys.* **29** (1957) 454; J. A. Wheeler, *Rev. Mod. Phys.* **29** (1957) 463 and *Geometrodynamics*, Academic Press, New York, 1962, p. 75; B. S. De Witt, 'The Everett-Wheeler Interpretation of Quantum Mechanics', in *Battelle Rencontres: 1967 Lectures in Mathematics and Physics* (ed. by C. De Witt and J. A. Wheeler) W. A. Benjamin, New York, 1968.
[71] Appreciation is expressed to Paul Van der Water for this quotation from William James.
[72] For the clarification of these issues of interpretation the prime desideratum would seem to be a fully developed mathematical formulation of quantum geometrodynamics. Already from the history of wave mechanics one learned that the favorable order of developments was physical idea first; mathematics second; and philosophy of measurement third. The same lesson appeared in the work of N. Bohr and L. Rosenfeld on the measurability of the electromagnetic field quantities: *Kgl. Dan. Vidensk. Selsk. Mat.-Fys. Medd.* **12** (1933) 8 and *Phys. Rev.* **78** (1950) 794. For the development of the mathematics needed for quantum geometrodynamics, three promising clues offer themselves. One is the extensive knowledge one already has of the mathematics of superspace. Another is what one can say on physical grounds about collapse. The third is the requirement of the quantum principle for fluctuations in the geometry and topology of space.
[73] For a brief account of the concept of 'monad' see for example the *Encyclopedia Brittanica*, article on Leibniz.
[74] Compare with the principle so well-known in biology, 'Ontology recapitulates phylogeny'.
[75] For conformal invariance in electrodynamics, see for example K. Johnson, R. Willey and M. Baker, *Phys. Rev.* **163** (1967) 1699 and earlier work cited there; and for conformal invariance in geometrodynamics, see for example section 8 of the chapter of R. Penrose, 'Structure of Space-time' in *Battelle Rencontres: 1967 Lectures in Mathematics and Physics*, (ed. by C. De Witt and J. A. Wheeler) W. A. Benjamin, New York, 1968.
[76] Cf. also 'Particles and Geometry', [58] for the closely related but not identical view that particles are residuals of primordial turbulence and are to be identified with the lower less rapidly decaying end of the spectrum of this turbulence. The difference is the difference between the Leibniz monad principle and the Misner mixmaster mechanism for bringing about similarity between all particles.
[77] Appreciation is expressed to V. Bargmann for suggesting this epitome of the quantum principle.
[78] In this connection, see especially G. W. Leibniz and B. Carter, note 65.

PART V

OBJECTIVITY AND ANTHROPOLOGY

(*Chairman:* JOHN M. ROBERTS)

JUDITH BUBER AGASSI

OBJECTIVITY IN THE SOCIAL SCIENCES

1. Introduction and Summary

In the present paper I shall attempt to deal with objectivity in the social sciences, particularly in sociology, on which I shall concentrate, because I am frankly disturbed by a trend current on the American scene. The trend may be characterized by its causing bitter division amongst social scientists and upheaval in their gatherings. What disturbs me about this trend is not so much the upheaval, as the ready tendency to despair of the basic precondition for objective social science, namely the assumption of the unity of mankind – both intellectual and moral. The radicalist social scientists who belong to this trend claim that the long established goal of objectivity in social sciences is a chimera and a subterfuge which has served the powers that be for too long already.[1] In the name of instant peace and liberation they imply that rational discourse between social scientists of different persuasions – the Establishment and the Revolution – is no longer possible. Sociologists of the women's liberation movement and black militant sociologists broadcast the idea that nothing can take the place of first-hand experience: only women can understand women's problems[2] and only blacks can understand blacks. All whites, including sociologists, are racists, at least subconsciously.[3] All those who do not join the Movement belong to the Establishment, at least subconsciously.[4] True, the group which advocates the jettisoning of the aspiration for objectivity is marginal. Yet I am concerned because exactly the most dangerous aspect of their activity, their attack on objectivity, is rather condoned and tolerated by most social scientists who see their good intentions and moralistic preoccupations and social conscience, and only complain about their bizarre and unseemly conduct, especially of the young ones among them. In my opinion the bizarre and unseemly, though offensive to the sensibility of one's colleagues and not very conductive to the scientific enterprise, is much less significant than the irrationalism they advocate, which may put an end to the enterprise altogether.

R. J. Seeger and R. S. Cohen (eds.), Philosophical Foundations of Science, 305–316. *All Rights Reserved*

In the present paper I shall present the following points:

(1) There are obstacles to objectivity common to all sciences; we attempt to overcome them as best we can. The obstacles special to the social sciences are caused by the special involvement of the investigator with his topic of study, which relates to both his interests and his emotional make-up.

(2) Methods to overcome these obstacles on the way to objectivity in the social sciences were suggested in the nineteenth and early twentieth century. The Marxist tradition had only a marginal following here until recently. The major American sociological schools prevalent in the forties, fifties, and early sixties, were the Warner school [5] and the functional analysts. [6] Both aimed at the elimination of the individual investigator's bias, but caused the establishment of a massive bias in favour of the status quo. Already in the fifties, C. Wright Mills used semi-Marxist ideas to ridicule functionalism [7] and a group of sociologists debunked the sacred cow of the middle class by the series of studies of suburbia. [8]

(3) In the mid-sixties Marxism became fashionable in sociological circles; instead of the emphasis on social equilibrium came the emphasis on change-inducing social conflict. Marx himself considered mankind as divided into hostile camps – the class-camps – yet he claimed that objectivity is possible (due to his basic law of social evolution), and he decidedly considered the possibility of individual intellectuals of the wrong class-camp to see the objective truth. [9] These aspects of the Marxist tradition are being jettisoned by considerable numbers of the present generation of left-wing social scientists. Indeed, the ideal of scientific rationality has become much dimmer in this group. The forces of irrationalism dispense with such items as a rigorous economic analysis of the existing system, they are vague about who are the potentional revolutionary social groups; they are influenced by anarchistic irrationalism, by Che Guevara's emotionalism, and by Mao's primitive collectivism. Those of us who still hope for rational discussion may well put this new phenomenon on the agenda as an urgent item for study within the community of social science.

2. Special obstacles to objectivity in the social sciences

It is my contention that though complete objectivity in science is an impossibility, aiming at it, or attaining as much of it as reasonably possible, is a necessary condition for the conduct of all scientific inquiry. Why should we consider objectivity so important that we should pursue it even when admitting it to be inaccessible? In my opinion, viewing inquiry as subjective, or as an entirely individual matter, would be the exclusion of all criticism; and this would be the exclusion of rational debate; and this would be the denial of the thesis of the intellectual or rational unity of mankind. It thus opens the door to irrationalism and elitism, whether social or racial.

The general obstacles to scientific objectivity in any field concern the fact that every human is heir to some intellectual preferences and standpoints. The individual is also heir to a social and cultural tradition as a result of his being a member of a specific group of national, religious, and ethnic characteristics. I do not wish to dwell on man's limitations qua man, since this is the topic of much philosophic disquisition. Rather, I wish now to move from the obstacles to all human attempts at objectivity, to the obstacles specific to the social sciences. These are, we are told, the values of the individual researchers, values meaning here preferences and judgments in the very field of human endeavour which is the topic or the subject-matter under investigation. For example, a social anthropologist may easily tend to evaluate and judge the practices and mores of people belonging to alien cultures in terms of his own. This is the well-known danger of ethnocentricity, so-called. There is no inherent difference between ethnic and class centricity. The investigator's individual experience may result in either negative or positive dispositions towards all sorts of groupings of people. He may identify with a group of people which seem to him to resemble his own group or, on the contrary, especially free of his own people's shortcomings to which he is most sensitive. The literature is more emphatic on the first kind of prejudice – due to observed similarity – but the opposite kind of prejudice – due to rebellion against one's own group – has already been noticed by Bacon in 1620 and is very prevalent amongst intellectuals: those who pin their hopes on the downtrodden as a counterfoil to characteristics of their own class and thus tend to misrepresent them quite grossly.

The prejudices resulting from politico-ideological convictions are, of course, commonplace; they occur in the natural sciences too, but are less serious there. Here we have both authorities demanding certain preconceptions, and scientists who represent these authorities either voluntarily or out of terror, especially in monolithic cultures. Even in pluralist societies, however, politico-ideological convictions play a significant role in distorting social realities. It is a commonplace that personal economic self-interest or the economic interest of the scientist's group may bias his judgment.

It is not possible to overcome these obstacles once and for all. Yet it is of the greatest importance that each individual investigator should make the effort to become aware, as much as he reasonably can, of those of his value judgments that are relevant to his studies. This is no easy task, even when, as I recommend, we let sleeping subconscious motivations lie. Every individual possesses layers and patchworks of values, acquired from different social milieux and during different phases of his development; they may easily be inconsistent and ambivalent and ambiguous. All that is required of the investigator is not psycho-analytical self-knowledge, but plain honesty and the readiness to be conscious of whatever knowledge of himself which is readily accessible. One has to be willing to subject one's preferences, expectations, hopes, and pet aversions, to some measure of rational examination: one may try to be clear as to what these are; one may try to pin oneself down; and one may then try to find out about possible consequences of one's preferences. This may be done with the aid of history or of social analysis, or criticisms by one's peers. For my part, worse then any pet aversions, or pet sympathies is the incredible ease with which intellectual fashions spread in the world of the social sciences. The fashion spread may be not a particularly dangerous bias, but it shows that entire groups of social scientists lack this basic requirement of critical awareness, without which there is no attempt at objectivity at all. This soon leads to a severe disillusionment, and the disillusionment destroys the fashion, but it does not create self-critical awareness, at least not necessarily; and so one fashion can lead to another and so on without much improvement.

I want to make it quite clear that I do not mean to say that the individual investigator should be an aseptic or neutral or disinterested party or that he should lack social concern or avoid social activity. Only that he try to be conscious and critical of his interests and preferences; which includes his being conscious of those moral options he takes which he does not subject to rational examination. In my opinion he will do better to declare openly both those preferences of his which he assumes to have survived rational examination and those which he frankly took as moral or aesthetic decisions not subject to such examination. Of course, this will make it easier for your student or reader to detect your bias and distortion in case you are not particularly cautious to avoid them; which, of course, is the better option. All this is fairly much in accord with the spirit of Max Weber's value-free sociology.[10] To which we come soon. Let me conclude, however, that it is quite advisable for anyone to study those problems which do carry moral import, according to one's own judgment of what is of moral import, but on the condition that one's criteria remain open to modification – especially as the result of such a study.

3. Traditional American Attempts to Solve the Problem of Value

The most important writers on our topic, Marx and Weber had little or no influence in the early stages of development of American sociology. The earliest American attempt to grapple with the problem originated in psychology and claimed psychology to be the totality of social science.[11] This school, behaviorism, tried to exclude from its research anything that is not observable objectively, i.e. people's conscience and awareness, feelings and values. While the behaviorists clearly avoided getting involved in the problems of the values of their objects of research, they also made the explanation of social phenomena impossible.

The American sociologists were very concerned with attempts to render sociology scientific by eliminating personal bias and partiality. Two attempts followed the classical Chicago school's attempt, the social stratificationists of the Warner school and the functional analysts of the Talcott Parsons school. Both offered criteria for objectivity, and both criteria introduced strong systematic biases in favour of the status quo, presumably unintentionally.

The Chicago school[12] undertook in the twenties and thirties the study of communities, neighbourhoods, and deviant social groups. They followed largely the rules of classical social anthropology of Malinowski, by viewing these microcosmic systems as Malinowski would view a primitive microcosmos – as social wholes. They accepted the value system of any whole when assessing any part or aspect of that whole – well in accord with classical functionalism. In particular they avoided all value ranking and upheld the principle of equality of all social systems which leads to moral relativism.

This attitude led to the neglect into the enquiry into the social and economic conditions which often had caused the peculiar degraded living-patterns of the groups under study.

The Warner school of social class[13] or stratification attempted to find an objective scale of measuring the social standing of any member of any American community. Warner was convinced that his method was utterly objective since it enabled any person, even a total layman, to determine the fairly exact social standing of a given subject as accurately as any set of lengthy interviews and participant observation could have. And from the determination of a person's position on the Warner scale a lot of predictions about him could be made – about his consumption patterns, his social associations, and even his taste in interior decoration. The claim that all American communities are divided into the same five or six discrete classes, on which all these predictions were based, was, of course, completely unfounded. It introduced, which is more important, a strong bias for the acceptance of fairly static and rigid social stratification as the standard, natural, normal, American condition. Many of Warner's less sophisticated followers in fields like education, welfare, and marketing, understood his middle-class, especially his upper middle-class (Class 3), to be the norm to whose values children in schools were supposed to be socialized, etc.

In most sociology departments across the country functional analysis was accepted in the fifties and up to the mid-sixties as the only possible way of doing scientific sociology. Parsons dealt with the entire American society as if it were one social system, i.e. a whole, and all its parts, all the social phenomena within the system, were claimed to be objectively explicable within the system by showing the function, i.e. positive role, which they play within the system in the system's working to maintain balance and stability. All this was well in accord with Durkheim's general ideas as presented over half a century earlier. The conservative aspect of this philosophy is too obvious to require separate statement.[14] Admittedly, in its later phases it found adherents who attempted – along with even Parsons himself – to allow for social conflict as an expression of instability and the cause of social change (in classical functionalism the function of conflict is to preserve stability).[15] But this compromise is not enough to remedy the bias, and is not consistent with the basic tenets of functionalism.

During the reign of functional analysis one major critic of it was heard – C. Wright Mills.[16] He pointed out the conservative bias of functionalism and ridiculed it as barely more than pompous jargon. C. Wright Mills himself attempted a social analysis of American society along economic-interest class lines. He certainly was not too successful in avoiding letting his personal bias (an extremely pessimistic view of the public as passive and contempt for the then Eisenhower administration for its subservience to big business) distort his image of American society and government. He was a convinced liberal socialist who often used Marxist methods of analysis, but also utilized the insights of later sociologists of elite phenomena. He was greatly concerned with objectivity, for which he had a most peculiar recipe:[17] first, relate your personal troubles to public issues, thus avoiding both personal bias and the bias of your own group (endorsing problems which are fashionable in your scientific community is such a bias); second, see your problem from all possible viewpoints; thirdly, verify your conclusions. Of course, it is the correlation of private troubles with public issues which catches the eye. It is a very dubious method of avoiding bias: it is all to easy and often rather cheap to blow-up one's private frustrations into social protest, and it is even easy enough to view the fashionable target of criticism as the source of one's private troubles; to call this objectivity is at best dubious.

A group of Mills' disciples[18] mounted in the early sixties a critical attack on conventional American sociology and especially challenged the functional analysis school's use of Max Weber's value-free sociology as a cover for hard-hearted defence of the status-quo – raising a generation of amoral social technicians.[19] They declared themselves advocates of the basic aim of scientific objectivity; they demanded the open and honest declaration of the researcher-teacher's values which motivated his choice of problems for research to his students and readers – (this I heartily endorse); they tagged on to this a somewhat dubious value-judgment of their own: values that involve the sociologist in social action are more valuable than other values.[20]

I have no time to discuss the studies in the fifties of suburban life, which followed no single strict system of thought but which had one common theme and common result contrary to both the Warner and the functionalist schools: the American middle class which had previously been respected by the American sociologists and whose values were upheld by many as the norm for American society, this class was shown up as unattractive, rediculous, even pathetic – and culturally unsophisticated. This, naturally, opened the way to the recent all-out attack on the so-called middle-class and its values by the new radicals. There is little doubt that this literature is of some value in spite of its having since been debunked as the 'myth of suburbia'. The myth is somewhat childish in making the suburb the scapegoat for a number of different and partial processes of social change; but it did raise controversy and it did bring about some more analytic studies of socio-economic changes in America.

4. Contemporary rebellion

In the mid-sixties, between 1965 and 1967 to be precise, the movements against the war in Viet-Nam on the one hand and for black power on the other formed a discernible new left ideology. This is a body of sharply negative and critical views of American society. American government was declared no longer a functioning democracy but an instrument of C. Wright Mills' military-industrial complex and according to the more extreme spokesmen of the movement, a fascist government ruling a progressively 'rotten' society. The hope for successful social reform, using the existing party-political and governmental institutions, was declared a dangerous illusion; all those still supporting these institutions were lumped together as 'the Establishment' and branded as the enemy. Confrontations with the Establishment became the major tactic and the slogans of revolution became the norm.

The new radical movement was largely based in the campuses of liberal arts colleges and major universities. The departments of sociology and of political science were most seriously affected it seems, in that a good number of faculty who had adopted this philosophy, attempted to radicalize their discipline. The novelty is not in the introduction of a new viewpoint, of a left-wing viewpoint, or even of an extremist viewpoint into social science; the novelty is that rational discourse, the possibility to debate social issues somewhat objectively among social scientists, which was previously assumed to exist also by most Marxists,[21] is now fully denied by many radicals. Name-calling is nothing new, but the new attitude is one which divides all sociologists and social philosophies into the camps of the Children of Light and Children of Darkness, thus encouraging name-calling as the only means of communication available.[22] From now on it seems no longer necessary to spend time in pedantic studies, in social and economic, minute and careful analyses of society or of the revolutionary forces. These activities, which Marx had considered absolutely essential, are now dismissed as a part of 'bourgeois crap'. What is needed, we hear, is the gathering of the immediate ammunition for the battle. There is no time for analysis, there is no need for analysis; in this period the only way to know social reality is by active participation in the struggle. Every intellectual activity, then, must have immediate relevance to the struggle, with the accent on immediacy.

Admittedly, the revolt against established conventional American sociology is understandable; admittedly it contains also some valid and useful criticism. Yet the revolt turned out to be chiefly a revolt against any aspiration for objectivity. It became, immediately, a revolt against any intellectual systematic endeavour.[23] The most dangerous element of such a move is the jettisoning of the idea of the rational unity of mankind, as well as the idea of the moral unity of mankind.[24] In the eyes of the radicals, not only general humanity is now divided decisively along the lines of the barricades of the revolution, or the stand on Viet-Nam, or along race- and sex-lines; social scientists are now also supposed to be irrevocably divided along the same lines.[25] The social problems of oppressed groups can no longer be studied by all honest scholars, but only by scholars of their own group.[26] Both the intellectual tradition and learning from history are gravely neglected. We have neo-Marxists ignorant of *The Communist Manifesto*[27] and anarchists ignorant of Bakunin[28] and a revolutionary movement that deem it superfluous to examine the great revolutions of the immediate past and their outcomes.[29]

The danger of this phenomenon is bad enough; worse is the fact that liberal sociologists of the left, akin to C. Wright Mills, are indulgent to this new danger. Irrationalism in the home base of rational attitudes towards social problems takes roots very quickly. My intention is not to offer as yet another sociological analysis of the phenomenon – I have not found a satisfactory one. Rather I wish to challenge all those who still cherish the idea of social science to talk to the young rebels, and to invite them to rational debates on whatever they feel concerned about. I also wish to propose that we challenge the older rebels systematically and publicly on this issue of objectivity, rationality, and the unity of mankind.

NOTES

[1] See, for example, Irwin Sperber, 'The Road to Objective Serfdom: a Comment on Oliver's 'Scientism and Sociology'', *Berkeley Journal of Sociology* (1969), pp. 111–121: "The champions of objective ideology are just as acquiescent to the prevailing social and economic pressures 'off the job' as when they are 'on the job'. The segregation of these roles is useful for the academician who will not in any case take a strong stand as a private citizen and who strives for the bonus of disguising such a noncommittal position as scientifically respectable.... It also justifies the opportunistic indulgence of doing the trivial, facile, publishable research irrelevant to the crises of one's own society" (p. 118). See also Marvin Surkin, 'Sense and Nonsense in Politics', *P.S.* **II**, No. 4 (1969) 573–581: "My purpose in this paper is to show that the rigorous adherence to social science methodology adopted from the natural sciences and its claims to objectivity and value neutrality function as a guise for what is in fact becoming an increasingly ideological, non-objective role for social science knowledge in the service of the dominant institutions in American society" purpose in this paper is to show that the rigorous adherence to social science methodology (p. 573).

[2] Marylin Goldberg, 'On the Exploitation of Women', *Liberation*, quoted in Edward Grossman, 'In Pursuit of the American Woman', *Harper's*, February 1970, p. 56. At the 1969 Convention of the American Sociological Association (San Francisco, September 1–4) there appeared on the list of the generally reasonable and useful demands of the Women's Caucus the claim for the teaching of women's sociology by women sociologists to women students in all sociology departments.

[3] The tendency of black political leaders, black social scientists and also some white radical social scientists to insinuate that any analysis of some aspect of the life of American negroes which they reject, is the result of the thinking of a *white* social scientist who necessarily does not understand *blacks*, has no feeling for them, is a 'subtle racist' – and worst of all, belongs to the category of 'White Liberals'. This tendency came first to the fore in the controversy around D. P. Moynihan's Report on the Negro Family in 1965.

A more recent and more violent example is the controversy about Black Culture in the *Berkeley Journal of Sociology*, 1967–1968; Douglas Davidson, 'Black Culture and Liberal Sociology', pp. 164–175, and his 'The Dilemma of the White Liberal: a Rejoinder to Kaplan', *op. cit.*, pp. 181–183.

[4] At the 1969 Convention of the American Sociological Association (San Francisco, September 1–4) the spokesman of the Radical Caucus denounced the executive committee of the association as 'lackeys of the Establishment'; (the committee had not accepted the

radicals' demand to come out with a resolution for immediate, unilateral withdrawal from Viet-Nam, but had appointed a committee on the problem of the political involvement of the association and had decided to poll the membership). Thereafter the spokesman of the Black Caucus denounced the association in its entirety as 'racist'.

[5] See W. Lloyd Warner, *Social Class in America*, 1949; Harper Torch, New York, 1960; the bibliography at the end includes the major titles of his school. See also next note.

[6] See Talcott Parsons, *Essays in Sociological Theory*, revised edition, Free Press, Glencoe, 1959, for a bibliography of Parsons to 1949. See also Robert K. Merton *et al.*, *Sociology Today, Problems and Prospects*, in 2 volumes, New York 1959; Harper Torch, New York, 1964.

[7] C. Wright Mills, *The Sociological Imagination*, Oxford Univ. Press, 1959; paper, Grove Press, New York, 1961, Chapter 2, Grand Theory, pp. 25–49.

[8] Surveys of the literature on suburbia are, for example, Dennis H. Wrong, 'Suburbs and Myths of Suburbia', in *Readings in Introductory Sociology* (ed. by Dennis H. Wrong and Henry L. Gracey), MacMillan, New York, 1967, pp. 358–364, and Maurice R. Stein, *The Eclipse of Community*, Princeton 1960; Harper Torch, New York, 1964, Chapter 9, pp. 199–226.

[9] Harold J. Laski, *On The Communist Manifesto*, 'An Introduction Together with the Original Text and Prefaces by Karl Marx and Friedrich Engels', Foreword For the American Edition by T. B. Bottomore, Pantheon Books, Random House, New York, 1967, pp. 145, 146: "Finally, in times when the class struggle nears the decisive hour, the process of dissolution going on within the ruling class, in fact within the whole range of the old society, assumes such a violent, glaring character, that a small section of the ruling class cuts itself adrift, and joins the revolutionary class, the class that holds the future in its hands, just as, therefore, at an earlier period, a section of the nobility went over to the bourgeoisie so now a portion of the bourgeoisie goes over to the proletariat, and in particular, a portion of the bourgeois ideologists, who have raised themselves to the level of comprehending theoretically the historical movement as a whole."

[10] Max Weber, ' 'Objectivity' in Social Science and Social Policy', in *Max Weber on the Methodology of the Social Sciences* (ed. by E. A. Shils and H. A. Finch), Free Press, Glencoe, 1949, and 'Science as a Vocation', in *From Max Weber: Essays in Sociology* (ed. by H. H. Gerth and C. Wright Mills), Oxford Univ. Press, New York, 1946. See also Hans Albert, 'Theorie, Verstehen und Geschichte', *Journal for General Philosophy of Science* **1** (1970) 1–23, especially Section V, for a more detailed treatment of Weber along the same lines.

[11] R. M. McIver, *Society, Its Structure and Changes*, Long & Smith, New York, 1931, pp. 529–530: "Incompetent to deal with the subjectivity of experience the behaviorists would discard it altogether. Seeking to get rid of subjective terms they get rid of the social fact, since it is fact only as created by and known to experience."

[12] For a survey of the Chicago school see M. R. Stein, *The Eclipse of Community, op. cit.*, Chapter 1, pp. 13–46; for a comment on a more modern variant of it, the neo-Chicagoans concerned chiefly with deviant sub-cultures, see Alvin W. Gouldner, 'Anti-Minotaur: the Myth of a Value-Free Sociology', in *Sociology on Trial* (ed. by Maurice Stein and Arthur Vidich), Prentice Hall, Englewood-Cliffs, 1963, pp. 35–52, especially pp. 46, 47.

[13] W. H. Warner, *Social Class in America, op. cit.*, p. 42.

[14] See, however, E. Gellner's thoughtful 'Time and Theory in Social Anthropology', *Mind* **67** (1958) 182–202. See also C. Wright Mills, *The Sociological Imagination, op. cit.*, p. 42. See also Daniel Foss, 'The World View of Talcott Parsons' in *Sociology on Trial* (ed. by Stein and Vidich), *op. cit.*, pp. 96–126.

[15] Robert K. Merton, 'Manifest and Latent Functions' in his *Social Theory and Social Structure*, Glencoe, Free Press, 1949, pp. 21–81.
[16] C. Wright Mills, *The Sociological Imagination, op. cit.*, p. 226 (last page).
[17] *Op. cit.*, p. 125ff.
[18] Alvin W. Gouldner, *op. cit.*, p. 52: "To do otherwise is to usher in an era of spiritless technicians.... If we today concern ourselves exclusively with the technical proficiency of our students and reject all responsibility for their moral sense, or their lack of it, then we may some day be compelled to accept responsibility for having trained a generation willing to serve in a future Auschwitz."
[19] Gouldner, *op. cit.*, p. 51.
[20] Daniel Foss, *op. cit.*, p. 126.
[21] See note 9.
[22] See, for example, Irwin Sperber, *op. cit.*, p. 113. New left sociologists and political scientists denounced with special vehemence the pluralistic models or theories of the U.S. economy, society, and government. The new left social science literature abounds in abusive terms like 'imperialist', 'fascist', 'oppressive', 'exploitative', 'racist', 'genocidal', 'colonizing', etc., not to mention mild terms such as 'decadent', 'one-dimensional', 'alienating', or 'manipulative' which claim to be an exhaustive characterization of 'The System' or 'The Establishment'.
[23] The major philosophical and methodological foundations of the new left attitudes which deny the possibility of objectivity and of communication with opponents are the following. Maurice Merleau-Ponty, *Phenomenology of Perception*, London 1962; *Sense and Nonsense*, Evanston 1964; *Humanism and Terror*, Boston, Beacon Press, 1969. Ernest Becker, *Revolution in Psychiatry: The New Understanding of Man*, Free Press, New York, 1964; *The Structure of Evil: An Essay on the Unification of the Science of Man*, Braziller, New York, 1968.

The only understandable contention in the whole of this literature is that objectivity belongs to the domain of the natural sciences, not to the human sciences – which is as old as German Romanticism.

What the movement has taken from Marx is chiefly the idea of the unity of theory and praxis. It means for them, however, that proper thinking is only possible after one is committed, in the existentialist's sense, to *the* Revolution.
[24] Marvin Surkin, *op. cit.*, p. 578.
[25] *Ibid*, p. 579.
[26] Herbert Marcuse, *An Essay on Liberation*, Beacon Press, Boston, 1969, p. 8.
[27] According to a newspaper report on Mark Rudd's visit to Boston University immediately after the Columbia revolt, this young revolutionary leader was then leafing through a copy of the Communist Manifesto which somebody had given to him, and which he then was seeing for the first time.
[28] For instance the authors of the magazine *Anarchy*.
[29] Thus for instance all the authors represented in *The New Left Reader* (ed. by Carl Oglesby), Grove Press, New York, 1969, still advocate the 'abolition of capitalism' as a matter of course; their attitudes toward the Soviet government vary from mild humorous (?) remarks on dogmatism (Fidel Castro) to severe criticism for being 'politically irresponsible' (C. Wright Mills). But none even attempts a rational analysis of what went wrong with the Soviet revolution, the first revolution which successfully 'abolished capitalism'.

I. C. JARVIE

ON THE OBJECTIVITY OF ANTHROPOLOGY

To raise sceptical doubts about the objectivity of the results of anthropology – or any other social science – is enough to tempt many people into offering a defense that is justificationist; which seeks, in other words, to make arguments that show anthropology *is* objective, which *justify* its claims. This strategy is doomed to defeat in the face of a determined sceptical assault. The only strategy which will avoid the trap[1] is one that goes on the offensive, which says "I hold anthropology to be as objective as any other science. If you do not, give me your arguments and I will try to answer them; I will not justify, and, if you agree anthropology is objective and merely want to see how I justify it, I will set aside your questions as idle." This will be my strategy here, and I apologise to those who would like to see a full-blooded justification of the claims of anthropology.

To simplify this immense topic I shall take 'objective' to mean the opposite of 'subjective, personal biased'; I shall take anthropology to mean social anthropology done by means of fieldwork (or its more intimidating name 'participant observation'). There is some connection between objectivity and truth. But it is not clear what it is.

Objectivity has a somewhat broader use than truth. Only statements can be true or false; but discussions, investigations, methods, judgements, people and many other things besides statements, can be classed as objective. That a person, or an investigation, or a trial, has been objective is not a sufficient condition for it to issue in a truth. Moreover, since we can reach the right conclusion by the wrong methods, objectivity is not a necessary condition of truth either. Words which are troublesome like this can be pinned down in various ways: by scrutiny of the myriad ways in which they are used – but this may be unenlightening; by imposing formal definitions of them on discussion – but this cannot but be arbitrary; by seeking to define the essence of their meaning – but this can lead to circularity or arbitrariness; and finally by embedding them in a context which provides rules for their interpretation as well as their use –

R. J. Seeger and R. S. Cohen (eds.), Philosophical Foundations of Science, 317–324.

such a context is known as a theory. The theory I adopt, which is Popper's, sees the aim of scientific investigation as the pursuit of truth,[2] and the interference of bias or personalities as vitiating progress towards that aim. Therefore, objectivity consists in placing checks on bias. No one is perfect, and no one should be uncritically trusted; the checks therefore must be social.[3] What must be institutionalized in the society of investigators is criticism; all work must undergo the ordeal of public criticism. This is an idealized description of the institutions of science.

Had we access to the truth, and were we able to prove that we had it, we might then have a method without bias (objective) and a result without error (truth). Only the tautologies of logic and mathematics can be proved to be true. They are true of all logically and mathematically possible worlds, whatever position we adopt on the relation of those worlds (subjective or objective) to the minds discussing them. The objectivity, or lack of bias or personal involvement, of logic and mathematics is gained precisely because they are vacuous: they say nothing about these worlds, only about their possibilities; they define them and their properties.

As soon as we try to describe, discuss, or explain which of the logically and mathematically possible worlds is realized in fact, then methods of proof and hence objectivity are lost. We preserve our perfect objectivity only by saying nothing.

Attempting to say something non-tautologically true brings out a connection of some sort between objectivity and truth; namely, in our theory that science makes *progress* in its descriptions, discussions, and explanations. As we test, refine, and modify our descriptions, discussions, and explanations we are implicitly maintaining the metaphysical view that there is a direction to the change, that we are improving them by moving towards the truth. What this amounts to is very difficult to specify fully, but Popper has proposed with his theories of depth and verisimilitude a partial explication of it.[4]

According to the theory of objectivity adopted here, the objectivity of the discipline lies not so much in its capturing what the objective truth about things is, but rather in whether it adopts means to foster maximum progress in our descriptions, discussions, and explanations, namely, subjects our ideas to the standard of intersubjective and repeatable testing. The character of antropological fieldwork is such that this standard of intersubjective testability is uniquely difficult to meet. Anthropology is usually an individual enterprise, conducted at considerable expense, in some inaccessible part of the globe, among people with an esoteric language. Only with the greatest difficulty can this research be repeated. Indeed, in the strictest sense, it cannot be repeated at all. Fieldwork takes place at a particular time, and within the unique network of social relationships established by *that* anthropologist, *then*. In such fluid human situations which change over time, and which differ markedly from person to person, no man's network of social relations, point of view and, hence, information can be duplicated. Fieldwork is a very private affair. It thus seems not to meet the demand for public criticisability, and the anthropologist may have to be uncritically trusted.

This argument need not carry us away. After all, in natural science no two scientific tests are truly identical since their numerical distinctness ensures that they take place in different space-time areas and it is only a hypothesis – and a metaphysical and therefore unfalsifiable one at that – that this makes no difference. ('Physics as we know it is indifferent to space-time parameters' is not a falsifiable statement because any counter-example could be interpreted as not belonging to objectively true physics.)

Can or does anthropology operate with a similar hypothesis to the effect that its research is repeatable to all intents and purposes, and that different time and different investigator does not matter? Is it possible to argue that the descriptive material could have been collected by anyone, and so the fieldwork is as near as no matter repeatable? In other words, that the better part of anthropological results are indifferent to time and the observer? This too would be a metaphysical hypothesis. While it may be unfalsifiable metaphysics, I would contend that most anthropologists would declare it false. Anthropological fieldwork does not involve one in the forging of social relationships in the field contingently, and that ideal of social science which feigns indifference to the fact that it is by men as well as about men is absurd.

A little more deserves to be said about this. Some philosophies conceive of the machine as more objective than the man. This looks plausible, but is a mistake. Machines only do what men have built into them: they are no more objective than their builders make them. Some of the biases, preconceptions and interpretations of their makers will be built-in. More important, whatever result they produce then has to be filtered back through the biases, preconceptions and interpretations of their operators. Science is made by man with the aid of machines; not *vice versa*. Because machine operations are mechanical they make less mistakes; but this is only because, recognizing his own mechanical limitations, man has devised in machines ways of correcting himself. A social science like anthropology, however, cannot possibly be mechanized. Cameras and tape recorders can only record what goes on around them, and that something is going on around them to be recorded can only be decided, evoked, or disclosed by a human being. A machine cannot explore a human society because it cannot enter into social relationships like talking. Either by getting the people to talk to him, or by them letting him overhear what they are saying, anthropologists gather data. Recording conversations on film and tape might eliminate some errors of memory, transcription, etc., but those machines might distract from, distort, or even prevent promising social relations which will in the long run be more enlightening.

Further than this, however, all social scientists are studying themselves: man. They are there to give human accounts of human beings. Martian accounts of human beings, arachnid accounts of human beings, mechanical accounts of human beings, may be very interesting and very different, but they are not part of the aim of the social sciences. The metaphysical and interpretative categories in which social science must be written are given: they are those which preserve the human meaning and significance of what we are privileged to be studying.

Human meaning and significance are properties of poetry too, however, which suggests that the social sciences are in danger of becoming private, or at least not objective and progressive.[5] The difference seems to be in that the poet captures moments, preferably unique, while the social scientist captures not moments, which are private and unique, but typical decisions and events for which he reconstructs a rationale. He concerns himself only with those decisions and events he considers to be typical of a great deal of human actions, and his rationale is thus of a general nature. So, while an individual anthropologist's experiences are not repeatable, similar sorts of events are, and attempts can be made to reconstruct their rationale and see if it checks out.

This must lead us to look rather closely at what goes on in the course of field research. What does an anthropologist do in the field that should put him on the road to objective descriptions, discussions, and explanations? To tell the truth, what fieldworkers do is wander around observing, learning the language, getting the feel of the place, chatting with people, interviewing them about their relatives, etc. A short name for this activity is 'gossip'. Anthropological fieldwork consists to a considerable extent in gossip – sometimes even structured or directed gossip.

Do I not, in admitting this, jeopardize my previous argument for the objectivity of anthropology? I don't see it: gossip is one of the principal means we have to find our way around in our own society. In many ways it provides us with information sufficient to cope with the world in which we find ourselves. Gossip is the principal means by which a foreigner like myself finds out things like what is the AAAS and what significance is to be attached to a session such as this. The source of my information is independent of its truth; in seeking the truth I must use whatever means seem appropriate.

The answers I would give then, were someone to impugn the objectivity of anthropological fieldwork because it has a profound personal equation, and is heavily dependent on gossip, are these. Martin has argued, convincingly enough, that fieldwork is neither a necessary nor a sufficient condition for successful anthropology.[6] This does not say that its history and its problems can be understood without reference to fieldwork. So a stronger argument is required and that is this. The personal equation and gossip do no more damage to the objectivity of anthropology than does the coffee or whisky imbided by Nobel Prize winners. All such things are *irrelevant* to the objectivity of anthropology. The objective truth, or lack of it, of anthropological results, the progress or lack of it in anthropological descriptions, discussions, and explanations, cannot be undermined by *ad hominem* arguments. All our procedures together suggest to us that if our results have been subjected to rigorous tests, the best we can devise, then we have done all we can to maximise the verisimilitude of those results. But we in fact do very little to test anthropological descriptions like field reports; that is the whole point – is this a blunder? No, I would say, we no more look over an anthropologist's shoulder than we do over a natural scientist's. Why should we? If he lies or misperceives so what? So he will be found out sooner or later? Not necessarily; the Piltdown forgery, or such incidents as C. P. Snow discusses in *The Affair*, may be the rare cases where the lie is exposed. He will mislead and misdirect a lot of others? Very possibly; but we either can't, or don't approve of, policing the whole world. Trust and independence are the least that are compatible with human dignity, including the dignity of that uniquely human activity scientific investigation. More than this, though, I suggest we neglect the policing of scientists because we know it is unimportant. Where we know incompetence or fakery might do obvious harm we tolerate some discipline, like in medicine or engineering,[7] usually by professionalizing. In anthropology we rely on people doing their best, and we can ask, what else is compatible with their dignity?

So much, then, for argument to the effect that anthropology is as objective as any other science. Some counterattack on those who deny its objectivity might also be in place. The most disastrous consequence of denying its objectivity is not so much that it makes a science of society impossible, but rather that it leads to relativism. The denial of a sience of society – advocated by those action-new-leftists who only allow as legitimate changing society not studying it – leads to absurdities. You can't change what you don't know, you can't know what you can't study, and you certainly can't know whether what you have tried to change is changed if you do not know how it was in the first place.[8] A similar denial for different reasons by the followers of Wittgenstein leads them either into relativism or into the view that a culture can only be understood in its own terms, from within, and so cultures are incommensurable.[9]

The seriousness of relativism is that – unlike denying the possibility of social science which runs in the teeth of the obvious existence and to some degree success of social science – it appears to be a conclusion forced on us by study of the way things are, and thus has a glossy surface of rationality to cover its deep and underlying irrationality. In the name of honest ignorance, and decent respect for other people, anthropologists declare it impossible to evaluate cultures as a whole, the one against the other or them all against some absolute standard. Arguing that all cultures differ profoundly on fundamental questions of value, and that all systems of value are culture-bound, they conclude that there are no standards universally applicable which would resolve these sorts of conflicts. Any adoption of standards by an anthropologist would make him no better than a missionary, who evaluates all cultures and especially their religions by how closely they approximate to the Judeo-Cristian values he is armed with. Such arbitrary and therefore irrational standards of value have allowed men to commit criminal and atrocious acts against those men and cultures found wanting.

My counter to this is that arbitrary evaluation is indeed irrational, but so is drawing the conclusion from the chaos of competing values that all values are equal. This is both against common sense and against good logic, for it is a *non-sequitur*: it does not follow. The excesses of absolutism are no longer, if they ever were, an excuse for shirking the fundamental human task of making up one's mind on moral issues. Elsewhere, I have tried to argue that our duty to try to behave decently towards others is not abrogated when we become anthropological fieldworkers, still less is it our duty to condone conduct in others we in other circumstances would condemn.[10] True, part of our duty is behaving decently by showing respect for others and their differences of opinion with us, not to mention our duty of self-preservation and the caution it entails. But research shows that to a surprising extent given their relativistic utterances anthropologists do behave decently by the standards they have been brought up in and will return to. This is moral common sense, and a blow against it involves the abrogation of all standards.

York University, Toronto

NOTES

[1] See J. O. Wisdom, 'The Sceptic at Bay', *British Journal for the Philosophy of Science* **IX** (1958) 159–63.
[2] See K. R. Popper, 'The Aim of Science', *Ratio* **1** (1957) 24–35; and *Conjectures and Refutations*, London and New York 1963, Chapter 10.
[3] K. R. Popper, *The Poverty of Historicism*, London 1957, pp. 155–56; *The Open Society and Its Enemies*, London 1962, Vol. II, pp. 216–23.
[4] These are to be found in the references given in note 2.
[5] I take it that much poetry aims to be intensely personal and subjective; that there is no progress in poetry in the sense of one poet's work superseding or replacing that of another. The question of whether this makes poetry more complex than art – which can be interpreted as a process of 'making and matching' – is puzzling. The descriptive and expressive resources of art constantly expand, but this is the only sense in which art progresses. Cf. E. H. Gombrich, *Art and Illusion*, London 1960.
[6] Michael Martin, 'Understanding and Participant Observation in Cultural and Social Anthropology', in *Boston Studies in the Philosophy of Science*, *Vol. V* (ed. by R. S. Cohen and M. W. Wartofsky), D. Reidel Publ. Co., Dordrecht-Holland, 1969.
[7] See J. Agassi, 'The Confusion between Science and Technology in the Standard Philosophies of Science', *Culture and Technology* **VII** (1966) 348–66.

[8] This is Popper's argument to show the irrationality of wholesale radicalism, see *The Open Society*, *op. cit. supra*, Vol. I, pp. 157–68.
[9] See my 'Understanding and Explanation in Sociology and Social Anthropology' in *Explanation in The Behavioural Sciences* (ed. by R. Borger and F. Cioffi), Cambridge 1970.
[10] 'The Problem of Ethical Integrity in Participant Observation', *Current Anthropology*, Vol. 10, 1969.

JACOB W. GRUBER

ACQUIRED MODELS AND THE MODIFICATION OF ANTHROPOLOGICAL EVIDENCE

Anthropologists have not traditionally concerned themselves with those epistemological questions concerning their investigative enterprise of which the problem of objectivity is one. While there has been a continuing interest in methodology, the concerns of method have been those which related to the kinds of data to be collected and their relevance to an understanding of particular life ways of particular peoples or social entities. While there is a continuing and increasing concern with the problem of the significance of the data for an understanding of the human condition, there still seems to be no overriding anxiety over the question of objectivity of knowledge in the way in which that has become a significant question for the philosopher of science except, of course, as that question may be translated into one which can deal with the precision with which problems are formulated and data gathered.[1] The reasons for this neglect of the epistemological questions which have so exercised segments of the physical sciences for the past half century are probably varied. The long identification, emerging from its own origins, of anthropology with 'science' (as opposed to 'social science') may have been a strong contributing factor in the inhibiting of any serious kind of philosophical introspection particularly in view of the fact that, in the nineteenth century, the designation 'science' did (as it still does in the minds of many) imply an objective, value-free activity quite separate from philosophy whose conclusions were assumed to be testable only in the mind of man rather than the reality of nature. This assumed relationship between 'science' and 'objectivity' is a function, however, of the particular history of science rather than that of anthropology. In anthropology itself, the undisciplined manner in which the field itself arose may also have contributed to the immediate concerns of data collection rather than those more subtle ones of data integrity. The press for ethnographic salvage and the need to preserve data which were already disappearing imposed an urgency for collection, description and preservation which allowed little time for questions of objectivity even were the scepticism of

R. J. Seeger and R. S. Cohen (eds.), Philosophical Foundations of Science, 325–336.

the process of science available to permit it.[2] Anthropology as a discipline with an emerging set of problems grew out of the development of the physical and, more particularly, biological sciences during the first half of the nineteenth century. As a field strongly identified with these sciences, it took on their assumptions, was heir to their prestige, and, in searching for its own theoretical success, used them as models.[3]

From the seventeenth century on, there are at least two continuing notions of science – its nature, function and process. Each carries with it a baggage of criteria for the determination of objectivity and the consequent establishment of Truth. Whatever their more subtle sources, Buchdahl suggests that their persistence in the traditions of western science is in part at least the result of the tremendous influence which Newton played in the establishment of both scientific attitudes and norms.[4] We should speak more properly of the influences of Newton, for Buchdahl regards the Newton of the *Principia* to be quite different from the Newton of the *Opticks*.

The former epitomizes 'the spirit of systematic reason' in its push for system and the mathematisation of natural knowledge in such a magnificent fashion as to suggest "that the whole physical universe in some sense mirrored the logical structure of a mathematical theory." The science of the *Opticks* on the other hand reflected an emphasis upon that which was in Nature and upon its observation. "It emphasized observation and experiment, it officially denounced the employment of hypothesis, it stressed the aim of investigating nature at close range, and it ended up by operating *against* the very spirit of systematization"[5] which was so important a goal of the seventeenth and eighteenth century savant. One can see, in passing, how much more congenial the spirit of the *Opticks* was to the Romanticism from whose natural passion much of modern natural science – including the science of man – found its initial impetus and support.

In short, the science of the *Opticks* is the clear development of a body of knowledge, a body of data, sensible in the literal meaning of that word, the product of observation, the inevitable and incontrovertible result of controlled experience. It was what a later generation was to call Baconism in the desire possibly to secure for it a more ancient authority.[6] For such a *science*, the canons of objectivity are those which relate to the establishment of proper controls in the observation of phenomena and the development of more sensitive instruments of experience or observation so that the elements of a gross Nature can be reproduced in the microcosm of the human mind or experience with an increasingly predictable fidelity. In such a science, objectivity is in a sense a function of the methods of investigation and the recognition of the variables which play some part in the distortion of perception of whatever sort.

There was – and there is – a science of man based upon this latter conception of what science was and of what it was supposed to do. During the first half of the nineteenth century there was a 'natural history of man' which sought to construct an ethnology, a history of the varieties of man, upon such a foundation. Such an anthropology saw as its essential task the establishment of a body of knowledge, a *science*, through an objective description of the varieties of man as they appeared, as they were perceived and sensed, in the variety of their physical structures, their customs, and in the babel of the many languages they spoke. Like the more general Natural History of which it was a part and from which it derived its methodology and goal, it sought to add to the body of natural knowledge; and new customs or new languages or new racial varieties were described as new to *Science*. As Prichard noted:

> The strict rule of scientific scrutiny exacts, according to modern philosophers, in matters of inductive reasoning an exclusive homage. It requires that we close our eyes against all presumptive and extrinsic evidence, and abstract our minds from all considerations not derived from matters of fact which bear immediately on the question... What is actually true, it is always most desirable to know, whatever consequences may arise from its admission.[7]

This was simply an extension into ethnological researches of an attitude which had already been, for a generation, the norm in such a new science as geology. In describing the 'new geology' which arose in the first decades of the nineteenth century – and which proved so successful and exciting – Lyell noted the insistence on 'observation':

> To multiply and record observations, and patiently to await the result at some future period, was the object proposed by them... all must be content for many years to be exclusively engaged in furnishing materials for future generalizations. By acting up to these principles with consistency, they... disarmed all prejudice, and rescued the science from the imputation of being dangerous, or at best but a visionary pursuit.[8]

That tradition of observation as knowledge and knowledge as understanding has been a continuing one both in the natural sciences and in anthropology; and its persistence as well as its strength and its persuasiveness is derived in great part, I think, from the assumption that such descriptions can be value-free.[9]

Of course as one reads these accounts – especially the older ones which are a part of an intellectual context so much different from our own – one is aware how precarious a guide is perception itself; how unconscious the describer was of the influence of his own culture, his own traditions, in the determination of what he saw, how he saw it and how he described it. Anthropologists have come to see how pervasive is the influence of a culture upon the perception – the reality – of its practitioners. And yet, there is a tendency to ignore its effect upon the observations of science or, more properly, the assumptions which underlie them. Such is the particular value of science itself and the heritage of its earlier definitions.

Even Prichard followed a continuing tradition of comparing Negro and European, black and white, on the implicit assumption that these represented the ultimate extremes of human variation – the former more and the latter less 'animal-like'. His *Natural History*, in fact, was a brilliant *tour de force* – its brilliance a product of his own sagacity – designed to support his firmly held and unconcealed belief that

> the Sacred Scriptures, whose testimony is received by all men of unclouded minds with implicit and reverential assent, declare that it pleased the Almighty Creator to make of one blood all the nations of the earth, and that all mankind are the offspring of common parents.[10]

Whether Prichard's monotheism and its corollary, the notion of the unity of mankind, could have been so persuasively argued in the absence of such a strong scriptural tradition is a question which must remain moot. However, as both Stanton[11] and Jordan[12] have shown, the social, political and economic conditions of slavery in the United States led equally able observers to opposite conclusions on this side of the Atlantic.

The manifest reality of a material progress whose results were everywhere apparent to a rapidly emergent middle class was also an important component of the intellectual background as the nineteenth century moved on; and the very existence of so many evidences of betterment, when coupled with an older tradition of the ladder of being, had significant, if hardly realized, effects upon a descriptive natural history whether in Taylor's early *Natural History of Society* or a more sophisticated version of the same theme – Morgan's *Ancient Society* in 1877. Note in the following passage from Taylor the number of evaluative expressions in what passes for an objective description of the different types of human culture:

When we attempt to take a comprehensive survey of the actual condition of humanity, our attention is not less forcibly arrested by the moral than by the physical differences which offer themselves to our view. One race is in a state of continuous and *progressive improvement*;... every day of its existence produces some new discovery tending to *increase the comforts and conveniences of life; intellectual advancement* seems to keep pace with material improvements; ...to want [i.e. lack] an amount of knowledge which would once be esteemed a glory, is now regarded as a disgrace. In fact, a *progressive advance* is manifest, to which imagination can scarcely assign limits... A second race appears to have set bounds to itself; the evidences of former progress are abundant, but no traces of a tendency to *further and future improvement* can be discovered... Passing over many intervening varieties, we arrive at a race which appears *little raised above the brute creation*; it has few evidences of having ever made progress, and none either of the power or will to advance itself beyond the present condition. There is neither memory of the past, nor *foresight of the future*... (italics mine).[13]

Here is our old friend 'progress'; and here is that familiar trinity of civilization, barbarism and savagery which is still so much with us. In the acceptance of the reality of the former, the efficacy of the latter is assured. And yet it is this assumption – something of a novel instance in the intellectual experience of mankind – which *colors* – and from a later vantage point – taints, the *science* of man which results.

Even more subtle however than the rather obvious impact of such a pervasive notion as that of progress upon the definition and organization of human experience was the effect of the investigator's own intellectual ancestry. Where could one find the telling sign of man's nature, the significant criterion for judgment or for classification? How was this complex phenomenon, this chaotic mass of variety to be reduced to the simplicity required by the new systematics? What was the mark of man? Language – or Biology – or Custom? For those who emerged from a background of medicine and anatomy, it was, of course, those traits so obviously exposed in the dissection room which defined man and his varieties; their persistence beyond custom and society argued for an absence of evaluation which insured their utility in objective classification. For those however who still fought the eighteenth-century battles over mind, it was in language that the essence of man lay for it was here that mind was capable of being perceived in its purest form; and when early ethnologists saw in the study of language the 'true' basis of ethnology, they were claiming for a classification based upon linguistic differences an objectivity hardly beyond argument. And for the traveller and voyager, the ethnological analogue of the zoological naturalist, it was the customs of living groups which served as the essential locus of differentiation for they saw in the *activities* of *groups* of men the important features of his nature. Each perceived – and often with intense precision – the world to be described; each ordered his descriptions with seriousness of intent and purpose; but each perceived and described within the frame of his own experiences and that of the particular traditions to which he was subject. Were these not the blind men describing the elephant? Each described, each contributed – and contributed significantly – to *Science*; but in whose description lay objectivity? Where lay Prichard's truth?

Or, to move on to subtler levels of bias, what of the reality even of our traditional notions concerning the organization of culture itself? In the traditional institutional scheme which orders our view of cultural reality and which serves as the threshold of analysis – that breakdown into social organization, religion, economics, etc. – do we not also impose a reality derived from our own particular tradition – one which goes back at least to Herodotus and the Greek geographers? Is the objectivity of the *Notes and Queries* (or any other similar guide to the collection of data) so clear as the specificity of its questions or the bareness of its outline would make it seem? Does not its organization and the very limitations imposed by its questions mask a bias derived from the very tradition which provides its rationale but from which it would escape?[14] Is there, can there be, an *ethnography* which is objective in the sense that it transcends the cultural tradition of the ethnographer?

Or, to cite one more example, is human behavior so highly skewed in the direction of maleness as both its traditional investigation and results would seem to suggest? Has not the institutionalized role of the male in western culture – first as priest, and then as philosopher and finally as scientist – both obscured and distorted the reality of culture in such a way as to have made objectivity in description and in analysis impossible? In such an instance, Morgan's misreading of the female role in Iroquois society is as instructive as the controversies over the question, in mid-nineteenth century, of female membership and attendance in the meetings of the Royal Geographical Society and both the Ethnological Society of London and its rival, the Anthropological Society.

What I am suggesting here through the use of these few examples, the numbers of which can be extended in quantity and range for the past and the present, is that the effect – or the effects – of our own deep-seated assumptions – assumptions which arise out of our own cultural traditions – is such that the traditional notion of objectivity in viewing other cultures is itself so difficult to achieve as to approximate an impossibility. It is the uncertainty principle compounded; for it is not only the observer as object which intrudes upon and distorts the interactional situation he observes, but it is his culture and tradition – that necessary system which is necessarily precedent to his action – which acts to define, to segregate and to preform the situation, the system, which he investigates and seeks to understand.

Anthropologists, however, are not so naive as not to have recognized the forming notions of their own socialization. Traditionally the anthropologist attempts to protect himself from the more obvious blindness by his training, much as the psychoanalyst does through the search for his own identity through the probing of his own psyche by the instruments he uses for that of others. The initial field work as well as the continuing ethos of field work are designed as the search for a cultural identity which permits a judgmental dispassion; and the older notions of cultural relativism served not only as a concept for the understanding of human variety, but also as a heuristic device for the investigation of particular cultures. Culture shock, so often experienced in the field, is not so much a reflection of the imperfection of the training as it is the sudden awareness of how different the culturally defined perceptions of the real world can be – and how helpless one is to enter into that world without the map provided by his own culture.

One significant aspect, then, of the 'objectivity' problem is this form of 'culturocentrism' – which I would distinguish from the more common ethnocentrism by its lack of an evaluative base. The effect of the observer's own cultural position upon his observations – to say nothing of his analysis – of other cultures, the way in which his own situationally determined sense of problem selects what he sees, can be the subject of an extended and informing lesson in objectivity in anthropology particularly as it intrudes itself into the innocence of the nineteenth century.

There is a more particular and more patent effect of the European or Western cultural background upon anthropology. It is that of a self-conscious science – and the explanatory systems it was constructing – upon the manner in which students of man were organizing an increasing mass of data whose transformation into a theoretical system – as against a collection of curiosa – provided anthropology with its form.

The nineteenth century – particularly its middle third – was a period during which scientists saw themselves as professionals and their science came to be institutionalized as a way of knowing. Expressions of this form of functional specialization include the development of scientific societies as audiences for the communication of esoteric information, the use of scientists *qua* scientists as expert consultants in government service, and the emergence of both science texts and science courses and curricula. One effect of the greater, if limited, control, the greater precision of knowledge *per se* which resulted from this particular kind of specialization, was the establishment of fixed and ordered systems of explanation drawn from the data of observation. Such observations themselves had purpose in that they contributed to the construction of elegant systems of explanation. Such *scientific* systems seemed more real – perhaps more objective – in that they rested upon an objective body of data, realities, objects perceived rather than thought or felt. The success and persuasiveness of such systems provided the guides for future activity. In an expanding public of professionals, the systems thus created served as 'models' for the organization of data which still existed as a mass of unordered observations. With their universes seemingly simpler and more remote than those of man, the physical sciences early established both methods and theories which, in their simplicity and apparent validity, could serve as useful models in anthropology. Like a language in the culture as a whole, a theory as model serves within a science as an essential structure which not only provides the scaffolding for the ordering of data but directs both perception and intellection as well. The efficacy of the model, however, depends upon its fit, upon the actual correspondence between the model itself as a set of known points and relationships and the data which it seeks to organize in its image. If such a correspondence does not exist, if the model does not fit, its continued use results in distortion. Anthropology, in its attempt to order its extremely complex data, has continually drawn upon theoretical constructions from other 'simpler' sciences under the assumption that such correspondences did, in fact, exist; and, in failing to recognize the lack of 'fit' between model and data, it has consistently drawn pictures of the human condition which have simplified to the point of caricature.

One case may illustrate the point:

Cuvier may be said to have invented the comparative method – at least as it came to be used in palaeontology and, by genealogical extension, in anthropology. Faced with the problem of reconstructing life forms from the fossil fragments alone available to him, he noted:

> Comparative anatomy, when thoroughly understood, enables us to surmount all these difficulties, as a careful application of its principles instructs us in the correspondence and dissimilarity of the forms of organized bodies of different kinds, by which each may be vigorously ascertained, from almost every fragment of its various parts and organs. Every organized individual forms an entire system of its own, all the parts of which mutually correspond, and occur to produce a definite purpose, by reciprocal action, or by combining towards the same end. Hence none of these parts can change their forms without a corresponding change on the other parts of the animal, and consequently each of these parts, taken separately, indicates all the other parts to which it has belonged... Thus, commencing our investigation by a careful survey of any one bone itself, a person who is sufficiently master of the laws of organic structure, may, as it were, reconstruct the whole animal to which that bone had belonged.[15]

The technique and his use of it were strikingly successful both in palaeontology and in the development of a scientific *mystique* in the lay community. He and his successors were able to populate the past geological epochs not just with the fossil fragments which the strata produced but with living animals and their environments – organic communities.

The key to Cuvier's success, however, lay in his own knowledge of anatomy and of comparative anatomy, a knowledge of the whole system of a particular organism in which and to which each of the individual parts was functionally and structurally related. It rested upon the validated assumption that an organism was itself an entity, that it was in effect, the simplest unit of organic behavior.

Faced with a similar problem, a half century later, those who attempted to reconstruct human prehistory, borrowed Cuvier's system (because they had been trained in it) and used it as a model for the solution of their own problems. For the prehistorians, the lithic artifacts, stratigraphically segregated, were like Cuvier's fossil fragments; and the organism to which they belonged was a system of behavior – a culture or a civilization. Lubbock makes the relationship specific:

> The archaeologist is free to follow the methods which have been so successfully pursued in geology – the rude bone and stone implements of bygone ages being to the one what the remains of extinct animals are to the other.[16]

And while it was part of his wisdom to recognize the exceptions and to be concerned in accounting for them, E. B. Tylor, early could verbalize the position more precisely:

> If the development of civilization were a uniform process, the discovery of a few objects made and used by a prehistoric tribe would be a sufficient index to determine the exact place of that tribe in civilization. For instance, the implements and carvings of the cave dwellers of Central France would enable us to conclude at once as to the domestic life, arts, customs, government, and religion of this old and interesting people; and even the rude flint implements of the quaternary or drift period would be an exact criterion of the general culture to which their makers had attained.[17]

Here, of course, is Cuvier's principle of correlation applied to a cultural system on the easy assumption – and one which had not been validated by any kind of cultural anatomy – that a culture – or more particularly, a cultural type – represented a real correspondence with the organism as system and that the relationships of its parts were, therefore, equally unvarying. Furthermore, in the absence of any sense of correspondence between the organism as a functional system and the culture as one, it was easy to exaggerate the importance of lithic systems in human cultures since such systems loomed so large in the palaeo-archaeological remains; and, in the application of the comparative method which worked so well in the palaeontological model for the explication of human history, it was equally easy to exaggerate the role or the importance, for purposes of investigation, of those existing cultures with a lithic technology. And, perhaps most significant of all, it was not unreasonable then, to create the temporal hierarchy in the palaeontological image – a hierarchy which, fitting the commonly accepted notion of progress, was so intrinsic a part of nineteenth century anthropology. Despite his own humanism and his own sense of a common humanity, Tylor, using the palaeontological image, could still consistently refer to 'higher' and 'lower' races within an evaluative frame for which a historical rather than a functional criterion was the most important.

The point is, of course, that human cultures are not organisms in themselves nor do present systems necessarily represent unchanged, earlier ones. They are not living fossils in that sense. To investigate them as such is, as many critics of nineteenth century anthropology pointed out, to proscribe the data and to distort understanding.

In summary then, the anthropologist has been – and I suspect is always – faced with two kinds of proscriptive systems: one is that formed of the assumptions which are part of his own tradition; such a system acts as a kind of template which he imposes through his own involvement upon the cultural systems he examines to provide that initial order without which he is completely directionless. The other is the more consciously realized explanatory system which has proved successful elsewhere and which he uses as the tested, the simpler, model for the meaningful organization of his own data. Both the hardly realized template and the consciously utilized model distort the search for data and the systems which result. I do not see how anthropological inquiry can be otherwise. Our problem is not to deny the existence of such devices but to recognize that they do in fact exist and distort, that each new model, far from providing us with *the* Truth, or even a deeper truth, provides us only with a different truth. Objectivity I think is *approached* – I think it presumptuous to suggest that it is attained – when we become aware of the extent of the distortion to which we are subject and its source – rather than through its elimination.

Dept. of Anthropology,
Temple University

NOTES

[1] See, for instance, Jacques J. Maquet, 'Objectivity in Anthropology', *Current Anthropology* **5** (1964) 47–55, in which his concern is primarily that of the effect of the history of colonization on the collection and interpretation of anthropological data in Africa.

[2] Jacob W. Gruber, 'Ethnographic Salvage and the Shaping of Anthropology', *American Anthropologist* (1970) in press.

[3] While one can describe an anthropological concern which is continuous in the western intellectual tradition, i.e. a continuing interest in the *varieties* of man, I refer here to that self-conscious interest in the human species as a part of the natural world which emerges after mid-nineteenth century and which establishes its own set of problems, develops a methodology and attempts a theory about man's origins, his nature, and his place in a changing nature. Such a discipline is already well established and recognized by the 1880's.

[4] Gerd Buchdahl, *The Image of Newton and Locke in the Age of Reason*, Sheed and Ward, London, 1961.
[5] *Ibid.*, p. 12.
[6] See George H. Daniels, *American Science in the Age of Jackson*, Columbia University Press, New York, 1968, especially Chapter 3. For a contemporary defense of Newton vis-à-vis Bacon see David Brewster's *Life of Sir Isaac Newton*, John Murray, London, 1831, especially Chapter 19.
[7] James C. Prichard, *The Natural History of Man*, 2nd ed., Bailliere, London, 1845, pp. 7–8.
[8] Charles Lyell, *Principles of Geology*, 9th ed., John Murray, London, 1853.
[9] Note, for instance, the statement which prefaces *A Manual of Ethnological Inquiry*..., prepared by a subcommittee of the British Association for the Advancement of Science in 1851, London 1952: "We are seeking facts, and not inferences; what is observed, and not what is thought."
[10] Prichard, *op. cit.*, p. 5.
[11] William Stanton, *The Leopard's Spots*, University of Chicago Press, Chicago, 1960.
[12] Winthrop D. Jordan, *White over Black: American Attitudes Toward the Negro, 1550–1812*, University of North Carolina Press, Chapel Hill, 1968.
[13] W. Cooke Taylor, *The Natural History of Society in the Barbarous and Civilized State*, Doubleday, New York, 1841, 1: 2–3. The actual presence of a recently immigrant Irish population in Great Britain and the continuing experience with a disorganized American Indian and an enslaved Negro in the United States served only to reenforce the notions of an evaluative hierarchy of human conditions. Of the Irish in London, a parliamentary report in 1836 could say "The Irish immigration into Britain is an example of a less civilized population spreading themselves, as a kind of substratum, beneath a more civilized community; and, without excelling in any branch of industry, obtaining possession of all the lowest departments of manual labour." (Quoted in Lynn H. Lees, 'Patterns of Lower-Class Life: Irish Slum Communities in Nineteenth-Century London', in *Nineteenth-Century Cities* (ed. S. Thernstrom and R. Sennett), Yale University Press, New Haven, 1969, pp. 359–385).
[14] See Margaret T. Hodgen, *Early Anthropology in the Sixteenth and Seventeenth Centuries*, University of Pennsylvania Press, Philadelphia, 1964, especially Chapter 5: 'Collections of Customs: Modes of Classification and Description'.
[15] Georges Cuvier, *Essay on the Theory of the Earth*, Blackwood, Edinburgh, 1817, p. 90.
[16] John Lubbock, *Pre-Historic Times*, 2nd ed., Williams and Norgate, London, 1869, p. 416.
[17] E. B. Tylor, 'The Condition of Prehistoric Races, as Inferred from Observation of Modern Times', in *Transactions of the Third Session of the International Congress of Prehistoric Archaeology, Norwich 1868*, Longmans, Green, and Co., London, 1869, p. 11. For a discussion of the scientific or intellectual syndrome of palaeontology, prehistoric archaeology and a historically oriented cultural anthropology see Jacob W. Gruber, 'Brixham Cave and the Antiquity of Man', in *Context and Meaning in Cultural Anthropology* (ed. by M. E. Spiro), Free Press, Glencoe, 1965, pp. 373–402.

PAUL W. COLLINS

THE PRESENT STATUS OF ANTHROPOLOGY AS AN EXPLANATORY SCIENCE

1. Introductory remarks

The enterprise undertaken in this paper is a brief exploration of the epistemological status of the various kinds of sentences which typically appear in the body of anthropological studies. Since this analysis presupposes a view of what science is, a few remarks should be made by way of preface in order to make this view explicit. No attempt will be made, however, to defend it.[1]

In the last two decades the objective nature of science in general has come under increasing attack. In this paper, however, the old-fashioned view that science is an objective, on-going enterprise controlled by empirical enquiry is taken for granted. In other words, for the purpose of the present analysis, we accept what Israel Scheffler (1967) calls the 'standard view of science'.

The standard view of science generally is associated with the following contentions:

(1) As analyzed formally, science can be defined as a set of true, universal sentences systematically related.

(2) Well-developed sciences can be analyzed formally as basically consisting of two types of sentences having different epistemological status: namely, theoretical sentences and experimental laws.[2]

(3) The sciences can be fruitfully regarded as explanatory disciplines. Accordingly, the analyses of explanation developed by scholars such as Carl Hempel (1965, Chapter 10) and Ernest Nagel (1961, Chapter 2) are accepted in their major details. It will be important, however, to make somewhat more explicit the nature of scientific explanation in order that the subsequent discussion of anthropological sentences may be made more intelligible.

2. Explanation in science

When we ask a question of any subject matter, the answer given, if it

R. J. Seeger and R. S. Cohen (eds.), Philosophical Foundations of Science, 337–348.

meets certain criteria, is an explanation. In this context 'explanation' is a technical term, and to be more explicit the term 'deductive-nomological explanation' will be used. A deductive-nomological explanation is one in which one or more law sentences appear in conjunction with sentences describing the conditions to which the law or laws apply, and on the basis of purely formal or logical operations, derive a sentence describing the phenomenon to be explained, i.e., a sentence in the declarative form of our question sentence. Carl Hempel has schematized the deductive-nomological type of explanation in the following way (Hempel 1965, p. 249):

	$C_1, C_2, \ldots, C_k$	Statement of antecedent conditions	Explanans
	$L_1, L_2, \ldots, L_r$	General laws	
Logical deduction	———	———	
	E	Description of the empirical phenomenon to be explained	Explanandum

This model of explanation will form the basis of future discussion, and some of the key terms need to be briefly discussed.

(a) The term 'deductive' appearing as part of the designation 'deductive-nomological' indicates that the conclusion of such explanations (the 'explanandum') must follow deductively – i.e., according to the rules of logic – from the premises (the 'explanans').

Accordingly, when this criterion is met, and the sentences forming the premises of an explanation are true, we know not only that the explanandum is true, but also that it could not be otherwise than true.

(b) The term 'nomological' in the designation 'deductive-nomological' indicates the requirement that at least one law of nature must appear as a working premise in any explanation.[3]

3. The Systematic Nature of Science

The most important characteristic of a scientific law lies in its interrelatedness with other laws. It has been indicated earlier that science, formally conceived, is a set of systematically related universal sentences which meet accepted criteria of empirical confirmation. By 'systematically related' is meant that the laws belonging to the set possess related scientific terms so that none of the laws are isolated from the others. Terms can be related in many ways, but basically laws are related through the medium of other laws and by their common relationships to either the defined or primitive terms appearing in a theoretical calculus. Carl Hempel (1966, p. 94) describes this characteristic of scientific laws somewhat metaphorically as follows:

> Scientific systematization requires the establishment of diverse connections, by laws or theoretical principles, between different aspects of the empirical world, which are characterized by scientific concepts. Thus, the concepts of science are the knots in a network of systematic interrelationships in which laws and theoretical principles form the threads. ... The more threads converge upon, or issue from, a conceptual knot, the stronger will be its systematizing role, or its systematic import.

Accordingly, an isolated 'law' has very little or no explanatory import. For example, the disconfirmation of the 'law' that all swans are white if of little interest to us because of this circumstance. On the other hand, we increase our knowledge when we are able to transform an unrelated universal into a law by finding systematic relations between it and other laws. When laws are thus systematically related, they confer 'indirect evidence' upon each other, thereby strengthening their empirical content. For example, the laws of the pendulum and the laws of planetary motion are systematically related so that direct evidence for one constitutes indirect evidence for the other.[4]

In short, it seems necessary that any science possess a network of interrelated laws in order to confer adequate power to its explanations.

4. Probabilistic Laws

Anthropologists, by the very nature of their science, must begin their inquiries with subject matters which are already highly organized. Man and his ways exhibit behavior patterns which are possible only as a consequence of the development of complex biological and psychological structures, and the uncertainties and variations associated with such organizations are part and parcel of the very subject matter which the anthropologist must face. Thus he finds himself dealing with the logical complexities associated with probabilistic laws, questions of strength of evidence, and the limitations of system-specific regularities.

Thus anthropologists must be content with probabilistic explanation since its generalizations can themselves seldom be more than probabilistic. Probabilistic laws, however, can belong to an interrelated network, and thereby attain indirect evidence and provide explanatory power. Indeed, contemporary physical theory is characterized by probabilistic laws with a high degree of systematic unity. For example, it is commonly known that the state-description of quantum theory is associated with a statistical interpretation. Accordingly, the derivative laws of quantum mechanics will themselves be statistical, and will possess derivative probabilities calculated on the basis of the mathematical theory of statistical probability. As a consequence, these laws are systematically related in explicitly statable ways, and direct evidence for any of these laws constitutes indirect evidence for the others.

In the case of probabilistic laws in the social sciences, indirect evidence should also be available. However, it is generally the case that the social sciences do not possess a body of laws which are explicitly formalized in such a way that the availability of indirect evidence can be assessed. Many generalizations are loosely related, which no doubt does provide some indirect evidence. But until such relationships are more adequately formulated and made more precise, it will remain impossible to assess the status of such generalizations.

This situation is especially unfortunate in the social sciences where available samples for establishing probability relations are frequently limited. If a formalized theoretical structure were available to provide indirect evidence for statistical probabilities involving limited samples, the probabilistic laws involved would be considerably strengthened. As it is, the law stating that almost all societies with Crow-type terminology also have some form of matrilineal descent illustrates this shortcoming. Although we may intuitively feel that other probabilistic laws are involved (such as the probability that people will structure their linguistic categories on the basis of their social organization), these laws have not been formalized and established.

However, since it is after all the case that generalizations in the social sciences are more or less loosely related in various areas of inquiry, there is every reason to expect that future developments will serve to gradually 'tighten' these relationships.

5. System-specific laws

In general, philosophers of science direct their attention to generalizations having the status of nomic universals or probabilistic laws since there are good reasons for taking the position that such generalizations are essential for explanation. Nevertheless, in the literature of anthropology there is a large body of generalizations which are important for anthropological analysis but which fail to meet the requirements established for explanatory purposes. These generalizations appear in more or less rudimentary form in the ethnologies of various populations: descriptions of marriage customs, trading relationships, residence patterns, and so on. Such sentences describe the behavior of all or of a class of the members of a given population given certain specified conditions, and can be conveniently referred to as *population-specific laws*. These limited regularities form the basic materials of anthropological analysis. They are descriptions of behavior patterns of a specific population, and may or may not be formulated in probabilistic form, although in principle they always admit of exceptions.

Population-specific laws, as we have indicated, constitute raw material for anthropological research and theory construction. Specifically, in the present context, they form the basis for the analysis and discrimination of cultural systems, and the formulation of *system-specific laws*. In systems analysis we begin with a number of variables discriminated according to a preliminary hypothesis. For purposes of illustration, let us assume that the following variables selected with reference to a given population constitute a system: the size of pig herds kept by members of the population, the crop return per acre of land under cultivation, the number of participants in a specified religious ritual, and the frequency with which the ritual is performed. With respect to these variables, then, let us assume that over a given period of time, as crop return decreases, the pig herds increase. As pig herds increase, furthermore, participation and frequency of rituals increase. Then, as ritual frequency and participation increase, pig herd sizes decrease, and so on. Relationships such as these, if properly refined, may be able to form the basis for framing system-specific laws, perhaps statable in mathematical form.

5.1. *Explanatory Value of System-Specific Laws*

By the very nature of the notion 'system-specific' it is evident that laws of this type do not qualify as genuine laws. As a consequence of our general knowledge concerning the universality of social change, we know that system-specific laws have a limited lifetime. However, the temporal period during which these laws are operative can be quite extensive, and during this time, the laws do have predictive and explanatory power.

The limited predictive and explanatory power which a system-specific law possesses derives from the fact that although it can be inferred from the context of its assertion that instances in the scope of its predication exist and are finite, the number of instances is not fixed for the temporal period for which the law holds. In other words, even though members of a given population do not remain constant, the system-specific laws will hold for some time. This situation differs from the obviously accidental universal 'all the screws in Smith's car are rusty', where the instances falling in the scope of predication (the screws in Smith's car) are limited to a finite fixed number of existing instances at a specific time (presumably at the time the universal is asserted).

Accordingly, although a system-specific law cannot serve as a premise in a deductive-nomological explanation, it can serve as a premise in an explanatory model whose conclusions will be true for a limited temporal span. If one is unable to explicitly state the duration of this period, however, the use of such laws is weakened.

5.2. *Heuristic Value of System-Specific Laws*

In spite of the fact that system-specific laws cannot qualify as nomic universals, they deserve our careful attention for at least two reasons: the practical value which can be derived from a knowledge of their systemic properties and the heuristic value which they may possess in directing our attention to more general laws of which the system-specific laws may be an instance.

To establish the first point, a few remarks by way of illustration should suffice. The discrimination of specific systems has been for some time the major concern of ecological approaches in biology. Relationships of energy transfer between various biota of a given region have been studied with a great deal of success with respect to the discrimination of specific ecological systems, and, as is well known, the results of these studies have had important practical and economic significance with respect to human populations. The possession of system-specific laws for such contexts has made it possible to control various ecological relationships to the advantage of human populations. Furthermore, even in the physical sciences attention to system-specific relationships has been essential to the extension of engineering techniques from the construction of such simple systems as the common thermostat to the more complex homeostatic systems employed in gun-aiming mechanisms and automatic pilots.

Similarly, the understanding afforded by the discrimination of specific systems in which human behavior taken collectively plays a significant role will no doubt be essential to the contributions which can be made by applied anthropology.

With respect to the heuristic value of system-specific laws, there is always good reason to believe that important steps towards the discovery of general regularities are made when it is possible to discriminate a specific system having a high degree of regularity in the relation of its variables, even though these latter regularities have a limited application.

To give a historical illustration which might illuminate this point, we can turn to the early career of physics. As proposed by Kepler, the celebrated three laws of planetary motion were system-specific laws relating to our solar system.[5] Kepler's own interpretation of planetary behavior was a rather esoteric mixture of mathematics and mysticism, whereby the sun possessed a world soul regulating the planets according to mathematical principles. He identified the sun with the First Person of the Christian Trinity. In short, as Kepler conceived his planetary laws, they were limited to a specific system and had their basis in the rationality of God.

Of course, the Keplerian laws have subsequently been shown to be derivable from the Newtonian laws of motion, and as a consequence have lost their system-specific character.[6] But at the time that they were framed, the scope of their predication was limited to a given finite number of bodies having a finite history. Still, the possession of these laws led, through Galileo, to the Newtonian formulation of the four laws of motion. These latter laws had extraordinary explanatory power, and elevated the science of physics to a well-established discipline.

It would be foolhardy to claim that the discrimination of system-specific laws in either biological ecology or anthropological analysis will lead so dramatically – or at all – to general laws having the explanatory potential of the Newtonian laws of motion. An entirely different route may turn out to be more effective. Nevertheless, such discrimination may easily be a step in the right direction.

6. Explanation in Anthropology

Our discussion to this point has, by implication, made it clear that accounts which appear in anthropological literature seldom, if ever, meet the strict requirements of scientific explanation. It is true that generalizations and probabilistic laws are implicit and even explicit in many analyses, but most often the terminology employed in these generalizations is too vague or too general to have much empirical import, and where some precision has actually been achieved, interrelatedness with other such generalizations remains unclear. At the present stage of anthropological inquiry, therefore, we are unable to readily assess the status of a generalization with respect to its possession of indirect evidence.

It is of course true that considerable advance has been abundantly evident in the concern for uniformity in the operationalizing of anthropological terminology, and the scope for the applications of such concepts is continually being broadened. Nevertheless, the complex nature of anthropological subject matter presents a continuing challenge to the investigator and it is not surprizing that a great deal more remains to be done to organize the principles of cultural behavior into an interrelated network of laws.

7. Theories and Hypothetical Orientations

Although we have indicated that in its present stage of development anthropology has not achieved a formal structure generally sufficient for the exercise of explanatory power, there is no reason to hold that this power will not be forthcoming in the future. The career of the physical sciences has demonstrated that the achievement of a formal structure has come by slow steps culminating in a formal theory. Since theories function to show how existing experimental laws are related, to indicate how they can be made more precise, and to guide inquiry to the discovery of new laws, a few final remarks on the status of anthropology from this perspective are called for.

7.1. *The Ideal of Science*

The ideal of any science is the development of a single theoretical framework which can relate and explain all phenomena within the scope of its subject area. This ideal has never been completely achieved by any science, although physics has had some dramatic successes, and has continued its inquiries in this direction.

Newton's great achievement, for example, was the unification of terrestrial and celestial mechanics into a single theoretical framework. From the axioms of his theory, plus various special assumptions, all the established laws of these two sciences could be derived. Nevertheless there were other subject areas regarded as falling under the scope of physics which could not be subsumed under Newton's postulates, such as thermodynamics and optics. Further developments, however, showed that with a few additional assumptions, the postulates of thermodynamics could also be derived from the axioms of mechanics. In this way, more and more phenomena gradually became included within the scope of a single theory, although, of course, the theories involved in this unification become altered in the process. And even though the process has by no means been completed, it is in this direction that physical inquiry proceeds.

The guiding principle in the attainment of this ideal of science is the formalization of an axiom set permitting the derivation (with special assumptions) of all known laws in its subject area.[7] When this has been achieved, and when the axioms and theorems have been related in various ways to the empirical world, these axioms and its theorems become known as a 'theory'. Hence the term 'theory' as used here is a technical one, and is not used in the loose sense of a vaguely formalized conjecture.

At the present stage of its development, anthropology clearly has no theory in the sense described above. However, insofar as anthropology is guided by the ideal of science in its experimental performance, it shares in the basic enterprise of all science.

7.2. *Hypothetical Orientations*

Basically, a theory is a complex postulate system possessing certain additional features not present in a non-empirical postulate system such as uninterpreted Euclidean geometry. These extra features serve to relate a theoretical postulate system to the empirical world. Nevertheless, the structure of Euclidean geometry is a good illustration of the formal aspects of any theory. First of all, there is an axiom set from which, by the use of 'rules of transformation', countless theorems can be generated. The axiom set of Newtonian mechanics are the 'laws' of motion, and derivable from this set as theorems are such laws as the law for the period of a simple pendulum and the law for freely falling bodies. By having a theory, the significant variables for analysis are delineated, indicated by the primitive and derived terms appearing in the axioms and theorems. In Euclidean geometry, for example, among the primitive terms are 'point', 'line', and 'plane'. Among its many derived terms are 'triangle', 'square', and 'circle'. Primitive terms of Newtonian mechanics include 'mass' and 'energy'. Among its derived terms are 'pressure' and 'acceleration'. The primitive terms of a theory are relatively few in number, but the derived terms, which are defined in terms of the primitives, are frequently extremely numerous.

The axiom set of a theory explicitly formalizes the relationships which hold between the primitive terms (and, sometimes, derived terms). This formalization has the ultimate function of integrating existing knowledge into a unified deductive system. But the fruitful aspect of a theory lies in its ability to derive relationships which had heretofore not been suspected, leading to the formulation of new laws of nature. Furthermore, it enables one to make existing laws more precise, and to relate previously unrelated laws into a single system.[8]

Anthropology, as we have indicated, at present has no formalized theory that can function with precision in the ways described above. At its present stage of development, it is in a state similar to that of physics prior to Galileo. However, just as physics prior to Galileo had various more or less precisely formalized 'approaches' to its subject matter – some wrong, some misleading, some fruitful – anthropology also has these 'approaches' to its subject matter. These approaches we will call 'hypothetical orientations'.

Hypothetical orientations function in very much the same manner as theories, although in a much less precise way. Accordingly, they tend to suffer – sometimes badly – from internal ambiguity. Nevertheless, insofar as they are usable, they serve the following roles: (1) they discriminate the relevant variables for analysis, (2) they provide a loose network of more or less vague or unformalized propositions justifying the relevance of the variables discriminated, and (3) they provide data from specific field studies which, in more or less determinate ways, support the loose network of propositions. Because of the looseness of the formalization, however, it is often difficult to access the relevance of alleged supporting data.

As anthropology stands today, the scope of the hypothetical orientations available may be limited to a specific population. Others, however, such as the attempt to explain culture patterns on the basis of child-raising practices, have universal scope. Nevertheless, it is in the gradual weeding out of the fruitless and misleading hypothetical orientations and the strengthening of the fruitful ones that anthropology proceeds in its attempt to attain the ideal of science.

Dept. of Philosophy,
State University of New York at Oneonta

NOTES

[1] A recent and excellent defense can be found in Scheffler (1967).

[2] Scheffler calls this position the 'two-tier picture of science', 1967, p. 47.

[3] For a comprehensive analysis of the characteristics of laws, see Nagel (1961), Chapters 4 and 5.

[4] For an explication of the concept of indirect evidence see Nagel (1961) p. 64–65.

[5] These laws were: (1) each planet describes an ellipse with the sun at one focus, (2) a line drawn from the sun to a planet sweeps out equal areas in equal times, and (3) the square of the time a planet requires to complete its orbit is proportional to the cube of its mean distance from the sun.

[6] This conclusion, however, is not altogether free from doubt. Kepler's laws make specific note of the situation that the conic section traversed by the planets is an ellipse. Newtonian laws alone, however, cannot serve as premises in deriving this conclusion. As Nagel observes, "additional premises seem to be unavoidable – premises which state the relative masses and the relative velocities of the planets and the sun." (Nagel, 1961, p. 58fn). It is not inconceivable, therefore, to claim that the need to specify specific features of our solar system to derive the Keplerian laws renders them system-specific after all. Indeed, if we consider the clearly system-specific law that all the screws in Smith's car are rusty, it is evident that if we are given certain conditions taken from the history of Smith's car (e.g., that it was driven into the ocean), we should be able to derive the accidental universal about the screws in Smith's car as an instance falling under the scope of various nomic universals (such as laws concerning the oxidation of metals).

[7] The relationship between theories and laws is complex, and is highly oversimplified in this discussion.

[8] A formal consideration of the nature of theories is beyond the scope of this work. The interested reader should consult Nagel (1961), Chapters 5 and 6; Hempel (1965), Section III; and N. R. Campbell (1920), Chapter 6.

BIBLIOGRAPHY

Campbell, Norman R., *Physics, the Elements*, Cambridge, England, 1920.
Hempel, Carl G., *Aspects of Scientific Explanation*, New York, 1965.
Hempel, Carl G., *Philosophy of Natural Science*, Englewood Cliffs, N.J., 1966.
Nagel, Ernest, *The Structure of Science*, New York, 1961.
Scheffler, Israel, *Science and Subjectivity*, New York, 1967.

ANTHONY LEEDS

'SUBJECTIVE' AND 'OBJECTIVE' IN SOCIAL ANTHROPOLOGICAL EPISTEMOLOGY*

I

Anthropology has often been accused, especially by social science disciplines which have modelled their epistemologies and methodologies on those of the physical sciences,[1] of being not only without rigor but especially of not being objective. Being subjective is, of course, bad.

The content of the accusation appears to involve the following allegations:

(a) that assertions about the reality under observation are impressionistic and emotionally-based, hence shaky at best and of no ultimate scientific value;

(b) that observations of that reality are non-replicable or, if they replicate, it is more or less by accident since there are no sound epistemic foundations for the observations;

(c) epistemic foundations are absent because the observations are made by an individual (especially through participant-observation procedures) without appropriate instrumentation (with the assumption that instrumentation solves the problem);[2]

(d) individual observations or judgements are inherently subjective, i.e., unique, personnal, *a function of the personality of the observer rather than of the reality observed.*

This set of allegations represents a very wide-spread conception of the nature of the subjective-objective opposition, one in which, plainly, the subjective and the objective are treated as being in the same domain of discourse.[3] This conception of the opposition underlies much of the interdisciplinary sniping and back-biting which is so often heard between sociologists and anthropologists, especially in joint departments, and which indeed reflects quite different ontological and epistemic viewpoints.

In what follows, I wish to argue that the opposition is a false one and to present some quite different considerations in the matter.

R. J. Seeger and R. S. Cohen (eds.), Philosophical Foundations of Science, 349–361.

In brief, I shall argue, first, that what have been treated as two poles within a single sphere of discourse are indeed subject matters of two quite different spheres of discourse – the subjective is in the sphere of axiology, the objective in that of epistemology. Second, I argue, therefore, that the relevance of the two to each other is quite different from that presented in most discussions of the matter. Third, I assert that objectivity does not *per se* have anything to do with individualness, multiplicity, or collectiveness of observation. Fourth, instrumentation is, *per se*, irrelevant to objectivity, and, fifth and last, that anthropological observational field techniques of individual participation and other forms of observation have, in themselves, always been adequately "objective" forms of observation.

II

That which the subjective basically boils down to, in the most unsophisticate form of the attack against it, involves a proposition that an observer undergoes a unique, unreplicable, *internal* experience which is, at best, conveyed to others through a personal metaphoric language.[4] In its more sophisticated form, the proposition is that a unique, unreplicable, personal interpretation is given to sensation or experience by the observer or 'subject', thus representing more his internal state than the external object of sensation. Notice, for example, how this is built into the very conceptual structure of experimental psychology which works with 'subjects' – and how this contrasts with anthropology, which works with 'informants'.[5]

'Objective', on the other hand, is conceived to involve the proposition that some of sensations or experiences or some set of interpretations of sensations or experience are, in principle, indefinitely replicable and repeatable – i.e. invariant.[6] The test of such replicability is 'intersubjectivity' or repeated, like observations by different 'subjects' or 'subjective individuals' or their instrumental substitutes. This 'objectivity' – and with it, ostensibly, scientific truth – is created, on one hand, by eliminating those subjective sensations and experiences which are supposedly unique to the individual by a kind of consensus process – what we all agree on is true and objective – and, on the other, by getting invariance in observation.

III

A number of considerations appear to me to make this most widely held conception of objectivity or of subjectivity-objectivity untenable.

First, we are given the fact that, epistemologically, in an ultimate sense, one can never empirically *know* that the sensations or experiences in one observer over several observations or in two or more observers observing the same event, or similar events, or each of them observing discrete events, are the 'same', 'isomorphic', or even similar. This fact has almost universally been connected with the idea of subjectivity. But it only has the connection if one makes the assumption that objectivity is congruent with invariance. Such as assumption is borrowed from classical scientific epistemologies and cannot be held for the historical sciences nor for historical dimensions of physical sciences (such as stellar evolution).

Second, all sensation and experience, even when mediated through the most refined instrumentation, must ultimately pass through interpretation by individuals.

Third, conceptually, this passing through individuals – the interpretive process – is closely associated with the unique and unreplicable.[7]

Fourth, as a result, what is called 'objective', is, in fact, essentially convention, or as I said above, consensus, in our western scientific methodology.

Fifth, therefore, the locus of what is ordinarily conceived of as 'objectivity' is actually in the sphere of our meta-language, rather than in the sphere of perception, sensation, and experience, or even in the sphere of our object language, although it has generally been thought to lie in the sphere of sensation.

Sixth, insofar as conventions, consensus, and meta-languages are subject to socio-cultural determinations, the entire body of what may allegedly be, and usually, is conceived as 'objective' may have little real relation to sensation or experience and may simply be replicated metaphores of 'unique' culturally-determined interpretations and private worlds – a kind of collective subjectivity by which 40000000 Frenchmen can, and indeed sometimes are, wrong. The notion of 'democracy' as frequently used in political science is a case in point; the 'culture of poverty' is another; the 'impoverishment of Negro linguistic facility' is another; the 'radicalization of rural migrants to the city' is another; 'the ether' and 'phlogiston' still others.

IV

Let me elaborate on the foregoing from another direction. I make the proposition that *all* sensation and experience is objective in the sense that some thing is observed by someone: characteristics and forms of an object are perceived. That the object be observable by more than one, or only one, observer or instrument only once does not make the thing more or less real nor its observation by an observer more or less 'objective' or 'subjective'. It only means that the thing is more or less accessible to observation. Low degrees of accessibility to observation have standardly been considered as subjective, especially when the accessibility is restricted to one observer, a point returned to below.

Thus, two classes of events become available for interpretation in our scientific work. First there are those which could be observed by more than one observer but are observed by only one because of the absence of other observers at the time of the event. This is the characteristic situation of the field anthropologist, although, in fact, there *are* other observers always present: the participants in the events who are ordinarily accessible, at best, only to other anthropologists, and that rarely.[8]

Second are intracorporeal events which are not accessible to any other observer under presently known technical conditions, but are keenly observed by the observer inside whom they occur.

Thus, dreams, headaches, and heartaches are quite *observable*, and objectively observed in the sense of the proposition stated above. Virtually no one (outside the operationalist experimental psychology laboratory) doubts their existence, their reality. No one asserts that these entire classes of phenomena do not exist or are figments of individual and collective imagination, like hippogryphs. What is more, we virtually *all* even accept the *description* of the dreams, headaches, and heartaches as *factually* correct within different degrees of refinement. We accept as correct, the characteristics and forms attributed to the intracorporeal object under observation, i.e. we take the description as objective.

Still further, there are entire bodies of praxis built upon the postulation that these sensations and experiences on the part of individuals are objective, i.e. correct observations of real objects – specifically medicine and the psychotherapies (not to mention human relating in general). The degree of success of the praxes may, indeed, be a measure, or even a function, of the degree of accuracy of the objective description using the unique and unreplicable sensations and experience.

I think that these kinds of events provide us an extreme case which permits us to see the real locus of the subjective-objective problem clearly. It is not in the uniqueness and individuality of the experience – because no one really doubts the epistemic validity, the reality, of what was observed, nor the individual as observer, nor even the reliability of the observer.

Indeed, the locus of the problem called 'subjectivity' is in an entirely different sphere – that of the axiological-interpretive, rather than the epistemic-ontological. The essential problem referred to in discussions of subjectivity is the question of evaluation – interpretation, meaning, value attached to things observed. Meanings appear to have little or no basic anchoring in objects as such or in sensation. No logical routes from sensation to meaning or vice versa have satisfactorily been established, although the attempt to do so is repeatedly made. The best or only route is still insight, involving Einstein's 'free creations of the human mind'. But even such insight presupposes other, anterior attributed meanings. Some of these insights as to meanings and the nature of things – like ether – evaporate. This is especially true of those of an evaluative nature, as, over time, the socio-cultural foundations of value judgements slowly transmute (whereas established ontological truths, including mathematics, continue).

In this regard, what is interesting but generally overlooked in the context of the discussion of objectivity-subjectivity, is that the axiological can have as locus *both* individuals and aggregates or collectives – even entire societies – in that an entire society's understanding of some thing may be based not on sensation and experience but on evaluative criteria or purely fictional meanings (like 'god'). So based, such an understanding can, in principle, be totally false from an epistemic point of view, though axiologically true (again, like 'god'). More commonly, we find meanings of the evaluative and fictional kinds attached to sensory and experimental content by links which appear plausible until carefully scrutinized – I have cited examples above.

The conclusion of this section is the assertion that what is usually referred to as being subjective is this axiological dimension attached to sensation and experience and not the sensations and experience themselves. The latter are essentially "objective", always, and on sound epistemic grounds. The problem of objectivity-subjectivity arises not in observation as has so often been asserted, not in the form of observation, but *in the acts of giving meaning*, especially evaluative meaning, to experience and observation.

V

A number of points of particular interest to anthropology are raised by the preceeding considerations.

First, the view of what is objective taken here broadens the range of epistemically valid experience usable by social scientists – and, in fact, always used by anthropologists. That experience includes unique events and inward individual observation, either their own or as reported to them by others, in other words, discrete historic incidents and cognitions of self and others. We need not here inquire as to what it is that the discrete historic incident tells us about the reality we are investigating other than to say that it provides a kind of index of process, of the confluence of repeatable ('lawful') and unrepeatable ('accidental') – objective – events and processes of the society under study. It gives us clues as to the linearity of process rather than the fixity of process more typically studied by sociology – the purportedly humanly universal, lawful, infinitely repeatable forms of action and interaction treatable, ostensibly, by the methods growing out of a probabilistically-conceived ontology, borrowed from physics.

At the same time, the objective, as here discussed, makes available the entire range of individual public and private experience as material which is epistemically valid for observation and interpretation. This too has been reflected in anthropological tradition in the form of collecting biographies, dreams, visions, and the like. The assumption has universally been made and universally been accepted that, as *data*, these materials are objective, real, accurate, valid, and functions of the societal realities in which they occur.

A second consequence has been the rather notable unconcern for an epistemology related to a probabilistically-conceived ontology. It is not that anthropologists have eschewed statistics and other mathematical forms – they have even innovated some – but, as one would expect from what I have said above as to what is conceived as objective that the use has been restricted to only those kinds of things which are intrinsically multiple and distributed. Even then, this use has usually been made only after some assertion has been made as to the relationship of such things to other things. The use of statistics in anthropology appears mostly to have been quantitative statements about previously determined structures which have developed from sequences of historical causes rather than as a methodology for the discovery of order in a world of chance. The position appears to be that a descriptive statistical statement is as much of an abstraction as is a relational statement but provides much less information and has less meaning-content, i.e. uninterpreted, it is of little interest to theory which is concerned mainly with qualitative relationship.

A third consequence relates to instrumentation. Anthropologists have been quite resistant to instrumentation – especially of the kind called questionnaires. In part, this follows from the epistemological viewpoint concerning statistics but also it arises from a conception – still largely implicit but reflected in the discussions of 'cultural relativism' and the current tidal wave of 'ethno-science' – that any observer external to a system observed is an incalculable interference in the system. Anthropologists have consistently maintained the necessity of minimizing the observer effect – hence participant-observation, a technique of attempting to fade into the background when you stand out like a sore thumb. Therefore, anthropologists learn the native language, native categories, etc., and use native informants and helpers wherever possible, even, say, in applying questionnaires. Any instrumentation is an interference with the observed system. This is particularly true with human populations which amplify the effect of the interference because they are also observers who can generate new responses and feed them back into the observer-system relationship in an endless regression of effects (see MacKay). The more pre-fabricated the questionnaire, and the more generalized the population to which it is to apply, the greater the interference effects, and the less interpretable the feedback, especially if the questionnaire is the main or only investigation technique being used.

Feedback from the interference effect can be of two kinds: (a) one, consisting of evaluative or axiological material, going into the responses which constitute the material of observation (e.g. giving answers to the questionnaire items that the respondents think the investigator wants to hear – anticipating his values – rather than their own perceptions and experience) thereby deforming the desired response material in the questionnaire, or, (b) a feedback to the researcher about the response by personnel of the system under observation to the interference as such, without a feedback into the responses to the content of the instruments. Combinations are of course also possible. Both interference effects give information about the system but the first cuts off or distorts the information desired in the instrument. The second gives more positive information since it gives data about the meaning of the questionnaire and its items to the respondents but does not necessarily distort response to the questionnaire items.[9]

In sum, I think that the anthropologists' skepticism about instrumentation for use among human populations is epistemically well-founded and healthy and that the very common field procedure of a lengthy period of intensive participant-observer ethnographic field work prior to instrumented quantitative work is epistemologically correct. Instrumentation does not resolve problems of objectivity, an expectation growing out of irrelevant mechanistic models and the assumed necessity for invariance.

A fourth consequence relates to the use of the anthropologists' emotional responses in the field as a source of data. Although publically this is eschewed, all of us know right well – and say so in the intimacies of the reminiscense sessions – that our emotional reactions – our 'subjective' evaluative responses to field situations – have often been exceedingly important in discovering objective aspects of the society studied. Epistemologically, this should indeed be so – the emotional response is almost always *to* some thing 'out there' and provides a cue to that thing, even to its characteristics. That is, the emotional response's 'objective' dimension. Where, because of the strength of the emotionality attached to the thing, I may doubt the validity of the *meaning* I attach to it, I can and must then compare my experience with that of others in order to attach a non-evaluative, non-fictional, or objective meaning to it, in other words, to sort out the ontological dimension from the axiological. It may be that similar emotional reactions will be discovered among a great number of differentiated observers, native and foreign alike (e.g. disgust and fear of the sexual tortures currently occurring for political reasons in Brasil) – and thereby also become a datum about the society (e.g. first, the feeling itself as datum; second, the fact of the deliberate use of terror and threat for control purposes in Brasil). This consideration leads me to recommend to all anthropologists in the field the constant examination of that set of *objects* that is comprised of their own emotional responses which often gives differential clues (e.g. why *this* emotion rather than another?) to the objective extracorporeal things in the society under study.

Finally, the peculiar field of operation of anthropologists – exotic cultures – has tended to immerse them much more frequently, forcibly, and ineluctably, for prolonged periods of time, in strange and intractable empirical worlds than members of any other discipline. The significance of this fact is that the anthropologists have constantly been exposed to sensations and experiences, to immediate levels of observation, which are outside the range of common experience and observation, encoded in common-language representations, to which evaluative and fictional meanings, individual or conventional, have, through time, become attached.

This anthropological field experience forces a constantly greater epistemic clarity as to what the nature of the data actually is – as nearly prior to interpretation as any human knowledge ever gets, and certainly prior to value-oriented interpretation. In principle, it should have forced anthropologists into great clarity about their central axioms and postulates, although they have rarely engaged in the endeavor of describing either. The lack of such endeavour and the consistent exposure to new, raw, exotic field data, resistant to culturally imposed interpretation deriving from precepts of their own culture, have led to a kind of philosophico-scientific naiveté (even sometimes a kind of dispraisal of methodological sophistication!) on the part of anthropologists which has often been commented on, slightingly by members of other disciplines, rather self-abasingly by anthropologists. Nevertheless, this naiveté has, in a way, represented the great strength of anthropology – its epistemic solidity, its 'objectivity' – since it has, until perhaps very recently, resisted having its panhuman materials encoded into the axiological – or subjective – frames of reference of our western cultural presuppositions that so badly afflict the other social sciences, especially political science and sociology.

Put another way, anthropology, because of its operational commitment to the study of the widest possible range of cultures, has tended to avoid the ethnocentrism and the axiological infusions into epistemology and ontology of the other social sciences. It has tended to express its science in ever-more generalized value frames, somewhat in the manner that Rapport considers essential (1969: 182–3) while some anthropologists, myself included, hold that it is, in principle, possible to extirpate *all* value interpretation from the science, leaving only the problems of non-value-oriented meanings and theoretical interpretation for exploration.

Boston University

NOTES

* I wish to thank Michael Kosok for a reading of this paper and for the discussion which followed; it illuminated many points. I am particularly indebted to him for clarifying the problem of observer effects on socio-cultural systems and the question of the utility of perturbation observation.

[1] The physical sciences, even today, have perhaps not satisfactorily resolved the multibody problem, almost the entire corpus of theory and method having, therefore, based itself on a two-body model. Two-body models have been carried over into social science methodology in the form of sociological and political-science work framed in terms of 'independent' and 'dependent' variables related by purportedly mathematico-deductive logic. Anthropology has rarely used this sort of model either as one of causal relations in the real world or as an underpinning for methodological procedures; it has tended to emphasize notions such as the 'complex', 'Kreis', 'pattern', 'configuration', 'system', etc., more or less in that historical order, all of them multibody notions. Again, other social sciences have been pervasively committed to statistical approaches, apparently operating with certain assumptions about the virtual *de facto* infinitude (or at least vastness) and repetitiveness of the universe treated so that probabilistic methods are deemed appropriate, again on a physicalist model. Anthropology has characteristically eschewed such assumptions and with them, most of the forms of inferential statistics based on the laws of chance. As another example, in America, Percy Bridgeman's operationalism did not originate, but widely influenced and gave impulse to, various forms of operationalism and behaviorism in psychology (e.g. B. F. Skinner and followers), in sociology (e.g. Lundberg, Schrag, and their followers), and in anthropology (E. Chappell, and briefly, C. M. Arensberg, with few or no followers), while, in the 1960's, a form of operationalism has hit political science, advocates of which call themselves 'behaviorists' (see Boring, 1929/1950, 'Objective Psychology', pp. 631–63, especially pp. 653–9 on 'operationism'). Strict operationism tends to put emphasis only on 'reliability' and to eschew 'validity', (see, also, Woodworth, 1931/1948, pp. 116–9), i.e., rules of correspondence for empirical propositions

(a weakness very characteristic in operationalist sociology and political science) to establish truth value of assertions. Anthropologists have always, in a vague way, been concerned with the latter, that is, with the interaction of observer and what there is to be observed, a concern consistent with the emphasis on participant-observation. Another model carried over to the social sciences is that of measurement; Rapoport (1969, pp. 180–1) points out that, though measurement is an important epistemic device for the biological and social sciences as it is for the physical sciences, even more important is the epistemic device of 'recognition', with respect to which measurement takes a rather secondary position. Finally, most of the models used have borrowed a mechanistic orientation from pre-twentieth century physics; this is particularly important as regards the observer-observed system and its feedback loops in the social sciences to which the mechanistic model is not applicable. The anthropologists' concern with 'cultural relativism' and 'participant-observation' has saved them to a considerable extent from commitment to this orientation.

[2] The Heisenberg principle, in one of its implications, is a statement to the effect that instrumentation does not *per se* resolve epistemic problems. Harris (1964, pp. 31–5) has made a similar point, in dealing with epistemic questions in anthropology.

[3] See, in this connection, the citations from Boring listed above, especially, for ex., pp. 631–2, in which consciousness (and introspection) is obviously equated with 'subjectivity' and both are opposed to 'objectivity'. Connected with this conception of the nature of the subjective-objective opposition appear to be ideas such as Theodore Abel's 'Verstehen', the problems revolving about the role of insight, operationism, behaviorism, and the extremely widespread notion that objectivity equals measurement and *only* measurement. Contra to this last notion, see Rapoport, as cited above.

[4] Note in this connection, the famous anthropologist Claude Lévi-Strauss, many of whose criteria for interpretation rest upon inner personal conviction (see Lévi-Strauss, 1967, pp. 14, 16, etc.). This personal epistemic establishment of truth is reflected in his language, difficult to understand. Leach (1970, p. 8) says: "... the whole corpus of Lévi-Strauss' writings is packed with oblique references and puns... which recall Verlaine's Symbolist formula... no color, nothing but nuance." Davy has remarked that the Symbolist poets "insisted that the function of poetic language and particularly of images was not to illustrate ideas but to embody an otherwise indefinable experience. ... Readers who find the precise meaning of Lévi-Strauss' prose persistently elusive should remember this part of his literary background." One suspects that much of Lévi-Strauss' work is indeed internal experience, only expressable through personal metaphor, because it is not precisely linked in an epistemic way with extracorporeal objects. It is, therefore subjective 'social science'.

[5] Generally, the roots of anthropological epistemology appear historically to be different from those of psychology. The latter appears to have grown more out of British traditions and their offshoots on the continent, rather anti-Cartesian; the former more out of the German idealist tradition, from Descartes, through Kant, and German idealism, and the various offshoots of these as they affected the disciplines of geography and cultural geography and the early ethnography in Germany of the 19th Century. These spread to Tylor in England and through Boas to the United States. Comtean positivism, with its antagonism to introspection (see Boring, 1929/1950, pp. 633–4) and emphasis on social facts, as an epistemology, seems to have passed down to the British social anthropologists.

[6] See Boring, 1929/1950, p. 634: "[Operationism] has come into some acceptance in psychology because actually the introspective method of Külpe and Titchener did fail to produce preinferential undebatable results, and because it was seen that introspection is subject to errors of observation, that the givens of experience, being the takens of science, may include a bias to taking." The invariance assumption is quite clear here. Why

should not the takens of science include the fact that the very variation in observation is a significant datum? The model is one derived from physics and excludes a whole range of possible experience involving constant variation within some parameters. This variation is then interpreted, in the discussion, as *errors* of observation. Because they were 'errors', a whole world of what I would include in objective reality was thrown out by a large part of psychology and segments of other disciplines.

[7] In this regard, Einstein's comments (1951, p. 13) are illuminating: "The concepts and propositions get 'meaning', viz., 'content', only through their connection with sense-experiences. The connection of the latter with the former is purely intuitive, not itself of a logical nature. The degree of certainty with this connection, viz., intuitive combination can be undertaken, and nothing else, differentiates empty phantasy from scientific 'truth'. The system of concepts is a creation of man together with the rules of syntax, which constitute the structure of the conceptual systems. Although the conceptual systems are logically entirely arbitrary, they are bound by the aim to permit the most nearly possible certain (intuitive) and complete co-ordination with the totality of sense-experiences...."

[8] The historian often finds himself in related type of situation, where, for example, a significant event is described in a single document; where, even though there were plural observers, there existed also a heavy selectivity for certain kinds of observers (e.g. events at a royal court); where he is engaged in 'oral history'. Of course, in all cases, except perhaps the last, no new observers can replicate the observation. Nevertheless, on the whole, the 'objectivity' of the historical data is taken for granted. It is accepted that, say, a Frey Sahagun or a Duarte Barbosa represented what was actually there to see.

[9] The Ethnoscientists in contemporary anthropology conceive that they have reduced the observer effect by the methods of ethnoscience: elicitation frames, asking native questions, and the like. Nowhere have proponents of the 'New Ethnography' made it clear how they explain to their informants what they are doing in the informants' society in the first place. Their arrival there as a foreign object is itself a major interference in the system and one whose effect no one has yet been able to assess. In fact, it has become more difficult to assess under the field rules of ethnoscience because the perturbations created in the system by the anthropologist practicing ethnoscience are more subtle and also restricted, ideally, to one type – the ethnoscience procedure. With, say, only one type of perturbation, one gets only the most limited feedback from the system as to what the effect of the perturbation may have been (the same problem as with the use of questionnaires without other data-gathering techniques). The field worker who does not only ethnoscience but also several other forms of data gathering is likely to produce several different forms of perturbation from all of which he can gather more information about the system. The argument, here, is that all observation involves an interaction; that it is mainly the interaction that the observer observes; that, given this fact, the best scientific strategy is to increase the number of types of interaction in order to maximize the information coming in as interaction feedback. It is illusion to assume that one can so reduce one's effect on the system as to see the system 'as if one weren't there'. It is illusion to think that instrumentation has this effect, thereby creating 'objectivity'.

BIBLIOGRAPHY

Boring, Edwin G.: 1929/1950, *A History of Experimental Psychology*, 2nd edition, Appleton-Century-Crofts, New York.

Einstein, Albert: 1949, 'Autobiographical Notes', in P. A. Schilpp (ed.), *Albert Einstein: Philosopher-Scientist*, Tudor, New York.

Harris, Marvin: 1964, *The Nature of Cultural Things*, Random House, New York.
Leach, Edmund: 1970, *Claude Lévi-Strauss*, Viking, New York.
Lévi-Strauss: 1967, *The Scope of Anthropology* (1959), Jonathan Cape, London.
MacKay, D. M.: 1955, 'Man as Observer – Predictor', in H. Westmann (ed.), *Man in His Relationships*, Routledge and Kegan Paul, London.
Rapoport, Anatol: 1969, 'Methodology in the Physical, Biological, and Social Sciences', *General Systems* **XIV**, 179–86.
Woodworth, Robert S.: 1931/1948, *Contemporary Schools of Psychology*, Ronald, New York.

PART VI

COMPARATIVE HISTORY AND SOCIOLOGY OF SCIENCE

(*Chairman:* JOSEPH AGASSI)

KARL H. NIEBYL

SCIENTIFIC CONCEPTS AND SOCIAL STRUCTURE IN ANCIENT GREECE

Awareness of the need for critically analyzing modes of thought as to their appropriateness to the problems to be analyzed is characteristic of modern times. In its critical character it contrasts, at least on the surface, with the emphasis on a basic consensus during the preceding millenia.

For some time now critical scholarship has begun to explore not only the relation between modes of thought, but particularly between basic methodologies underlying major modes of thought and their corresponding social, economic and general cultural conditions.[1] Modern Western science had its origin in Greece. Therefore it is here that one must first seek this relationship. This paper presents some thoughts on the major characteristics of the relation between the structure of society in ancient Greece and the perceptions and conceptions in terms of which man began to act – which, indeed, man created for that purpose. It was in Greece that the first prerequisite for scientific thought, the conceptualization of language, developed. However, before this conceptualization of language could become science as it has been understood in modern times, society needed to develop the capability to measure natural – and that means in the last instance social – phenomena. As this quantification of man and Nature occurred only between the 12th and 16th centuries, its history and analysis lie outside our present task.

It will have to be kept in mind that designs in history, as they might be inferred from the presentation of an historical development, are always, *ex post*, and can never be *ex ante*. Thus historical analysis designed to shed light on the present situation always in fact must start with the present and proceed backwards into the past. The present to which this paper is addressed finds modern science in a deep-going crisis of its basic methodology. It is hoped that the analysis to follow will make a contribution towards the eventual resolution of that crisis.

R. J. Seeger and R. S. Cohen (eds.), Philosophical Foundations of Science, 365–381.

1. The historical setting of the relation between thought and material reality

The analysis of thought and its relation to the social reality to which it responds presupposes knowledge of the way in which it came into existence. The type of thought characteristic of Western civilization developed in ancient Greece. Unless we wish to shut our eyes and uncritically assume that the particular form of rational thought characteristic of Greek society – and through it of Western civilization – arose simply and directly by parthenogenesis, it is incumbent upon us to inquire into its societal origin.

The term 'social organization' as employed in this paper refers to an aggregation of individuals who, in the face of a niggardly Nature, began by cooperating in collecting activities and much later integrated into social organisms at first with agriculture, and finally with industry, as the organizing focus. Thus, there is one significant difference between the past, encompassing Western antiquity and early feudalism, and modern times, and that is that man depended during the long initial period primarily if not exclusively on Nature, *i.e.*, on forces beyond his control. This condition is being modified and eventually changed through industry – a manner of maintaining social existence in which the critical emphasis is on the contribution that *man himself* makes.

Nature, in impressing its overall *necessity* upon man, unified thereby his attitude into a basic consensus in which contrary expressions remained relatively mute. The historical fact of man's subservience to Nature provides a background against which to view the developing industrial age with its fundamentally antagonistic and individualistic units of social and economic existence, and begins to throw light on the difference in the attitudes exhibited in modes of thought: scientific as much as religious, practical as much as artistic. These frames of reference were and are of a most general character within which man attempted to find his place in the world, as much as the terms through which social activity, indeed social life, was only possible. We shall refer to these most general and at the same time inescapable and fundamental frames of reference – in extension of the use Professor Kuhn has made of this term – as the major, and for the West fundamental, paradigms.

Yet within the period of this paradigm there are further meaningful divisions. Delimited at the time nearest to the present by the industrial phase of western man's existence, and going backwards in history, feudalism represents the transitional process by which the productive and social relations of Greek and Roman antiquity can be seen to change from their slave base to one in which eventually the peasant and craftsman becomes a free laborer, responsible to himself for his existence and eventually the primary creator of social existence. And just as the contradistinction between the slave and free labor base of social life serves to circumscribe a process of societal development with a discernable and analyzable structure, so the genesis of classical antiquity can be seen in the transformation of primitive forms of social life, with the Homeric period serving as the transformation phase, and the Homeric epics providing illustrations of these changes. The aspect of this transformation period that interests us in the present connection, is that of the interrelationship between changes in language and language structure and therewith changes in the nature of thought, and the structure of society during this period.

2. The Relation of Thought and Social Structure Preceding Homer

Language had arisen in primitive society as a means of coordinating first collecting – and then division of labor activities. The form of the human larynx has provided the possibility for producing articulate sounds with which to communicate in the evolving social structure and thereby to reflect and form a corresponding language structure.

During the process of changing from the earliest collecting aggregations to what properly might be called social organization, certain changes in the function of individuals and groups occurred within the emerging social context. By forming social organizations, the individual member of the social unit became separated, cut off from any *direct* individual linkage to Nature. The fact that now only a collective endeavor was capable of sustaining the sum total of the individuals within such social unit provided man with a basically new condition. Later, man not only became estranged from Nature, but with the further developing various forms of division of labor, he also became estranged from the total social product and thus the meaning of the share of the social product that he himself worked on and thus from himself. This process of progressive estrangement *produced*, along with the estrangement *the collective necessity* to generate *societally* a means to overcome its effects, the *ability* of *social man* to become *aware* of the social, interdependent character of his work.[2] This awareness took the form of models of his social experience in his mind by which he was able to see himself as a constituent and operative part of the whole, and through which he became able to engage in his activities. The type of mental model generated by estrangement produced a frame of reference within which man could see himself not only in particular actions and action situations, but in which the reflected thought and the action perceived remained as yet *undifferentiated*, remained *entities, whole*. Just as 'action' then was inseparable from mental perception, so images came into existence only as a reflection of or directions towards actions, *i.e.*, either as reflections of impending actions, or in the forms of magic and ceremonials which were teaching and learning devices for the oncoming generations. This type of thinking is here designated as 'adequate' abstract thinking.

The analysis of the genesis and structure of language based on comparative archaeological, anthropological and ethnological evidence and economic, sociological and cultural-anthropological analysis demonstrates that the distinction between head and hand is not given in Nature.[3] While at the earliest stage, the collecting stage, we observe a predominance of the hand over the head, the formation of social organization, and with it the formation of language and the concomitant forms of estrangement lead to an interpenetration between societal practice and an emerging self-consciousness, the emergence of man's mind. Continuing to concentrate on man's increasing role in relation to the objective conditions given in Nature, we observe a developing separation between head and hand, mainly in terms of age, between the youth who have to learn the terms of social existence and having learned produce social existence through their labors, and the group of elders who, having lived and perceived that existence, are able to teach the youth and to advise the members of the social organization on its activities. In this adequate, abstract thinking the final point of reference remained the whole, the social organization. Against it, as the background, there is communication in the case of action and indoctrination, both of them wholistic or, as it has been called, magic.[4] At this stage, language had as yet no generic terms; the terms employed referred to situations which included in each case the totality of their social frame, *i.e.*, precisely that frame of reference that under conditions of alienation was separated and became hidden in grammar.[5] Verbs, in this context, referred still to particular situations; they were designed for them and for no others. It was the activity, the actual function as it operated with the social whole that counted, and not its immediate and to us today apparent interdependence. Thus, the living body was perceived at that stage only as an aggregate of operative functions in a given situation, and not an entity in itself. Homer had no word for 'mind', or for that matter for 'soul'. As Professor Snell has pointed out, the word for the latter ψύχη (psyche) in Homer's time still means 'the force that keeps the human being alive'.[6] Another term which later connoted 'mind' was θύμος (thymos), which to Homeric man was the generator of motion; and still another was νοῦς (nous) which was then thought of as the cause of images.

3. The transformation of the relations of social production in ancient Greece

The long lasting migrations into the Greek peninsula initiated a basic social structural change.[7] While the coordinated collecting activities of earliest social formations and early agriculture had resulted in a social surplus that had made it possible for a larger population to sustain life than the same number could have individually in a one-to-one relationship with Nature, the absence of additional cultivatable soil,[8] even the use of pasture for grain cultivation, left the food supply far below the needs ever increased through further immigrations. Yet grain was available on the fertile border region of the Black Sea, and was reachable without great difficulty through island-hopping in the Aegean Sea, through the Hellespont and by coastal shipping on the Pontus Euxinus. As the initial direct piracy of grain was unable to guarantee a steady supply, it became necessary to develop an exchange. Exchange presupposed, however, a corresponding tradeable surplus of the types of goods desired by those who were to be persuaded to part with their surplus, *i.e.*, to exchange olive oil and Greek wine for the needed grain. The labor force able to produce directly, or indirectly by freeing others for the task, this new type of surplus had to be able to produce more than it consumed. This, again, could be accomplished, in the absence of technology, only by force, by compelling that labor to consume less than it produced. That force was economic, political and cultural, and its form was slavery, first developed by the transformation of the *barbaroi*, the immigrants, into slaves, and later by piracy and importation.[9] The surplus now produced was in the first instance appropriated by the οἶκος (oikos), the household. Transport of the surplus in its appropriate form to the port where the actual exchange for the grain was taking place, was under the supervision of the head of the household. As these exchanges became institutionalized, it was no longer the community represented by its head that concluded the exchange transaction with the foreign importers, but the head himself – at least in the initial and here significant period – who appeared now as the appropriator and trader in one person. The appropriation and the trade of the surplus by the head of the household constituted the functional emergence of private property along patrilineal lines.[10] The all-penetrating character of this new and eventually fundamental economic and political institution provides us with a clue to an understanding of the changes in thought form to which we referred above.[11] It drove a decisive cleavage into society. A new class had emerged which had ceased to participate in the material production of life, and which not only arrogated to itself the surplus, the land and the slaves as private property and transformed formerly free members of the social organization into *de facto* slaves by means of debt,[12] but also claimed the political and administrative power in the community. As even independent craftsmen and free peasants were slaveowners – however small in number their slave holdings may have been in the individual case – they, too, participated in the alienating effects, if in different degrees, of the appropriation and the trading of other people's labor.

4. The Transition to Conceptual Thought

In this transitional phase of the development from Homeric to classical forms of thought, we meet with the so-called materialist Greek philosophers, and especially with Heraclitus. Heraclitus wrote about the soul in a way very different from Homer. When Heraclitus wrote that "you could not... find the ends of the soul though you travelled the whole way: so deep is its Law (*logos*)",[13] the *logos* was thought to pervade everything. The *logos* in Heraclitus is the organizing force which begins to gain its own abstract quality. Yet in spite of this changing attitude Heraclitus remained a materialist philosopher who included the non-physical as well as the physical in his thought. With him concept and matter never became fully severed. Thomson interprets the *logos* of Heraclitus as 'the interpenetration of opposites'[14] in which the abrupt counterposing of words and clauses conveys the continuing dialectics of thought and reality, though enveloped in an hieratic style.[15]

In Homer we find no such abstract general concept. Two men, for instance, cannot have 'the same mind'. Thus in Homer the direct relation of the particular to the social whole still prevails, while in Heraclitus we observe the emergence from a mere awareness of the social whole into abstract consciousness. The image as part of the language-term begins to separate out and to reappear vaguely though no less definitely as underlying all meanings and expressions, as *logos*. The thus perceived interconnection between words and meaning, between words and a language *sui generis*, reflects at one and the same time a step in the alienation away from productive participation, and thereby the emergence of the intellect. The intellect, the mind, was in the process of becoming a social product.

With Democritus concepts became ideas emptied of their material content. At that point in history science came to be defined by the unequivocal separation of the physical and material from the non-physical, the ideas, which in turn involved the distinguishing between the mover and the moved, between matter and force, between thing and its qualitative aspects, echoing thereby the cleavage in society that had brought forth the social class whose alienation had permeated its language and thought. This separation had been initiated in the last instance by a change in the relation between man and Nature in that man modified his dependence on Nature through exploiting his fellow man, thereby producing the new tradeable surplus, individual economic property, and a class structure. The effect of this new development upon the structure of society had been quite different from the estrangement of man from Nature which we discussed earlier. The abstract perception of the wholeness of society in the minds of individuals in the case of estrangement had caused no divergence between the mental image and the actual conditions it reflected. Under the now developing conditions of class alienation it became no longer possible for the appropriating class to change its mental images in line with the ever-changing productive and hence social relations. As it had been severed from the former, it did not perceive its change in terms of the latter. And the greater the social distance of the appropriating class from the productive process, the greater the alienation of its thought, the more it was to empty its language, and the images it was used to convey, of those lingering references to material reality that were still characteristic of Heraclitus' thought. They were finally left with 'pure' concepts, ideas the nature of which could then be expressed in geometric terms and relations, which were now asserted *a priori*. As Plato wrote over the door to his academy: Only those who know geometry may enter here.

5. The Relation of 'Utopian' Thought to Science

We have, so far, confined our historical analysis to tracing the changes in the structure of language, and thereby of the mind, in relation to changes in the structure of society from pre-Homeric times to classical Greece by concentrating on those changes experienced by the dominant, appropriating class. In the same process, however, the material producers of social life, the working people in their various social forms, while experiencing directly the structural change they were bringing forth, were cut off from command over what they were producing and were excluded in whole or in part from administrative activities and political life. Thus they, too, no longer experienced the wholeness of their social existence. But the form frustration took in their minds retained the materiality of the images, in contradistinction to the emptying out of the images of the appropriators of all references to material reality.

As the very material forms of the productive process were ever changing, from communal cooperation to the use of foreign slave labor and to the loss of communal freedom through debt, it was this experienced and self-created continuity of change that in the truncated perception of their lives acquired the quality of a deeply felt confidence in the future. Δίκη (dike), to them, remained 'the right way of life', temporarily interrupted by their servitude, and to be restored in a confidently expected future.[16] For the purpose of this discussion, this form of thought will be referred to as *utopian* thought, with a meaning different from that given to the term by Plato. While this type of thought is different from that of the class appropriating the labors of the former, to whom indeed δίκη (dike) had turned into an abstract and static concept of justice, the two thought forms are yet to be seen in the long run as interdependent. Though considered utopian in the early phases of the long process leading to eventual emancipation in modern times, it was the material experience of change and of the problems accompanying that change voiced in that thought that provided the ever renewed challenge to a dominant class which in the very same historical process was becoming equipped with abstract concepts that were beginning to lend themselves to a scientific albeit static analysis of these problems. Indeed, it was these challenges accompanying major changes in the modes of production, and within the major ones minor social-structural changes, that provided the sources for the eventual reconstruction of major, minor and overlapping paradigms.

6. The nature of the conceptual tools of science

We had said that the terms by which in unalienated society communication had been effected referred always to total situations. Mental images and action were then still undifferentiated. Just as the initial estrangement had produced the need for mental images involving in their background always perceptions of the whole, so now the alienating experience of the dominant group deprived these images of their immediacy and vitality. It made them static. And being static, and having thereby no longer the possibility of forming thought that expressed social-structural, qualitative change, their concepts were in fact denying such change. It is important, however, to note that it is the nature of the concepts as they had historically emerged in the course of the development of classes that provided the possibility for denying change, and not, at least not in the first place, the need or desirability to produce a type of concepts that would deny the possibility of change to those chafing under conditions of exploitation. It was the static character of the concepts that lent itself to the latter use.

Static representations of total situations do not permit their use in directing effective action appropriate to conditions of social-structural change. The 'still-life' quality of such images always refers to a past situation and past actions. To make them capable of directing conduct in terms of this basically static frame of reference it was necessary for the particular picture, as distinct from its underlying frame of reference, to become moveable. This effect was, historically, accomplished by selecting out of the total picture, and that meant out of the language structure handed down from the past, those parts that could serve as universals. These, then, were used as moveable parts in the composition of a variety of mentally pictured situations expressed in a composed language which conveyed meaning just as colored pebbles used in the composition of a mosaic make a picture. These parts had lost their material functional content, or, as we called it earlier, were emptied of all material functional content once alienation and class differentiation had set in. The use of these generalized moveable parts made it now possible to direct society and to evolve a philosophical frame for guidance in spite of the static character of the concepts involved.

It is this process that provided the dynamic force for the changes in the structure of the Greek language from the times of Homer to those of classical Greece. The phenomenal aspects of this process of change can be seen, according to Snell, after adjectives and verbs had been formed in pre-Homeric times, in the substantivation of adjectives and verbs by means of adding the definite article to them and thus transforming them into nouns. The emerging alienation of the dominant class found its expression in the language of that class in the transformation of the concrete noun first into the generic noun, and finally into the abstract noun. The result was the accomplishment of a first, though fundamental, step in the development of an ability to form and convey abstract concepts, and thereby to reason scientifically.

In line with what we had just said about the nature of utopian thought, we find, as we should expect, that Hesiod, the one ancient writer who might be considered in the 'utopian' tradition previously referred to, does not use the article characteristic of abstract concepts usable for scientific analysis.

When in later stages also the utopian myths of the people became permeated by the static character of the new language structure, they were deprived of their originally forward-looking and restoration-anticipating character and thus transformed into what became known as legends.[17] To the same end, we see wholism treated statically, for instance in the Oedipus trilogy where it serves to demonstrate man's supposed inability to know and direct his own life. While Snell suggests that the development towards the use of the definite article is very slow and "even Aeschylus does not yet employ it with abstractions",[18] I would like to suggest that it is not so much the slowness of the introduction of the definite article that explains its absence in Aeschylus' work, but that Aeschylus wishes to create a favorable predisposition in his audience by appealing to its utopian predisposition by means of a use of language appropriate to that purpose. The near-contemporary of Aeschylus, Heraclitus, speaks already of 'the art of thinking', 'the universal', and 'the *logos*'.[19] The substantives formed here no longer refer to the same order of things as ordinary concrete nouns. Yet the formation of a concrete noun is the first step towards the appearance of the abstract noun in that the concrete noun contains within itself the potentiality of forming generic nouns and thereby abstractions.[20]

The process of substantivation, and thereby forming abstractions and in turn making the formerly inherent parts of the total image independently moveable, extended *pari passu* to verbs, predicates, prepositions, copulas, *etc.* It was in this manner that the sentence, and thereby grammar, was formed – a further step in the elaboration of the intellectual tools that would serve as the prerequisite for scientific analysis.

7. The Implications of the Static Character of Scientific Concepts

The development of the ability of man not only to form concepts and to construct a logical grammar had had its price: its static character. In the process of formation, the attitude of the social class that created these forms of language and thought had become more and more that of an observer.[21] The type of language structure appropriate to the observer became characterized at the same time by the separation of the physical from the non-physical, by the separation of thought and action.

The attitude of the alienated observer became visible in the thought forms of Democritus who, in the words of Diogenes Laertes, "throws the qualities overboard" and to whom, therefore, properties appear to be "in reality nothing but a variety of *ideai*, of forms, as he sometimes calls his atoms, arranged in geometric patterns."[22] Thus it is space and numbers, *i.e.*, concepts deprived of any reference to material reality, and hence of qualitative change that become the means of *learning* (μαϑημα, mathema) about the reality that is behind the appearance of it (and not gaining knowledge by experience). The entirely abstract nature of this learning process transforms, then, sensory shadings (properties) into the only analytical expression possible under these conditions, *i.e.*, the measurement of shadings. The means for such measuring had to be derived at this time entirely arbitrarily as, for instance, in the abstract use by Pythagoras of the tonal scale, or of the Egyptian technique of land measurements in his famous theorem. Snell correctly observes in this connection that "the Greeks were not interested in observing the infinite transitions within chord and pitch; they were content to record the constant relations responsible for the harmonies, and to use figures... as integer numbers."[23] Indeed they would have been unable to develop the concept of a discrete quantity, as the social conditions for its development arose only with the quantification of man and the conceiving of it abstractly in and after the 14th century.[24]

The high point in this development of categories capable of scientific, if static, utilization is thus reached just before Plato. The Periclean Age constitutes in fact the first major crisis in the social-structural development of Graeco-Roman antiquity, the decisive crisis for ancient Greece. The establishment of monopoly in international trade, as much as in landholdings in an economy based upon an increasingly brittle slave base was leading Athens to ever more military and imperialistic adventures. The growing uncertainties of its economic and politically privileged position led the dominant social group into a mental frame in which it became preoccupied with the problem of maintaining its *status quo ante*. While in the past it had been its concern to promote an increasingly abstract frame of reference within which to provide the rationale for conduct, as well as to understand itself within its alienated position, it became now preoccupied with the need to explain, nay, to indoctrinate those whose lack of vested interest in the prevailing conditions made them prone to threatening revolutionary change, with the defeatist thought that indeed their hope for real change was vain.

It was to this purpose that the abstract conceptions developed up to Phythagoras and Democritus were now bent. Heraclitus had employed adjectives of sensation to express his philosophical thought while Democritus had used adjectives of form, quantity and size as scientific vehicles for this thought. But the problems that Socrates and Plato were concerned with demanded adjectives such as 'fair', 'good' and 'just' which inferred teleological judgments. We can fully concur with Snell when he points out that these types of adjectives are not appropriate in a system of scientific concepts. We had tried to show that it had become the very characteristic of science in antiquity to exclude qualitative, and thereby moral aspects from its consideration of man. To illustrate the point: both Democritus and Plato were concerned with the 'good'; but while with Democritus the good is based upon what is pleasant to the senses, and thereby "is amenable to a calculus based upon mechanical motion",[25] Plato's adjectives point intentionally towards "a goal which always lies just beyond the present horizon of possibilities."[26] In Democritus the soul atoms move mechanically and that is passively after their initial push separated them from the whole. Their motion is thus free of all ethical implications. Plato, on the other hand, places the responsibility to act on the active individual as he has to emphasize – for reasons indicated before – the inherent and supposedly unsurmountable limitations of man's actions. The static motion in the system of Democritus accounts for the fact that the verb is passive, and that Democritus knows only one person, the third. While thus the language forms as used by Democritus continue the path towards alienation of the observer, and thereby, towards the formation of a scientific language, the thought forms of Plato referred to here represent a deviation from this development. They are the beginning of a transitory stage that gave expression to the height of the crisis of Greek society.

With Aristotle these conditions began to settle down again, though not in the direction of renewing a progressive ascent in the formation of scientific concepts which indeed with the conditions generative of the thought forms used by Pythagoras and Democritus had to essentially come to an end, but in terms of the uneasy and long-lasting economic and political stagnation that eventually led to the change of the focus of social life from Greece to Rome, from the emphasis on the exploitation of the eastern Mediterranean to that of the western Mediterranean.

What this stagnation seemed to make permanent was not only the alienation of the dominant social group from the active productive process, but from a productive process that itself was increasingly exhibiting paralysis. The economy and society in this period were not any longer based so much on the work of slaves, but became supported by supposed alliances, conquests and tributes appropriating directly the products of whole nations. Though the appearance was thus created as if the 'glory that was Greece' was continuing, the decomposition of the body economic and politic ran its course. Yet it was at the beginning of this period that the class schism and the consequent abstract use of language terms, and therein the consciousness and mind of the dominant class, reached its height in that the intellect became not only distinguished as such in Aristotle, but severed from the 'body' as the 'mover' was separated from the 'moved'.

Though this science in the form of philosophy had since the days of Plato begun to acquire an ideological, that is apologetic character, the application of the scientific concepts won so far, to mechanical problems celebrated its first triumphs during the subsequent hellenistic period.

8. Aristotle and the Utopian Element in the Science of Antiquity

Yet this is not the whole story of the genesis of scientific concepts, and thereby science in antiquity. We have traced the continuity of social life which knows decline of one form of social relations only as coincident with a newly emerging one in the thought and thereby in the thought forms of the dominant class. In the beginning of this paper, however, reference was made to the fact that while this paper was concerned with the relations of science and social structure, that is, science as enunciated by the dominant social class, there were also the activities of the working people, slave and free, and the developing changes of the mode of production produced and hence experienced by them, that reflected in their minds in the form of an awareness of what they were doing, though not in the abstract terms developed by the dominant class. The form of this awareness of change was confidence in continuing change, and thereby also in the eventual change in their conditions, in an eventual reestablishment of social justice. We had called this attitude 'utopian'. It was reflected, if dimly, in some of Aristotle's formulations. Motion as conceived by Democritus was passive. The frame of reference in Plato's thought, while active, was static. Aristotle, however, distinguished between qualitative and quantitative changes, as well as statics. The source for this apparent return to earlier qualitative forms of thought is to be found in the contemporary experience of social and economic developments which ever more clearly contradicted the assumption of mere quantitative increases and decreases. They compelled Aristotle to go in his conceptual formulations beyond Plato's active denial of the possibility of change, in order to be able to account for qualitative changes at least abstractly while the reality of it impressed itself upon him and his time with ever greater force in actions and voices that proclaimed these changes not only in utopian hopes but in a deeply felt if not yet abstractly, theoretically formulated conviction that lent them strength.

After Plato's introduction of teleological concepts, it was this *abstract* conception of qualitative change that marked the beginning of the long turn away from the further gestating of conceptual forms useful eventually to Western scientific analysis. While the implications of this development were at first hardly visible – and indeed seemed to be contradicted by the flowering of scientific inquiry and application during the hellenistic period – they did come to the foreground in the revival of Aristotelianism in the 13th and 14th centuries. Only when, in the late middle ages, man began the arduous road to shift the providing of the conditions for his existence from Nature to himself[27] and through the knowledge gained in this process of producing himself gained the capability of consciously directing it and through it to overcome the alienation that had marked the beginning of the entire and lengthy process, only then did it become possible to engage upon the construction of a new fundamental paradigm, one of the basic characteristics of which is the overcoming of the static quality of the scientific tools the genesis of which and thereby the implicit limitations of which it had been the task of this paper to indicate.

An important implication in the analysis presented in this paper – though not its immediate objective – has been to at least indicate that one of the historical sources for the eventual formulation of a new paradigm which, though almost mute in the historical records during the period of the emergence of the first major methodological frame of reference, was the utopian, qualitative mode of thought emerging at the same time. Just as utopian thought, giving voice and demonstrating through its actions the qualitative social-structural change which it was instrumental in bringing forth and of which both Plato and Aristotle had to take cognizance, so it was the utopian thought generated in the Peasant Wars and the English Revolution formulated in the Socinianism that influenced Newton in his completion of the major paradigm, that gave the modern classical methodology to the science of the West. While this major paradigm was still based on static concepts and quantitative measurement, the struggle for an adequate accounting of qualitative change has intensified since Leibniz and the emergence of industrial democracy with the result that the transition from utopianism and classical science to qualitative science has become one of the tasks of today.

Temple University,
Philadelphia

NOTES

[1] Thomas S. Kuhn, *The Structure of Scientific Revolutions*, The University of Chicago Press, Chicago, 1962. See also, Karl H. Niebyl, 'The Need for a Concept of Value in Economic Theory', *The Quarterly Journal of Economics* **54**, No. 2, February, 1940.

[2] G. Thomson, *Studies in Ancient Greek Society*, Vol. II: *The First Philosophers*, Lawrence and Wishart, London, 1955, pp. 35ff.

[3] Cf. George Thomson, *Studies in Ancient Greek Society*, Vol. I: *The Prehistoric Aegean*, Lawrence and Wishart, London, 1954; Friedrich Engels, *Die Dialektik der Natur*, in: *Marx-Engels Gesamtausgabe*; G. Clark, *From Savagery to Civilization*, London 1946; E. A. Smith, 'Myths of the Iroquois', *2nd Annual Report of the Bureau of Ethnology for 1881/1882*, U.S. Government Printing Office, Washington, D.C., 1883, pp. 47–116.

[4] Thomson, *The Prehistoric Aegean*, pp. 435ff.

[5] We are discussing here, as will be obvious, the social genesis of the 'presettings' that relate subject to object etc. which Noam Chomsky thinks may be found in a deep structure 'far below'. Cf. his *Language and Mind*, Harcourt, Brace and World, New York, 1968 and *Syntactic Structures*, Mouton, The Hague, 1968; see especially the former, p. 83.

[6] Bruno Snell, *The Discovery of Mind*, Harvard University Press, Cambridge, 1953. I am indebted to Professor Snell for his analysis of language in ancient Greece, though he is in no way responsible for the analysis to which I relate it. Cf. p. 8.

[7] Cf. V. Gordon Childe, *Prehistoric Migrations in Europe*, Oosterhout, The Netherlands, Anthropological Publications, 1969, pp. 65, 149, 158–160. We are omitting here discussion of Mycenae and/or oriental influences on Greek life in the second millenium minus as of minor significance in the given context. See: A. Bartoněk, 'Zur sozialökonomischen Struktur der mykenischen Gesellschaft', *Neue Beiträge zur Geschichte der alten Welt*, Vol. I, *Alter Orient und Griechenland* (ed. by E. C. Welskopf), Akademie Verlag, Berlin, 1964, pp. 149–162. Also: A. Winspear, *The Genesis of Plato's Thought*, The Dryden Press, New York, 1940, pp. 13–18.

[8] A. French, *The Growth of the Athenian Economy*, Routledge and Kegan Paul, Ltd, London, 1964, p. 8.

[9] At times members of immigrating tribes became metoikoi (μέτοικοι) *i.e.*, inhabitants without citizens rights and without ownership of soil.
[10] Cf. Childe, *op. cit.*, p. 59.
[11] It would go far beyond the time allotted to this paper to follow these changes in detail. Cf. the fundamental research into the nature of the productive relations in Greek antiquity of E. C. Welskopf, *Die Produktionsverhältnisse im alten Orient und in der griechisch-römischen Antike*, Akademie Verlag, Berlin, 1957; and M. I. Finley, 'The Servile Statuses of Ancient Greece', in *Revue internationale des droits d'antiquité*, 3rd ser., July, 1960, pp. 165–189; and the collection of essays edited by the same author, *Slavery in Classical Antiquity*, H. Heffer and Sons, Ltd, Cambridge, England, 1960. See also for a different interpretation: Fr. Vittinghoff, 'Die Bedeutung der Sklaven für den Übergang von der Antike in das abendländische Mittelalter', in: *Congrès international des sciences historiques*, Rapports, Vol. 6, Stockholm 1960, pp. 71–73, and by the same author, 'Die Theorie des historischen Materialismus über den antiken "Sklavenhalterstaat"', in *Saeculum* **XI**, No. 1-2.
[12] The development and expansion of debt-slavery only broadens our argument. See also M. I. Finley, 'Between Slavery and Freedom', in: *Comparative Studies in Society and History* **VI** (1963/64) 233ff.
[13] K. Freeman, *Ancilla to the Pre-Socratics*, Harvard University Press, Cambridge, 1957, 45, p. 27; Snell, *op. cit.*, p. 17.
[14] Thomson, *The First Philosophers*, p. 132.
[15] *Ibid.*, p. 135.
[16] Winspear, *op. cit.*, pp. 39–40.
[17] Similarly, cf. Giorgio de Santillana and Hertha von Dechend, *Hamlet's Mill*, Gambit, Boston, 1969, especially Chapter IV.
[18] Snell, *op. cit.*, p. 229.
[19] *Ibid.*, pp. 112–113, 2, 114, 50.
[20] Professor Snell sees another source for abstract nouns in abstracts that began their careers as mythical names. Mythical name, like *phobos*, have no generic significance *ab ovo* either however, as fear became generic only in the same manner as other concrete nouns that lent themselves to the formation of abstractions when there were people in appropriate conditions who formed abstractions of these terms, whether nouns, etc., or names that in fact in the given conditions acquire generic significance. There seems to be no reason to infer that φόβοσ (phobos) had such generic potentiality in a nonexploitative society.
[21] The relevance of these facts has impressed itself upon us in our own time with force. This is, however, not the occasion to argue the significance of this phenomenon. It must suffice here to record its origin in classical antiquity.
[22] Snell, *op. cit.*, p. 238 (Diog. Laert. 9, 72).
[23] *Ibid.*, p. 239. Also, M. R. Cohen and I. E. Drabkin, *A Source Book in Greek Science*, Harvard University Press, Cambridge, 1958, p. 1.
[24] Cf. Anneliese Maier, *Die Vorläufer Galileis im 14. Jahrhundert*, Roma 1949, pp. 26ff.
[25] The calculus of pleasure of Democritus is a materialistic vestige.
[26] Snell, *op. cit.*, p. 240.
[27] The limiting implications of the former had been indicated earlier in this paper.

ROSHDI RASHED

ALGÈBRE ET LINGUISTIQUE: L'ANALYSE COMBINATOIRE DANS LA SCIENCE ARABE

Si l'on met à part le calcul des probabilités, c'est sur le terrain de deux disciplines que l'analyse combinatoire s'est le plus souvent exercée: l'algèbre et les études linguistiques, fussent-elles de la langue en général ou de la langue philosophique.[1] Nul n'ignore que c'est surtout depuis le début du 18ème siècle, notamment avec Jacques Bernoulli et Montmort,[2] que l'analyse combinatoire a pris son essor, pour les besoins du nouveau calcul, dans la mesure où il s'agissait de problèmes de partition d'un ensemble d'événements et non exclusivement de nombres. Chacun sait par ailleurs qu'avant cette rencontre favorable au développement sans précédent de l'analyse combinatoire, algébristes et linguistes produisaient et utilisaient déjà certaines de ces méthodes: c'est ainsi du moins que mathématiciens et linguistes arabes ont découvert l'analyse combinatoire.

A bien regarder, on s'aperçoit toutefois que les savants arabes, comme c'est encore le cas d'une certaine manière au 16ème sinon toujours au 17ème siècle,[3] séparaient ce que nous réunissons, depuis une date d'ailleurs récente, sous le concept d'analyse combinatoire. Alors que l'algébriste ne voyait guère dans l'instrument invoqué par le linguiste le sien propre, ce dernier s'efforçait pour sa part de réinventer ce dont l'algébriste possédait déjà les éléments. Cette conscience théorique fragmentée est en outre discrète dans la science arabe et on n'éprouva pas alors, comme au 17ème siècle, le besoin de désigner par un nom particulier l'analyse combinatoire. Le linguiste semble découvrir des méthodes combinatoires tout naturellement, pour peu qu'il entreprenne une exploration raisonnée de certains phénomènes linguistiques. Quant à l'algébriste, il dénomme des procédés mais non encore une activité qui, en tant qu'activité organisée, exigerait l'attribution d'un titre. Or, s'interroger sur la fragmentation et la discrétion d'une conscience théorique – l'unité de l'analyse combinatoire – c'est s'engager à différencier entre projets spécifiques du linguiste et de l'algébriste. Nous verrons ainsi que si pour le premier, l'analyse combinatoire est le moyen de rationaliser

R. J. Seeger and R. S. Cohen (eds.), Philosophical Foundations of Science, 383–399.

une pratique ancienne, pour le deuxième, elle n'est, en dernier lieu, qu'un instrument technique pour fonder un problème théorique: une autre conception de l'algèbre ou encore le projet d'une algèbre autonome. Instrument ici et là, sans doute reste-t-il à souligner qu'elle est une fois un moyen de résoudre théoriquement un problème pratique et une deuxième fois un moyen forgé au cours de la solution d'un problème théorique. La différence des projets est, croyons-nous, responsable de la méconnaissance où fut tenue l'unité de l'analyse combinatoire, tandis qu'une fois cette différence oubliée, les portes sont largement ouvertes à des interprétations abusives de l'activité combinatoire, unifiant ce qui n'était pour les savants qu'épars et multiple.

Encore faut-il relever que ces deux sens de l'analyse combinatoire, si différents soient-ils, ont en commun au moins une condition de possibilité qui peut être résumée schématiquement par le changement des rapports des deux termes: science et art. Fonder l'autonomie de l'algèbre c'est se donner pour but sa constitution comme science, mais ceci revient à admettre qu'une science serait aussi un art, qu'elle pourrait donc se donner sans l'assurance d'un objet parce qu'elle en a plusieurs – arithmétique et géométrie – bref à concevoir une science sans affirmation de l'être. Le linguiste, par sa conception du traitement théorique d'un art, celui du lexicographe, abolissait lui aussi une ancienne distinction entre science et art dans la mesure où il entendait attribuer le statut d'une science à une connaissance conçue dans ses possibilités de réalisation pratique et dont le but lui est extérieur. Or, si une meilleure compréhension de ce changement renvoie au moins pour une part à une sociologie de la connaissance,[4] il reste que, pressenti mais jamais saisi, il a été le prétexte de jugements sur l'esprit pragmatique de la science arabe opposé à l'esprit théorique de la science hellénique, jugements qui, depuis Renan et par la suite Duhem et Tannery, ont été souvent repris.

Il est évidemment tout à fait impossible de faire dans les brèves limites de cet exposé une histoire, fût-elle peu détaillée, de l'analyse combinatoire. Mais s'il nous a paru important d'en parler, ce n'est pas seulement par suite de l'intérêt du problème mais aussi pour marquer une distance, dans ce domaine particulier et peu connu de la science arabe, avec une certaine histoire des sciences dont l'érudition ne dissimule point le préjugé de continuité, lequel interdit souvent la reconstitution d'une activité rationnelle historiquement datée et géographiquement localisée. L'analyse combinatoire est à plus d'un égard un cas exemplaire car, en dehors de la tradition hellénique, les tenants de la continuité l'ont souvent considérée comme une déviation, à négliger ou à réduire à une prétendue 'pensée analytique, atomistique, occasionaliste et apophtegmatique...' propre aux savants arabes. Mais si par reconstruire nous voulons dire essentiellement comprendre, il nous faut multiplier les références, c'est-à-dire reconsidérer cette activité à partir de deux ordres de questions: celles que les savants arabes se sont posées à eux-mêmes – scientifiques et extra-scientifiques – et les autres, celles auxquelles une science confirmée répondra plus tard. Nous allons donc suivre successivement l'apparition de l'analyse combinatoire en algèbre et en linguistique.

Le premier recours à l'analyse combinatoire en algèbre est souvent daté du 11ème siècle et plus précisément attribué à un ouvrage non encore retrouvé d'al-Ḫayyām (1038–1123): telle est l'opinion qui prévaut chez les historiens des mathématiques. En fait, on voit apparaître dès la deuxième moitié du 10ème siècle un intérêt particulier pour l'analyse combinatoire conçue pour améliorer et étendre le calcul algébrique, dans le domaine particulier des équations algébriques et des problèmes d'extraction de racines, comme en témoignent les titres des essais d'Abu al-Wafā (940–998) et ceux du célèbre astronome-mathématicien al-Bīrūnī (973–1048). Ce fait d'histoire n'a cependant pas reçu l'explication qu'il mérite. Pourquoi le développement de l'analyse combinatoire dans la science arabe au 11ème siècle? Cette question ne reçoit aucune réponse, soit que l'on n'y pense point, soit que l'on évoque l'effet d'une influence heureuse – toujours indémontrable! – de la science chinoise ou hindoue ou les effets de la fortune et du hasard.

A bien regarder cependant, on peut voir que c'est à cette même période qu'est élaborée l'idée d'autonomie et de spécificité de l'algèbre, autonomie ne signifiant pas seulement séparation d'avec la géométrie mais encore – et surtout – arithmétisation de l'algèbre.[5] Pour résumer brièvement ce programme: on applique l'arithmétique à l'algèbre en sorte que celle-ci conserve pour des variables $x \in [0, \infty]$; $x \in [-\infty, 0]$ est introduite par la définition $x = -y$; $y \in [0, \infty]$ – les opérations essentielles de l'arithmétique: $+|-$, $\times|\div$. Encore ne faut-il point oublier que l'algèbre se présente au 11ème siècle principalement comme la science des équations algébriques. Le fait le plus surprenant est ici que ces algébristes qui s'efforcèrent plus que d'autres de réaliser l'autonomie de l'algèbre sont ceux-là mêmes qui développèrent les méthodes combinatoires. Ce développement apparaît lui-même au terme d'un retour *délibéré* de l'algébriste à l'arithmétique, par suite des exigences du nouveau projet et afin d'en rechercher les moyens nécessaires. Pour expliciter ces affirmations, il nous faut rappeler rapidement comment l'algèbre se développa, après al-Ḫawarizmī au 9ème siècle, mais aussi contre lui.

Que l'on hésite à attribuer la paternité de l'algèbre à Diophante pour la réserver à al-Ḫawarizmī, se justifie dans la mesure où, contrairement au premier, le deuxième a considéré l'algèbre pour elle-même et non plus comme un moyen de résoudre des problèmes de la théorie du nombre. L'algèbre a désormais pour objet principal, comme on le répétera plus tard, "le nombre absolu et les grandeurs mesurables (étant) inconnus mais rapportés à quelque chose de connu de manière à pouvoir être déterminés." L'objet principal de la connaissance algébrique est donc de déterminer des opérations "au moyen desquelles on est en état d'effectuer le susdit genre de détermination des inconnues soit numériques, soit géométriques." Devant la diversité des 'êtres mathématiques' – géométriques, arithmétiques – l'unité de l'objet algébrique est fondée seulement par la généralité des opérations nécessaires pour ramener un problème quelconque à une forme d'équation ou encore de préférence à l'un des six types canoniques énoncés par al-Ḫawarizmī

$$\begin{array}{lll} (1)\ ax^2 = bx & (4)\ ax^2 + bx = c & \\ (2)\ ax^2 = c & (5)\ ax^2 + c = bx & a, b, c > 0 \\ (3)\ bx\ = c & (6)\ bx\ + c = ax^2 & \end{array}$$

d'une part, et des opérations pour dériver des solutions particulières, c'est-à-dire un 'canon',[7] d'autre part. C'est dans la mesure où al-Ḫawarizmī a aboli, comme nous l'avons dit, l'opposition entre science et art que cet objet – les opérations – peut être considéré comme l'objet d'une science. Une opération est objet d'une connaissance théorique sans se référer toutefois à une théorie de l'être algébrique. Elle est aussi objet de la connaissance d'une activité qui a sa fin en dehors d'elle, puisque conçue dans ses possibilités, soit de réduire un problème à une certaine forme, soit de dériver d'une manière parfaitement réglée des solutions particulières. Un algébriste du 12ème siècle, as-Samaw'al, semble avoir saisi cette situation: pour lui, contrairement à la géométrie, dans l'algèbre, "le début de la connaissance est la fin de l'action et la fin de la connaissance est le début de l'action." Mais si cette abolition de l'opposition entre science et art – fût-elle dialectique ou exclusive pour chaque genre – a été à l'origine d'une science de l'algèbre, trouver la spécificité de cette science, c'est en définir l'autonomie. Or l'algèbre d'al-Ḫawarizmī se heurtait encore à l'obstacle de la démonstration géométrique: cherchant à déterminer les conditions d'existence des racines pour résoudre des équations du 2ème degré, sa démonstration est géométrique et ses règles de solution ne donnent que la racine positive.[8]

Les successeurs d'al-Ḫawarizmī, tout en poursuivant ses recherches, ont réagi, comme nous l'avons déjà dit, contre l'insuffisance de la démonstration géométrique en algèbre. Cependant, la nécessité pressentie d'une démonstration numérique n'a été elle-même possible qu'au terme d'une extension du calcul algébrique et de son domaine, puis de sa systématisation. Les successeurs immédiats d'al-Ḫawarizmī se mirent à cette tâche sans tarder: Abū Kāmil (850–930)[9] intègre les irrationnelles comme objet du calcul à part entière en tant que racines et coefficients. On développe des opérations exigées par la solution de systèmes d'équations linéaires à plusieurs inconnues et pour l'extraction des racines de polynômes algébriques... La systématisation, surtout de la théorie des équations, prendra place précisément au 11ème siècle: on tentera alors, al-Ḫayyām par exemple, une classification complète des types canoniques des équations cubiques.[10]

L'extension et la systématisation du calcul algébrique ont permis de formuler l'idée de démonstration algébrique dans la mesure où elles ont fourni les éléments d'une réalisation possible. Au début du 11ème siècle, c'est-à-dire peu de temps avant al-Ḫayyām, un des savants les plus actifs dans ce domaine, al-Karagi, s'engage à donner, outre la démonstration géométrique, une autre démonstration, celle-là algébrique, des problèmes qu'il considère. Al-Ḫayyām ne se contente pas de réaliser la coexistence des deux démonstrations mais en dégage la raison dans un texte-programme. Après avoir donné une solution de l'équation du 3ème degré à l'aide des propriétés des sections coniques, il écrit: "et sachez que la démonstration géométrique de ces procédés ne remplace pas leur démonstration numérique, lorsque l'objet d'un problème est nombre et non pas une grandeur mesurable."[11]

Dans le même esprit, au 12ème siècle, as-Samaw'al semble exiger une démonstration algébrique dans la mesure où l'algèbre, différente de la géométrie, est une approche analytique des problèmes mathématiques ou comme il l'écrit: "l'algèbre est une partie de l'art de l'analyse, tandis qu'en géométrie on peut déterminer l'inconnue sans recours à l'analyse."[12]

C'est précisément au cours de l'extension et de la systématisation soulignées plus haut, pour une algèbre dont la pièce maîtresse est la théorie des équations, que les algébristes sont revenus à l'arithmétique pour développer l'analyse combinatoire. On comprendra alors pourquoi les recherches de techniques pour extraire les racines de degrés supérieurs quelconques ont pris pour eux une importance particulière. C'est au cours de l'élaboration de ces techniques qu'ils se sont tournés vers l'analyse combinatoire pour découvrir, d'une part, le tableau des coefficients binômiaux et sa règle de formation et, d'autre part, la formule binômiale énoncée verbalement et pour des puissances entières. On sait enfin par as-Samaw'al qu'al-Karagi a construit pour ce calcul le triangle de Pascal. Dans ce texte on trouve en effet le tableau des coefficients binômiaux, sa loi de formation

$$C_n^m = C_{n-1}^{m-1} + C_{n-1}^m$$

et le développement

$$(a+b)^n = \sum_{m=0}^{n} C_n^m a^{n-m} b^m$$

pour n entier.[13] Ce texte est le premier, à notre connaissance, où ces règles sont énoncées d'une manière aussi générale. Il est probable qu'elles ont été également formulées par al-Ḫayyām dans un ouvrage non encore retrouvé. Il écrit en effet dans son algèbre:

Les Indiens possèdent des méthodes pour trouver les côtés des carrés et des cubes, fondées sur une telle connaissance d'une suite de nombres peu étendue, c'est-à-dire la connaissance des carrés des neuf chiffres, à savoir du carré d'un, de deux, de trois, etc.... ainsi que les produits formés en les multipliant l'un par l'autre, à savoir du produit de deux en trois, etc.... J'ai composé un ouvrage sur la démonstration de l'exactitude de ces méthodes et j'ai éprouvé qu'elles conduisent en effet à l'objet cherché. J'ai, en outre, augmenté les espèces, c'est-à-dire j'ai enseigné à trouver les côtés du carré-carré, du quadrato-cube, du cubo-cube, etc.... à une étendue quelconque, ce qu'on n'avait pas fait précédemment. Les démonstrations que j'ai données à cette occasion ne sont que des démonstrations arithmétiques, fondées sur les parties arithmétiques des éléments d'Euclide.[14]

Plus tard, au 13ème siècle, on retrouvera les mêmes résultats à cette différence près que la formule binômiale s'écrit toujours verbalement

$$(a+b)^n - a^n = \sum_{m=1}^{n} C_n^m \, a^{n-m} \, b^m.$$[15]

On les retrouvera également dans 'la clé de l'arithmétique' d'al-Kāshī au 15ème siècle.[16]

Tandis que l'analyse combinatoire suivait cette voie en algèbre, un développement parallèle s'esquissait en linguistique. Moins importante ici par ses résultats mathématiques, l'analyse combinatoire indique un domaine extérieur aux mathématiques où elle peut toutefois s'exercer. C'est cette tentative, négligée par les historiens des sciences que nous allons maintenant considérer.

L'intérêt constant des Arabes pour leur langue a frappé tant les orientalistes occidentaux les plus modernes que les historiens arabes les plus anciens. Inséparable des progrès de la linguistique actuelle, l'étonnement des premiers est provoqué non seulement par la multiplicité et la diversité des recherches linguistiques des Arabes mais aussi par l'orientation structuraliste avant la lettre qu'on peut y trouver, peut-être d'ailleurs commune à tous ceux qui ont tenté d'analyser leur propre langue, comme le montre, entre autres, l'exemple hindou. Les historiens arabes classiques y voyaient, pour leur part, un événement aussi important et original que la constitution de la logique.[17] En effet, outre les linguistes eux-mêmes, juristes,[18] théologiens-philosophes,[19] classificateurs,[20] se sont toujours attachés, pour des raisons diverses, à une réflexion sur la langue, les uns pour une exploration rationnelle des phénomènes du langage, les autres pour résoudre la question épineuse de l'éternité ou de la création de la parole divine, les derniers enfin pour présenter selon un mode raisonné une classification de leurs matériaux empiriques: plantes, substances curatives, etc.... Pour les linguistes, cet intérêt d'origine vraisemblablement religieuse s'est rapidement laïcisé. Suscitée par la double nécessité de créer un conservatoire des mots et des significations et d'élaborer les règles syntaxiques de la parole divine afin de présenter le sens original de la révélation transmise dans la langue des 'païens', cette tâche s'imposa, à cause de l'extension rapide de la nouvelle religion, en l'absence d'une institution particulière pour veiller à l'interprétation conforme de la parole sacrée, source première d'unification doctrinaire de peuples de langues, traditions et cultures diverses. Ces motivations reléguées à l'arrière-plan, une laïcisation devait bientôt permettre aux premiers linguistes de traiter au même titre aussi bien de la parole divine que de la poésie païenne préislamique. Il reste cependant que les grammairiens devenus lexicographes n'entendaient au début par lexique qu'un glossaire particulier à une matière ou à une région, explicitant des mots désuets ou de sens difficile. Chez les Arabes, comme dans les autres cultures, il s'agissait là de lexiques dont le domaine était limité et l'arrangement incertain. Dans ces glossaires, le principe de composition ou d'arrangement des mots est essentiellement sémantique.

L'idée de remplacer ce travail de monographie lexicographique par un lexique de l'ensemble des mots de la langue s'est manifestée pour la première fois chez al-Ḫalīl Ibn-Aḥmad: c'est précisément pour résoudre ce problème pratique que la langue fut proposée comme l'objet d'une analyse combinatoire.

Ḫalīl entend en effet rationaliser la pratique empirique du lexicographe ou mieux encore résoudre théoriquement le problème pratique de la composition d'un lexique de l'arabe. La tâche n'est nullement immédiate dans la mesure où le principe sémantique de classification propre aux anciens lexiques est difficilement généralisable et, par conséquent, peu efficace. Une telle généralisation aurait exigé un système de concepts fondé et précis. Dans l'état des recherches sémantiques au 9ème siècle, pour ne pas parler de la situation actuelle, un tel système ne pouvait être élaboré. La composition d'un lexique de la langue ne pouvait donc être qu'une recomposition et la langue sera désormais soumise à l'analyse pour une énumération exhaustive de tous les mots.[21] C'est à cette seule condition que tout mot trouvera sa place dans le lexique et y figurera une – et une seule – fois. Le projet de Ḫalīl se précise: énumération exhaustive, d'une part, application bijective de l'ensemble des mots dans les cases du lexique, d'autre part, en sont les conditions, dans la mesure où elles sont les contraintes auxquelles doit se soumettre tout principe de composition du lexique. A ces deux contraintes internes, il faut en ajouter une troisième externe: la nécessité de rendre le dictionnaire accessible et facilement maniable à un usager éventuel.[22] Or, ces contraintes étant de toute évidence formelles, le principe de composition doit être de même nature. Sa constitution requiert donc l'élaboration préalable sinon d'une théorie du fonctionnement idéal de la langue, au moins d'une doctrine de l'ensemble du phénomène linguistique à partir de la seule reconstruction du vocabulaire, à savoir les éléments linguistiques qui demeurent identiques à travers la variation des significations des mots. Mais pour élaborer cette doctrine, le lexicographe se doublera du phonologue. Seule leur collaboration préparera efficacement l'avènement de l'analyse combinatoire.

La doctrine de Ḫalīl peut être ramenée à une proposition essentielle: la langue est une partie phonétiquement réalisée de la langue possible.[23] Si en effet l'arrangement r à r des lettres de l'alphabet – avec $1 < r \leqslant 5$ selon le nombre des lettres de la racine comme on le verra – nous donne, dit Ḫalīl, l'ensemble des racines – et par conséquent des mots – de la langue possible, une seule partie, limitée par les règles d'incompatibilité des phonèmes des racines, formera la langue. Composer un lexique revient donc à constituer la langue possible pour en extraire ensuite selon les dites règles, tous les mots qui s'y soumettent. Thèse importante dont la formulation a toutefois exigé une étude phonologique que Ḫalīl entreprendra dès l'abord. Il exploitera pour cette étude, comme il l'a déjà fait pour constituer la métrique, sa connaissance de la musique. Une différenciation entre deux niveaux d'analyse – signes et significations – lui permettra de prétendre à une reconstitution de la langue à partir du seul niveau des signes. Cette différenciation devait aussitôt en suggérer une autre entre son périodique – musical – et son irrégulier, apériodique, c'est-à-dire entre voyelle et consonne. Les consonnes sont ensuite ordonnées en classes selon leurs points d'articulation. Commençant par les laryngales pour aboutir aux labiales, il donne les classes suivantes:[24]

1 – ʿ ḥ h ġ
2 – k q
3 – g s ḍ
4 – ṣ s z
5 – ṭ d t
6 – ḏ̣ ḏ ṯ
7 – r l n
8 – f b m
9 – w alif yā'

Pour certaines classes, il distingue entre lettres sourdes et sonores: ainsi, dans la première classe ʿ est sonore tandis que h est sourde et dans la cinquième, on a pour d sonore, t sourde.[25] Un examen de la classification de Ḫalīl et des explications données dans Kitāb al-ʿAyn, à la lumière de la phonétique moderne, montre aisément que la répartition des sons en classes suivant les points d'articulation, d'une part, et l'opposition sourde/sonore, d'autre part, est dans l'ensemble correctement approchée. L'ordre des consonnes à l'intérieur de chaque classe reste cependant approximatif et les élèves de Ḫalīl – Sībawayhi par exemple – reprendront son analyse pour la perfectionner.

Avant d'appliquer sa connaissance à la tâche qui l'a sollicitée – la lexicographie – le phonologue va d'abord l'exploiter en vue d'une étude morphologique de l'arabe qui facilitera considérablement la démarche du lexicographe.[26] Il découvre ainsi une caractéristique morphologique de l'arabe, et des langues sémitiques en général, c'est l'importance des racines dans la dérivation du vocabulaire et le nombre suffisamment réduit de ces racines. La racine comme groupement de consonnes et seulement de consonnes, signifié auquel s'attache le plus souvent un signifiant générique, ne pouvait apparaître comme unité théorique d'analyse avant les distinctions précédentes entre sens et signification, d'une part, voyelle et consonne, d'autre part. Ces racines sont en outre des formes limitées, au plus quinquilitères et en grande majorité trilitères, de telle sorte qu'il suffit de calculer les permutations d'un groupe de lettres au plus égal à 5 pour avoir le nombre de toutes les racines de la langue possible. Ce calcul effectué, Ḫalīl va procéder enfin à la composition de son lexique. La méthode est simple: il faut calculer d'abord le nombre des combinaisons – sans répétition – des lettres de l'alphabet pris r à r avec $r = 2, \ldots, 5$, et ensuite le nombre des permutations de chaque groupe de r lettres.[27] En d'autres termes il calcule $A_n^r = r! \cdot C_n^r$, n étant le nombre des lettres de l'alphabet $1 < r \leqslant 5$.[28] Pour $r = 3$ par exemple il a par cette méthode toutes les racines trilitères de la langue possible. Ces considérations de phonologie et de morphologie l'amènent cependant à considérer un problème que les linguistes arabes après lui comme Abū ʿAlī ibn-Fāris, Ibn Ǧinni, as-Suyūṭī ne cesseront de développer: les incompatibilités entre phonèmes à l'intérieur d'une même racine. Les règles d'incompatibilité[29] permettent d'extraire de la langue possible un certain nombre de racines et de concevoir ainsi celles qui doivent être comprises dans le lexique. Nous ne saurions donner ici le détail de ces règles d'incompatibilité. Résumons-les en gros de la manière suivante: les deux premières consonnes de la racine ne peuvent appartenir ni à la même classe de localisation, ni souvent à des classes de localisation voisine. Les deux dernières consonnes de la racine subissent la même règle mais peuvent être semblables. La dérivation des mots à partir des racines se fait par schémas finis, eux-mêmes objet d'une combinatoire. Ni ces schémas ni leurs combinaisons ne sont encore reconnus explicitement par Ḫalīl. Ils le seront seulement quand la phonologie comme la morphologie de l'arabe seront considérées pour elles-mêmes, et non du point de vue local du lexicographe. Cette oeuvre sera celle des élèves et successeurs de Ḫalīl, tandis que dans Kitāb al ʿAyn la dérivation des mots reste sans règle apparente.

En guise de conclusion, rappelons quelques points:

(1) L'analyse combinatoire, développée différemment par les linguistes et les algébristes, répond donc à deux projets différents dont l'un est de résoudre théoriquement un problème pratique, tandis que l'autre entend au contraire fonder une conception théorique.

(2) La conscience fragmentée et discrète de l'unité de l'analyse combinatoire se manifeste par l'absence d'un concept particulier pour la désigner et renvoie à la différence des projets.

(3) Dans les deux cas cependant, l'analyse combinatoire se produit au terme d'une transformation essentielle de la conception des rapports entre art et science, dans une tradition en quelque sorte extérieure à celle de l'hellénisme arabe. En effet, cette transformation éclaire au moins partiellement l'apparition ici de deux disciplines scientifiques qui se proposent comme terrain du développement et de l'exercice de l'analyse combinatoire.

Ces hypothèses, de nature épistémologique, nous ont permis sur le terrain même de la reconstruction historique, d'une part, de réintégrer dans l'histoire des sciences une branche dont les anciens Arabes n'imaginèrent pas l'exclusion hors du champ de l'activité scientifique et, d'autre part, de revenir pour les raisons déjà expliquées, au début du 11ème siècle afin d'y découvrir des textes traitant de l'analyse combinatoire et d'avancer ainsi de deux siècles la date d'apparition du premier texte connu en ce domaine. L'histoire épistémologique peut donc permettre, et de comprendre une activité rationnelle datée et localisée, et d'en assurer une meilleure reconstruction historique.

CNRS – Institut d'Histoire des Sciences, Paris

NOTES

[1] Comme par exemple l'alphabet philosophique proposé par 'l'épître naïrouzienne' d'Avicenne et surtout les tentatives de Raymond Lulle, notamment dans l'Ars Magna, voir E. W. Platzeck: *Raimund Lull, sein Leben – seine Werke, die Grundlagen seines Denkens* (Prinzipienlehre), t. I, pp. 298 ff., Düsseldorf 1964. Pour l'histoire de l'école lullienne, voir W. Risse: *Die Logik der Neuzeit*, t. I, Stuttgart 1964, pp. 582 ff.

On a voulu voir dans cette combinatoire l'origine d'une tendance qui, de Lulle à Leibniz, aurait conduit à la fondation du calcul logique. Mais on sait aujourd'hui, comme Risse l'a parfaitement montré, qu'il n'y a pas continuité des questions et des solutions de Lulle et de son école à la pensée leibnizienne. La tentative de Lulle est le point de départ d'une métaphysique, bien plus que d'une logique.

[2] Il s'agit de l'*Ars Conjectandi* de Jacques Bernoulli dont la deuxième partie concerne 'La doctrine des permutations et des combinaisons', Basel 1713, pp. 72–137 et du *Traité des combinaisons*, rédigé par Montmort pour la deuxième édition de son *Essai d'analyse sur les jeux du hasard*, Paris 1713, pp. 1–72.

[3] Pour l'histoire de l'analyse combinatoire dans la première moitié du 17ème siècle, voir la thèse très bien informée d'E. Coumet: *Mersenne, Frenicle et l'élaboration de l'analyse combinatoire dans la première moitié du XVIIème siècle*, 2 tomes, dactylographiée.

[4] Trois orientations se partagent pour l'essentiel le domaine de la sociologie de la connaissance. La première veut expliquer la configuration du savoir scientifique, par sa liaison avec les structures des moyens et rapports de production: c'est la thèse marxiste. La deuxième trouve cette explication au terme de la constitution de représentations collectives, elles-mêmes, manifestations et éléments constituants de la conscience collective, fût-elle transcendante comme chez Durkheim ou immanente comme chez Gurvitch, d'une totalité sociale. La troisième ne reconnaît pas plus à la sociologie de la connaissance qu'à toute autre sociologie le droit d''expliquer' mais seulement celui d'interpréter des significations au moyen d'une réduction éidétique. Il s'agit ainsi pour M. Weber, comme pour beaucoup d'autres après lui, de soustraire la connaissance à son histoire, pour la comprendre ensuite comme la projection quelque peu pâlie d'un type idéal.

Si pour la première orientation, les réalisations ne dépassent malheureusement pas encore le niveau d'affirmations trop générales pour être vraiment démonstratives – telles que la naissance de la bourgeoisie commerçante et les débuts de la science classique ou le développement 'technologique' à l'époque de la renaissance et le commencement de la mécanique, comme l'affirment certains non-marxistes ayant subi l'influence de la pensée marxiste – les deux autres orientations sont dangereuses. La tendance durkheimienne propose comme instrument d'analyse un concept encore vague qu'elle donne pour défini: la conscience collective. La tendance weberienne ne satisfait ni le savant ni le philosophe: le premier refusera le statut de science à une sociologie qui confond la tâche du savant – la construction de modèles théoriques – et celle du philosophe – l'interprétation de significations; le deuxième réclamera des garanties que ni les weberiens ni les phénoménologues ne sont en mesure de lui donner: comment en effet affirmeraient-ils que les liaisons éidétiques ne sont pas, surtout dans ce domaine, les produits de contingences?

S'il est donc nécessaire de surmonter les incertitudes internes à la sociologie de la connaissance, elle-même, avant d'en tenter l'exploitation dans l'histoire et la philosophie des sciences, il est indispensable de commencer par reconstruire l'histoire de l'activité scientifique que l'on veut expliquer. Cette histoire fait souvent défaut et particulièrement dans le domaine des sciences arabes. Or c'est seulement à ce prix que le discours du sociologue cessera d'osciller entre des affirmations programmatiques sans aucune portée de réalisation et des assertions générales dénuées de toute valeur explicative.

[5] Ce fait n'a pas été suffisamment souligné et distingué d'une autre tendance qui, à la suite de Diophante, poursuivait une amélioration et une extension de l'application de l'algèbre à l'arithmétique. En réalité, les deux orientations existent conjointement chez les mathématiciens arabes. Quant à l'arithmétisation de l'algèbre, elle se manifeste particulièrement depuis al-Karağī et après lui. Woepcke en introduisant le livre d'al-Karağī – al-Faḫrī – fait remarquer: "l'Auteur fait souvent observer qu'on doit être préparé à l'intelligence des règles du calcul algébrique – Ḥisāb al-Mağhūlāt par les règles de l'arithmétique vulgaire – Ḥisāb al-Ma'lūmāt." V. Woepcke: *Extrait du Fakhrî: Traité d'algèbre par Abou Bekr Mohammed ben Alḥaçan Alkarkhi*, Paris 1853, p. 7. Or cette remarque n'exprime que très superficiellement la tâche d'al-Karagī. Il s'agit dans ce livre d'introduire d'une manière à la fois systématique et délibérée les opérations de l'arithmétique en algèbre, de telle sorte que l'on ait les opérations suivantes $\times / :$, $+ / \div$, non seulement pour les nombres mais aussi pour les termes algébriques. Cette application de l'arithmétique se révèle aux algébristes – al-Karağī par exemple – comme un moyen nécessaire d'organiser et d'étendre l'exposé algébrique. On découvre alors la spécificité de la démonstration algébrique.

Ainsi, après avoir étudié les puissances de l'inconnue où il examine en même temps $x, x^2, x^3, \ldots$ et $1/x, 1/x^2, 1/x^3, \ldots$ al-Karağī poursuit en introduisant dès le départ les opérations de l'arithmétique pour les expressions algébriques rationnelles. V., chapitre 3–8, *op. cit.*

[6] V. Al-Ḫawarizmī: *Kitāb al ğabr wa'l muqābala* (ed. by A. M. Musharrafa et M. M. Ahmad), Le Caire 1939, pp. 16–17.

[7] L'idée de canon est une des idées centrales de l'ouvrage d'al-Ḫawarizmī. Elle suit d'une manière systématique la solution de chacun des types d'équation et à peu près dans les mêmes termes. Pour l'équation $x^2 + bx = c$ par exemple elle correspond à ce que nous pourrions transcrire de la manière suivante:

$$x = \frac{b}{2} + \sqrt{\left(\frac{b}{2}\right)^2 + c}$$

et pour l'équation

$$x^2 + c = bx \quad \text{à} \quad x = \frac{b}{2} \pm \sqrt{\left(\frac{b}{2}\right)^2 - c}.$$

Mais dans ce dernier cas, si $(b/2)^2 < c$, c'est-à-dire si $x \notin \mathcal{R}$, le problème est alors, tant pour al-Ḫawarizmī que pour les algébristes qui lui succèdent jusqu'au 18ème siècle, 'impossible'. Al-Ḫawarizmī formule le problème – pour l'équation $x^2 + 10x = 39$ par exemple – de la manière suivante: "Un carré et dix fois sa racine égale trente neuf." Il écrit: "Prenez la moitié des racines, elles deviennent cinq. Multipliez-les par elles-mêmes, elles deviennent vingt-cinq. Ajoutez-les à trente-neuf, elles deviennent soixante-quatre. Prenez sa racine qui est huit et soustrayez la moitié des racines qui est cinq. Il reste trois. C'est la racine du carré demandé qui est donc neuf." Ce qui revient à écrire:

$$x = \sqrt{\left(\frac{10}{2}\right)^2 + 39} - \frac{10}{2} = \sqrt{64} - 5 = 8 - 5 = 3.$$

Il faut remarquer ici que faute de symbolisme, un vocabulaire bref et limité exprime l'idée de canon par la répétition de termes à peu près semblables.

[8] La démonstration d'al-Ḫawarizmī est géométrique. Pour l'équation précédente $x^2 + 10x = 39$, il prend deux segments $\perp AB = AC = x$. Il prend ensuite $CD = BE = 5 = (10/2)$. La somme des surfaces de $ABMC$, $BENM$ et $DCMP$ est égale à 39, celle du carré $AEOD$ est de $25 + 39 = 64$ donc $x + 5 = 8 = \sqrt{64} \rightarrow x = 3$.

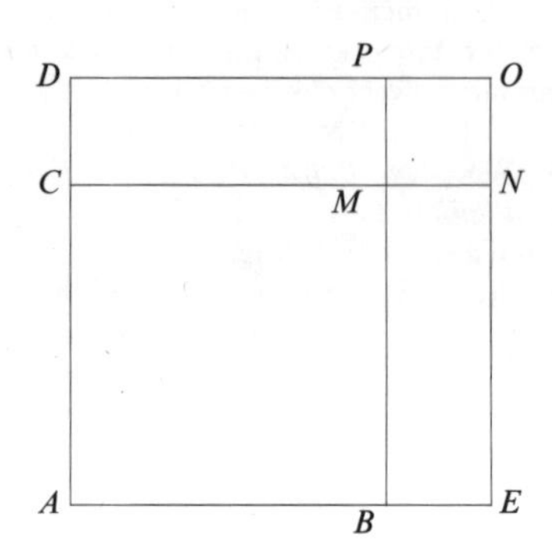

[9] V. H. Suter: *Das Buch der Seltenheiten der Rechenkunst von Abū Kāmil al-Miṣrī*, Bibl. Math. **11** (1910–1911) 100–120 et également la traduction de M. Levey du *Kitāb fi'l ǧabr wa'l muqābala d'Abū Kāmil*, Univ. of Wisconsin Press, 1966.

[10] V. la traduction de F. Woepcke: *L'algèbre de Omar Alkhayyâmi*, Paris 1951.

[11] V. Woepcke: *op. cit.* p. 9. Nous avons substitué ici à la traduction de Woepcke 'ne rend pas superflu', 'ne remplace pas' qui exprime avec plus de rigueur la phrase d'al-Ḫayyām.

[12] Pour as-Samaw'al, nous avons consulté le manuscrit no. 2718 Aya Sofia d'*al-Bāhir*, 113 feuillets, v. p. 27^{2-v}.

[13] Il est possible de démontrer que le quadrato-cube de tout nombre se divisant en deux parties est égal au quadrato-cube de chacune de ces parties, à cinq fois le produit de chacune d'elles par le carré-carré de l'autre et à dix fois le produit du carré de chacune d'elles par le cube de l'autre. Et ainsi de suite dans un ordre croissant.

Rappelons maintenant un principe pour connaître le nombre nécessaire de multiplications de ces degrés les uns par les autres, pour tout nombre divisé en deux parties. Al-Karǧi a dit que si l'on veut y parvenir il faut poser sur un tableau un et un au-dessous du premier, déplacer le (premier) un sur une autre colonne, ajouter le (premier) un à celui qui est au-dessous de lui: on obtiendra ainsi deux que l'on posera sous le un (déplacé) et on posera le (deuxième) un au-dessous de lui. On aura donc un, deux et un. Ceci montre que pour tout nombre composé de deux nombres, si l'on multiplie chacun d'eux par lui-même une seule fois – car les deux extrêmes sont un et un – et si l'on multiplie chacun d'eux par l'autre deux fois – parce que leur moyen terme est $\overline{2}$ – on obtiendra le carré de ce nombre. Si on déplace ensuite le un de la deuxième colonne sur une autre colonne, qu'on ajoute le un (de la deuxième colonne) au deux (au-dessous de lui), on aura trois qu'on inscrira sous le un (de la troisième colonne); si on ajoute alors le deux (de la deuxième colonne) au un au-dessous de lui, on aura trois qu'on inscrira sous trois, puis on inscrira un sous ce trois; on obtiendra ainsi une troisième colonne dont les nombres sont un, trois, trois et un. Ceci nous apprend que le cube de tout nombre composé de deux nombres est donné par le cube de chacun d'eux et trois fois le produit de chacun par le carré de l'autre. Si on déplace encore le un de la troisième colonne sur une autre colonne, qu'on ajoute le un (de la troisième colonne) au trois au-dessous de lui, on aura quatre que l'on inscrira sous un; si on ajoute alors le trois au trois au-dessous de lui, on obtiendra $\overline{6}$ qu'on inscrira sous quatre; si on ajoute ensuite le deuxième trois au un au-dessous de lui, on aura quatre qu'on inscrira sous six, puis on déplacera un sous quatre: il en résultera une autre colonne dont les nombres sont un, quatre, $\overline{6}$ quatre et un. Ceci nous apprend que la composition du carré – carré d'un nombre constitué de deux nombres est donnée par le carré – carré de chacun d'eux – puisqu'on a un aux deux extrêmes – puis par quatre fois le produit de chacun des nombres par le cube de l'autre – parce que quatre succède aux deux extrêmes un et un – car la racine du cube est le carré – carré et enfin par six fois le produit du carré de chacun d'eux par le carré de l'autre – car le produit du carré par le carré est le carré-carré. Si on déplace ensuite le un de la quatrième colonne sur une cinquième colonne, qu'on ajoute le un au quatre au-dessous de lui, le quatre au six, le six au quatre et le quatre au un, puis qu'on inscrive les résultats sous le un déplacé de la manière susdite, qu'on inscrive enfin le un qui reste, on aura une cinquième colonne dont les nombres sont un, $\overline{5}$, dix, dix, $\overline{5}$ et un. Ceci nous apprend que pour tout nombre divisé en deux parties, son quadrato-cube est égal au quadrato-cube de chacune des parties – car les deux extrêmes sont un et un – à cinq fois le produit de chacune d'elles par le carré-carré de l'autre – puisque le cinq succède aux deux extrêmes sur les deux côtés et dix fois le produit du carré de chacune d'elles par le cube de l'autre – parce que le dix succède aux deux cinq. Chacun de ces termes appartient au genre du quadrato-cube car le produit de la racine par le carré-carré et celui du cube par le carré donnent l'un et l'autre le quadrato-cube, on peut ainsi connaître le nombre

x	x^2	x^3	x^4	x^5	x^6	x^7	x^8	x^9	x^{10}	x^{11}	x^{12}
1	1	1	1	1	1	1	1	1	1	1	1
1	2	3	4	5	6	7	8	9	10	11	12
	1	3	6	10	15	21	28	36	45	55	66
		1	4	10	20	35	56	84	120	165	220
			1	5	15	35	70	126	210	330	495
				1	6	21	56	126	252	462	792
					1	7	28	84	210	462	924
						1	8	36	120	330	792
							1	9	45	165	495
								1	10	55	220
									1	11	66
										1	12
											1

(nous avons reproduit ce tableau en remplaçant l'énoncé, racine, carré, cube, ... par les symboles x, x^2, x^3 ...)

de carrés et de cubes pour une quelconque limite (ou quelconque puissance). V. As-Samaw'al *op. cit.* pp. 45 ff.
[14] V. Woepcke: *op. cit.*, p. ??.
[15] Ce texte est traduit en russe par Ahmadov et Rosenfeld, v. *Istor.-Mat. Issled.* **15** (1963) 431–444.
[16] Cf. P. Luckey: *Die Rechenkunst bei Ǧamšīd b. Mas'ūd al-Kāšī*, Wiesbaden 1951, pp. 24 ff. et du même auteur: 'Die Ausziehung der n-ten Wurzel und der bionimische Lehrsatz in der islamischen Mathematik', *Math. Annalen* **120** (1948) 217–274.

La thèse fondamentale de Luckey est qu'al-Kāšī au moins a retrouvé le tableau des coefficients binômiaux en utilisant le procédé connu plus tard sous le nom de procédé de Ruffini-Horner. L'idée mérite une discussion approfondie que nous poursuivrons ailleurs.
[17] Cf. Faḫr al-Dīn al-Rāzī: *Manāqib al-Imām al-Šafi'ī*, p. 99:

Le rapport d'al-Šafi'ī à la science des 'uṣūl al-Fiqh' – la jurisprudence canonique – est ce rapport même qui lie Aristote à la logique et al-Ḫalīl ben Aḥmad à la métrique. En effet, avant Aristote, les hommes argumentaient et s'opposaient à partir de leurs dispositions naturelles en l'absence d'une loi universelle pour ordonner les définitions et les preuves; il va sans dire que leurs mots étaient confus et vagues car la nature seule sans référence à une loi universelle, ne peut parvenir à autre chose. Ayant constaté cela, Aristote s'isola pendant une certaine période et présenta sa science de la logique où il proposait aux hommes une loi universelle à laquelle on pût se référer pour la connaissance des définitions et des preuves. De la même manière, les poètes, avant al-Ḫalīl ben Aḥmad, composaient des vers en se fondant sur la seule nature quand al-Ḫalīl présenta sa science de la métrique qui consiste en une loi universelle des mérites et des défauts du vers. De la même manière encore, avant l'Imām al-Šafi'ī, les hommes traitaient de questions relatives aux uṣūl al-Fiqh, argumentaient et s'opposaient sans pour autant posséder de loi universelle à laquelle se référer pour la connaissance des preuves juridiques, du mode de la controverse et de la démonstration. Al-Šafi'ī déduisit la science des uṣūl al-Fiqh et proposa aux hommes une loi universelle à laquelle se référer pour la connaissance des degrés de la preuve juridique.

On distinguait d'une manière générale dans la tradition classique arabe entre deux classes de sciences: les sciences antiques – 'ulūm al-awā'il – et les sciences arabes – 'ulūm al-'arab – c'est-à-dire les sciences linguistiques où les auteurs ne reconnaissaient aucune priorité à la science hellène.
[18] Cf. à titre d'exemple: *Kitāb al-Mu'tamad fī uṣūl al-Fiqh d'Abu'l Ḥusayn Muḥammad al-Baṣrī*, Institut Français de Damas, t. 1, 1964, pp. 15 ff.
[19] Cf. à titre d'exemple dans 'la grande encyclopédie théologico-philosophique', *al-Mu'nī*, t. 1 d'Abu'l-Ḥasan abd al-Ǧabbār, concernant la création de la parole divine, Le Caire 1961.

Dans les ouvrages des Mu'tazilites de même que dans ceux des linguistes, des discussions théoriques développent longuement les problèmes de l'origine et de la nature de la langue. V. par exemple l'ouvrage du linguiste al-Suyūṭī- *al-Muzhir* (ed. par al-Mawla *et. al.*), t. 1, pp. 7 ff., Le Caire (non datée).
[20] Cf. à titre d'exemple, Meyerhof-Ṣobḥī: *The abridged version of 'the book of simple drugs' of Aḥmad ben Muḥammad al-Ghāfiqī by Gregorius abu'l-Farag*, Cairo 1932 et Ibn al-Baytār: *Gam'al-mufradāt, Traité des simples* (ed. par L. Leclerc), Paris 1877–83.
[21] Pour l'histoire de la lexicographie et afin de situer la portée de l'étude lexicographique, v.: A.-B. Keith: *A History of Sanscrit Literature*, London 1924; Müller: *Handbuch der klassischen Altertums Wissenschaft*, vol. II, 1913; J. Collant: *Varron grammairien latin*, Strasbourg 1923; K. Krumbacher: *Geschichte der byzantinischen Literatur*, München 1896.
[22] Nous lisons ainsi au début du *Kitāb al-'Ayn*:

Voici ce que composa al-Ḫalīl ben Aḥmad al-Baṣrī... des consonnes alif, bā', tā', ṯā'... au moyen desquelles les Arabes parlaient, pivot de leurs mots et de leurs expressions, en sorte qu'aucun n'en soit excepté. Il voulut ainsi connaître les vers et les discours des Arabes afin que rien ne lui en soit irrégulier. p. 52.

[23] Cette théorie, présente dans le texte attribué à al-Ḫalīl, a été reprise et développée plus tard par Abū 'Alī ben Fāris, Ibn Ǧinnī et as-Suyūṭī... Ce dernier rapporte en outre dans son ouvrage cité plus haut les opinions d'Ibn Fāris et d'Ibn Ǧinnī. V. *op. cit.*, pp. 240 ff.

[24] V. *Kitāb al-'Ayn*, pp. 52–53 et p. 65.

[25] *op. cit.*, p. 64.

[26] *op. cit.*, p. 55. Al-Ḫalīl a dit:

"Aucune construction du nom ou du verbe ne dépasse chez les Arabes le nombre de cinq consonnes. Quelles que soient les adjonctions dépassant cinq consonnes qu'on peut trouver dans un verbe ou dans un nom, sache qu'il s'agit là d'une adjonction à la construction qui n'appartient pas à la racine du mot."

[27] "Sache qu'un mot bilitère se permute de deux manières comme – qad, daq; šad, daš – qu'on mot trilitère se permute de six manières et se nomme sextuple comme – ḍaraba, ḍabara, baraḍa, baḍara, raḍaba, rabaḍa – qu'un mot quadrilitère se permute de vingt-quatre manières, car ses consonnes au nombre de quatre se multiplient par les modes du trilitère normal qui sont au nombre de six et donnent ainsi vingt-quatre dont on écrit ceux que la langue utilise tandis que l'on écarte ceux qu'elle n'utilise pas... et qu'un mot quinquilitère se permute de cent-vingt manières, car ses consonnes au nombre de cinq se multiplient par les modes du quadrilitère qui sont au nombre de vingt-quatre et donnent ainsi cent-vingt manières dont la plus petite partie est utilisée et la plus grande négligée."

Autrement dit, pour chercher la permutation de r consonnes, on cherche la permutation de $r-1$ consonnes multipliée par r, ou encore $r! = r(r-1)!$

[28] Le calcul attribué par Abū Ḥamza à al-Ḫalīl et reproduit par as-Suyūṭī – *op. cit.*, p. 74 – est correct, à savoir A_n^r pour $n=28$ et $r=2,\ldots,5$. En plus de ce calcul, la composition même du lexique permet d'avancer la formule correspondante. La méthode est souvent citée et beaucoup plus tard, on la trouve encore dans les Prolégomènes d'Ibn Ḫaldūn, reprise comme un élément du fonds culturel. En effet, pour chercher C_n^r pour $r=2$ par exemple, il procède empiriquement, prenant la première consonne qu'il combine avec les autres et obtient ainsi 27 mots. Il prend ensuite la deuxième consonne qu'il combine avec 26 pour avoir 26 mots, et ainsi de suite. Il additionne ensuite ces combinaisons et multiplie la somme par 2 pour avoir les arrangements. Pour $r = 3$, il procède de la même manière, mais en considérant les mots à deux consonnes comme une seule consonne que l'on combine avec les 26 consonnes restantes. Avec les 27 racines à deux consonnes, il forme 26 racines à trois lettres, et ainsi de suite. Il additionne et multiplie par 6 pour passer aux arrangements, et ainsi de suite pour $r = 4$ et $r = 5$.

V. Ibn Ḫaldūn: *Prolégomènes*, pp. 548 ff., chap. 'La science du langage'.

[29] V. *Kitāb al-'Ayn*, p. 63 et pour la consonne '.

Ḫalīl a dit: "Le ' ne se combine pas avec le ḥ en un seul mot par suite de la proximité de leurs points d'articulation." p. 68.

Ou encore: "Mais le ' est négligeable avec les consonnes suivantes g, h, ḥ, ḫ." p. 69.

Il reste cependant qu'après al-Ḫalīl ces problèmes deviendront l'objet d'études systématiques.

STEPHEN TOULMIN

SCIENTIFIC STRATEGIES AND HISTORICAL CHANGE

I

There are moments when we become aware that the tide of thought is turning, and that the natural flow of men's proper concerns is carrying us in a new direction. Only a few years back, the 'multidisciplinary' topic of this particular session would have been regarded as hardly serious. Most philosophers of science then saw their tasks as centered on formal logic, and they happily set aside the history, sociology and psychology of science as being 'of merely empirical interest'. (In this they agreed with Descartes. As he put it: historical enquiries, like foreign travel, may broaden the philosopher's mind, but they have no relevance to his professional problems.) Their indifference was requited. Historians, sociologists and psychologists of science were equally content to have philosophy leave them to their own devices. For some half-a-century, the accepted maxim in our corner of the academic world could have been put in the words of the hymn:

> ... We must shine,
> You in your small corner
> And I in mine.

So if we are to treat the Comparative History and Sociology of Science as having any genuine relevance to *philosophy* of science, we must – surely – begin by facing the preliminary, Kantian question, *Wie ist eine solche Wissenschaft überhaupt möglich?*: How is such a multidisciplinary enquiry possible at all?

That is the first of the three issues I shall take up in my remarks today. For reasons that I shall analyse as concisely as I can, philosophers, sociologists and psychologists of science alike have been taking an essentially *static*, or *structural*, view of science; and, as a result, the history of science has been reduced (in Collingwood's phrase) to a 'scissors- and-paste' catalog or chronology. Yet (I shall argue) suppose we adopt an

R. J. Seeger and R. S. Cohen (eds.), Philosophical Foundations of Science, 401–414.

alternative, more *dynamic* approach to science, we shall at once begin to see more clearly how its historical and philosophical, psychological and sociological aspects are interrelated. A more dynamic account turns out to be, also, a more integrated account; and in the central part of my remarks I shall discuss one class of examples that illustrate this point. Finally, I shall survey very briefly the new kinds of question that will become important for us if we do adopt this alternative, more integrated approach to the analysis of the sciences.

II

Until recently, those of us who wished to study the sciences in a consistent, integrated way found ourselves frustrated at the very outset. In order to learn about the historical development of scientific ideas, we turned to one set of books and authors; in order to study the psychology of the individual scientists whose ideas they were, we had to turn to others; still another set dealt with the social structure of the professions through which scientific ideas find institutional expression; while others yet again analysed philosophically the arguments by which such ideas were – and were to be – validated. Each set of books was written in its own terms, and their authors showed little inclination to peek over the fence into the garden next door. By some masterpiece of initial abstraction, apparently, the scholars concerned had defined the central question of each specialized sub-area in ways that kept them permanently isolated.

What was the nature of this initial abstraction, by which the history and sociology, philosophy and psychology of science became differentiated off, specialized and eventually separated from one another? The core of the answer to that question is, I believe, as follows. Some fifty years ago, the philosophers and behavioral scientists concerned stopped thinking of the sciences as essentially *historical and developing* human enterprises, and became preoccupied instead with *single temporal cross-sections* of those historical enterprises. So the philosophers studied only the intellectual content of each science – conceived of as a logical system of propositions current at a given moment in time; the sociologists considered only the structure of scientific institutions – conceived of as forming a social system of groups, and roles, operative within the total system of society at that time; while the psychologists concentrated on the supposedly-universal processes by which the child developed into the scientifically-mature adult – whatever the scientific ideas current among adults in his particular culture or society.

The historical reasons for this switch of attention are intriguing, but too complex to go into here. Its effects are evident enough. It led sociologists and anthropologists to disregard the *historical*, or *diachronic*, processes of social and cultural change, and to focus on *contemporaneous*, or *synchronic*, factors and categories – groups, social forces and equilibria, cultural functions, etc. It led psychologists to ignore the relations between *cognitive development* and *enculturation*, as though all children must inevitably develop the same conceptions of (say) causality or conservation, regardless of manifest differences in the concepts generally current in different epochs and cultures! And it led philosophers to equate the *rationality* of a science with the *logical* character of the systems of propositions in which, supposedly, one could express its entire intellectual content. Thus began the period of intellectual history in which we have all grown up, and with which we are all – I nearly said, all too – familiar.

Now, I am not going to argue today that the Emperor had no Clothes. This half-century of intellectual specialization has, no doubt, been an active and interesting period in all the fields concerned. But I do insist that this specialization was achieved only at a price. Having begun by abstracting particular temporal cross-sections from the historical development of science, and having confined their attention to relations discoverable *within each single cross-section considered separately*, philosophers and behavioral scientists alike found it impossible to reconstruct, out of the relations so discerned, any satisfactory account of *historical change* – whether in the intellectual content, the social institutions or the learning-procedures of science. For the terminology of 'logical structure' gives us no way of characterising the 'rationality' of conceptual changes in science; the terminology of 'social structure' gives us no way of characterising the 'dynamic processes' of institutional change; and so in every case.

Let me remind you, at this point, of what Aristotle said about the concept of 'geometrical points'. Such *points* (he argued) are a fiction or artefact, which we arrive at, and define, by the intersections of *lines*, rather than the other way around. If we imagine that our task is to start with *points* and then construct *lines* out of *points*, we shall end in confusion: there *is* no procedure for taking an infinite number of points and making a line out of them. Nor, I would add, is there any intellectual procedure, either, for taking a temporal sequence of propositional systems or institutional structures, as customarily defined, and reconstructing out of these static structures the historically-developing rational enterprise of science, from which they were arrived at by abstraction in the first place.

Let me come to the heart of the matter. It was always a mistake to identify 'rationality' and 'logicality': to suppose, that is, that the *rationality* of a science could be explained in terms of the *logical* attributes of the propositional systems intended to express its intellectual content at one time or another. For questions of rationality are concerned, precisely, not with the particular intellectual position that a man (or professional group) adopts at any given time, but rather with the *conditions on which*, and the *manner in which*, he is prepared to *criticize and change* that position. The rationality of a science is embodied, that is, not in the propositional systems current *at particular times*, but in its procedures for discovery and conceptual change *through time*. So, when inductive logicians declare, "There is no logic of discovery," we can happily agree; and we can at once go on to stand this dictum on its head. For logic is concerned with the *internal articulation* of intellectual systems, considered either at particular times, or apart from time; so the *rationality* of scientific discovery – the intellectual procedures by which scientists arrive at *conceptual changes*, in the course of time – will necessarily elude analysis and judgment in formal terms alone.

It is time, accordingly, to carry the philosophy of science back to its true starting-point. This is (I have said) not a logical system, or temporal sequence of logical systems, but rather an *intellectual enterprise* whose rationality lies in the procedures governing its historical development. We may perhaps, for certain purposes, represent the provisional end-product of that enterprise in the form of a propositional system; but to do so remains an abstraction. The 'system' so arrived at is not the primary reality: it is a fiction or artefact of our analysis, and the results we can achieve by this method are limited. The true starting-point remains the living, historically-developing intellectual enterprise of science, and it is to this that all our results must be referred back for validation.

Certainly, this is the direction in which one feels the new tide of theoretical discussion flowing today. Sociologists and anthropologists are at last turning away from the static categories of 'social structuralism' and 'cultural functionalism', and are taking up, once again, long-neglected historical problems about social development and cultural change. Philosophers of science, too, are reawakening that historical sense that they cultivated naturally in the time of William Whewell, and are paying closer attention to what Russ Hanson called 'the patterns of discovery'. Given this new direction, we can acknowledge the constraints that the 'specialist abstraction' has been placing on us all these years, and begin posing new, and more lifelike questions about the sciences: questions that cut across the accepted boundaries between philosophy and psychology, sociology and history. My second task here will be, to illustrate how these new questions arise.

III

Regarded as an historically-developing enterprise, any well-developed natural science has two faces. We can think of it as an intellectual 'discipline', or as an institutionalized 'profession'. If we consider it in disciplinary tems, its temporal development will then be a topic for the history of scientific ideas: we shall have to enquire (e.g.) how concepts which were at one stage entirely 'speculative' subsequently became 'well-established', only to fall later into the category of 'mere approximations' or even of 'superstitions'. By contrast, if we consider it in professional terms, its temporal development will then be a chapter in the history of scientific institutions and procedures: we shall now have to enquire (e.g.) how professional groups and technical procedures which at one stage had no standing, or authority, subsequently won general authority and acceptance, only to be relegated later to a subsidiary role if not completely discredited. But now these will no longer be two separate, independent histories. The *selection-criteria* by which novel concepts win an 'established' place within the scientific discipline will now have to be seen in relation to the actual procedures used by those crucial individuals or 'reference-groups' within the profession whose current role it is to pass judgment on conceptual innovations 'in the name of' the science concerned. The philosophical story and the sociological story thus overlap, becoming two complementary aspects, two interrelated ways of looking at a single, larger enterprise.

By way of illustration, let me say something about these interrelations: i.e. about the ways in which the intellectual development of scientific disciplines parallels, and is paralleled by, the sociological development of scientific professions – notably, the respects in which the activities of the judges and reference-groups in a profession reflect, and are reflected in, the selection-criteria governing conceptual developments in the corresponding discipline. One can too easily take it for granted that the relationship between a scientific discipline and its professional exponents is a one-way affair: that the logical demands of the inductive method exercise an unambiguous authority over scientists, so that scientific judgment demands of the men concerned only that they should keep their heads and remain 'logical'. This (I shall argue) is an illusion. At any point in the development of a science, questions of *intellectual strategy* can arise for which there is *no* uniquely 'logical' decision-procedure; and in these situations rational decision indeed becomes – in the familiar, colloquial sense of the phrase – a 'matter of judgment'. To this extent, the relationship between the conceptual development of a science, and the institutional development of the profession that embodies it, will be more complex: a two-way affair, in which the history of scientific *ideas* can no longer be disentangled from the history of the individuals and *groups* whose task it is to formulate, criticize and judge those ideas.

To borrow an idiom from legal parlance: in science as in the law-courts, one can distinguish between 'clear' and 'cloudy' cases, or between 'routine' issues and 'moot' points. Let me first say something about 'routine' cases. The scientists working in any particular discipline commonly share very much the same intellectual aspirations or ideas. They have (that is to say) an agreed, or sufficiently agreed conception of 'explanation': or, to speak more exactly, their *differences* about what would represent a 'complete' solution to their shared intellectual problems are for the time being of only marginal significance. One effect of this consensus is to determine correspondingly well-defined selection-criteria, for deciding between the many conceptual variants, or innovations, that are circulating within the discipline at any time. For the operative question can then be stated simply enough: as being, how far the adoption of any proposed conceptual innovation will improve our conceptual grasp in the field concerned – how much nearer it will take us towards the collective intellectual goal of the science, as for the time being understood and agreed.

The fact that the operative question in such cases can be *stated* simply does not, of course, imply that it can necessarily be *answered* simply. Even here, scientists are often faced with invidious choices. A proposed theoretical change may appear highly attractive in one respect, retrograde in another – there are always multiple criteria, and these may well point us in opposite directions. Even where the current intellectual strategy of a sience is not in question, therefore, the tactical choices may still be hard to make. (Just how much price is to be paid in terms of coherence, or elegance, in exchange for a small improvement in the predictive power of our computational procedures? Just how are we to balance off a gain in simplicity of one kind, against a loss in simplicity of another?) Still, to the extent that there is general agreement about broad intellectual goals, it does become possible to appraise, in rational terms, the critical activities of individual scientists or reference-groups working within the science. To that extent, the accepted rational demands of scientific judgment will have a clear authority over the professional scientists working in that field. Thus, looking back at 19th-century physical science, we may think in retrospect that William Prout's desire to press beyond atomic theory to the sub-atomic constitution of matter was, within his historical and professional content, genuinely 'premature'; that Michael Faraday's commitment to the visual imagery of 'tubes of force' and the like was 'unorthodox', and so on. The existence, for the time being, of agreed criteria for judging between novel ideas carries with it, implicitly, criteria for judging also the decisions of those individual scientists and reference-groups who are, at that point in historical time, the *custodians of rationality* within that science.

IV

Yet this is not always so. Alongside these clear and straightforward issues there are other more difficult ones: cases which are *intrinsically* cloudy, just because in certain respects the intellectual strategy of the relevant science is temporarily in doubt. Let us touch on three examples: (1) By the late nineteenth century, the program for physics first propounded at the end of Newton's *Opticks* had been largely completed, and the period from 1900 to 1914 was a time of uncertainty. The new conceptions of relativity and quantum theory would establish themselves firmly only after 1919, and for the time being physicists were casting around for a way ahead, undecided what new intellectual goals they should set themselves. Against this background, it is worth re-reading the exchange of papers between Max Planck and Ernst Mach printed in the *Physikalische Zeitschrift* for 1910–11. In a long essay called 'The Unity of the Physical World-Picture', Planck surveys the whole historical development of physical thought, and finds in it a constant direction, involving the progressive elimination of 'subjective' sensations and other 'anthropocentric' elements, in favor of quantitative, 'intersubjective' magnitudes and theoretical 'invariants'. This done, he criticizes Mach's philosophy of science, arguing that Mach's 'sensationalism' willfully reintroduces into the heart of physical theory those very 'subjective' elements that the main stream of physical thought had been aimed at eliminating. In due course, Mach replied with an equally elaborate paper – partly an intellectual autobiography and partly a restatement of his methodological program, which calls for a physics committed to the avoidance of metaphysics and based solely on 'observables' or (as he regards them) 'sensory observations'.

Now, whether Planck's or Mach's conclusions were, in the event, the sounder is not my foremost question. The two men differed quite fundamentally over questions of strategy, and in some respects the quantum physicists of the Copenhagen school have, rightly or wrongly, preferred the strategy advocated by Mach. What is significant for our purposes is, simply, the character of the argument between them: notably, the clear-sighted manner in which Planck analyses the changing explanatory demands that have guided the development of physical theory. As he sees it, the new strategic principles appropriate to the problems of theoretical physics in his own day must make it the 'legitimate heir' of previous physical investigations. They therefore have to be formulated and judged, not in formal, abstract terms, but with an eye to the whole historical evolution of physics. (2) The second example, from the late 1940s, concerns the emergence of the phage group and the 'take over' of theoretical biology by men trained originally in physics, which formed the background to the development of molecular biology. (The whole episode has been admirably documented in a recent article by Donald Fleming.) When, in 1944, Avery and his colleagues published their classic demonstration that D.N.A. was the carrier of a particular hereditary trait in a single bacterium, they were constrained from claiming too much, by their commitment to the currently-accepted attitudes of classical genetics. Within classical genetics – which had been one of the great success stories of modern biology – biochemical questions about the material nature of the gene were unimportant, if not entirely irrelevant. As a result, the 1944 paper was, in Fleming's words, 'muffled and circumspect': the authors were 'almost neurotically reluctant' to identify genes with D.N.A. Watson and Crick were subject to no such constraints; but the extent of their success should not lead us to overlook the strategic battle had gone on before. For they were the self-confident heirs of a new approach that had been hammered out, in the years between 1944 and 1953, by men like Szilard and Delbrück. Avery and his colleagues exemplified an attitude that the new physicist/biologists were to reject completely. Delbrück has said that biology, as he found it, was a 'depressing' subject: the accepted styles of biochemical interpretation "stalled around in a semidescriptive manner without noticeably progressing towards a radical physical explanation." And, once again, what is significant here is the nature of the considerations on which the new approach was based. Szilard brought to biology (Fleming quotes him as saying) "not any skills acquired in physics, but rather an attitude: the conviction which few biologists had at the time, that mysteries can be solved." This attitude, shared by Delbrück and the phage-group, enabled them to bring a fundamentally new strategy to bear on the problems of virology and genetics, of which Crick and Watson's molecular biology was to be the fruit. (3) My third example is taken from contemporary physics. Here again, there are theoretical difficulties which call, not for better mathematics or more ingenious experiments, but rather for a strategic re-appraisal of the basic aims of the subject. Thus, some physicists today – such as Geoffrey Chew – believe that "developments of the past three decades suggest that the capacities of the elementary particle idea may finally have been exhausted, without the identification of an ultimate set of primitive entities." Many physicists concede, for instance, that none of the known set of hadrons could be elementary. This being so (Chew suggests) it may be that "an end to the elementary particle road really has been reached," and in that case physics "must find an alternative."

In the present situation, therefore, we can proceed in either of two directions. On the one hand, we can refuse to accept that 'the end of the road' *has* been reached, and we can continue "a vigorous search for new entities which might conceivably be identified as fundamental constituents of matter." Unfortunately (as Chew points out) "the elementary particle idea never gets to an end... If there are elementary particles, why those particular particles?... It always leaves you with the problem of understanding the last particle you have identified." Chew's own alternative is to look for a theory of an entirely different kind, in which the existing sets of particles are explained, not by appealing to other, yet smaller particles, but by showing that "the particles are as they are because this is the only possible way they could be." Unfortunately, this so-called 'bootstrap' idea has its own difficulties. As Chew concedes, "By its very nature, it cannot be formulated through equations of motion in the time-honored tradition of all previous physical theories, because in principle there are no entities which could conceivably appear in the equations of motion": so its acceptance would manifestly require us to modify our ideas about the form a satisfactory 'physical explanation' should take. Whether such an alternative approach will prove sound, only history can show: if it does (Chew concludes) "the dilemma with the hadrons" will turn out to have been "the precursor of a wholly new kind of science."

These three examples differ substantially in several respects. For instance, only one of the three can really be said to involve a 'crisis' in the development of a science, though in all three the consequences of the strategic changes involved are equally far reaching. There was no crisis in theoretical biology in the mid 1940s. If men like Dulbrück and Luria argued for a new strategy and a new conception of 'explanation', they did so out of a desire to bring physical and biological understanding into closer relation, not to rescue biology from any particular threat of intellectual breakdown. But the examples do have one feature in common. In each case, the conditions that normally apply cease to hold good. There is no longer a collective, agreed conception of 'explanation' in the science concerned and so unambiguous criteria for judging between new ideas.

Correspondingly, there is room in such cases for individual differences, and for the exercise of individual judgment, of a kind that does not exist over routine issues. In each case, the fundamental theoretical question at issue ceases to be, "What conceptual innovations will best solve our outstanding problems, and help us towards our agreed intellectual goals in this field?" Rather, it becomes, "What general intellectual goals *should* we be aiming at here? What explanatory tasks *are there to be* tackled in this field?"

Questions of this form, of course, cut too deep to be answerable by resort to any routine procedure. They require us not so much to continue playing to an old familiar set of rules, as to reappraise the goals of our whole theoretical game. Such 'moot' cases thus bring us up against questions that can safely be begged in routine cases: questions about the nature of strategic choice in a scientific discipline, and about the functions of the men whose judgments carry authority in the relevant profession. This being so, they cannot be handled unless one stands back, and views current intellectual problems against a longer historical background.

V

Up to this point in my remarks, I have been speaking about scientific judgment in legal parlance – (moot points, cloudy issues, etc.) – simply as a matter of convenience. Now I want to go one step further. For I believe that we can understand better the questions that arise about scientific strategies, and the sorts of judgments they call for, if we compare them with one particular class of judicial decisions: namely, those crucial cases in constitutional law in which (e.g.) the U.S. Supreme Court is compelled to reinterpret the provisions of the Constitution, and to *reanalyse* the social functions of law itself in their application to novel historical situations.

Let me remind you of Oliver Wendell Holmes' discussion of legal reasoning in his book, *The Common Law*. In the court of last resort (Holmes argued) we are faced with constitutional cases for which no definitive and unambiguous decision-procedure is available, or in which the accepted principles, as hitherto interpreted, lead to evident anomalies and inequities. When this happens, the judicial task is no longer simply to apply pre-existing procedures to fresh cases. Rather, it is now to stand back and reconsider the overall justice of the provisions in question against a larger socio-historical background. In the final juridical context, logic thus becomes the servant of fundamental human purposes: jurisprudence and judicial practice alike are ultimately based on our developing understanding of the historical sociology of law.

Now, I realize that the sociological jurisprudence of men like Mr. Justice Holmes and Professor Roscoe Pound is itself open to dispute. I cite it here, only because Pound and Holmes put their finger very precisely on the point at which legal reasoning ceases to be a *formal* or *tactical* matter, of applying rules and principles to new situations, and becomes concerned with *strategic* issues: the point (that is) at which the *just* way ahead can be determined, only by reappraising the fundamental social purposes of the law in a new historical context. Over such strategic issues (they insisted) it is no longer possible to regard decisions, clearly and definitively, as 'right' or 'correct'; but they are *none the less rational* for all that. At a point of this kind, the Supreme Court judge can no longer speak for the law *as it is*: instead, he passes judgment in the light of a longer-term vision, both of what the law has been and is, and of *what it should become*. To that extent, the resulting judgment will inevitably be, not 'the judgment of *the law*', so much as (say) Holmes' or Frankfurter's *best personal judgment* of the way in which the law should now develop, if it is, at this point in historical time, most completely to fulfill its basic social ideals of equity, humanity and certainty. For this is a 'rational frontier', at which fallible individuals, acting in the name of the human enterprise that they represent, have to open up new possibilities in order to deal with novel and unforeseen problems. Here, accordingly, we can no longer separate the rational procedures of the law from the men who are reshaping them or from the historical situations in which these men find themselves.

It is the same (I argue) in science. At moments of strategic uncertainty, when the intellectual goals of science have to be reappraised, the rational procedures can no longer be stated in formal rules, and the decision-criteria for judging conceptual innovations cease once again to be definitive and unambiguous. We cannot, for instance speak of the decision of a Mach, a Delbrück or a Chew as being, by any standards available in their times clearly and definitively 'right' or 'correct'; but they are *none the less rational* decisions for all that. At a point like this, a Delbrück no longer speaks for biology *as it is*: instead, he does his best to pass judgment in the light of a longer-term vision, both of what biology has been and now is, and of *what it should become*. The resulting decision will inevitably be, not 'the judgment of *biology*', so much as Delbrück's own *best personal judgment* of the way in which biology must now develop, if it is most completely to fulfill our basic intellectual ideals of understanding, at this point in human history. Here, too, we are at a 'rational frontier', where novel problems demand new methods of thought; and where we can no longer separate the rational methods of science entirely from the men who are reshaping them, or from the historical situation in which those men are working.

VI

I have given here an illustration – and it is merely *one* illustration – of the kinds of scientific situation in which it is impossible to consider men or ideas, professional institutions or rational procedures, in isolation from one another; and in which it is necessary to put them all into the context of an integrated, historical account of science. While there may, indeed, be some questions about science that we can legitimately state in sociological rather than logical terms, or in philosophical rather than historical terms, many of the most significant questions overlap these traditional boundaries. So, if the historical development of science really is (in my sense) the story of a human 'rational enterprise', the re-synthesis implied in the title of the present symposium is then not only overdue but inescapable. For in that case, in the very act of separating the philosophy of science from its sociology, its history from its psychology, we shall have been missing the essential character of science.

Consider, for instance, the notion of 'authority' in science. At any time, there are certain 'authoritative' individuals in any branch of science, whose opinions carry special weight, when it comes to judging the merits of theoretical innovations, as solutions to outstanding problems. The backing of (say) a Kelvin or a Helmholtz, a Sherrington or a Jacques Loeb, can get a new idea taken seriously – or ignored – in a way that the advocacy or opposition of a lesser man could never do. The intellectual 'authority' that such men exercise is both a *personal* authority, and also the authority of the discipline *in whose name* they have come to speak. Yet how (we may ask) do these particular men acquire the right, and the power, to speak 'in the name of' (say) neurophysiology or physical theory? And how, in successive generations of a scientific profession, does intellectual authority pass from one such reference-group of scientific judges to another? Here is a group of questions that we can never answer satisfactorily, either in terms of the theory of social structure alone, or by appeal to inductive logic alone.

It is at points like these that the customary interdisciplinary boundaries most seriously obstruct us. When it comes to considering (e.g.) how individual scientists or scientific institutions serve, and shape, the changing rational purposes of science, we can no longer take refuge in 'pure' philosophy, or in 'pure' sociology, or answer our questions in terms drawn from only one of these familiar specialisms. The 'rationality' of science is a highly complex affair. It is not a mere internal property of propositional systems, and so a matter for formal logic, but involves, rather, an interaction between scientists and their ideas, their institutions and the historical situations with which they have to deal. Any account we give of scientific rationality must be rich and complex enough to do justice to all these different aspects. True: we have not yet reached a stage at which we can give such an account. But it is as well for us to know what we are entitled to aim at, and to demand, so that we can decline, in the meantime, to be satisfied with anything second-best.

University of Chicago

BIBLIOGRAPHY

Cairns, John, *et al.* (eds.), *Phage and the Origins of Molecular Biology*, Cold Spring Harbor, New York, 1966.

Chew, Geoffrey F., 'Crisis for the Elementary-Particle Concept', Wellesley College Science Symposium, January 1968, *Publications of the University of California Radiation Laboratory*, No. UCRL-17137.

Fleming, Donald, 'Emigre Physicists and the Biological Revolution', in *Perspectives in American History*, Vol. II, 1968, pp. 152-189.

Planck, Max, 'The Unity of the Physical World-Picture'; and Mach, Ernst, 'My Scientific Theory of Knowledge and its Reception by my Contemporaries', together with Planck's reply: originally printed in *Phys. Z.*, for 1910 and 1911. New translations of these three papers appear in an anthology of papers on philosophical aspects of twentieth-century physics: *Physical Reality* (ed. by S. E. Toulmin), Harper Torchbooks, New York, 1970.

ERNAN MCMULLIN

LOGICALITY AND RATIONALITY: A COMMENT ON TOULMIN'S THEORY OF SCIENCE*

Stephen Toulmin's discussion of historical change in science makes use of a number of dichotomies that have become increasingly popular in the decade since the appearance of Kuhn's *Structure of Scientific Revolutions*. One of these, that between 'logicality' and 'rationality', he sharpens more perhaps than anyone else has done, with far-reaching consequences not only for the historical sociology of science but also for the philosophy of science and for the general theory of man as a knowing being. If he is right, not only have most people been looking in the wrong direction in their attempt to circumscribe and understand the structures of human rationality, but it is not even clear that it *has* any structures, of the kind they were looking for, at least.

I sympathize with Professor Toulmin's emphasis on the historical dimension of science, and his unhappiness over the way in which this dimension was often excluded from philosophical consideration in the dominant positivist philosophies of science of the 1930–1960 period. I agree that a close collaboration between historians, sociologists, psychologists and philosophers is required if the processes of change in science are to be properly understood. I would agree that the rationality of science as a human activity cannot be a matter of 'logic' alone, if by 'logic' is understood "the internal articulation of intellectual systems." [1]

But I will argue that the dichotomies drawn by Toulmin, particularly that between logicality and rationality, are too sharp, and that this leads him into a position just as one-sided as the one he is criticizing. Although he wants to retain the notion of science as a rational enterprise, by defining rationality in terms of "an interaction between scientists and their ideas, their institutions and the historical situations with which they have to deal" (Section 6), he thins it out to a dangerous (and I think, unjustifiable) extent. To make the *rationality* of the acceptance of a proposed new theory in science depend upon sociological, psychological or historical factors, runs the risk of either separating rationality and justification, or confusing causes with reasons.

R. J. Seeger and R. S. Cohen (eds.), Philosophical Foundations of Science, 415–430.

1. FOUR THESES

It will be useful first to outline Toulmin's argument in terms of four interconnected theses, each corresponding to a section of his paper (Sections 2, 3, 4, 5 respectively).

Thesis 1: Logicality and rationality are distinct features of science. The former is "an internal property of propositional systems" (Section 7), "considered either at particular times or apart from time." It is 'static', 'structural', 'synchronic', discovered by making 'a temporal cross-section'. Rationality, on the other hand, has to do with "the intellectual procedures by which scientists arrive at conceptual changes, in the course of time". It is 'dynamic', 'diachronic', exhibited only in the continuous development of science, and not in a temporal cross-section or even a sequence of such cross-sections, since one cannot reconstruct "out of these static structures the historically-developing rational enterprise of science." Where philosophers (and others) went wrong was in:

> equating the rationality of science with the logical character of the systems of propositions in which, supposedly, one could express its entire intellectual content.... For the terminology of 'logical structure' gives us no way of characterizing the 'rationality' of conceptual changes in science.

Thesis 2: The selection criteria in terms of which conceptual novelties in science are evaluated can be understood only by referring to the individuals or groups within the profession (the 'custodians of rationality') whose "role it is to pass judgement on conceptual innovations 'in the name of' the science concerned." Thus, the philosophical and the sociological accounts of these innovations are necessarily intertwined; they reveal two complementary aspects of the same enterprise. It is an 'illusion' that "the logical demands of the inductive method exercise an unambiguous authority over scientists." These demands may be muted by the historical particularities of the scientists themselves, considered as a professional group in a state of psychological and sociological change.

Thesis 3: One must distinguish in any science between periods of 'routine', when there is 'broad agreement' about intellectual goals, selection-criteria, and the like, and periods of 'reappraisal', when these goals and criteria are being questioned. These latter are not necessarily crises, in the strict sense; they are times when 'unambiguous criteria for judging between new ideas' do not exist, and there is, for example, no "collective agreed concept of 'explanation' in the science." There is thus a need for 'the exercise of individual judgement', until the period of uncertainty ends. The distinction between the sorts of issue that dominate discussion in 'routine' as against 'reappraisal' periods is that between 'tactics' (puzzle-solving, where the rules and principles are given), and 'strategy' (long-range problems, where the level of questioning goes right back to the foundations).

Thesis 4: An exactly similar problem arises in those (fairly rare) judicial cases when judges can no longer rely on precedent or written constitution but are forced to "reappraise the fundamental social purposes of the law in a new historical context". Here, either no definitive decision-procedures are available, or else the application of the accepted norms 'lead to evident anomalies and inequities'. Thus, logic must "become the servant of fundamental human purposes." "The judge can no longer speak for the law as it is; instead, he passes judgement in the light of a longer-term vision, both of what the law has been and is, and of *what it should become.*" The resulting judgement will inevitably represent not 'the judgement of the law', but the 'best personal judgement of the judge'. It cannot be called 'right' or 'correct', in the sense of being governed by accepted rule, but it is none the less 'rational'.

I have thought this rather detailed summary worthwhile, both because these theses draw together the threads of Toulmin's own work,[2] and because they are recurrent in the recent literature in philosophy of science. T3 is, of course, a variant of Kuhn's influential distinction between 'normal' science and 'revolutionary' science. T2 is the sociological theme which Kuhn, Polanyi, Habermas, and many others, have been pressing, from a variety of by-no-means identical points of view. T1 is implicit in the recent work of Feyerabend, and many thought they detected it also in Kuhn.[3] T4 is, it would seem, peculiar to Toulmin;[4] it reflects a suspicion of disciplinary boundaries and a corresponding reliance on interdisciplinary analogies of the sort that characterized even his earliest work.

The crucial thesis is, of course, T1. The other three theses are introduced as support for it. I shall argue that the notion of 'logicality' utilized by Toulmin has to be modified; it is a straw-man, as he defines it. T3 importantly qualifies T2, as Toulmin himself notes. If the notion of logicality be enlarged (as I think it must), then T2 no longer supports T1. The analogical argument from T4 to T1 does not work either. It will be argued that though T1, as it stands, is wrong, it errs by oversimplification; behind it there is an often-overlooked (though less dramatic) truth.

2. Logicality

What is the 'logicality' that Toulmin here (and elsewhere) contrasts with rationality, to the point of almost making them opposing? It appears as a property of propositional systems, akin to formal validity. It is generated by an accepted set of rules or procedures, and expresses itself in the logical relations between the various propositions of a science, considered as a single proof-structure. It requires that the conceptual system in terms of which the propositions are expressed should also be given, or that it can at least be postulated.

What sort of logic is required in order to make science, considered as a set of propositions, work as a proof-structure? Clearly, deductive logic alone will not suffice, as the classical Aristotelian theory of science supposed it should. There are deductive sequences in science, of course, beginning usually either from hypotheses that are being tested or from theories that are being applied to specific cases. But this is only a small part of the regular routine of the scientist. What about inductive logic? Here we encounter a major difficulty. There *is* no completely acceptable formal theory of inductive logic; indeed, many (like Popper) would reject the possibility of such a logic entirely. A great deal of effort has gone to the formulation of an abstract theory of confirmation, and with some success, but here again it would not be correct to say that there is a theory available to logicians which would enable them to evaluate the probability of theories advanced by scientists.

At this point, a choice is open to us. We can either content ourselves with the rather straitened notion of logicality as deductive character, or else accept a broadening of it to bring it closer to the actual procedures of the scientists. To accept the former alternative would be to trivialize Toulmin's argument entirely. T1 would still be false, strictly speaking, since deduction is clearly one of the rational procedures employed in science (in the assessment of new concepts, *inter alia*), and thus logicality would at the very least be *part* of rationality and not distinct from it. But more significantly, the narrow concept of logicality has not been seriously proposed by *any* philosopher of science in our century as the proper mode of reconstruction of scientific proof. In particular, it would have been rejected by the Vienna Circle, who are after all the main target in this context of Toulmin's criticism. If we are to be guided by what *they* were looking for (and this ought to be a first clue, at the very least), then the concept of logicality has to be construed much more broadly.

How broadly? If it is to include all those procedures which are generally accepted as means of establishing the reliability of scientific assertions, then it will clearly be very broad indeed. There would, for example, be the different statistical and curve-fitting techniques in terms of which empirical generalizations are formulated and tested. One would presumably want to take account of a variety of criteria commonly used in the evaluation of theories (predictive success, simplicity, empirical content, fertility, etc.). These are *logical*, in the sense that they are means of assessment. They allow one to evaluate an assertion in terms of the evidence advanced on its behalf. They are not cogent; their application does not give a definitive result. There is no single formalized theory uniting them together which would allow the calculation of the probability or likelihood of one theory as against another, on the basis of given evidence. Thus, there will be room for disagreement among those who use their procedures. Non-logical factors will play a role.

But before discussing these, it is necessary to be more specific about what qualifies a procedure as a 'logical' one here. After all, scientists make use of all sorts of procedures, at least some of which are certainly not 'logical' in any helpful use of the term. For instance, one would want to exclude experimental techniques of testing apparatus, or taking measurements. Nor would the replication of experiments to check on the reliability of data be called a logical procedure. Logic has to do with the relations between two sorts of proposition, one counting as assertion to be tested, the other as evidence. Science is necessarily a propositional activity; it may also rely on non-propositional insights or on manual skills. But until a set of propositions has been articulated, one cannot yet speak of *science*. Thus, the rationality of these pre- (or non-) propositional activities cannot be equated with the rationality of science, even though it may be a part of the latter, or (more exactly) a condition of it.

Much has been made (by Feyerabend and Kuhn, in particular) of the incommensurability problem, the difficulty of comparing two theories where exact translation from one to the other cannot be made, and where the 'evidence' for one may thus not be linguistically the same as that for the other. But as has frequently been pointed out,[5] there does not seem to be any reason why logical modes of comparison could not be followed, where the degree to which one theory is supported by *its* evidence could be compared with the degree of support for the other. Once again, such a comparison will not be coercive, but in science there is nothing unusual about that, as we have seen. Disagreement is likely, of course, to prove much more intractable in such cases, not so much because of specific inter-translation problems as because of the larger role likely to be played by non-logical factors. The rejection of the once-canonical (and undoubtedly convenient) distinction between theoretical terms and observational terms, and of the associated 'foundationalist' assumption of the logical independence of the 'evidence' (common to all classical philosophies of science) entails that we can no longer separate *probanda* from 'data' as neatly as the logician would wish. But this does *not* mean that logical procedures (in this broader sense) cannot be used in evaluating the relation of *probanda* and data, or even in comparing rival *probanda* where the data are 'theory-laden'.[6]

But there are still some problems as to just where the boundaries of 'logicality' should be drawn. Two questions, in particular, arise. Must the rules constituting logicality in an instance of scientific assessment be *explicit* ones? Clearly not, at least not in the sense that scientists are operating from an explicit list of them. But ought they not be at least *capable* of verbal formulation, rules of the sort that could be written down in a handbook on method? To say no to this would appear to call into question whether they can properly be called 'logical' (if an implication from 'logical' to 'formal', to 'independent of experience with this particular material', holds good). Yet to say yes seems to run counter to how assessments are made in science and how people are trained to make them. I will opt for the latter answer, but must leave to another occasion an adequate defence of it. Thus, we will assume that logicality is dependent on procedures and criteria that are, at least in principle, capable of being formulated verbally as rules.

In what sense must these rules be 'agreed upon'? Is it sufficient that they represent the common practice of scientists? Or ought one require something more secure, namely that they be capable of some sort of theoretical justification? The rules of deductive logic are sanctioned by something more reliable after all than the circumstance that they are used by logicians. (It must be admitted that the exact nature of the sanction is a matter of notorious dispute!) To answer this, we have to inquire further into the implications of the first alternative. How *does* a particular procedure come to be 'common practice' among scientists? Is it simply a matter of the consensus of a professional group, to be understood by tracing the historical and sociological peculiarities of that group? Clearly it *does* represent the consensus of what is without doubt a very peculiar social group. But how was this consensus arrived at? By a gradual process of testing, by the pragmatic success of the science based on this procedure, by an aesthetic conviction of its 'rightness', by its making sense as an appropriate mode of logical assessment of claims about nature.... There is no single answer here. But all of the answers have this in common that the consensus is based on *reasons*, on test; it develops gradually, not primarily because of some socio-psychological changes in the professional group, but because of the continuing challenge of a reality that is over against the scientist, which is not subject to his sociopsychological demands, which forces him to adapt, not only his theories but even the very methods in terms of which the theories themselves are assessed. This is not to say that his methods and his goals are not affected by his biographical peculiarities; rather it is to claim that the effect of these peculiarities is progressively decreased because (to the extent that they *are* peculiar) they will not be successful in helping him attain the basic goals of science. And what are these? It is easy to answer: successful prediction and explanation, but that is not so much an answer as a series of further questions.

Perhaps the above may be enough, however, to indicate that when we speak of the logicality of science as rooted in the 'common practice' of the scientist, this is not intended to be a sociological remark, nor is our quest on the whole much aided by sociological analyses. To put this concretely, we are implying that if natural science were to develop in a culturally very different setting, with different historical antecedents, there is no reason to suppose (and good reason to deny) that the modes of assessment used would be significantly different; to the extent that they *do* differ, the exigencies of fitting thought to a reality it did not create will force them closer together over the course of time. The objectivist argument is much stronger, of course, for the second-level *procedures* of science than it is for the first-level *theories* of science, on which history has a much more individualizing effect. But even the latter have such an inductive pressure on them, a pressure which has grown as science has developed, that one has to be wary. The testimony of 'common practice' in the case of *science* (by contrast with other disciplines) is one drawn from several centuries of remarkable pragmatic success in predicting and controlling the processes of Nature, and of growing insight into the structures of Nature, from nebula to nucleus.

These remarks, which would need much more elaboration in order to constitute a satisfactory treatment of logicality, suggest that Toulmin's T4 is incorrect, and that this his T2 needs to be put more restrictively. But first, let us take a look at T3.

3. Routine and Reappraisal

It would be generally accepted that reappraisals occur every now and then in any branch of science. Kuhn's distinction between 'normal' science and 'revolutionary' science has been attacked on several scores, among the most important for our purposes being that it suggests too sharp a cut between the two, and that it speaks in terms of *periods*, as though there were no 'normal' activities going on in revolutionary science or revolutionary challenges simmering below the surface of the most 'normal' science. Instead of a sharp dichotomy, let us assume a spectrum from the most radical revolution (the Galilean one) to a reorganization of a field in the light of a new model, like the Watson-Crick model in biology. And let us, following Toulmin, avoid words like 'revolution' and 'crisis'.

But one distinction must still be drawn. Reappraisal can occur at two rather different levels. One is at the level of method, of procedure, of goal. The other is at the level of specific theories in 'first-order' science. The distinction is not quite as sharp as it seems (think of the concept of force proposed by Newton in his mechanics, which also challenged the accepted notion of mechanical explanation of the day). Yet in most contexts, it can be drawn easily enough. In T3, Toulmin is clearly only referring to second-order or critical reappraisal; the instances he gives have one feature in common, he says; "there is no longer a collective agreed conception of 'explanation' in the science concerned, and so no unambiguous criteria for judging between new ideas" (Section 4).

First-order reappraisals, where there is no significant criteriological disagreement, are thus implicitly classified as part of 'routine' science. Such reappraisals (when a theory is failing in some respect, and the need for a new or modified theory is widely felt) are far more common than the other. And much of the conceptual development in science (e.g. the development of the concept of energy in the 19th century, or on a more limited scale, the development of the concept of a spin quantum-number after 1926) is associated with them, i.e. with theoretical changes which do not involve significant methodological change. But Toulmin admits that T2 (emphasizing the importance of the sociological approach to science as the activity of a specific social group) does *not* apply, in any important way, to 'routine' science:

To the extent that there is general agreement about broad intellectual goals, it does become possible to appraise, in rational terms, the critical activities of individual scientists.... To that extent, the accepted rational demands of scientific judgement will have a clear authority over the professional scientists working in that field (Section 3).

This means, then, that much (perhaps most) conceptual development in science is governed by the relatively straightforward logicality of common practice. The T2 suggestion would apply at most, therefore, only to developments associated with second-order reappraisals. But if this be so, the rationality of science has a very large admixture of logicality indeed, and T1 is largely eroded. In short, T3 and T2, if taken together, do not constitute an argument for the distinctness of rationality and logicality. *At most*, they would show that the rationality of conceptual developments associated with *critical* reappraisals cannot be reduced to logical terms.

4. Logicality and Temporality

One of the governing metaphors of Toulmin's work is the contrast expressed in the dichotomies of T1, between two ways of analysing science, as static-structural-temporal cross-section, on the one hand, or as dynamic-developmental-continuous flow, on the other. Rationality cannot be disclosed by the first approach, only by the second:

The rationality of science is embodied, not in the propositional systems current *at particular times*, but in the procedures for discovery and conceptual change *through time* (Section 2).

Now one might well want to ask why a successful theory should not testify, in some fashion, at least, to the rationality of science. But let us leave this objection aside to concentrate on the dichotomy on which the assertion rests. In one important respect, at least, this dichotomy is spurious. It derives its plausibility from the suggestion that the 'temporal cross-section' is a neatly-laid-out deductive system, lacking any element of change or dynamism. But if one looks at this suggestion more closely, its arbitrariness becomes apparent. If one takes a 'cross-section' in the history of science, one may just as easily find two theories in competition with one another, with the evidence telling in one direction (say) rather than the other. Or one might find a long-accepted theory faced with some awkward anomalies. But even if one were to find a single established theory in undisturbed possession, it would be misleading to regard it as a deductive system from the logical standpoint.

In an HD system, it is true that deductive relations exist between the hypothesis and the various laws or individual predictions that can (with the aid of state-descriptions) be derived from it. But it is crucial to recall that the logical relations of *evidence* are working in the opposite direction. It is the deducibility of a verified consequence that attests to the validity of the hypotheses, not the truth of the hypothesis that attests to the truth of the consequence. Thus, the system, despite appearances, is *not* a deductive one, provided it be regarded as a claim about the world. (Once again, it is on the never-far-from-mind analogy of a purely *formal* system, a deductive calculus, that the plausibility of this static-dynamic cut mainly relies.) Whether one considers a single system or alternative ones, the same broadly inductive criteria are at work in either case.

Assessment goes on constantly in science. The cross-section analogy is not a helpful one in understanding it. What is discovered by 'taking a cross-section', i.e. by considering the principles governing the assessment of a theory at a given moment in time, ought not differ in any important respect from what one would learn by tracing the assessment over a period of time. But now an important ambiguity is revealed. Does the 'cross-section' above refer to the act of assessment or to the history of the theory itself? To answer this, let us see which the logicist philosophers Toulmin is criticizing would choose. It would *have* to be the former, though the metaphors employed by Toulmin could easily suggest the latter. That is, they endeavoured to isolate the logical criteria governing the act of theory-assessment. What they did *not* do, however, and it is crucial to underline this, was to limit their consideration to a moment in the history of the theory supposedly being assessed. They did not, as it were, 'freeze' this theory at an instant, and ask what could be learnt from that instant.

This can be demonstrated by a glance at the sort of logical criteria they were concerned with. They talked not only about empirical content, but about predictive success over a period, about fertility in guiding research, about unexpected confirmations, about ability to absorb anomaly by internal modification, and so on. These are *historical* questions, not to be answered by a simple inspection of theory-and-evidence, considered as a timeless propositional system. Professor Toulmin is quite right in suggesting that from the logicist standpoint it is much *easier* to restrict oneself to such an inspection, if one wishes to estimate degrees of likelihood of theories numerically. But it would manifestly *not* represent the practice of scientists; it would leave out of account some of the most important inductive criteria, those which have to do with the *career* of the theory (model) or theories under consideration. Once again, let it be emphasized that one is *not* dealing with a formal deductive system here, but with the far more complex inductive assessment of the way in which a given theory shows itself capable of suggesting guidelines and meeting challenge over the course of time.

This is a *logical* issue, as we have defined it. The procedures are for the most part agreed upon, and their purpose is the assessment of likelihood. This assessment is a tentative affair; it does not issue in a numerical probability. The criteria will not be seen in quite the same way by different scientists, so that disagreements may arise. But this does not militate against its being a *logical* matter; we have become so used to limiting 'logic' to the automated steps of purely formal systems that we may too easily overlook the procedure-governed character of what scientists are doing, even when they are disagreeing. The procedures used have been developed and tested by centuries of service. They can therefore be discussed and empirically evaluated by reference to the history of science.

There are no valid grounds, then, for contrasting logicality and temporality in the way Toulmin does in T1. To make a logical assessment of a temporal system does not 'freeze' the system; one is not forced to lay aside the dynamic features of the system, no more than one is forced to do this in mechanics. There too, an 'assessment' (a mechanical theory) not only can take the change in the mechanical system into account but can in fact base itself on the changes directly. Plato tended to disjoin logicality and temporality just as Toulmin does, supposing that since logical relations are atemporal, there could not be a logical account (in terms of the Forms) of the changes in things. Change and the associated temporality were regarded as incapable of reduction to the intelligibility of form. But astronomy and later mechanics showed that it *could* be done, and that there is a confusion involved in the assumption that the representation of a changing reality must itself be changing, (or that an unchanging, i.e. 'scientific' representation can only be of an unchanging reality):

For logic is concerned with the internal articulation of intellectual systems, considered either at particular times or apart from time (Section 2).

It could be Plato speaking. But when Toulmin goes on immediately to add:

So the rationality of scientific discovery – the intellectual procedures by which scientists arrive at conceptual changes in the course of time – will necessarily elude analysis and judgement in formal terms alone

the resemblance with Plato abruptly vanishes. For he makes rationality a matter of how scientists *arrive at* conceptual changes; it is "embodied in (the) procedures for discovery."

5. Discovery and Assessment

One is reminded here of the distinction drawn by Reichenbach and Popper and so much insisted on by the logical positivists, between the contexts of discovery and of justification. They dismissed the former from philosophical consideration, and claimed that rationality attaches only to justification. How a theory is hit upon is irrelevant in an assessment of its scientific character; what matters is what evidence can subsequently be brought on its behalf. Toulmin has quite evidently, and deliberately, stood this distinction on its head (the metaphor he himself uses in this passage to describe what he is doing). He withdraws rationality from the context of justification altogether, and limits it to the context of discovery.

One could charge him with rhetorical overkill, with an incautious use of paradox. There is nothing in the history of the notion of rationality in Western thought that could warrant one's denying rationality to the very process (justification or proof) which has always been taken as its clearest realization. But instead of urging this, perhaps sufficiently obvious, criticism it may be more to the point to recall the attacks that have been launched against this discovery-justification dichotomy by Hanson and others with whose views Toulmin is ordinarily sympathetic. Is it not an oversimplification to separate the two as though assessment played no part in discovery nor discovery in assessment? Clearly they are continuously interwoven. It is wrong to exclude logic from the "procedures by which scientists arrive at conceptual changes," whether one does this in order (like Popper) to characterize these procedures as irrelevant to the rationality (i.e. logic) of science, or (like Toulmin) in order to make these procedures *define* the rationality of science and thus exclude logic from that rationality. Assessment is not a separate act that somehow succeeds discovery; the process by which "scientists arrive at conceptual changes" has within it a continuous dialectic of conjecture and assessment that controls the direction of conceptual change, as of the other elements that matter to scientific progress (new models, modified generalizations, new tests, and so on).

Toulmin appears to identify change in science with conceptual change in order to underline his thesis that the rationality that is peculiar to science is found primarily in the processes of conceptual change, and that 'logicality' plays no part in these processes just because they are *processes*, and because logic requires an agreed concept-system. But some of the most important kinds of change in science (those associated with the discovery of new laws, for example) do not necessarily involve conceptual change. Further, Toulmin speaks of concepts as though they were independent entities, capable of separate variation.[7] It is easier for him to divorce logicality and rationality if he focusses science (and thus rationality) on the shift in individual concepts, which on the face of it does not seem to be a subject for logical assessment. Such assessment is proper rather to propositional systems, and he has tried to exclude these from the dynamic of science in the making.

But the fact is that the units of change in science are *not* concepts but concept-sets, i.e. theories. It is rarely the case that concepts alter in isolation, without effect on one another. The degree of interconnection of the concepts in a theory is not often as tight as it is in mechanics, where an alteration in one would mean an alteration in all. But on the other hand, there is enough loose interconnection in even the most modestly-elaborated theory that one does not attribute changes to the individual concepts in the theory without first checking the theory as a whole. And of course, it will in any case be the logical assessment of the theory as a whole that will serve as justification – and therefore as part of the *dynamism* – even of individual conceptual changes that occur within the scope of the theory.

6. Rationality

There is still much more that should be said on the topic of rationality, but the scope of this article does not permit it. I have argued that logical factors play a dominant part in the reappraisals that go on constantly in active parts of science. But I have also conceded that the 'logic' here is not a coercive one which comes up with unequivocal answers. What other factors enter in, and how do *they* relate to rationality? This is really the central issue for contemporary philosophy of science. One has to admit socio-historical influences of all sorts. These explain (causally) how certain conjectures come to be made, and (in part, at least) why certain positions are maintained. But they are not (I would submit) part of the *rationality* of the science. They are secondary to the hard challenge of the empirical, which in the long run (sometimes only the *very* long run) will moderate their influence at both the theoretical and the meta-theoretical levels. I am arguing, then, that sociological and historical reconstructions may help us to understand why science took the course it did in a particular case. But this does not imply that these sociological and psychological factors help define the rationality of science as a human activity.

But there are still other factors involved, which are non-logical in the sense that they do not include definable and communicable procedures, and yet which have a certain degree of objectivity to them. A trained scientist has a variety of intuitive skills, of ways of patterning, which take years to acquire and which do not come from any manual. These 'personal' factors (to use Polanyi's term) are very much a part of human rationality. They are checked and controlled in very different ways in different disciplines. In science, failures in skill or insight are usually easy to detect: the prediction will come out wrong; the classification will break down; the analogy will not work. The good scientist will, over the long run, show his mettle; the untrained one likewise will not go undetected. The patterning skills of a genius for a time may be hard to separate from those of a crank, but once again the empirical demands made on scientific hypotheses are severe enough that as a rule the genius and the crank can, after a time, be confidently separated.

It is less easy to do this in other fields. And this brings us back finally to T4. The analogy between the constitutional lawyer and the scientist is not a helpful one. The sources of dynamism in law and in science are basically different. And assessment of success or failure of a change in law is an altogether different affair than is evaluation of a theory-change in science. The first and simplest criterion in the latter case is correct prediction; if anomalies in prediction arise, a warning light must immediately show. It is difficult to find any analogue to this in the case where a Supreme Court judge reinterprets the provisions of the Constitution. What would count as prediction or as anomaly? The relations of law and science to the empirical order are so different that it appears unwarranted to use the extension of the one as a means of understanding the extension of the other.

A serious constitutional problem clearly makes heavy demands on the interpretive skills of the judge. It will be more difficult to establish just who possesses the requisite skills than it would be in the corresponding case in science. If two judges disagree, there is no direct means of arbitrating their difference. And when the Supreme Court makes a decision 5–4, they *decide* what the law should become, and the law automatically does become what they make it. Scientists do not have matters quite so much their own way! When we say that 'best personal judgement' plays a significant part in constitutional decisions in law and in fundamental theoretical reappraisals in science, we mean that in these contexts logical procedures are often secondary to questions of trained skill and personal insight. Thus, in such cases it is true (as Toulmin reminds us) that "we cannot separate the rational procedures from the men who are reshaping them" (Section 5).

But the consequences of this are very different in the two cases. The rationality of science demands that we progressively, and in quite specific ways, render theoretical conjectures independent of the authority of the men who made them. When drastic and controversial suggestions are made by a Delbruck or a Chew concerning the restructuring of an entire field of science, we do not *immediately* dub these 'rational' (except in the not unimportant sense in which *all* conjecture is the product of man's rational power). The rationality of *science* is rather more demanding than that. Only when these conjectures are considered in the context of the evidence, only when they gradually test out over the course of time, do we attach the label 'science' to them. What distinguishes science from other fields, what marks off scientific rationality from other kinds, is the nature of this testing process, rather than the manner in which conjectures are made and the role of authority in their initial acceptance as worthy of test.

The analogy with law is even less close in this respect than it would be with other disciplines, like history. For when the Supreme Court judge makes a decision this *constitutes* his finding as law; whereas in science it is improper to speak of the scientist who proposes a conjecture as 'making a decision', particularly not a decision which would bind others. His proposal is merely the *first* step, and before a decision can be reached on its merits, there will be a period of assessment. It is surely the structure of assessment, therefore, that we must principally investigate when we are concerned with the nature of scientific rationality.

University of Notre Dame

NOTES

* Comment on S. Toulmin, 'Scientific Strategies and Historical Change', *AAAS Symposium on Comparative History and Sociology of Science*, Boston, December 1969.

[1] Section 2, *op. cit.*, References to Professor Toulmin's paper will be by section number. The remainder of the footnotes below were added in proof (1973).

[2] This is all the more true since the publication of his *Human Understanding*, Princeton 1972. In fact, the article commented on here contains the nucleus of the argument of the later book. The only major theme not alluded to explicitly in the earlier article is the analogy between organic evolution and conceptual development in science, on which the book relies so very heavily.

[3] In the Postscript to the new edition of SSR, Kuhn appears to reject it (pp. 199, 205-6), though not using either of the terms in which it is stated.

[4] Kuhn rejects it, or at least is distrustful of this sort of analogy, *op. cit.*, pp. 208–9.

[5] See, for example, the discussion after Feyerabend's paper in *Minnesota Studies in the Philosophy of Science*, Vol. IV (ed. by M. Radner and S. Winokur), Minneapolis 1970.

[6] The rejection of the foundationalist thesis does, of course, count against the ambitious logicist program of reconstruction of induction and confirmation at one time backed by logical positivism, but not (it now seems to me) so as to constitute a quite *new* sort of argument against it.

[7] This become a central motif in his *Human Understanding*, and is indispensable to the evolutionary metaphor he relies on so heavily there. Concept variation is taken to be analogous with genetic mutation in a group of organisms, and he tries to use the notion of selection in the same way in both cases in order to explain how 'adaptation' occurs. But the same three questions remain: (1) ought one identify change in science with conceptual change? (2) are not conceptual changes inseparable parts of a broader sort of change, theoretical change? (3) does not logical assessment play a central and ongoing part in theoretical changes, thus making logicality quite central to rationality in science?

JOSEPH AGASSI

ON PURSUING THE UNATTAINABLE

The aim of this note is to criticize the view that it is never rational to attempt the impossible; it is not, however, the aim of this note to advocate all impossible aims. The ideal of positivism, which positivists deem obviously attainable, namely the unity of science in rationality and the rational unity of mankind, is here viewed as very worthwhile, but quite possibly impossible and certainly not obviously possible. Yet, to repeat, not all impossible aims or unattainable goals are reasonable to pursue.

I

There is a simple and straightforward sense in which pursuing the unattainable is palpably irrational: it is the pursuit of what we know to be unattainable while ignoring this relevant knowledge. However, this does allow for rational or reasonable error: when speaking of the unattainable we do not mean what we take to be attainable by some error or another. It is quite obviously rational to attempt to attain an aim on some reasonable assurance, and this remains so even if later on it turns out to be unattainable. After the destruction of a work of art, but before the news of the destruction is broadcast, the search for it may be reasonable even though it is unattainable. A fortiori, it may be rational to try to attain the almost impossible; for example, if it is one's only chance for happiness or if the stakes are sufficiently high: the young hero may rationally try to marry the princess, even though all his associates know his venture to be hopeless – but only insofar as he still retains hopes, however faint, and insofar as he feels that the stakes are high enough for him to make the odds fair enough. Otherwise, if the error on which the pursuit is easily detectable, or if chances are grossly miscalculated, or if the cost of possible error is wilfully ignored, if the pursuer refuses to consider the possibility that he may err and the cost of his error, then we may rightly view him as rather irrational.

There is the claim of the psychologist to muse about, which is that

R. J. Seeger and R. S. Cohen (eds.), Philosophical Foundations of Science, 431–444.

some person will permit himself to fall in love only hopelessly: only when assured of the hopelessness of attaining this end will he dare pursue it. Moreover, the psychologist will assume that in such cases, the hopeless pursuer will refuse to consider the possibility that his pursuit is hopeless, and he will argue obsessively to prove to himself that he has high hopes. Let us assume all this to be true. Is the end, hopelessly pursued, the unattainable end, simply the one which the pursuer is really after? Not in the least; his real end is to avoid attaining it, says Freud. If so, why not simply avoid pursuing it? There may be different answers to this. For example, the flesh is conditioned to chose an end that the spirit wishes to avoid attaining. It is relieving a pressure to let the flesh, in its ignorance, to pursue what only the spirit knows to be better unattained. Then the flesh is rational and pursuing the princess, while the spirit is rational and pursuing avoidance of love plus relief of the pressure of the flesh. Freud's theory of the id is a variant of this. Usually Freud suggests that the id and ego stand for flesh and spirit; however, the ego is conscious and often the flesh is taken account of by the conscious and the fear of attaining this aim by the subconscious (as in the case of the groom who 'forgot' to go to his own wedding).

The Freudian theory of conflict will present matters even more subtly. Conflicting aims may remain unresolved, Freud indicates, yet their pressure may be relieved by playing against the odds; a person in conflict may wish to retain his conflict, or more precisely, is in conflict as to whether he wishes to retain his conflict. Here, then, relief of pressure from conflict while retaining the conflict may, indeed, be the true end of one's action. Whatever is our opinion about the truth or falsity of Freud's views, we can learn from it that possibly it is within one's aim to pursue another aim, and yet these two need not be in a simple hierarchic order; there is a difference between hopeless love aimed at the avoidance of consumation and courtship aimed at consumation. Regardless of any given facts, this may help us develop a more sophisticated approach to all goal-directed behavior: the real goal, the final end, may conflict (truly or seemingly) with the partial goal, with the more immediate end.

Let us take a more extended, long range, abstract end – any promised land, private or tribal, perhaps even religious, scientific, or aesthetic – any end beyond achievement within one lifetime. What is the rationality of an individual's pursuing it, even though it is admittedly unattainable for him and perhaps even unattainable for his descendents, or for the tribe as a whole, or for the whole of mankind? Let us first discuss the case where the end is attainable for the species but not for the individual, and later the case where it is not even attainable for the species.

In the first case, where, say, the final goal is not given to one's own self, but is given to one's descendents, the rationality of the pursuit is not really problematic. Any act for the sake of posterity, be it one's children, or one's future honorable mention in the future textbook of science, etc., can be viewed as a special, slightly more complicated case of a long term project which lies within one's own reach. The rationality of working for one's posterity can be reduced, with little effort, from that relative to the goal attainable only to one's successors to that relative to some goal which one may well achieve. We can say a father works not for the good of his children; his true end is the knowledge of, even mere hope for, his children's good opinion of him while he lies on his deathbed, say. And, clearly, in order to gain their approval, he has to pursue their ends, which are, indeed, not within his reach, yet well within theirs.

Here we have a curious case, where the partial end is farther ahead than the true and final end. Usually the partial or subordinate goal is more immediate than the final or primary goal; the attainment of the partial goal is means for the final or true goal; as means, as a link in a causal chain, it is achieved first. One may offer a criticism of any reduction of an end to a part of a more immediate end by the suggestion that it is subjectivist, since self-deception will satisfy the reduced end just as much as real work for one's son's future interests. This criticism is clearly acceptable on occasion, but it may, generally, be not harmful to the reduction proposed. Some people can deceive themselves, and when pressed hard enough they prefer the self-deception that they labor for their children's future over actions to such an effect. Other people are less capable of self-deception and so have no option. They may be unable to deceive themselves, whether from habit against it, or from having frank and free exchanges with their children. In either of these cases they can only feel having done something for their children's future when they have reasonable grounds for such a feeling – by standards of reasonableness acceptable in their community, particularly to those people whom they except to be present when they might die. Of course, taking the moment of one's death as all-important is here a mere simplification. The real point being that some philosophers wish to reduce all aims regarding posterity to aims regarding one's own lifetime.

The debate, however, has here deteriorated from a general complaint to a particular instance of it, which may not be good enough. The general complaint was that the partial end, when allowed to be later than the final end, has no causal link akin to the partial end whose actual attainment is a causal step towards the final end. Thus, for example, if immediate price of a commodity, in money terms, or more abstractly, as the exchange of concern for appreciation – if the immediate price depends on future expectation, we have a funny infinite regress! This regress, however, may be admitted as rather innocuous. It becomes important, perhaps, when we wish to improve our expectations. But when explaining a father's concern for his son's remote future, all we have to notice is that he and his son have some expectations in common, and on the basis of these they trade concern for remote future as against a promise of appreciation in the less remote future. And the difficulty of a detailed and correct reduction, then, may well be rooted in our own ignorance of the exact terms of contract between father and son, as well as the exact area of common expectation, the influence of disagreements, of misunderstandings and self-deceptions, and similar complications – all of which are immaterial to the principle of reduction we have discussed here.

The reduction of the rationality of working towards one's children's future to the rationality of working towards one's own future, say one's own future peace of mind – particularly in one's old age – is one which we need not insist on. Some philosophers, following Spinoza's wake, insist on it. They reduce all unselfish motives to selfish ones, and by implication all motives concerning posterity to motives concerning a more immediate future. Now, quite obviously, the demand to reduce all motives to selfish motives can always be met – on the condition that selfishness is defined broadly enough, of course. The reduction, thus, is not too satisfactory: it may be made easy by assuming the following ploy: being morally conditioned, one can only attain happiness, peace of mind, or any other selfish goal, when acting morally – even if this means trying to attain some unattainable ends. The result is that any reduction is handy, from any unselfish goal to a selfish one, including an unattainable goal to an attainable one. This Spinozist ploy was repeatedly and systematically used by Freud, in the guise of the theory of the super-ego, its formation and content, and the guilt pressure that it effects.

Briefly, according to Freud, any goal, selfish or absolutely altruistic, reasonable or utterly mad, may on occasion be stored in one's super-ego during childhood (before one can either examine or protest) and thus be operative in the sense that a man may have to obey his super-ego without questioning it and without endorsing it – simply in order to attain no other end but an immediate relief of a painful sense of guilt. The immediate end – peace of mind – is here the final end, and the end of the super-ego is served as a means for it. This is how Freud achieved a causal explanation of purposeful action: by seeing the sense of guilt as a means of pressure, as a cause, a motive force, or a propellant. We need not view the theory as strictly causal, however; we may view the relief of pressure (of one's sense of guilt or of one's bladder) as goal-directed, as directed towards relief. We may agree, then, that Freud reduces with ease – too great an ease – all remote ends to an immediate one, namely that of relieving one's sense of guilt.

II

It seems, then, that all remote goals are now somehow taken care of, and with great ease; the goals attainable to oneself in the remote future as well as the goal unattainable to the species on principle, such as Heaven on Earth, some other Utopian dream, or the attainment of full rationality, or of knowledge of the whole truth and nothing but the truth, which is the discovery of the secret of the universe. There is no difficulty in reducing the unattainable goal to attainable ones in the manners described above. The attainable goal may be approaching the unattainable goal, or facilitating the approach to it; the end need not be the final one, but the coming to a point as near to it as possible. Also the goal may be the more immediate one than that of finally achieving the highest end possible: we may declare the goal to be the immediate pleasure of the search rather than the find, as in the case of going for the princess, described above. One may even declare the goal to be, as a matter of fact, the immediate pleasure derived from the search. This really is the easiest reduction: the benefit of the excercise might be proportional to the effort invested in the search; one may thus end up recommending the maximum investment with assurances against success! For this end one may require the pursuit of an impossible end plus the maximum desirability to approximate it as much as possible; one might proclaim that the greatest effort made by the greatest numbers may lead humanity only infinitesimally nearer to the unattainable goal, yet we should all put all efforts unsparingly to attain this slight improvement. Particularly when one is ambivalent about one's goals, one tends to favor such a philosophy. Thus, one serves a few and even conflicting ends when one aims hard for the impossible.

This is not to say that the only cause for the endorsement of unattainable goals is psychological conflict. Indeed, some such endorsement is advocated here, though psychological conflict is not. It is possible, however, to suggest that many thinkers, particularly in the nineteenth century, preferred remote ends to immediate ones, in typical pre-Freudian self-doubt. When John Stuart Mill asked himself, we remember, what if his end will be attained in his lifetime, he suffered a severe nervous breakdown. All this, however, is no argument for or against unattainable goals. It is interesting to notice that strong emotional drives may also stand behind the opposite view, which condemns all unattainable goals as chimerical.

As experience shows, philosophers, especially as graduate students, show sometimes enormous hostility to the idea of going for unattainable goals. The reason they often give (I am reporting from my own limited experience) is not so much that it is irrational to go for an unattainable goal – this they keep in reserve and use only when very hard-pressed – but that if the goal is admittedly forever unattainable it then simply does not exist, and so talking of pursuing it is plainly meaningless! This argument is not from the theory of rationality but from epistemology. If one analyzes it carefully enough, one finds it to rest on the positivist verification principles of meaning: what we cannot verify we cannot understand. That this principle has a terrific emotional import is obvious, and it may well merit some examination, however superficial.

The verification principle is so vague that philosophers could never quite get its meaning straight – if it has a straight meaning at all. (In my view it has only emotive import, but no cognitive meaning.) It is alright to say, what *I* do not verify *I* do not feel I fully understand. Faraday, the greatest fantasist of physics, said he could never feel he fully understood a description of an experiment unless he could call it his own – that is to say, unless he performed it with his own hands and saw its results with his own eyes. Faraday was no positivist; he only expressed a very strong positivist feeling, and he strictly confined this feeling to matters regarding experiment alone. No positivist, not even the most extravagant one, ever asserted, what *I* cannot verify, *I* cannot possibly accept. One reason why Descartes' philosophy sounds so frightening – Kierkegaard regarded his going through with his doubt heroic because it was so frightening – is that he began with this dictum; but, of course, once he established the veracity of God and of the Natural Light, things looked much less frightening. Perhaps the main reason why no philosopher, not even Berkeley, could fully consider the case of solipsism is that it really is frightening to believe no one around, to trust no one's testimony except one's own. (Indeed, it is quite mad to doubt that one is born to a human female, regardless of one's source of information concerning the facts of life or the reliability or otherwise of that source: do you remember who told you that you came from a womb and does it matter to you? Will you ever calculate the probability that indeed you came from a woman's womb?) Once verification means strictly the acceptance of only the evidence of one's own senses, then verifiability becomes too constraining a principle – too constraining even for any positivist to contemplate it as a serious option.

On the contrary, the positivists always stressed that verification was required only in principle, not in actual fact. Whatever this meant, it allowed for the ingenuous acceptance of testimonies of other scientists – without first calculating on the basis of evidence the degree of their veracity. In the heyday of positivism, when the moon was just coming to within man's reach, the example 'there are craters on the far side of the moon' illustrated what positivists meant by a statement verifiable in principle but not in fact. This really expresses the positivist ethos: When Man – *any* man – reaches the backside of the moon, then we – *all* of us – will know whether there are craters there. When a positivist speaks of Man or of people or of us, when a positivist uses the first person plural, he really means the scientific community. And he means to exclude from this, first and foremost, the obscurantists – who do not count since they have abdicated their rationality. Further, amongst the obscurantists he includes the speculative metaphysician. The ethos of the positivist is one which contrasts science with metaphysics by stressing that science is bold but only to the limits of the practicable: reach for the moon, not for the outer galaxies; certainly not for the outer reaches of the mind, where experiment possibly cannot follow. Thus, in my opinion, the emotional import of the positivistic verification principle is clear, and at least quite unobjectionable, if not also laudable, in spite of its branding the metaphysician as an obscurantist. When a metaphysician turns out to favor science and, more so, when his speculative system has a benign effect on science, then the positivist – at least Carnap and Weissmann – has no qualm in annexing the metaphysician into the community of science: he simply stresses that he requires verifiability only in principle. What, then, happens to the cognitive import of the verification principle?

Statements about the back of the moon, we said, are in principle verifiable; even statements about temperatures in the heart of the sun may be so declared. In spite of all criticism of logical nature, we can say that these were meant to be included. But what will positivists say about fantasies regarding outer galaxies, wild science fiction stories bordering on the metaphysical? Where and how will they draw the line between where experiment cannot follow as yet, but may one day, and where it cannot ever follow at all? This question was studied by Kant. His studies are now obsolete. There is none to replace them. It is fairly commonsense to admit that this question is beyond our reach: Kant could not suspect, but we can argue, that the limits of reason are beyond the limits of reason: we cannot find out what one day we may be able, and what we shall never be able, to find out. The answer to the question, posed by Kant, what goal is attainable, what is not, is, itself, quite unattainable by Kant's own standards and even by more lax ones. The very examples with which the positivist ethos is illustrated, indicate that positivists want commonsense more than an abstract formula, and by commonsense they wish to put before science challenges not too easy to ignore and not too hard to undertake; surely this is too dependent on circumstances!

The young positivist who rejects vehemently the idea of the rationality of pursuing some unattainable goals, is quite rational, and on two counts. On one count, he sees no need to start with the assumption that a goal – Utopia, the ultimate truth – is unattainable; and when early in his career he is told that they are beyond reach he sees no reason for believing this. On the other count he fears that the unattainability of the goal, if openly admitted, might be too discouraging. Yet one may reach a point where the faith in the attainability of a goal is but a mirage, retained by a voluntary act of self-deception, by the fear that the recognition of the unattainability of a goal may bring a halt to the search. But it may be, as I have argued, perfectly rational to go on searching the seemingly unattainable. The positivist's fear which leads him to self-deception is ungrounded.

The positivist Utopia is well worth retaining: the unity of mankind in rationality: the unity of science in humanity. For all we know it may even be attainable. Suppose we stress that the attainability of the goal need not be immediate, or even demonstrable here or now, but we may, in principle, hope that one day it will be. In that case the great fear that the goal is meaningless need not be so paralyzing. The very suggestion to the contrary – the claim that the positivist Utopia is obviously attainable – if at all seriously entertained, is self-defeating. Regardless of whether one thinks the aims of science are attainable or not, one can agree to the following central point. It is alright to leave the question open; but answering it in the affirmative as a matter of course renders the Utopia not very exciting; whatever we (nearly) have, as a matter of course, since we are so obviously poor, is hardly worth having. Whatever we obviously can find is not something very exciting to look for. But the positivist Utopia – the rational unity of mankind and of science – is the possibly attainable possibly unattainable goal, well worth pursuing.

III

One thing, then, a skeptically minded student may learn from positivism. The contrast should not be made between the attainable and the unattainable goal, and not (as the positivists suggest) between the obviously attainable and the rest, but rather between the obviously unattainable and the rest. This is but the application of skepticism to itself, quite central to the ancient skeptic doctrine of ataraxia. Yet it is not so obvious that we should apply it to the case of goals, and the positivist insistence of the obviousness of the verification principle may be a proper stimulus here.

There is, indeed, a slight bonus here. The doctrine of doubt, when applied to itself, hardly leads to new insights (though it may relieve some pressures). Yet when applying it to goals, it becomes more interesting.

At first, superficially, one may hastily conclude from doubt that at least one important goal is clearly unattainable. Since the aim of science is knowledge, and since the possibility of knowledge is seriously doubted, we may conclude (with Socrates at the end of the *Symposium*) that the end of science, knowledge, is quite unattainable. Yet, on a second thought we may view things differently. Strictly speaking, the positivist, when wishing to insure the attainability – or even the mere existence – of a goal, speaks not of any objectively attainable goals, but only of some of these, namely of those which are certainly attainable. This raises the question, are all *objectively* attainable goals *demonstrably* obtainable? This hinges on our question, what does the positivist mean when he says he requires verifiability only in principle? I do not think anyone has worked out this question. (Even if you declare that all true laws of nature are discoverable, you need not say all attainable goals are certainly attainable – unless this is a law of nature!) This is why I deny that the verification principle has cognitive meaning. Hence I also deny that the principle has ever been refuted, of course. (Only very simple variants of it were refuted with the phrase 'in principle' sufficiently clearly, but quite naively, construed for the sake of the argument.) And so, one chief reason for pursuing the seemingly unattainable may be, that we do not know whether in principle it is attainable or not. We can even say, we do not know what our goals are – whether individually (remember Freud) or collectively (remember the positivists). The princess may, and sometimes does, return love to the daring insolent unworthy suitor just because he is just this; with a happy or an unhappy ending possible. The unity of mankind in rationality and of rationality in science may be attainable in one sense of rationality or another not yet sufficiently explored, or even in a known sense of rationality yet by hitherto undreamt means. We do not know whether this is so, and the doubt itself may be a strong motive for the search. The word 'skeptic' means searcher, and it is a pity that the old skeptics stressed the value of peace of mind (ataraxia) rather than search, though they did say that peace of mind came, as an afterthought, as a result of search, and though, no doubt, many have found peace in the search. To conclude, if we doubt even our claims about our goals, our study becomes more interesting.

The question, I feel, still remains: why do so many agree to aim beyond reach only when they are ambivalent, and why do so many find the idea that their aim is utterly beyond reach so disturbing? It is not so counter-intuitive, after all, to think (as I tend to do) that it is a disaster to have no secret of the universe to hanker after, namely to think that hopefully the end of science will never be reached. After all, Lessing already expressed the sentiment, among other classical writers, in a manner which did win popular acclaim.

What both the ambivalent seeker of the impossible and the confident seeker of the possible share, is the view that seeking the impossible is highly frustrating. This view, as a personal expression of individual tastes, is not objectionable – merely regrettable. But all too often it is presented and even pressed hard as a principle of objective validity, psychological, methodological, even logical. This, obviously, has been empirically refuted. Though endless search can be frustrating, and though peace of mind can be achieved with little or no search, we do have empirical testimony of lives devoted fruitfully and happily to tasks not yet attained and perhaps unattainable. Kepler's search for the harmony of the spheres and Einstein's search for a unified field theory are such instances, and by no means the only ones.

Things might fall better into place, perhaps, if we rectify the reduction of all motive to selfish motive. No doubt this reduction is one which has a laudable and an objectionable aspect. Consider a true martyr, i.e. one who loves his life and would not see any merit in sacrificing it, yet who sacrifices it for the sake of preventing some worse catastrophy, such as the violation of the principles of humanity (in accord with his opinions, whatever these may be). One may say, and it has been said, that the martyr is giving up a long life of shame in preference to a moment of life with dignity, and that really in his act of martyrdom he only expresses his high preference for dignity. No doubt, there is a lot to this, yet we know that the martyr may not be thinking in this manner no matter how deliberate his action is. Take, more specifically, the Freudian theory, according to which the martyr is avoiding a supreme stress created by a sense of guilt which would crush him if he avoided the act of martyrdom. This theory is more comprehensible, yet there are more and clearer counter-examples to it.

The merit of all reduction of motive to selfish motive, however, is in its staunch individualism, in its incorporation of a blanket refusal to consider any action of the race, the tribe, or the General Will, through the individual. In brief, the strong part of the reduction is its incorporating the autonomy of morals, of the attribution of responsibility to the individual alone. But it does overshoot its mark, not only by denying altruistic act or act for the general interest, it even denies that one can construe the general interest and act in accord with it. We know for a fact that people, small and big, do have views concerning the general interest and act at times in accord with their views about it. They may sincerely act in accord with what (in their opinion) is no less than the interest of the whole human race. Every time we applaud a person who puts world peace before any other interest we acknowledge – rightly or wrongly – that he falls into this category.

IV

Can we, then, not act also in accord with an end which happens to be not achievable even by the human race? Surely this does happen; sometimes we do it knowingly, sometimes in ignorance. Inasmuch as science is taken for religion, it is no doubt a shallow religion; yet this happens regularly. And science as a religion-substitute contains its hagiography and folklore and mythology, these including some science fiction. And repeatedly science fiction, both in its less pretentious and in its more pretentious instances, speaks of a more general purpose, to which humanity as a whole is but a small contributor. We need not assent to this view; we merely must admit its legitimacy even while we wince at it. This, however reluctant, is an admission that in principle one may crazily devote one's action to what one deems that one dimly perceives as the end of a huge process to which we may be but a small party.

There is no doubt of the religious dimension, and even piety, of all this. The better science-fiction writers make no bones about it. What is unpleasant about it is its inherent pretentiousness, however, not its religiosity. Some science-fiction writers, philosophers, and religious leaders of various schools have declared (and others denied), that ofttimes the effort is laudable, and is imprinted on the cosmic book of history, quite regardless of its outcome. The positivists who found all this outrageous, were particularly outraged by the infusion of the religio-metaphysical into the domain of science.

Thus, it all falls into place. The positivists insisted on the unity of science as a fact, and on its dictating sobriety on us, not so much in order to declare that the millenium has arrived, as in order to prevent talk about the millenium altogether. This was to no avail. Once we hanker after the millenium, we can only claim that it is here, or that it is remote, or that it is the unattainable goal. The positivists did not mean to tell us that it is here; but to the hankering – not as a mere feeling which, like any feeling, of course, they allowed – they refused to allow the cognitively recognizable expression. They simply did not succeed, and the feeling, the hankering, the religious dimension of science, kept finding diverse expressions in our culture. And so, the positivist thesis of the unity of science itself became one expression of that hankering, plus the implicit and unintended suggestion that the paradise wished for is around the corner. In the darkest days for humanity, preceding World War II, this was the message – clearly the unintended message, of course. It was meant to preach sobriety, but it preached the millenium of sobriety.

This conclusion may be answered with ease. Suppose that positivism preaches sobriety and hence (in some sense of 'hence') the avoidance of a discussion of the scientific millenium (which is the uncovering of the secret of the universe); suppose that raising the question nonetheless does make positivism millenarian; why should this trouble the positivist who preaches the avoidance of the question? The trouble, however, lies in the positivists' change of this philosophy of science. Already Pierre Duhem, the last great positivist of the period preceeding that of Wittgenstein and the Vienna Circle, was caught in this same difficulty. He declared that we can speak the truth and nothing but the truth, yet that we cannot speak the whole truth (which is the secret of the universe). Now, as long as we can have a class of absolutely true statements and increase its content in time, we may say that the question, does the secret of the universe exist? is meaningless: we can say that the answer to the question is in principle unverifiable, unconfirmable, untestable, etc., and we can try to dismiss the question altogether. But suppose that science does not progress as the increase of content of the class of all known true statements.

Suppose, science is a series of different pictures of the world, such as the Newtonian and the Einsteinian picture in physics, or the Darwinian and neo-Darwinian in biology, etc. And suppose we agree that there is some preferability here, which is not oblivious of the value of truth.

Suppose, that is, that there are some scientific world-pictures which are nearer to the truth than others; then we have admitted, willingly, that there is a secret of the universe which may or may not be attainable to science.

Duhem himself was aware of all this. He usually declared that the preferability of one world-picture over another is judged with the aid of criteria other than the truth – such as utility and beauty. He could not, however, altogether relinquish the idea that the preferability also relates to the approximation to the truth. Louis de Broglie, the famous quantum physicist, remarked in his introduction to Duhem's *The Aim and Structure of Physical Theory* that this amounts to a declaration of bankruptcy.

Duhem also realized that having a succession of world-views destroys the idea that at least reports of observation are final and unalterable. Yet he insisted that we do have a class of verified statements of observation which increases in time. He somehow felt that though put in scientific language an observation report is made open to revision (with the revision of the language of science), when put in ordinary language a statement is vague enough to remain unchallengeable. This is a hypothesis concerning the unrevisality of ordinary language which is false, and whose reputation has led the positivists of the post World-War II to a new theory of the reform of ordinary language through the process of explication of concepts so-called. With this admission that both ordinary language and scientific systems are revisable, there is nowhere to turn, it seems; we have criteria of improvement, and these may point the way towards the very best, and the very best may, indeed, be the uncovering of the secret of the universe. All we need to solve so many traditional philosophical problems is, I think, to admit that the secret of the universe is quite likely the unattainable goal of all research.

ACKNOWLEDGEMENT

I am indebted to Lucien Foldes, Jay Hullett, Kurt Klappholz, and Sir Karl Popper for many discussions.

Boston University and Tel Aviv University

SCIENCES AND CIVILIZATIONS, 'EAST' AND 'WEST'

*Joseph Needham and Max Weber*** †

I

For some years now, I have been engaged in a series of 'case histories' and 'mental experiments' whose main aim has been to strengthen the foundations of an evolving comparative historical *differential* sociology of sociocultural processes and patterns, most notably sciences, in *civilizational perspective*. As I have conceived them, the 'case histories' call for the strict study of determinate 'civilizational complexes' with a view to analyzing the variations and the shifting balances of their changing cultural ascendancies resulting from the critical contacts. In this same spirit I am continuing to seek to chart the differential impacts and outcomes of specified 'intercivilizational encounters'.[1]

The title and subtitle of the present paper suggest clearly enough the wider settings to which the present 'case' and 'experiment' relate.

The working hypothesis on which I shall be proceeding herein is that sciences *as well as other cultural expressions* have taken many different shapes and enjoyed different saliences in the accredited frameworks of orientations and disciplines of different civilizational settings. Most important of all is the fact that the structures of consciousness and conscience associated with the varied sciences of great societies have been characterized by variable mixes of different cultural elements and emphases. This may be seen to best advantage in the study of the changing orientations of societies which have been the vehicles of world religions and the seed-beds of civilizational transformations.[2]

I further proceed on the postulate that the stratigraphy of these shifting ascendancies is especially available to view when two complex societies with deep historical imbeddedness come into conclusive contact with one another in respect to the ultimate structures of their traditions and commitments, above all the rationales legitimating their cultural ontologies and epistemologies. A large part in defining the eventual mixes of cultural elements is played by what I have been calling the 'moralities

R. J. Seeger and R. S. Cohen (eds.), Philosophical Foundations of Science, 445–493.

(and logics) of thought' and the 'logics (and moralities) of action' enshrined in their symbolic universes by their symbolic technologies.[3]

It is surprising how few have been the historians, sociologists, and philosophers of science who have placed due emphasis on certain basic contexts which must involve noteworthy considerations for all their diverse orientations, however specialized they be. Too few have recognized, for example:

(1) how large was the part played by the varied fortunes of science in shaping the different destinies of East and West;

(2) how large a part sociocultural, including religious, factors have played in the varied fortune of science.[4]

II

In the present essay I shall mainly refer to episodes in the East-West relations during two periods of critical intercivilizational encounters: the era of the so-called Scientific Revolution – 16th and 17th centuries – and the era of the so-called "Twelfth-Century Renaissance" and the Crusades – the 12th–13th centuries.[5]

I begin with the crucial case of 'East'–'West' relations in the era of the Western Scientific Revolution.

On a superficial view, the Scientific Revolution in the West was one continuous break-through from Copernicus to Newton. Actually, the story tells differently.[6] There was no easy road for either astronomy or physics in the 16th and 17th centuries.[7] The *De revolutionius orbium caelestium* published in 1543 as the great astronomer lay dying at his home in Fromburk (Frauenburg) in Poland, acquired a readership only by being presented to the world in an anonymous introductory Letter to Readers describing the new image of the universe as a contrary-to-fact "hypothesis". It was many years before the author of this fateful Letter was publicly identified as Andreas Osiander, German Protestant Church official and theologian.[8] Despite the masking of Copernicus' purposes and claims, his work did not have easy sailing in the first century after its publication either in the Catholic[9] or Protestant worlds.[10]

Galileo's early successes did not protect him from later opposition and disfavor. After 1616 he was under admonition to desist from arguing for the truth of the Copernican hypothesis.[11] From this point forward he walked in a labyrinth. Nonetheless, in 1623 he arranged to publish his *Dialogue concerning the Two Chief World Systems*. Soon he was to undergo trial and humiliation at the hands of the Inquisition.

Despite Galileo's forced self-denunciation, the Scientific Revolution pulsed forward. The new science of Copernicus, Kepler, Galileo, Newton triumphed and made especially rapid headway in the areas marked by a surge toward scientific enlightenment. The capstone came in 1687 with the publication of Newton's *Principia*.

What happened in China, Japan, India was another story altogether.

To tell this story we must turn first of all to the renowned mission of Matteo Ricci, S. J.[12] The Jesuits who were associated in the Ricci mission were familiar with the new developments associated with Copernicus. Ricci himself was a student of Christopher Clavius at the Roman College. Many knew Galileo and were themselves Galileo's fellow members at the Academy of the Lynxes. Once, however, the Congregation of the Index had spoken against the Copernican hypothesis, the Jesuits felt bound to obey. They made few references to Copernicus or his work. This situation lasted throughout the crisis associated with Galileo. Publicly the Chinese Jesuits made no reference to the *Dialogue Concerning the Two Chief World Systems* and to the sentence against Galileo.

In short, the Jesuits did not feel under obligation to inform their Chinese hosts of the very stormy developments in Europe. As a result of this silence, the Chinese were kept from learning about the great new breakthroughs in astronomy and physics. Indeed, it was not until the late 18th century that the Chinese and Japanese were appraised of these developments. During most of the early 17th century the astronomy which was taught them and described in books by the Jesuits was that of Tycho Brahe, which represented a third way, the geoheliocentric system which respected the Biblical teaching.[13]

Professor Nathan Sivin has indicated that some Chinese astronomers early saw the limitations of the Tychonic system.[14] The net effect of the Jesuit diffidence, however, was to delay the Chinese reception of Copernicus and Galileo until the late 19th century. The most powerful estimates of the effect of this delay will be found in the following statements by Sivin and Duyvendak, respectively:

The Jesuits were also unable to discuss the wider repercussions of the Scientific Revolution, in particular Galileo's central idea that the only firm basis for knowledge of nature was the work of scientists themselves. The Church's injunction of 1616 against the teaching of heliocentrism was meant to reject this notion. To the very end of the Jesuit scientific effort in China, the rivalry between cosmologies was represented as between one astronomical innovator and another for the most convenient and accurate methods of calculation, rather than between the scholastic philosopher and the mathematical scientist for the most fruitful approach to physical reality. Thus the basic character of developing modern science was concealed from Chinese scientists, who depended on the Jesuit writings....

... The impossibility for the Jesuits, the mediators of Western science in China, to accept Galileo's heliocentric theory, is a matter of immense cultural significance.... It is really with the condemnation of Galileo that the paths of East and West diverge, not to meet again until the nineteenth century....[15]

III

My direct purpose herein will be to speak to the challenges posed by these developments and to review the historical and sociological theses set forth by Needham in his life-long studies of the dialogues and encounters of East and West. Toward this end I shall be drawing especially on four of Needham's works, his monumental *Science and Civilization in China*, his *Grand Titration*, *Clerks and Craftsmen in China and the West* and a series of popular lectures which offer many clues to Needham's personal attitudes, *Within the Four Seas*.[16]

To help illuminate the theoretical frameworks of Needham's hypotheses and challenges I shall mainly draw here upon clues culled from the pages of a small company of titans of sociology: Max Weber principally, but also Sir Henry Maine, Emile Durkheim, Marcel Mauss, and a number of more recent writers. I single out these authors because I am convinced that along with Needham they offer us building blocks on which to rear a comparative depth-historical *differential* analysis of sociocultural processes at the societal, civilizational and intercivilizational levels which we so sadly lack today and will more desperately need tomorrow.[17]

Among other contributions, these authors suggest cues on how to correlate factors involved in the movements toward and away from participations in communities of discourse, passages toward and away from wider rationality and fuller freedoms of entry into and exits from frameworks of inquiry; passages toward the elaboration of social and cultural milieux conducive to applications and validations of structures of rationales of evidentiary canons, procedures – expressed in languages marked by logical constancy, abstractions of form and universality of scope.[18]

IV

I begin with two expressions of Needham's challenges whose current importance for the comparative history and sociology of science and society it is impossible to exaggerate:

Perhaps the most far-reaching issue in Needham's challenge calls out to us from the pages of his monumental *Science and Civilization in China* which sums up the meaning of the famed Jesuit mission in China in the early 17th century. Here Needham asks whether the new astronomy which was in the power of the Jesuits to mediate to the Chinese needs to be thought of as a *peculiarly Western science* as opposed to *Eastern* science or as a *new, universally valid*, world science. Fully cognizant that the entire tenor of his vast undertaking hinges on his answer to this question, Needham replies:

> It is vital today that the world should recognise that 17th century Europe did not give rise to essentially 'European' or 'Western' science, *but to universally valid world science, that is to say, 'modern' science as opposed to the ancient and medieval sciences.* Now these last bore indelibly an ethnic image and superscription. Their theories, more or less primitive in type, were culture-rooted, and could find no common medium of expression. *But when once the basic technique of discovery had itself been discovered, once the full method of scientific investigation of Nature had been understood, the sciences assumed the absolute universality of mathematics, and in their modern form are at home under any meridian, the common light and inheritance of every race and people....*
>
> ...*And what this language communicates is a body of incontestable scientific truth acceptable to all men everywhere.* Without it plagues are not checked, and aircraft will not fly. The physically unified world of our own time has indeed been brought into being by something that happened historically in Europe, but no man can be restrained from following the path of Galileo and Vesalius, and the period of *political dominance* which modern technology granted to *Europeans is now demonstrably ending.*
>
> In their gentle way, the Jesuits were among the first to exercise this dominance, spiritual though in their case it was meant to be. To seek to accomplish their religious mission by bringing to China the best of Renaissance science was a highly enlightened proceeding, yet this science was for them only a means to an end. Their aim was *naturally to support and commend the 'Western' religion by the prestige of the science from the West which accompanied it. This new science might be true, but for the missionaries what mattered just as much was that it had originated in Christendom.* The implicit logic was that only Christendom could have produced it. Every correct eclipse prediction was thus an indirect demonstration of the truth of Christian theology. The *non sequitur* was that a unique historical circumstance (the rise of modern science in a civilisation with a particular religion) cannot prove a necessary concomitance. Religion was not the only feature in which Europe differed from

Asia. But the Chinese were acute enough to see through all this from the very beginning. The Jesuits might insist that Renaissance natural science was primarily 'Western' but the Chinese understood clearly that it was primarily 'new'.[19]

Precisely how the Jesuits or the Chinese defined the situation is no easy matter to decide from evidence made available by Needham and others. What matters to us at the moment is how *we* need to assess the elements at issue in this conclusive contact of ultimate rationales.

And now to the second set of Needham's challenges:

Between the first and fifteenth centuries the Chinese, who experienced no 'dark ages', were generally far in advance of Europe, and quite independent of the great ideas and systems of the Greeks. Not until the scientific revolution of the Renaissance did Europe draw rapidly ahead. Throughout those fifteen centuries, and ever since, the West has been profoundly affected not only in its technical processes but in its very social structures by discoveries and inventions emanating from China and East Asia. Not only the three which Lord Bacon listed (printing, gunpowder and the magnetic compass) but a hundred others – mechanical clockwork, iron-casting, stirrups and efficient horse-harness, the Cardan suspension and the Pascal triangle, segmental-arch bridges and pound-locks on canals, the stern-post rudder, fore-and-aft sailing, quantitative cartography – all had their earth-shaking effects on a Europe generally more unstable. Why then did *modern* science, as opposed to ancient and medieval science, develop only in the Western world since the time of Galileo?[20]

What shall we say to Needham's challenges and questions? How well does he himself answer them in the light of our present understanding of comparative historical sociology? How secure are the foundations on which Needham's questions rest? Is it truly the case, for example, that:

(a) Between the first and fifteenth centuries the Chinese, who experienced no 'dark ages', were generally far in advance of Europe, and quite independent of the great ideas and systems of the Greeks?

(b) Europe did not draw rapidly ahead until the scientific revolution of the Renaissance, more exactly the 17th century?

(c) Throughout the first fifteen centuries of the Christian era and ever since, the West has been profoundly affected not only in its technical processes but in its very social structures by discoveries and inventions emanating from China and East Asia?

I anticipate the discussions to follow by remarking that we will not be able to find a satisfactory answer to the issues Needham poses if we accept without check the statements he has proclaimed facts in the first two paragraphs I have just concluded. The issue of factual accuracy is no small or pedantic matter. The established assurance of an accurate comparative history of science will prove the clue to the analysis we shall describe as comparative historical differential sociology in civilizational perspective.[21]

V

Scholarly courtesy demands that I stand aside at this point and wait upon Needham to develop his own case on the basis of the evidence he knows so well. What does Needham himself see as the explanations of the phenomena and problems under investigation?

First impressions to the contrary apart, it will be noted that Needham is not content to rest his case upon a single or simple formula. Strewn throughout his works are many hypotheses offered with varying stresses scoring the orchestration of elements supposed to constitute the explanations of the phenomena. The main sources which he draws upon for explanatory paradigms are Marx and Engels[22], Karl Wittfogel[23], Edgar Zilsel[24], and other less renowned scholars who are believed to have advanced a so-called dialectical materialist, or Marxist interpretation of history.[25]

It quickly becomes apparent, however, that Needham feels free to range outside conventional Marxist canon. Some weighty and critical hypotheses suggested to him by the Chinese data are related to other sources. One such critical source is Alfred North Whitehead.[26]

Essentially Needham conceives his main challenge to be double-edged:

(1) Chinese science and technology markedly outstripped the West from the second to the sixteenth centuries.

(2) How, in the face of this evidence, explain the fact that the Galilean-Newtonian science did not emerge in China?

Seeking explanations in terms of the sociocultural directions embarked upon by the autochthonous civilizational complexes[27] he seems, in my reading, to mark out five interrelated systems or complexes in terms of which the changes in socio-cultural patterns can be explained:

(1) geographical-environmental settings
(2) socio-economic structures
(3) cultural ontologies and philosophical principles
(4) symbolic (in the narrow sense) technologies, notably linguistic patterns of the written and spoken languages

(5) community and associational patterns and values, including norms of conduct.[28]

The key features of these data ask to be arranged under these rubrics as follows:

(1) The *geographical-environmental* conditions of China were marked by an isolation of the Chinese civilization unknown among the Mediterranean or even the Hindu civilization. Furthermore, China was a gigantic land mass connected by rivers, rather than a series of coasts connected by a virtually enclosed sea, and, despite the similarity of the range of climates between China and Europe, only China experienced a monsoon season.[29]

(2) In its *socio-economic structures*, medieval China was a non-hereditary 'feudal bureaucracy' founded on an agrarian base, which dispensed with or lacked an organized system of slavery. Supervision of defense and the waterworks was reserved to the Imperial authority.[30]

(3) A central feature of *cultural ontologies* of China was the idea of *li*, the central idea shared by both the Confucians and Taoists, of harmony and inter-relatedness among all phenomena – mankind included – in the universe. The preference of the Taoists for withdrawal and 'wordless' contemplation of Nature (discerning 'The Way' and 'the One') contrasted with the heavy emphases among the Confucians on literacy, scholarship and learning.[31] Here Needham would also include the way in which the calendar – very closely linked to religious and political concerns – played a central role due to the interconnectedness of celestial and human affairs.

(4) *Linguistic-symbolic technologies* of China refers to the complex structures of its language – as written by the literati and as spoken – the restraints built into its language by its pictographic cast and the early innovation of a genuine decimal number system.[32]

(5) *Community and associational structures and values* comprise the pieties, obligations and norms of conduct deriving from traditional claims of family, clan, secret-society, and other claimant group structures.[33]

A few short pages will not suffice to show how these five complexes ground the explanation of so vast an amount of data as Needham has assembled. I would briefly indicate by way of illustration how their intersections help explain four phenomena which are equally critical for our purpose:

(1) The marked superiority of China over the West in technological inventiveness from the 2nd to the 16th centuries.

(2) The failure of empirical scientists and merchants to emerge in China as they did in Europe.[34]

(3) The failure of an idea of 'laws of nature' to emerge in China (Whitehead, Zilsel, and others).[35]

(4) China's failure to achieve a breakthrough to the experimental-mathematical mode of Galileo.[36]

The evidence can be given in synoptic form as follows:

(1) The absence of any systematic exploitation of slave labor was in Needham's view a key factor in the superiority of China's technology prior to the sixteenth century. The absence of slaves only secondarily gave incentive to the invention of labor-saving devices. Primarily, the absence of slave labor meant that the adoption of any such invention would pose little threat to the texture of society. Hence, China's socio-economic structure was more amenable to the propagation of inventions than, say, that of Rome.[37]

China's cultural ontologies were also important factors in its early technological successes. The Taoist practice of contemplation of Nature – to discern the 'Tao', which would reveal the proper (anti-Confucian) 'Way' for society[37a] – cleared the path for many advances in pharmacology, alchemy, medicine and some parts of physics. For example *wu wei*, the principle of non-intervention (the same principle which blocked the path to the notion of 'Law of Nature') made the idea of action-at-a-distance (that is, of non-mechanical action) unproblematic, leading to the discoveries of the seismograph and of many magnetic phenomena unknown in the West until quite late.[38]

On the other side of the coin, the Confucians, in the service of the feudal bureaucracy, lent 'orthodoxy', and thereby encouragement, to its projects – especially astronomy (for calendar-making) and hydrodynamics.[39] In the case of the latter two technologies, we see also the contribution of the environmental factors of monsoon and river systems. The agrarian-based feudal bureaucracy had to control seasonal floods which threatened both crops and the safe passage of tax-barges carrying grain as in-kind payment.[40]

(2) As for the *non-emergence of merchants and experimental scientists in China*: here, as Needham tells the story, the feudal bureaucracy plays the leading role. Unlike the late medieval European city, where the mayor or burgomaster shared responsibility with the guilds, the Chinese city was governed by the bureaucratic representative of the Emperor.[41] There was no gap of power in which the merchants could insert themselves. This situation was exacerbated by the fact that merchants held the lowest station in the society, after scholars, farmers and artisans in that order. As Needham puts it, every merchant's son desired to become a bureaucrat. Furthermore, continual interventions[42] by the bureaucracy in the name of the Emperor – e.g., sumptary laws – in the economic affairs of merchants inhibited their attainment of power and influence.

Just as the merchants, so also the artisans – who were next to the bottom of the ladder in the society, and according to Needham were the richest source of experiment and invention – were inhibited from coming to the fore. Like the merchants, artisans could achieve little status.[43] Consequently, bright young men were drawn away from the crafts to the bureaucracy, applying their intelligences to the literary studies necessary to pass qualifying examinations, rather than to invention. Furthermore, the lack of training in the language of the literati made it virtually impossible for artisans to make even their discoveries understandable or respectable among the powerful Confucian scholars, the advisors of the emperor.[44]

The connection Needham finds between the failure of the merchants to achieve power and the non-emergence of experimentalists becomes clear when he puts his critical question in a new way: What was there peculiar to the Western development which pushed it beyond the situation which had crystallized in China? Raising the question in this form allows Needham to bracket the importance of theoretical traditions from the Greeks until the end of the 14th century.[45] It also compels him to turn to stress changes in the economic and social realms. He writes:

It may well be that concurrent social and economic changes supervening only in Europe formed the milieu in which natural science could rise at last above the level of the higher artisanate; the semi-mathematical technicians. The reduction of all quality to quantities, the affirmation of a mathematical reality behind all appearances, the proclaiming of a space and time uniform throughout all the universe; was it not analogous to the merchant's standard of value?[46]

Needham also asks whether the lack of Chinese interest in exploiting technology in certain areas – and hence the failure to break through to Galilean science – could be attributed to the lack of a profit-motive due to the feudal bureaucracy. His answer on this issue is striking in its simplicity:

Put in another way, there came no vivifying demand from the side of natural science. Interest in Nature was not enough, controlled experimentation was not enough, empirical induction was not enough, eclipse-prediction and calendar calculation were not enough – all of these the Chinese had. Apparently a mercantile culture alone was able to do what agrarian bureaucratic civilisation could not – bring to fusion point the formerly separated disciplines of mathematics and nature-knowledge.[47]

(3) The very idea of *'law'* was radically different in China and the West. Whereas in Rome an abstract system of legal concepts and procedures had to be developed to encompass adjudication of the discrepant interests within and among Roman groups and the relations of Romans to the many 'peoples' of the empire (each having its own 'law') and the relative lack of internal differentiation, China's geographical isolation made this unnecessary.[48] As a result, the notion of 'laws' promulgated by a divine lawgiver was wholly foreign to the Chinese understanding of law. The latter started from the assumption that the entire universe of man and entities was a 'pattern', a 'ceaseless regularity' and it was not a commanded regularity; it was a spontaneous co-operation of all things according to their natures.[49]

(4) An issue which Needham himself describes as 'a focal point' in his grand design is put most powerfully in his restatement of his challenges in the context of the relations of mathematics to science. Needham writes:

If the foregoing pages have been numerous, we must reflect that from many points of view mathematics has always been a discipline of its own of equal rank with the whole of the natural sciences. The conclusion of the account of Chinese mathematics brings us to what might be described as a focal point in the plan of the present work. What exactly were the relations of mathematics to science in ancient and medieval China? What was it that happened in Renaissance Europe when mathematics and science joined in a combination qualitively new and destined to transform the world? And why did this not happen in any other part of the world?[50]

As we presently see, this form of the question is very close to Weber's form of the question.

Looking more carefully into the inner mechanism of this development, Needham writes:

... It has often been said that whereas previously algebra and geometry had evolved separately, the former among the Indians and the Chinese, and the latter among the Greeks and their successors, now the marriage of the two, the application of algebraic methods to the geometric field, was the greatest single step ever made in the progress of the exact sciences. It is important to note, however, that this geometry was not just geometry as such, but the logical deductive geometry of Greece. The Chinese had always considered geometrical problems algebraically, but that was not the same thing.[51]

Needham does acknowledge that certain of the formative ideas which were to find fruit in Galileo appeared as early as the 13th century among the philosophers at Oxford.[52] He hastens from this to explain that

these ideas came later to be associated with the University at Padua, where Averroism was strong and logic was studied as a preliminary to medicine, not law or theology. Discussions there, between the +14th and the +16th centuries, led to a *methodological theory which, except for the important element of mathematisation, showed some similarity to the eventual practice of Galileo.*[53] [My emphasis]

Needham comes close to anticipating our own view toward the end of his discussion where he declares:

Thus while the practice of the higher artisanate was akin to the second or experimental part of the Galilean method, the theorising of the scholastics foreshadowed the first or speculative part. But how widely they were aware that agreement with empirical fact was the ultimate test of hypotheses seems doubtful, nor is it clear that they always understood the importance of examining new phenomena ... which had not already been used as the source of the hypothesis under test. Moreover, they rarely succeeded in advancing beyond the primitive style in their hypotheses. Robert Grosseteste of Lincoln (+1168 to +1253) has been selected as the key figure in this natural philosophy, but the dual process of induction and deduction goes back to Galen and the Greek geometers, probably reaching Grosseteste through Arabic sources, such as the encyclopaedist Abū Yūsuf Yaʿqūb ibn-Isḥāq al-Kindī (*d.* +873) and the medical commentator 'Alī ibn Riḍwān (+998 to +1061). Though Grosseteste may have believed that organised experimentation beyond mere further experience should be used to verify or disprove hypotheses, it is not claimed that he himself was an experimentalist. He does seem however to have influenced the +13th-century group of practical scientific workers which included the Englishmen Roger Bacon (1214 to 1292) and Thomas Bradwardine (1290 to 1349) in physics, the Frenchman Petrus Peregrinus (*fl.* 1260 to 1270) in magnetism, the Pole Witelo (*c.* 1230 to 1280) in optics, and the German Theodoric of Freiburg (*d.* 1311) with his admirable theory of the rainbow. It is curious that during the period in which these men were working, China was the scene of a scientific movement quite comparable. But after the early years of the 14th century there was a marked regression, and verbal argument again dominated in Europe until the time of Galileo himself. ... There was thus a continuous line of experimentalists in Europe from Roger Bacon to Galileo, but after about +1310 the contribution of scholastic philosophy ceased, and for three centuries practical technology was the order of the day.[54] [My emphasis]

Needham proceeds to a careful dissection of the phases of the so-called Galilean method proceeding on the assumption that the new or experimental philosophy was characterized by the search for measurable elements in phenomena, and the application of mathematical methods to these quantitative regularities. He is obliged to acknowledge that the conscious experimental test of precise hypotheses which formed the essence of the Galilean method differed from instigative experimentation of technologists and craftsmen. He adds:

In such empirical ways it was possible to accumulate great stores of practical knowledge, though the lack of rationale necessitated a handing down of technical skill from one generation to the next, through personal contact and training. With due regard to different times and places there was not much to choose between China and Europe regarding the heights of mastery achieved; no westerners surpassed the bronze-founders of the Shang and Chou, or equalled the ceramists of the Thang and Sung. The preparations for Gilbert's definitive study of magnetism had all taken place at the other end of the Old World. And it could not be said that these technological operations were non-quantitative, for the ceramists could never have reproduced their effects in glaze and body and colour without some kind of temperature control, and the discovery of magnetic declination could not have occurred if the geomancers had not been attending with some care to their azimuth degrees.[55]

VI

If, as I have suggested above, we look into the pages of a number of comparative historical sociologists of social and cultural process whom Needham hardly mentions by name; if, that is, we look into Weber, Durkheim, Marcel Mauss and, last but not least, Sir Henry Sumner Maine, we will be rewarded by finding critical new clues in the way of approaches to answers to Needham's questions. The most rewarding of these bear upon two related clusters of concepts which are variously represented in the works of these men. The clusters and concepts may be described as follows:

(1) *rationalism, rationality, rationalization* and what I elsewhere describe as *rationale-structures* in their several and joint relations to the prospect

of promoting the fullest possible rationalization of intelligence[56]; (On the many issues involved here Weber serves as our main, but not our only, guide. A limiting aspect of this fact is that Weber's teachings on these heads are not entirely free of ambiguity.)

(2) the sources and outcomes of different passages to *universalism*, *universalities* and *universalizations* in the multiple contexts of thought and action in the varied spheres of social relations and symbolic cultures.[57] (None of our authors speaks about this second complex of themes expressly in any detail; the main contributors are Weber, Durkheim and Maine. Durkheim offers clues on the relations between changes in social structure to changes in the size and density of communities of discourse and persuasion.[58] Weber's hints can be elaborated to yield an important body of notions expressing the pattern of *'double dialectic'* of universalization processes in the movements and changes in the social and cultural spheres.[59] Of this we will speak at greater length later.)

To return to what I have called *the fullest possible or maximum rationalization of intelligence*: If decisive movement in this direction is to occur, it is requisite that great numbers of men and women be free to transcend the particularistic restraints of family, kin, class, caste and to allow their minds to wander freely in zones which are acknowledged to be immune against intervention. These zones are neutral zones protected against the insurgence of the political or theological censor and the group dictate.[60]

The promotion of a full rationalization of intelligence demands that substantial numbers of persons be legally empowered and psychologically disposed to carry on mental production at the highest level of operation without being called to a halt by disabling private or public inhibitions or barriers.[61]

We proceed now to set down some of the main answers to Needham's challenges.

VII

Although Marco Polo was astonished by what he saw on his visits to China, it is now evident that had he been oriented to a genuinely comparative analysis of cultural and social process he might have needed to report the giant strides which Europe had achieved during the twelfth and thirteenth centuries. The years between the first crusade and the Council of Vienne of 1311 were extraordinarily eventful for Western Europe.[62]

On the surface, China was a vast land Empire with extraordinary evidences of complex technologies of transport, communications, exchange, regulative legal and instructional systems, and so on. During the European High Middle Ages we are describing here, however, Europe underwent a succession of shocks which were to leave it stronger rather than weaker. The following developments will be recalled:

(1) The overcoming of Muslim power in the Mediterranean.[63]

(2) The development of self-governing cities by no means lacking in legitimate authority to rule on their own behalf, as a cryptic phrase of Weber mistakenly allows us to believe.[64]

(3) The spread of liberties in the several senses of both liberties *from* and liberties *to*. In the present instance the critical liberties are the liberties *from* the control of territorial powers and liberties *to* employ one's own abilities in the 'public's' or one's own behalf.[65]

(4) The spread of many centers of learning, especially universities, which were dedicated to expanding the scope and exploring the roots of knowledge as then conceived. Of critical importance here was the liberty to philosophize on matters not dogmatically established.[66]

(5) The systematic concordance of knowledge and learning. Most important here were the summing up of 'scientific' theology, natural philosophy, political philosophy, moral philosophy, and law – Roman, canon, common, municipal.

From the sociological side, the 12th and 13th centuries were the seedbed of the modern European society. It was exactly the differentiation into kingdoms, principalities, cities, estates (*stände*), professions, universities and so on which helps us understand the extraordinary pulse-beat of the developments of the 12th and 13th centuries.[67]

It was the West which was to explore the vast new uses of numerous techniques and implements which were originated by the Chinese, Hindus, and Muslims. An especially striking case is the developments in the West of the so-called Hindu-Arabic numeral system and the zero.[68] It was the West which was to give new force and meaning to the mathematical approach to natural philosophy and to move dramatically toward mathematical physics with the help of newly recovered Greek texts. In these centuries, the West was entirely to restructure the rationales of thought and action, knowledge, opinion and conscience.[69]

Only because we have been disinclined to study the evidence and traces of the comparative development of the symbolic technologies in East and West have we failed to see the most important elements in the development of the West.

VIII

In other writings I have emphasized the advance which Needham makes over many other investigators in his presentation of some of the key issues involved in his 'challenges'. I have, indeed, proposed that Needham needs to be included among the ranking pioneers in the comparative historical sociology of science and sciencing in civilizational perspective.[70]

The point has now come to enter some necessary demurrers. For all his expertise in science and sinology, Needham falls short of Weber, whom he hardly appears to mention, when it comes to the analysis of the interplays of science, civilizational setting, cultural orientation and life-ways of peoples – the Chinese and China being no exception in this regard.[71]

Weber became interested in studying variations and variabilities in the cultural patterns of the great civilizational complexes of West and East early in his career. The entire thrust of his research led him to become increasingly committed to what I am calling the *comparative historical (differential) sociology of social-cultural processes and civilizational patterns*.[72]

Everyone of Weber's works offers clues to his answers to Needham's questions. Weber's clues are easy to trace through the course of his writings from the essays of 1904–5 carrying the title, *The Protestant Ethic and the Spirit of Capitalism* to his possibly last essay, 'The Author's Introduction' to the *Collected Essays in the Sociology of Religion*. The turning point in this sequence came with his initiation of the studies which went under the name, *Wirtschaftsethik der Weltreligionen*.[73] As one might expect, some of the best formulations appear in his work on Confucianism and Taoism which was translated under the title, *The Religion of China*.

Here in the *Religion of China* Weber strongly elaborates the immense supports to archaic traditionalism given by the clan and family structure, the absence of a *polis* and city of the Occidental type, the importance of the mandarin elite and the Confucian ethic, the prevalence of the ideal of the scholar-gentleman and the denigration of trade and systematic disciplined work.[74]

Weber gives due stress to the fact that China far surpassed the West in many sectors of economy and polity but this does not lead him afield; he does not infer that the priority in developments of economy and technology will necessarily be paralleled in leadership in respect to pace and pattern in the passages to new phases of rationalized orientations and schemes of action. In the case of China it was precisely the enormous authority and hold of practical prudence and this-worldly rationalism which barred a full rationalization at appropriate levels of individual conduct, social organization, and cultural framework.[75] In the West, it was the thrusts to new transcendental expressions of the charismatic which broke down the old supports of traditionalism in the spheres of knowledge. Deeply sensitive to the great role played by transformative break-throughs in the West, Weber writes:

> If this development took place only in the Occident the reason is to be found in the special features of its general cultural evolution which are peculiar to it. Only the Occident knows the state in the modern sense, with a professional administration, specialized officialdom, and law based on the concept of citizenship. Beginnings of this institution in antiquity and in the Orient were never able to develop. Only the Occident knows rational law, made by jurists and rationally interpreted and applied, and only in the Occident is found the concept of citizen (*civis Romanus, citoyen, bourgeois*) because only in the Occident again are there cities in the specific sense. Furthermore, only the Occident possesses science, in the present-day sense of the word. Theology, philosophy, reflection on the ultimate problems of life, were known to the Chinese and the Hindu, perhaps even of a depth unreached by the European; but a rational science and in connection with it a rational technology remained unknown to those civilizations. Finally, western civilization is further distinguished from every other by the presence of men with a rational ethic for the conduct of life. Magic and religion are found everywhere; but a religious basis for the ordering of life which consistently followed out must lead to explicit rationalism is again peculiar to western civilization alone.[76]

Weber was deeply interested in the characteristics of Chinese science and magic. He was aware of the fact that Chinese religion did not set itself in opposition to new theological and scientific outlooks; he also perceived that passages to new science often arise from new cosmological and ontological visions and needs.[77] The very absence in China of a stress on transmundane God had a limiting effect on Chinese ontology.[78]

Weber's understanding of the strange turnings of rationalism in China and the inhibiting effects of the absence of transcendant points of reference are very great insights into the failure of the Chinese to proceed toward transformative breakthroughs. In this same connection Weber does appreciate the critical importance for the scientific development of the West of the Greek heritage, especially of the importance in that of Euclid. Weber is not as helpful as one might wish in an analysis of the issues of universalism and universalization, but he does help us in this terrain. Few discussions of Weber on the Chinese are as revealing as Chapter VI in his *Religion of China*, where, among other issues, he discusses the absence of natural sciences in China. Here he anticipates many of the points which reappear in his 'Author's Introduction' of 1920. At one point in the chapter Weber emphasizes the importance of the absence of powerful metaphysical interest in the Confucian tradition. He adds:

> The development of mathematics had progressed to trigonometry – but this soon decayed because it was not used. Confucius evidently had no knowledge of the precession of the equinoxes which had been known in the Middle East for a long time. The office of the court astronomer, that is the calendar maker, must be distinguished from the court astrologer who was both an annalist and an influential adviser. The former was a carrier of secret knowledge and his office was hereditarily transmitted. But relevant knowledge can hardly have developed, witness the great success of the Jesuits' European instruments. Natural science as a whole remained purely empirical. Only quotations seem to have been preserved from the old botanical, that is pharmacological work, allegedly the work of an emperor.[79]

He puts his twofold general question in his usual way:

Why did the distinctively *modern* science arise in the West, and in the West alone? And why did China "*not* enter upon that path of rationalization"? What factors may have operated to cause it to fail to achieve full height and crystallization in the East until its adoption in the West? What factors in the sociocultural environment, what factors in the political and familial structures worked to check the full rationalization of intelligence and of science and intelligence in the Oriental world?[80]

The reader wishing a full presentation of Weber's answers has no recourse but to read the entire corpus of the *Collected Essays in the Sociology of Religion*. At the very least he must examine the 'Author's Introduction' to these *Collected Essays* and Chapter VI of the *Religion of*

China. The reader will quickly recognize how strongly Weber stresses the enormous authority in the West as contrasted with the East of the universalizing and universalistic modes of thought and sensibility. Weber appreciated the prominent place accorded to systematic structure and other abstract formalisms in the West.

Weber allows us to see into his deepest intentions toward the close of his 'Introduction'. The *origin* of Western science, he explains, cannot be attributed to capitalistic interests:

Calculation, even with decimals, and algebra have been carried on in India, where the decimal system was invented. But it was only made use of by developing capitalism in the West, while in India it leads to no modern arithmetic or book-keeping. Neither was the origin of mathematics and mechanics determined by capitalistic interests. But the *technical* utilization of scientific knowledge, so important for the living conditions of the mass of people, was certainly encouraged by economic considerations, which were extremely favourable to it in the Occident. But this encouragement was derived from the peculiarities of the social structure of the Occident. We must hence ask, from *what* parts of that structure was it derived, since not all of them have been of equal importance?[81]

IX

There is no hope of achieving a truly reliable comparative sociology of cultural change without first having a thoroughly grounded comparative history. To help resolve the issues, therefore, which divide Needham and Weber and other comparative historical sociologists of science and civilization, we need to turn our attention to the main lines of socio-cultural development in Western Europe.

The first theme on which we must fix our attention is the extraordinary character of the era of the Crusades in World history.

As grateful as we must be to those who pioneered in the study of the 12th and 13th centuries – to Haskins, Gilson, Sarton and others – we are now aware that we have come a long way and that new reviews of perspectives and interpretations are needed. In our own day, R. W. Southern[82], Marshall Clagett[83], Dom David Knowles[84], M. D. Chenu[85], Pierre Michaud-Quantin[86], and others have helped us to see the 12th and 13th centuries in a new light.

Briefly, the more we come to know the European cultural and social transformations of the crusading era the more we find ourselves having to conclude that they represent a watershed in the international history of civilizations. Without wishing to be understood as implying that the breakthroughs effected in the 12th or 13th centuries were without precedent or parallel, we need to say that anything short of a comprehensive survey or analysis of other civilizational complexes and civilizational encounters would not suffice to allow us to suggest the massive innovations which occurred during that epoch. Indeed, it needs to be realized that a large part of the significance of those days is based upon the extraordinary responsiveness – negative as well as positive – of men of that era to strands of influence coming from other eras and civilizational centers.[87]

Eager appropriations and sifting of critical traditions and texts are the hallmark of the time. Not only did the men of this era recover neglected Greek and Roman texts – most notably Greek philosophical and scientific writings and the Roman law – they also labored to translate Arabic, Hebrew and classical works which had become part of the Islamic and Hebraic corpus. Thanks to Joseph Needham, Lynn White, Jr.[88], and others, each passing day brings new evidence of borrowings from China, India, and other lands of the East.[89]

A prime responsibility here must be to delineate certain central changes of existence, experience and sensibility which underlie the extraordinary productions of that time. To put my point briefly, the twelfth century is the era of the crystallization of new structures of consciousness which would have to be described as *rationalizing-and-rationalized structures of consciousness* which gain increasing dominance over *faith-structures* of consciousness and sacro-magical structures of consciousness.[90] It is the time when monastic theology gives ground to scholastic theology, the time when new images and horizons of conscience, self, person, society, the cosmos, action, justice, forms of rule, institutions of law and learning take on a cast that have ever since been distinctive and primary features of the Western European world.[91]

The key changes effected in this process are marked by the following emphases: in place of a sacro-magical, albeit sacramentalized, sense of the creation and creator, there appears a stress upon the need and ability of men to know and prove one's faith, to know and explain natural phenomena by the principles of natural philosophy, to offer rational justification of their acts and opinions.[92]

X

It remains to review these comparative historical efforts in wider terms. Only by so doing can we identify the differences in the axes of social and cultural development: the differences in the axioms which ultimately are expressed in the unfolding projective sociocultural geometries of the orientations and institutions of the two worlds. Needham and Weber need to be brought into direct contact with one another if we are to describe these different geometries correctly. Let us consider the differences in notions assumed to be the ground of institutional development.

The hallmark of the West in the twelfth and thirteenth centuries is the crystallization of orientations and institutions which simultaneously rest upon the two-fold commitment to the *concrete individual person* and *objective universal*.[93] The main notion on which the institutions are reared is that individuated persons are the bearers of rights and rationalized universals become the focal points of governing norms.

Such widenings of outlook imply new sanctions and new turning to the future. Universalities of this scope presuppose the passages beyond sacro-magical and faith-structures of consciousness to rationalized structures of consciousness which place high premium on universal rationales and rationalities.

I have elsewhere suggested the ways in which the faith-structures of consciousness issue in new contents which call for new understanding. Faith pursuing understanding of itself and its new contents, elaborates structures of rationality.[94] The widened communication and communions take shape in the appearance of new forensic contexts in which discrepant views on the nature of rationality, the proofs of claims, need to be correlated in a manner that invites and requires rational assent. Illustrations of this process abound in Anselm of Bec, Abelard, St. Thomas, and other scholars.[95]

The avenues by which the multiplicity of discordant perspectives and claims are brought into some harmony or concordance are dialogue and dialectic.[96] Where alternative pre-rational methods exist for the establishment of the norms of action and belief there seems to be limited recourse to these logical elaborations. Under certain sets of conditions, the new forensic contexts may be kept confined to limited areas and restricted domains.

History affords many illustrations of the limited institutionalization of universalism or rationality or dialectic within restricted spheres. We must not be surprised, therefore, if we find institutions apparently devoted to science or logic in many traditionalist societies. Here we need a distinction between a universalism of limited application which is the base for an intellectual elite or a restricted meritocracy and a societal commitment to universalization, that is, a commitment to encourage the expansion of the boundaries of participation to increase the avenues of communication and to develop skills in communication among those who have so far not truly entered into dialogue because of self-inhibition or the restraints imposed by others.[97]

XI

The times we are describing, the 12th and 13th centuries, were eras of explosive innovations and abrasive conflicts in all spheres and also of unrelenting efforts to find ways of balancing and uniting opposed positions and perspectives. Clearly there could be no going forward until – and unless – the unsolved conflicts of earlier days and incipient schisms of our own time were brought into new harmony and the contraries of *Yes* and *No* were in some way reconciled by a higher synthesis. The paradigms for these days were Abelard, Gratian, St. Thomas, and other summists.

A related and equally significant fact about the 12th and 13th centuries was the general conviction throughout the European Christian world that it was possible, with the exercise of dedicated will and unremitting intellectual labor, to discover an intelligible cosmos and to forge a rational world united in all its spheres, concordant in bringing complex harmony where appearance only presented dissonance.

Chenu and others who have perceived the critical importance of ideas of universality in the 12th and 13th centuries are correct in implying that there was a thrust toward universalistic perspective at every level of action, reference, and significance. Everywhere men sought to take hold of the universal to depict a universe of order in nature, in law and everywhere else.

Thanks to the help afforded us by special studies and researches in semantics such as those of Michaud-Quantin, we are able to avoid falling into traps. The seemingly higher concentration of references to universality in the field of natural philosophy should not distract our attention from the very great evidence that universality was pursued everywhere else, in jurisprudence, in moral philosophy, and so on. The approaches to concordances of discordant canons were efforts to achieve universality. The treatises on human action on the cases of conscience were movements in the same direction. Why did they devote themselves so arduously to the writing of Summae and Encyclopedias? All the thinkers of the 12th and 13th centuries were inspired by the hope that they might rear a great edifice which would house all the confusing and vexing particularities of circumstance.

Our own research makes clear a main locus of this effort was the pregnant notion of *conscience*[98], destined to be the spur and setting to an immense development of special interest to those interested in the comparative study of the shapes of diverse cultural patterns.

XII

Needham has clearly established that the Chinese attained very high levels of achievements in technology and protosciences from the 2nd to the 16th centuries.

The situation was quite different in the spheres of high-level science. Never having known the Greek heritage, the Chinese lacked an adequate base for the independent invention of Galilean science. The Chinese were always inhibited by restraints of various sorts – familial, political, social and intellectual.

Needham notwithstanding, Western science did not have to wait until the days of Leonardo and Galileo to have its first flowering. From time to time Western science did undergo a loss of animation and fell into a sort of limbo. At no time, however, were the grounds for it totally gone. The Greek heritage lacking in China persisted in the West.[99] Its revival and reinforcement in the 12th and 13th centuries were indispensable elements in the 14th century surge forward – especially at Merton College, Oxford, and Paris – of mathematics and mathematical physical and logical modes of orientation.[100]

Charles Haskins, Johann Nordström, and a number of more recent writers (Lynn White, Jr., William and Martha Kneale) have been right to ask whether 12th century developments in thought, art and life might not have been in many ways more critical for the West than the later and better known Renaissance of the 14th and 15th centuries. It was the 12th century which saw the critical breakthroughs in many avenues of the sciences and logic, natural philosophy, the arts, letters, style, sentiment, the scholastic rationales of conscience and opinion.[101] The present writer is pleased to find himself concurring with the views of the Kneales and Dom Knowles that Galileo had harbingers in the 12th and 13th centuries.[102] One of these was surely Peter Abelard; another was surely Adelard of Bath.[103]

How account for the extraordinary expansion of these novel interests? How explain the effort to find all one might know about the realm of the intelligible? Through what series of sociological and cultural changes would men come to develop such profound curiosities, skills and techniques? In my view there is need to grasp certain neglected facts about the changes in the structures of existence, experience, and expression.

Burckhardt and others apart, the 12th century witnessed the flowering of a wide variety of individuated selves and persons who were convinced that they had the ability to create arts, sciences, machines, laws, and even law codes, and were prepared to ameliorate their lot by deliberate innovation. The era sees an extraordinary widening of the horizons of consciousness and new strivings to absorb and explain new experiences and to map worlds in which they are involved.[104]

It is no wonder that the 12th and 13th centuries place such strong stress on critical notions and terms not before conceived as they then came to be. Presently I shall explore neglected facets of these notions and terms in order to make clear what they imply in the way of the expansions and differentiations of the structures of consciousness.[105] Thanks to M.-D. Chenu and P. Michaud-Quantin it is now possible to recover the aura and scope of certain recurring metaphors and terms, notably *universitas*, *civitas*, *communitas*, *persona*, *libertas*, *conscientia*, *aequitas*, *liber*, *machina*, about which I shall speak again later.[106]

With Maine's help we are able to grasp the fact that the twelfth and thirteenth centuries saw surges toward the expansion of the number of those qualified to be persons in their own right.[107] Europe in the 12th and 13th centuries underwent an accelerated passage from status to contract, from what Maine called a *stationary* society to a *progressive* society.[108] At every turn we see evidence that increasing numbers become ready and willing to ameliorate their lot by deliberate innovation. We see the breaking down of invidious ascriptive solidarities and the release of 'individuals' for independent action. Although the full expression of many of these developments comes in subsequent centuries, it may be noted that already in the 12th and 13th centuries we have the deliverance of domestic bondsmen from the yoke of their masters, the enfranchisement of sons from the power of their fathers, and the increased freedom of women to act as persons in their own right.[109] It would be folly to say that this triple juridical revolution occurred everywhere in Europe at the same pace and to the same effect. Clearly the greatest advances in the spread of liberties and personhood occurred in the new cities, where freedom was open to all.[110]

We are in debt to Max Weber for a number of notions which are most helpful in understanding sociocultural process in the 12th and 13th centuries. I will speak now of only two central ideas; the first is expressed most acutely in the 'Author's Introduction' (1920) to the posthumously published *Collected Essays in the Sociology of Religion*, which may have been written just before his death in 1920, where Weber offers clues of extraordinary value in the way of explaining the differences between the path and tempo of Western development as opposed to the development of China, India, and the East generally. What Weber recognized here was that at an early date the West became committed to scientific and juridical rationalism which was marked by the highest degree of striving for system, abstract formalism, and above all, universality in the terms of reference.[111] The universalities of Greek philosophy and science, the universalities of Roman law and political theory, were an enduring heritage which recurrently entered into new fusions in the Western world. The 12th and 13th centuries built upon these universalities, both in Greek science and Roman law and were thus committed to structures of consciousness different in critical respects from those which had prevailed in China, in India, in Islam, among the Hebrews, etc.[112]

In two of his writings, notably in his 'Types of Religious Rejections of the World and their Directions', and in his special monograph on *The City*, Weber perceived another phenomenon that had an unique history in the West. This phenomenon was called by the name of *fraternization*[113], which occurred at different times and under different stimuli. In proclaiming universal brotherhood for all mankind, Christianity bore the promise of being an universal solvent of caste, clan, tribe, kindred, family and of other structures incorporating elements of invidious dualism.[114] Also of grave importance were the fraternizations and confraternizations which Weber identified as the clue to the unique character of the medieval Western city. Nowhere else had so many cities come into being as a result of sworn brotherhoods which proceeded against prescriptive rules of dominion of ecclesiastical and temporal powers. This singular fact led to the appearance of new institutions which were havens for free men. Nowhere else would there develop so variegated a pattern of free workers, traders, professional men, notaries, patricians, clergy, etc. The maxim of the city was 'city air makes free'. Anyone found in the city or living in the city was presumed to enjoy freedom.[115]

It would be hard to understand the main features of philosophy, theology, and science during our era without making reference to the flows of experience and action in new urban settings marked by the great prominence of novel types of forensic settings. The cities and universities were places where men of varied and discrepant experience and viewpoints engaged in unending dialogue and dispute concerning all possible occasions, actions and events. It is this predominance of friction and dialogue that helps us grasp the paramount role of dialectic and scholasticism.[116]

The many new disputes and movements which have their source in the medieval city derive their strength from the fact that new groups and individuals continue to contend one against another in defining and shaping the ways in which they shall carry on their lives. Panofsky is not wrong when he links Gothic art and scholasticism; the two are comparable and conjoint efforts to build new fabrics which harmonize discordant points of view and perspectives.[117]

In their different ways, Greece, Rome, Israel, Christianity, the Medieval Church were all oriented toward universalism.

Weber does not enter into detail about the intervening processes, the dialectical conversions of changes in the social organization and in the cultural structure, but here, too, it is apparent that he perceives the critical importance of movements of fraternization and confraternization, all the conversions of brotherhood, including the passages from brotherhood to otherhood, from tribal brotherhoods to every form of universal otherhood.[118] Elsewhere in his writings he perceives the vast distinction between the Western city and the city of the Orient and especially of the Far East.[119]

Our main answer would then be that we cannot hope to cope with the challenges put to us by Needham without applying for aid to the cultural sociology of Max Weber and Henry Sumner Maine.[120]

It is now apparent that many philosophers and theologians of the twelfth and thirteenth centuries had a deep need to make sense of the whole universe, to discover its patterns and structures, and phases of its coming into being, the connection among all of its parts, the measure of its motions. It is no accident that new thinkers of the time proceeded on the assumption that the world, being in a profound sense the work of God's hand, was necessarily a *machina mundi*, which could be understood as a *machina* if only the explorations were deep enough.[121] Numerous treatises of the time are devoted to exploring the proofs and traces of the divine artificer in the world of nature. Many were no longer content to deduce knowledge of the natural world from the Book of Scriptures. Thus it is no surprise that from the beginning of our era, we encounter thinkers who draw heavily upon Muslim philosophers, scientists and commentators to understand the natural world.[122]

For many, everything that was made by the divine architect was necessarily intelligible and was itself a revelation. The usual suggestion was that there were *three* books – the *Book of Creation (or Creatures)*, the *Book of Nature*, and *the Book of Conscience*.[123] Knowledge concerning the world was deemed to be available in each of these books, and already in the 13th century we find authors who look to nature and natural philosophy, especially as made available through the writings of Plato and Aristotle and the Muslim thinkers for truth concerning nature.

The motives of the search were many, but the principal intention was to establish nature's workings. Particular stress on explorations of this sort occurred among the English Franciscans, and those closely associated with them. We need only mention in this connection the work of Robert Grosseteste, Roger Bacon, John Peckham, Peter Maricourt and others.[124] The flowering of this effort comes in the 14th century, especially among the logicians and philosophers associated with Merton College at Oxford.[125]

XIII

We may summarize the results of our survey thus far in the following ways:

(1) Although Needham evidently had a much greater knowledge of the sciences and of sinology than did Weber, the analysis he provided of the historical sociology of Chinese science does not match that of Weber. For reasons which need not be detailed here, Needham strongly tends to favor a hypothesis which inevitably simplifies a complex issue. Needham insists too much on the crucial importance for the development of modern science of the economic and cultural dominance of the so-called mercantile 'bourgeoisie' and 'merchant capitalism'.[126] A more carefully woven net than this is evidently needed to capture the data now accessible to us.

Needham seems not to have been aware of a point made most forcefully by Weber, namely, that the breakaway developments which are connected with modern science cannot be described as one or another form of rationalism – whether prudential, calculative, ethical, philosophical. Indeed, it was precisely against a mistaken thesis over the all important role of rationalism in helping to explain the character of the disenchanted cosmos of the 19th century which constitutes the point of departure for Weber in his *Protestant Ethic*.[127]

If I may be allowed to cite from one of my previous writings,

> The central object of Weber's interest was the uncovering of cultural, so-called spiritual, foundation of the distinctive bureaucratic enterprise, organization, and outlooks of the modern Western world. Prudent "rationalism." he insisted, was not the view which spurred the spread of the vocational, innerworldly asceticism, as the dominant *ethos* of the modern industrial era. Far from being smooth and straight, the roads to modernity were paved with "charismatic" breakthroughs of traditional structures. These breakthroughs, forged in an atmosphere of religious and social effervescence, had their issue in the overcoming of the invidious dualisms which had hitherto inhibited new rationalizations (and new rationales) affecting all patterns of action and conviction relating to work, wealth, welfare, regulation of self and society, political order, the sense, ultimate worth, the experience of justification, and so on.
>
> As the years passed, Weber was to strengthen his desire to discover whether and why the rationalization process had been more intensive and extensive in the West than anywhere else. Already, indeed in his first edition, however, the *civilizational* differences in the central orientations to "religion" and "world" proved to be of critical importance for his way of thinking.[128]

Needham so far seems to have missed the extraordinary importance for transformative development of the breakdown of traditionalist rationalism. His recurring identification of middle class membership with anti-traditionalism is again and again put in doubt by Chinese and non-Chinese evidence.[129] In China, as in the West, many of the most important encouragements to breakthrough came from movements beyond ascriptive solidarities toward universalizing and confraternizing associational, communal, and cultural structures.[130]

(2) In the West one of the greatest forces in the direction of universality was the Christian message as it extended the Jewish idea of tribal brotherhood of all to universal brotherhood of mankind. Another very powerful source of universalism was Greek philosophy and logic and Roman law. Wherever these had influence, movements to universality achieved great importance.

We may put our proposition in a partly paradoxical form. In many traditionalist societies the roads to universality were barred by the strength of the ethos of family and tribe, kindred and clan. At some moments in some societies there is a passing beyond these nuclear structures to new unions and associations. The passage originates in movements of wider social, civic and religious confraternization and in aspirations to universality in law and science.[131]

(3) In brief, the argument maintains that for a society to make advances in the direction of universality it is necessary that there be some levers spurring movement in that direction. In the cultural structures of the West, Christianity, Greek and Roman law were levers working to that end. In the Far East these did not serve as goads to change. Indeed, for this reason changes of a universalizing character in the modern East had to wait upon the influences of the cultural structures of the Western world.[132] Westernization needed to become universalization in order for secure advance on the road to modernity to be achieved. It seems highly probable that Weber was in a better position than Needham to appreciate the character and significance of these wider cultural and social movements. The relations of Weber and Needham to some aspects of the issues posed in Needham's challenges may be put as follows:

(a) For Weber the priority of China in technology by no means establishes China's superiority in science and social organization during the era up to Galileo.[133]

(b) It appears to have so far escaped Needham's notice that the 12th and 13th centuries in particular were eras of extraordinary social, political and cultural innovation in European and international history. Needham would need to enrich his theoretical resources to do justice to the effervescences and renewals of the 'medieval Renaissance'.[134]

(c) Needham neglects to pay enough attention to the extraordinary societal differentiations which distinguish the Western world from China. Where but in Western Europe could one find its incredibly diverse patterns of cities, parishes, countries, states, territorial sovereignities, universalities? Where but in the West could one expect to find the patterns of citizenship, the corporate structure of cities, the political philosophies?[135]

(d) Needham also understates the evidence for the fact that the separation of church and state in the West gave vast opportunity for widened inquiry into fresh horizons. Neither the Pope nor the King nor the Emperor were ever in a position wholly to dictate the thrust of reflection and criticism.[136] The case was different in China.

(e) Needham has so far failed to give due stress to the fact that religious and ethical *universalism* in Christian thought and the surge forward of Christian sentiment and expression were of the greatest importance in providing a climate for the development of distinctively Western thought with its surge toward *generality*, *nationality*, and *universality*.[137]

(f) Needham sometimes seems to assume that religion and theology could never have been spurs to scientific advance in the Western world. Religion did a great deal more than breed inquisition, as Needham in one place seems to imply. As a matter of fact, all of the major theologians of the Western world in the 13th and 14th centuries were curious about the natural world and trusted to be able to develop notions useful for exploring, describing and explaining the natural world and natural motions. The advances in the physics and mathematics in the West developed by medieval and early modern theologian-philosophers, especially in the 14th century, are now carefully documented, as we have noted, in the works of Clagett, Crombie, Grant, Alexandre Koyré, Annaliese Maier, John Murdoch, and others.

(g) Weber had a stronger awareness than Needham has so far shown that the very Western image of human action and world changed dramatically in the 12th and 13th centuries. Collective consciousness gave ground to the individual conscience around which there evolved comprehensive logics of opinion, belief and action. Interestingly, the treatises on human action were treatises on conscience and its cases, all of which provided a setting for the proliferation of highly developed structures and decision matrices of the moralities and logics of thought and action.[138]

(h) For my own part, I must add that these new outlooks and sensibilities found expression in an ever more intensive pursuit of disciplined experience and trustworthy methods for assuring certain knowledge of the books of creatures (or creation), conscience, and nature. We need not be surprised that it was a thirteenth-century theologian-philosopher-scientist – Roger Bacon – who was the first to sound the call for *scientia experimentalis* and to speak explicitly of 'laws of nature'.[139]

True, neither the men of the 13th century nor the hardy thinkers of the 14th century Oxford and Paris attained a Galilean level of scientific thinking and experiment. Weber and Needham are agreed on this point. Happily, scholars of the stamp of Koyré, Annaliese Maier, Beaujouan[140], Claggett, Murdoch, and others have taken us beyond the extreme claims of Duhem on this score in studies based on scrupulous control of the texts. And here I must stop this installment of our story.

The limits of the present essay do not allow me to go beyond the 14th century for fuller exploration of Needham's claims. Quite deliberately I have concentrated here on the moments of the civilizational turning points, the 12th and 13th centuries. In doing this I do not mean to suggest that everything of value in Leonardo and Galileo came from the Middle Ages through any simple or direct derivation. Clearly Galileo was very powerfully influenced by the new currents of science and sensibility which emerged in the era of the so-called Renaissance. The new editions of Ptolemy and Archimedes almost certainly were a very powerful inspiration for his work.[141] However, from the perspective here adopted these constitute steps forward within a milieu and ambience which had achieved the new crystallizations in the 12th and 13th centuries.[142]

Elsewhere in this series of volumes edited in the name of this Colloquium, I have described some of the central currents of thought and sensibility which achieve their culmination in the work of the early modern pioneers in science and philosophy.[143] There is no need to repeat this material here. The substance of the present discussion can be communicated in a single sentence: Without an Abelard, an Adelard, a Grosseteste, a Roger Bacon, there might have been no Galileo.[144] A most critical spur to the breakthroughs of the era from Copernicus to Newton was the great leap forward during the course of the High Middle Ages of *concrete universality* in all the spheres of existence and thought, in all the domains of life, public and private. The momentous – even revolutionary – achievements of the *medieval* Renaissance were vibrant elements in the sweeping surges of science and philosophy in the 16th and 17th centuries.[145]

New School for Social Research,
New York

NOTES

* Graduate Faculty, New School for Social Research, (N.Y.).

** Slightly variant versions and fragments of this text have been presented in the following settings: A session on the History and Philosophy of Science, sponsored by the AAAS and Boston Colloquium in the Philosophy of Science at the December 1969 meetings of the AAAS in Boston, Mass.; University of Pittsburgh Center for the Philosophy of Science, Pittsburgh University, Pittsburgh, Pa., March, 1972; Philip Merlan Lecture, Scripps College, Claremont, California, January, 1973; Symposium on the History and Philosophy of Science, Queens College, Flushing, New York, April, 1973; Annual Meeting, International Society for the Comparative Study of Civilization (U.S.), Boston, Mass., Feb., 1974.

† The author may be allowed to take this occasion to express his particular gratitude for the hospitality shown him by his Scripps College sponsors and hosts, Dr. Franceska Merlan and the Administration of Scripps College. I also wish to acknowledge the help from my research assistant, Patrick Byrne of Stony Brook, in reviewing the material in Section V.

The following list includes some works which are not cited by page in the foregoing footnotes. They have been listed here because they constitute critical backgrounds of some of the larger themes of the present essay.

In the interest of simplification I have decided to use abbreviations in citing Needham's writings. The two main abbreviations are:

GT – The Grand Titration. (See below under Needham, 1969a.)
(The extremely compelling – soon to become 'classical' – papers reprinted in this volume constitute the indispensable point of departure for the appreciation of Needham's large contribution to research in the differential comparative historical sociology of sciences and civilizations. Particular attention in this connection is called to chs. vi and vii, "Science and Society, East and West" and "Human Law and the Laws of Nature".)

SCC – Science and Civilization in China. (See below under Needham, 1954–1971.)

Readers wishing additional bibliographical leads are urged to review the references in three works by the present writer. (*cf.* below under Nelson, 1968b, 1973b, 1973e.)

[1] See Nelson (1973d, 1972a, 1974b).
[2] *idem* (1973a).
[3] *idem* (1964, 1965a).
[4] Cf: *idem* (1972a, 1974a).
[5] *idem* (1965b, 1968b, 1973a).
[6] Stimson (1917); A. R. Hall (1954); von Gebler (1897).
[7] Rosen (1959, 1965).
[8] Rosen (1959). It needs to be noted that a number of recent writers have begun to present Osiander's initiative in a very favorable light, cf. Koestler (1959); Wrightsman (1970); Christianson (1973). This is in the sharpest contrast to the views earlier taken by Kepler and Galileo. Kepler (1600–01); Galileo (1615–16); cf. Clavelin (1968).
[9] Writings newly recovered from oblivion and partially transcribed by E. Garin (1971) – the *De veritate S. Scripturae* and its attached *opuscula* by Giovanni Maria Tolosani, Florentine Dominican of San Marco – puts in doubt a number of points of view which have long prevailed among historians of science. An especially arresting and influential expression of the main element in this perspective will be found in Koyré (1964), p. 69: "The Catholic Church seems to have been not only quite unperturbed by Copernicanism before the advent of Giordano..., but altogether unaware of its 'inherent challenge'."

In sharp contrast to this statement we now learn from Tolosani that he undertook to complete a confutation of Copernicus' *De Revolutionibus*, which his friend, Bartolomeo Spina, "*magister Sacri et apostolici Palatii*," had been prevented from finishing by his death in 1546. Tolosani's work is represented in the opusculum *De coelo supremo immobile et terra infima stabili*..., which dates from 1546–1547.

Tolosani goes to considerable pains to argue the case against Copernicus from the point of view of a professional scholastic and theologian. To his mind, Copernicus failed to marshall the requisite disproofs of earlier positions and the necessary demonstrations of his largely Pythagorean view. Interestingly, Tolosani explains that Copernicus was learned in languages and knew a great deal about the astronomical and mathematical sciences but little about the "physical or dialectical sciences." Also, Copernicus gave offense by his intemperance and lack of learning in the divine sciences. To our surprise, Tolosani knew that Copernicus was not the author and did not share the view of the anonymous Preface (Osiander). Lastly, these materials were known to Tommaso Caccini, Galileo's scourge, who carried on an unrelenting campaign against the astronomer in Florence, Bologna, and elsewhere. I reserve fuller discussion of these materials for another occasion. I owe thanks to Edward Rosen for bringing Garin's paper to my attention.
[10] Schofield (1965); Westman (1971); Christianson (1973).
[11] For references to the texts and issues, see Nelson (1968b), pp. 37, n. 121; cf. *idem* (1965b, 1973b).
[12] Ricci (trans. Gallagher) (1953); *SCC*, III, 437–61; Nelson (1973b); Sivin (1973a); Schofield (1965); Hellman (1963).
[13] Boas and Hall (1959); Hellman (1963, 1970).

[14] Sivin (1973a).
[15] [a]Duyvendak (1948); cf. Nelson (1973b); [b]Sivin (1973).
[16] Needham (1969a, 1970, *SCC*, *GT*).
[17] Nelson (1972a, 1974b).
[18] *idem* (1974a).
[19] *SCC*, III, pp. 448–449.
[20] *GT*, p. 11.
[21] Nelson (1969c, 1975)
[22] See *GT*, pp. 123–124, 129, 192–194, 202–203; *SCC*, II, p. 291. Needham explicitly rejects fixed stage theories of evolutionary sequence.
[23] Needham acknowledges his dependence on Wittfogel's early work (1931); he is very critical of Wittfogel's later work on *Oriental Despotism*, cf. *GT*, p. 150, 192–194, 203–204.
[24] Zilsel (1942a, 1942b); cf. *GT*, pp. 193, 280, 302–310; *SCC*, II, pp. 534–543.
[25] Two scholars particularly praised by Needham are Jean Chesneaux and André Haudricourt; see, *e.g.*, *GT*, pp. 123–124, 192–194.
[26] See, esp., *GT*, p. 124, 323–326; *SCC*, II, p. 201, 291–292, 454. Whitehead is credited by Needham with having developed philosophical views which very closely resemble the central Chinese concepts of process. This leads Needham to ask a very striking question:

"The Chinese world-view depended upon a totally different line of thought. The harmonious cooperation of all beings arose, not from the orders of a superior authority external to themselves, but from the fact that they were all parts in a hierarchy of wholes forming a cosmic pattern, and what they obeyed were the internal dictates of their own natures. Modern science and the philosophy of organism, with its integrative levels, have come back to this wisdom, fortified by new understanding of cosmic, biological and social evolution. Yet who shall say that the Newtonian phase was not an essential one?" (*SCC*, II, p. 582.)

[27] *GT*. This stress is a critical element in Needham's comparative historical differential sociology.
[28] Needham's findings are summed up under four 'factors': geographical, hydrological, social and economic. *GT*, p. 150.
[29] *GT*, pp. 190–191.
[30] *GT*, pp. 26, 167–171, 206–208.
[31] *GT*, pp. 34–35, 138–162; *ibid.*, ch. 8; *SCC*, II, pp. 519, 528, 557.
[32] *SCC*, III, pp. 3–4, 21.
[33] *GT*, p. 207.
[34] *GT*, pp. 38–41.
[35] *GT*, ch. 8; *SCC*, II, ch. 18.
[36] *SCC*, II, pp. 540–543.
[37] *GT*, pp. 26, 167–171, 206–208.
[37a] *GT*, pp. 34–35, 158–162.
[38] *GT*, p. 210.
[39] *GT*, p. 186.
[40] *GT*, pp. 18, 32, 181, 195.
[41] *GT*, pp. 185–186.
[42] *GT*, pp. 39, 196.
[43] *GT*, pp. 28, 39.
[44] *GT*, p. 31.
[45] *GT*, pp. 190–191, 217.
[46] *SCC*, III, p. 166.
[47] *ibid.*, p. 168.

[48] *SCC*, II, pp. 518–519.
[49] *SCC*, II, pp. 557, 562, 564, 579.
[50] *SCC*, III, p. 150.
[51] *SCC*, III, p. 156.
[52] *SCC*, III, pp. 161–162.
[53] *loc. cit.*
[54] *SCC*, III, pp. 161–162.
[55] *SCC*, III, p. 159.
[56] Nelson (1968a, 1975).
[57] *idem* (1972a, 1974a).
[58] Durkheim (1915, 1933).
[59] Nelson (1975); Weber (1958).
[60] Nelson (1974a).
[61] *idem* (1974a).
[62] *idem* (1973a).
[63] Pirenne (1956).
[64] For reasons which remain to be discussed in detail on another occasion, Weber placed cities under the rubric, "non-legitimate domination (nicht-legitime Herrschaft)." Weber (1968), p. 1212.
[65] *ibid.* (1968), pp. 1249–1256.
[66] Sutton (1953).
[67] Weber (1968), ch. xvi, p. 1212ff., *ibid.*, ch. xv, *passim*; *idem* (1928), chs. xxviii, xxix.
[68] Sombart (1915); Weber (1928); de Roover (1937).
[69] Nelson (1969a, 1973a), Chenu (1968).
[70] Nelson (1973c).
[71] *idem* (1975).
[72] *idem* (1973c, 1974a).
[73] *idem* (1974a); this is the collective title given to the studies in the two last volumes of the *Gesammelte Aufsätze zur Religionssoziologie*; cf. under Weber below.
[74] Weber (1951), ch. v, vi.
[75] *ibid.*, pp. 151–152.
[76] *idem* (1920, ed. 1958).
[77] *idem* (1951).
[78] *ibid.* pp. 155–159.
[79] *ibid.*, p. 154.
[80] *idem* (1958); *idem* (1951).
[81] *idem* (1958).
[82] Southern (1953, 1970). A particularly helpful and strong discussion of the issues involved in the present essay will be found in Southern's striking title essay of his *Medieval Humanism* (1970), ch. 4, pp. 29–60.
[83] See, esp., Clagett (1959b, 1961, 1968).
[84] Knowles (1963b, 1964).
[85] Chenu (1968, 1971).
[86] Michaud-Quantin (1972a, 1972b).
[87] Haskins (1957, 1961).
[88] White (1971b).
[89] *idem* (1971a).
[90] Nelson (1973d).
[91] *idem* (1969b, 1973a); cf. Chenu (1969).

[92] Nelson (1973a).
[93] *ibid.* (1973a).
[94] *idem* (1973d).
[95] Chenu (1968); Jolivet (1969); Gerber (1970).
[96] Chenu (1968).
[97] Nelson (1974a).
[98] Nelson (1969, 1973a).
[99] Haskins (1960); Sarton (1968), II and III, *passim*.
[100] Crombie (1953), pp. 178–188; Dijksterhuis (1961), pp. 188–193; Crosby (1955), for Bradwardine; C. Wilson (1956), for William Heytesbury.
[101] Nelson (1973a).
[102] Kneale (1962); Knowles (1964).
[103] Abelard (trans. 1971); Adelard (ed. 1934).
[104] Nelson (1969a, 1973a); Chenu (1969).
[105] Close studies of historical semantics are an indispensable component of a comprehensive research program for a comparative historical *differential* sociology of sociocultural process. Such a program involves cross-civilizational investigations of the careers of clusters of key terms with a view to uncovering their workings, analogues, locations in different settings. In this spirit I have been at work for some time on the cross-civilizational analyses of a selected list of axial terms for Western civilization in the 12th and 13th centuries.
[106] Chenu (1968); Michaud-Quantin (1970a, 1970b).
[107] Maine (1861), ch. 2. For some interesting facets of the early history of the notion "person," see Mauss (1938).
[108] *ibid.*, ch. 5, pp. 164–165.
[109] *ibid.*, pp. 163–164.
[110] Pirenne (1956).
[111] cf. Weber (1951), chs. v and vi *passim*, esp. pp. 125–128 for suggestive observations on these issues. Weber notes the failure of Chinese philosophy to give birth to scholasticism "because it was not professionally engaged in logic, as were the philosophies of the Occident and the Middle East." Discussing Confucius he notices the "absence of *speech* as a rational means for attaining political and forensic effects, speech as it was first cultivated in the hellenic *polis*." He continues:

"Such speech could not be developed in a bureaucratic patrimonial state which had no formalized justice.... The Chinese bureaucracy was interested in conventional propriety, and these bonds prevailed and worked in the same direction of obstructing forensic speech. The bureaucracy rejected the argument of 'ultimate' speculative problems as practically sterile. The bureaucracy considered such arguments improper and rejected them as too delicate for one's own position because of the danger of innovations."

[112] Nelson (1973a, 1973d); Grunebaum (1971).
[113] Weber (1946), pp. 328–330; *idem* (1968), III, pp. 1241–1260.
[114] Nelson (1969), *passim*; Weber (1927), pp. 262–263.
[115] Pirenne (1956); Weber (1968), III, p. 1239.
[116] McKeon (1935); Chenu (1969); Le Goff (1964), pp. 422–428.
[117] Panofsky (1967).
[118] Nelson (1969a), esp. pp. xix–xxv; Weber (1928), pp. 262–263.
[119] Weber (1968), pp. 13–31.
[120] Nelson (1972a, 1974a).
[121] Nelson (1973–1974a); Gieben (1964), p. 144, for Grosseteste, *De machina universitatis*,

cf. Sternagel (1966); White (1969).
[122] Sarton (1968), II, *passim*; III, esp. pp. 109–152, 279–330, 399–405, 485–542, 561–569, 611–634, 709–720, 829–861, 985–1023.
[123] Nelson (1973–1974b).
[124] Crombie (1953), *passim*.
[125] *ibid.*; Murdoch (1968); Clagett (1967).
[126] *GT*, pp. 182–184.
[127] Nelson (1973c), pp. 72–74.
[128] *ibid.*, p. 78.
[129] *GT*, pp. 174, 184, 197, 211.
[130] Weber (1946), pp. 328–330; *idem* (1951), pp. 142–145.
[131] cf. Nelson (1974a).
[132] Bellah (1963) in Eisenstadt (1967), pp. 243–251; Nelson (1972b), pp. 125–126.
[133] Weber (1951), chs. v–vi, esp. pp. 150–152.
[134] Chenu (1968); Clagett and others (1961); Southern (1970), Nelson (1973a).
[135] Weber (1968).
[136] Weber (1951), esp. pp. 13–32; *idem* (1968), III, Chs. XV–XVI, *passim*.
[137] Nelson (1973d, 1974a).
[138] Nelson (1969a), pp. 239–246; (1969b); cf. Chenu (1969); Michaud-Quantin (1962).
[139] Few recent writers seem to give enough weight to the distinctions among the *several sources of the several notions* of the law and author of Nature; the Greco-Roman, the Hebrew, early Christian, patristic, scholastic and early modern sources. My own sense is that the more we intensify our research the less certain we become that medieval nominalism and Calvin were as decisive for the ideas of law and order of Nature as Whitehead (1925), Zilsel (1942b), Collingwood (1945, Needham (1962; *SCC*, II, p. 539 ff.), Oakley (1961) and more recently, Oberman (1973), have contended.

The main ground for my conviction arises from the substantial evidence I have already gathered and hope to discuss elsewhere that critical images and ideas of divine guidance and construction of the world of nature and creation long antedate the 14th century. Some bits of proof may be offered from the history of the images of world-machine and the divine geometer-engineer-architect in the medieval and early modern period:

The notion of the world-machine occurs again and again in the 12th and 13th centuries; cf. Sternagel (1966); Nelson (1973a, 1973–1974b). Thus Robert Grosseteste even wrote a treatise, *De Machina Universitatis*, which is briefly discussed in S. Gieben (1964), pp. 144–168; cf. esp. pp. 144, 151, 153 for Grosseteste's exemplarism.

Other sources of images of God as the Divine Artificer shown directly engaged in the geometric construction of the world were set forth with the authority – and through the citation – of scriptural passages: *Wisdom* xi. 20, "Thou hast ordered all things according to measure, number, and weight"; *Proverbs* viii. 27, "He set a compass upon the face of the depth"; *The Revelation of John* xxi, ff. These images and texts have a long history in the early and high Middle Ages. References to the spread of the first of these images will be found in L. White, Jr. (1971b), where emphasis is placed on the Winchester Gospel (*ca.* 1000 A.D.) (*ibid.*, p. 189); also see A. Blunt (1937–1938), who offers a spirited discussion of Blake's representation of Urizen setting compasses on the world.

It must also be noted that a goodly number of thinkers and writers applied ideas of God's omnipotence *to rule out notions of natural law*. Thus it became necessary for Galileo directly to controvert Urban VIII for his (the Pope's) arguments from Divine Omnipotence in the *Dialogue*.... How checkered were the afterlives of these images and proof-texts may be seen in one especially revealing episode. G. B. Riccioli wrote his *Almagestum Novum* at the behest of the hierarchy to defend the inquistorial and papal rulings on Galileo. Nonetheless it carries a frontispiece calling attention to the hand of God ordering all things by measure, number and weight. (Bologna, 1651; cf. ill. in Stimson.)

It is hard to bring this note to a close without calling attention to the provocative conclusion of Needham's last chapter on 'Human Law and the Laws of Nature' in the *GT*. There Needham raises exceptionally challenging questions;

"...But historically the question remains whether natural science could ever have reached its present state of development without passing through a 'theological' stage."

"...The exact degree of subjectivity in the formulations of scientific law has been hotly debated during the whole period from Mach to Eddington, and such questions cannot be followed further here. The problem is whether the recognition of such statistical regularities and their mathematical expression could have been reached by any other road than that which Western science actually travelled. Was perhaps the state of mind in which an egg-laying cock could be prosecuted at law necessary in a culture which would later have the property of producing a Kepler?" (*GT* p. 330)

On the first of Needham's questions above it is now urgent to read the recent essay of Helmut Koester (1968). Identifying Philo "as the crucial and most important contributor to the development of the theory of the natural law," Koester adds:

"Only a philosophical and theological setting in which the Greek concept of nature was fused with the belief in a divine legislator and with a doctrine of the most perfect (written!) law could produce such a theory, and only here could the Greek dichotomy of the two realms of law and nature be overcome. All these conditions are fulfilled in Philo, and the evidence for the development of this theory of the law of nature in Philo is impressive." (*ibid.*, 540).

[140] See, esp., Beaujouan (1957a, 1957b, 1963).

[141] Sarton (1955).

[142] Kneale (1962); Knowles (1964); Chenu (1969).

[143] Nelson (1968b).

[144] Just as this goes to press there comes to hand a suggestive essay by Joseph Levenson which appears to raise doubt on a major theme and conclusion of the present essay; see Levenson (1953). Levenson finds that a number of Ch'ing philosophers remind one of Abelard, but none came so far as Francis Bacon. Levenson writes:

"In European history, divergence from idealism could take the forms of the pre-scientific nominalism of Peter Abelard... as well as the form of Sir Francis Bacon's... inductive empirical science; our Chinese thinkers' affinity was with Abelard, not with Bacon" (pp. 158–159).

Levenson's view seems to me insufficiently morphological. He misses the extent of Abelard's breakthrough of the structures of conscience. Also, he fails to note the importance of Abelard's thrust in the direction of a new logic on intentionality. Lastly, Levenson exaggerates the role of Francis Bacon in the development of high science. In the same spirit, the philosophers described by Levenson hardly seem to parallel Abelard. Comparisons are, indeed, a prime necessity of comparative historical differential sociology; they also, however, regularly involve serious risks.

On the present issue, there is an excellent passage by the Kneales:

"...On the other hand, it is arguable that the exercises of the medieval universities prepared the way for modern science by sharpening men's wits and leading them to think about the methods of acquiring knowledge. For it is certainly a mistake to suppose that all the philosophers of the Middle Ages believed in systems of deductive metaphysics, and that experimental science began quite suddenly when Galileo or some other Renaissance worthy made an observation for the purpose of refuting a generalization of Aristotle or Galen, just as it is wrong to suppose that Luther was the first to suggest reform of the Church." (Kneale (1962), p. 226.)

[145] Nelson (1973a).

BIBLIOGRAPHY

Abelard, Peter (1079–1142): 1960, *Historia Calamitatum*, Paris.

Abelard, Peter (1079–1142): 1962, *Dialectica* (ed. with an intro. by L. de Rijk.), Assen.

Abelard, Peter (1079–1142): 1971, *Ethics* (ed. and trans. with an intro. and notes by D. E. Luscombe), Oxford.

Adelard of Bath (early 12th c.): 1903, 'De Eodem et Diverso' (written between 1105–1116), *B.G.P.T.M.* **IV**, Issue 1 (ed. by H. Willner), Münster.

Adelard of Bath (early 12 th c.): 1934, 'Quaestiones naturales', *B.G.P.T.M.* xxxi, Issue 2 (ed. by M. Müller), Münster.

Al-Ghazali (1058–1111): 1955, *Tahafut al-Falasifah* (Trans. by S. A. Kamah), Lahore, India, also cf. below under Averroes (1959).

Anselm of Canterbury, St. (c. 1033–1109): 1962, *Basic Writings: Proslogium; Monologium; Gaunilon's On Behalf of the Fool, Cur deus homo* (trans. by S. N. Dean), LaSalle, Indiana.

Anselm of Canterbury, St. (c. 1033–1109): 1967, *Truth, Freedom and Evil: Three Philosophical Dialogues* (ed. with rev. trans. by J. Hopkins and H. Richardson), New York.

Averroes (1126–1198): 1959, *Tahafut al-Tahafut* [*The Incoherence of the Incoherence*] (trans. by S. Van den Bergh), 2 vols., London.

Bacon, Roger (c. 1214–1292): 1859, *Opera quaedam hactenus inedita* (ed. by J. S. Brewer), London. Vol. I contains *Opus tertium*; Vol. II. *Opus minus*; III. *Compendium philosophiae*.

Bacon, Roger (c. 1214–1292): 1964 (*ca.* 1267), *The Opus Majus of Roger Bacon* (ed. and intro. by J. H. Bridges), 2 vols. and supplement in Latin, Frankfort.

Barker, Ernest (Sir): 1948, *Traditions of Civility*, Cambridge.

Barth, Karl: 1960, *Anselm: Fides quaerens intellectum: Anselm's Proof of the Existence of God in the Context of His Theological Scheme* (trans. by I. W. Robertson), Richmond, Virginia.

Beaujouan, G.: 1957a, *L'interdépendance entre la science scolastique et les techniques utilitaires (XIIe, XIIIe et XIVe siècles)*, Paris.

Beaujouan, G.: 1957b, *'La science dans l'occident médieval chrétien'*, in Taton, 1963, pp. 468–532.

Beaujouan, G.: 1963, 'Motives and Opportunities for Science in Medieval Universities', in Crombie, 1963, pp. 219–236.

Bellah, Robert: 1967, 'Reflections on the Protestant Ethic Analogy in Asia (1963)', in Eisenstadt (1967), pp. 243–252.

Bennett, Adrian A.: 1967, *John Fryer: The Introduction of Western Science and Technology into Nineteenth Century China*, Cambridge, Mass.

Bernard, Henri: 1941, 'Notes on the Introduction of the Natural Sciences into the Chinese Empire', *Yenching Journal of Social Studies* **III**: 2, pp. 220–241.

Blunt, Anthony: 1938–1939, 'Blake's "Ancient of Days": The Symbolism of the Compasses', *Journal of the Warburg Institute* **2**, pp. 53–63.

Boas, Marie and Hall, A. R.: 1959, 'Tycho Brahe's System of the World', *Occasional Notes of the Royal Astronomical Society* **3**:21, pp. 253–263.

Bochner, S.: 1966, *The Role of Mathematics in the Rise of Science*, Princeton.

Bonaventura, Saint (d. 1274): 1956, *Itinerarium mentis in Deum* (trans. with intro. by P. Boehner), New York.

Burckhardt, Jacob: 1958, *The Civilization of the Renaissance in Italy* (trans. by J. B. Middlemore, intro. by E. Nelson and C. Trinkaus), New York. (Originally 1860.)

Burtt, E. A.: 1932, *The Metaphysical Foundations of Modern Physical Science*, New York.

Chenu, M.-D., O. P.: 1968, *Nature, Man and Society in the Twelfth Century. Essays on New Theological Perspectives in the Latin West* (ed., trans. by J. Taylor and L. K. Little, Preface by E. Gilson), Chicago.

Chenu, M.-D., O. P.: 1969, *L'éveil de la conscience dans la civilisation médiévale* (Conférence Albert-Le-Grand 1968), Montreal and Paris.

Christianson, J. R.: 1973, 'Copernicus and the Lutherans', *Sixteenth Century Journal* **4**:2, pp. 1–10.

Clagett, Marshall: 1955, *Greek Science in Late Antiquity*, New York.

Clagett, Marshall (ed.): 1959a, *Critical Problems in the History of Science*, Madison, Wisconsin.

Clagett, Marshall: 1959b, *The Science of Mechanics in the Middle Ages*, Madison, Wisconsin.

Clagett, Marshall, Post, G., and Reynolds, R., (eds.): 1961, *Twelfth-Century Europe and the Foundations of Modern Society*, Madison, Wisconsin.

Clagett, Marshall: 1967, 'Some Novel Trends in the Science of the Fourteenth Century', in Singleton (ed.) 1967, pp. 275–303.

Clagett, Marshall (ed. and trans.): 1968, *Nicole Oresme and the Medieval Geometry of Qualities and Motions*, Madison, Wisconsin.

Clark, G. N.: 1937, *Science and Social Welfare in the Age of Newton*, London.

Clavelin, Maurice: 1968, 'Galilée et le refus de l'équivalence des hypothèses', *Galilée. Aspects de sa vie et de son oeuvre*, Paris, pp. 127–153.

Cohen, I. B. and Taton, R. (eds.): 1964, *L'aventure de la science: Mélanges Alexandre Koyré*, I (Histoire de la Pensée XII), Paris.

Collingwood, R. G.: 1945, *The Idea of Nature*, New York.

Crombie, A. C.: 1950, 'Galileo's Dialogues concerning the Two Principal Systems', *Dominican Studies* **3**, pp. 105–138.

Crombie, A. C.: 1953, *Robert Grosseteste and the Origins of Experimental Science, 1100–1700*, Oxford.

Crombie, A. C. (ed.): 1959a, *Medieval and Early Modern Science*, 2nd ed., 2 vols., Garden City, New York. (1st ed. 1952.)

Crombie, A. C.: 1959b, 'Commentary on Papers of Rupert Hall and Giorgio de Santillana', in Clagett (ed.) 1959, pp. 66–78.

Crombie, A. C.: 1959c, 'The Significance of Medieval Discussions of Scientific Method for the Scientific Revolution', in Clagett (ed.) 1959, pp. 79–102.

Crombie, A. C.: 1961, *Augustine to Galileo*, Cambridge, Mass.

Crombie, A. C. (ed.): 1963, *Scientific Change* (Symposium on the History of Science, Oxford, 1961), New York.

Crosby, H. Lamar: 1955, *Thomas of Bradwardine, his 'Tractatus de proportionibus'*, Madison, Wisconsin.

D'Elia, P., S. J.: 1960, *Galileo in China: Relations through the Roman College between Galileo and the Jesuit Scientist-Missionaries (1610–1640)*, Cambridge, Massachusetts.

de Rijk, L. M. (ed.): 1962, *Logica modernorum*, vol. I: *On the Twelfth Century Theories of Fallacy*, Assen.

De Roover, Raymond: 1937, 'Aux origines d'une technique intellectuelle: la formation et l'expansion de la comptabilité à partie double', *Annales d'histoire économique et sociale* **IX**, pp. 171–193, 270–298.

Dijksterhuis, E. J.: 1961, *The Mechanization of the World-Picture* (trans. from the Dutch), Oxford.

Drake, Stillman: 1957, *Discoveries and Opinions of Galileo*, New York.

Drake, Stillman: 1967, 'Mathematics, Astronomy, and Physics in the Work of Galileo', in Singleton (ed.) 1967, pp. 305–330.

Drake, Stillman: 1970, *Galileo Studies. Personality, Tradition, and Revolution*, Ann Arbor, Michigan.

Duhem, Pierre: 1905a, *L'évolution de la Mécanique*, Paris.

Duhem, Pierre: 1905b (1954), *La théorie physique son objet, sa structure* (trans. by Philip P. Wiener: *The Aim and Structure of Physical Theory*), Princeton.

Duhem, Pierre: 1906–13, *Études sur Léonard de Vinci*, 3 vols., Paris (reprinted 1955).

Duhem, Pierre: 1913–59, *Le Système du monde: Histoire des doctrines cosmologiques de Platon à Copernic*, 10 vols., Paris.

Duhem, Pierre: 1969, *To Save the Phenomena: An Essay on the Idea of Physical Theory from Plato to Galileo* (trans. by E. Doland and C. Maschler, intro. by S. L. Jaki), Chicago. (Original French printing, 1908.)

Durkheim, Emile: 1893, *The Division of Labor in Society* (trans. by G. Simpson 1933), New York.

Durkheim, Emile: 1912, *The Elementary Forms of Religious Life* (trans. by J. W. Swain 1915) London.

Duyvendak, J. J. L.: 1948, 'P. D'Elia's *Galileo in China*', *T'oung Pao* **38**, pp. 321–329.

Edgerton, Samuel Y., Jr.: 1966 'Alberti's *Perspective*: A New Discovery and a New Evaluation,' *Art Bulletin*, Sept.–Dec., **XLVIII**, pp. 367–378.

Eisenstadt, S. N.: 1965, 'Transformation of Social, Political, and Cultural Orders in Modernization', *American Sociological Review* **30**, pp. 659–673.

Eisenstadt, S. N.: 1968a, 'The Protestant Ethic Thesis in an Analytical and Comparative Framework', in Eisenstadt (ed.), 1968b, pp. 3–45.

Eisenstadt, S. N. (ed.): 1968b, *The Protestant Ethic and Modernization. A Comparative View*, New York.

Forest, Aimé, van Steenberghen, F., and de Gandillac, M.: 1951, *Le mouvement doctrinale de IXe au XIVe siècle* (Histoire de l'Église depuis les origines... jusqu'à nos jours, ed. by A. Fliche and V. Martin, vol. 13), Paris.

Foscarini, P.: 1615, *Lettera sopra l'opinione de' Pittagorica e del Copernico della mobilita della terra*, Naples.

Foster, Michael B.: 1934, 'The Christian Doctrine of Creation and the Rise of Modern Natural Sciences', *Mind* **43**, pp. 446–468.

Foster, Michael B.: 1936, 'Christian Theology and the Modern Science of Nature', *Mind* **45**, pp. 1–28.

Galileo: (1615–1616) 1932, 'Considerazioni circa l'opinione Copernicana', *Le Opere di Galileo Galilei*, vol. V, Ristampa della edizione nazionale, Firenze, pp. 349–70. (Originally 1615–1616).

Galileo: (1632) 1953a, *Dialogue concerning the Two Chief World Systems – Ptolemaic and Copernican* (trans. by Stillman Drake, foreword by Albert Einstein), Berkeley, California.

Galileo: 1953b, *Opere*. A cura di Ferdinando Flora, Milan (La letteratura Italiana. Storia e testi, vol. 34, tomo I.)

Galileo: 1965, *Sidereus Nuncius. Nachricht von neuen Sternen* (ed. and intro. by Hans Blumenberg), Frankfurt am Main.

Garin, Eugenio: 1971, 'A Proposition de Copernico... (Schede III)', *Rivista di critica di storia di filosofia* **XXVI**, pp. 81–96.

Gerber, Uwe: 1970, *Disputatio als Sprache des Glaubens*, Zürich.

Ghellinck, Joseph de: 1946, *L'essor de la littérature latine au XIIe siècle*, Museum Lessianum, Section historique, iv–v, Brussels.

Ghellinck, Joseph de: 1948, *Le mouvement théologique du XIIe siècle. Nouvelle édition*, Museum Lessianum, Section historique, x, Brussels.

Gieben, Servus, O. F. M., Cap.: 1964, 'Traces of God in Nature according to Robert Grosseteste', *Franciscan Studies* **24**, 144–158.

Glacken, Clarence, J.: 1967, *Traces on the Rhodian Shore*, Berkeley and Los Angeles, California.

Grant, Edward C.: 1962a, 'Hypotheses in Later Medieval and Early Modern Science', *Daedalus* **91**:3, pp. 612–616 (cf. B. Nelson, 1961).

Grant, Edward C.: 1962b, 'Late Medieval Thought, Copernicus and the Scientific Revolution', *Journal of the History of Ideas* **23**, pp. 197–220.

Grant, Edward C.: 1971, *Physical Science in the Middle Ages*, New York.

Grosseteste, Robert: 1963, *Commentarius in VIII Libros Physicorum Aristotelis* (ed. by Richard C. Dales), Boulder, Colorado, (Originally *ca.* 1228–1233).

Grossmann, Henryk: 1935, 'Die gesellschaftlichen Grundlagen der mechanistischen Philosophie und die Manufaktur', *Zeitschrift für Sozialforschung* **4**, pp. 161–231.

Grunebaum, G. E. von: 1962, *Modern Islam*, Berkeley, California.

Grunebaum, G. E. von (ed.): 1968, *The Islamic World*, London.

Grunebaum, G. E. von: 1970a, *Logic in Classical Islamic Culture*, Wiesbaden.

Grunebaum, G. E. von: 1970b, *Classical Islam*, Chicago.

Grunebaum, G. E. von: 1971, *Mediaeval Islam: A Study in Cultural Orientation*, 2nd ed., Chicago and London.

Hall, A. Rupert: 1954, *The Scientific Revolution. 1500–1800*, London. (2nd. ed., 1962).

Hall, A. Rupert: 1959, 'The Scholar and the Craftsman in the Scientific Revolution', in M. Clagett (ed.) pp. 3–21.

Hall, A. Rupert: 1963a, *From Galileo to Newton: 1630–1720*, New York.

Hall, A. Rupert: 1963b, 'Merton Revisited or Science and Society in the Seventeenth Century', *History of Science* **2**, pp. 1–16.

Hall, Marie Boas (ed.): 1970, *Nature and Nature's Laws. Documents of the Scientific Revolution*, New York.

Haskins, C. H.: 1957, *The Rise of the Universities*, Ithaca, New York.

Haskins, C. H.: 1960, *Studies in the History of Mediaeval Science*, New York.

Haskins, C. H.: 1961, *The Renaissance of the 12th Century*, New York (originally published 1927.)

Hellman, C. Doris: 1963, 'Was Tycho Brahe as Influential as He Thought?', *British Journal for the History of Science*, **1**, pt. IV:4, December, pp. 295–324.

Hellman, C. Doris: 1970, 'Tycho Brahe', *Dictionary of Scientific Biography* (ed. by C. C. Gillispie), vol. II, pp. 401–516.

Hessen, B.: 1931, 'The Social and Economic Roots of Newton's "Principia"', in *Science at the Crossroads*, London, pp. 147–212. Reprint, with new Introduction by R. S. Cohen, New York 1971.

Hine, William L.: 1973, 'Mersenne and Copernicanism', *Isis* **64**:221, March, pp. 18–32.

Hofmann, Rudolf: 1941, *Die Gewissenslehre des Walter von Brugge O. F. M. und die Entwicklung der Gewissenslehre in der Hochscholastik* (Begründet von Clemens Baeumker), *BGPTM* **XXXVI**, issue 5–6, Münster.

Holton, Gerald (ed.): 1965, *Science and Culture*, Boston.
Holton, Gerald (ed.): 1973, *Thematic Origins of Scientific Thought: Kepler to Einstein*, Cambridge.
Horton, Robin: 1971, 'African Traditional Thought and Western Science', in *Rationality* (ed. by B. Wilson), pp. 131–171.
Husserl, Edmund: 1970, *The Crisis of European Sciences and Transcendental Phenomenology* (trans. and intro. by David Carr), Evanston, Illinois.
Jolivet, Jean: 1969, *Arts du langage et théologie chez Abelard, Paris.*
Jones, Christine: *See* under Schofield.
Kapp, Ernst: 1942, *Greek Foundations of Traditional Logic*, New York.
Katzenellenbogen, Adolf: 1959, *The Sculptural Program of Chartres Cathedral: Christ Mary, Ecclesia*, Baltimore.
Kepler, Johannes: (1600–01) 1967, 'Apologia Tychonis contra Ursum', Landmarks of Science New York. (Readex Microprint from *Joannis Kepleri astronomia opera omnia* **I**, Frankfurt, 1858, pp. 216–287.)
Kirk, Kenneth E.: 1925, *Ignorance, Faith and Conformity*, London. (Reprinted 1933.)
Kirk, Kenneth E.: 1927, *Conscience and Its Problems. An Introduction to Casuistry*, London.
Kirk, Kenneth E.: 1937, *The Vision of God*, London.
Kline, Morris: 1972, *Mathematical Thought from Ancient to Modern Times*, New York.
Kneale, William and Kneale, Martha: 1962, *The Development of Logic*, Oxford.
Knowles, David, Dom: 1963a, *The Historian and Character*, New York.
Knowles, David, Dom: 1963b, *Saints and Scholars: Twenty-five Medieval Portraits*, Cambridge.
Knowles, David, Dom: 1964, *The Evolution of Medieval Thought*, New York.
Koestler, Arthur: 1959, *The Sleepwalkers*, New York.
Koester, Helmut: 1968, 'ΝΟΜΟΣ Φ ΥΣΕΩΣ: The Concept of Natural Law in Greek Thought,' in *Religions in Antiquity. Essays in Memory of Erwin Ramsdell Goodenough* (ed. by Jacob Neusner), Leiden, pp. 521–541.
Koyré, Alexandre: 1949, 'Le vide et l'espace infini au XIVe siècle', *Archives d'histoire littéraire et doctrinale du moyen-âge* **24**, pp. 49–91.
Koyré, Alexandre: 1957, 'Galileo and Plato', in P. P. Wiener and A. Nolan (eds.), pp. 147–175.
Koyré, Alexandre: 1964a, *see* under I. B. Cohen and R. Taton.
Koyré, Alexandre: 1964b, 'The Exact Sciences', in R. Taton, ed. (1964), pp. 11–104.
Koyré, Alexandre: 1965, *Newtonian Studies*, London.
Koyré, Alexandre: 1968, *Metaphysics and Measurement: Essays in the Scientific Revolution*, Cambridge, Massachusetts.
Kwok, D. W. Y.: 1965, *Scientism in Chinese Thought: 1900–1950*, New Haven, Connecticut.
Le Goff, Jacques: 1964, *La civilisation de l'occident médiéval*, Paris.
Lenoble, Robert: 1943, *Mersenne et la naissance du mécanisme*, Paris.
Levenson, Joseph R.: 1953, 'The Abortiveness of Empiricism in Early Ch'ing Thought', *Far Eastern Quarterly* **XIII**: 1, pp. 155–165.
Lindberg, David C.: 1970, *see* under J. Pecham.
Lopez, Robert S.: 1971, *The Commercial Revolution of the Middle Ages, 950–1350*. Englewood Cliffs, New Jersey.
Lottin, Dom O.: 1942–1960, *Psychologie et morale aux douzième et treizième siècles*, 6 vols., Louvain.
Maier, Annaliese: 1964–1967, *Ausgehendes Mittelalter. Gesammelte Aufsätze zur Geistesgeschichte des 14. Jahrhunderts.* (vol. I, 1964; vol. II, 1967), Rome.

Maier, Annaliese: 1956–1968, *Studien zur Naturphilosophie der Spätscholastik* (2nd ed., enlarged), Rome. (vol. I: *Die Vorläufer Galileis im 14. Jahrhundert*; vol. II; *Zwei Grundprobleme der scholastischen Naturphilosophie das Problem der intensiven Grösse die Impetustheorie*: vol. III: *An der Grenze von Scholastik und Naturwissenschaft. Die Struktur der materiellen Substanz das Problem der Gravitation die Mathematik der Formlatituden*; vol. IV: *Metaphysische Hintergründe der spätscholastischen Naturphilosophie*; vol. V.: *Zwischen Philosophie und Mechanik. Studien zur Naturphilosophie der Spätscholastik*.

Maine, Henry J. S. (Sir): 1963, *Ancient Law*, Boston. (Originally published 1861.)

Mathias, Peter (ed): 1972, *Science and Society: 1600–1900*, Cambridge.

Mauss, Marcel: 1938, 'Une categorie de l'esprit humain, la notion de personne celle de "Moi"', in *Sociologie et Anthropologie*, Paris. (English trans. by L. Krader, 1968, in *Histories, Symbolic Logics, Cultural Maps*, Special issue of the *Psychoanalytic Review*, B. Nelson (ed.), **55**, pp. 457–481.

Mauss, Marcel: 1960, *Sociologie et anthropologie* (ed. with intro. by C. Lévi-Strauss), Paris.

Mauss, Marcel: 1968–1968, *Oeuvres*, 3 vols. (ed. by V. Karady), Paris.

McKeon, Richard: 1935, 'Renaissance and Method in Philosophy', *Studies in the History of Ideas*. **III**, pp. 37–114.

McMullin, Ernan: 1967, 'Empiricism and the Scientific Revolution', in Singleton (ed.), 1967, pp. 331–369.

McMullin, Ernan (ed.): 1968, *Galileo, Man of Science*, New York.

Merton, Robert: 1968, *Social Theory and Social Structure*, New York. (Originally Published 1949.)

Merton, Robert: 1970, *Science, Technology and Society in Seventeenth-Century England* (with a new intro. by the author), New York. (Originally published 1938.) *See* also under B. Nelson (1972a).

Michaud-Quantin, Pierre: 1962, *Sommes de casuistique et manuels de confession au Moyen-Age*, Montreal.

Michaud-Quantin, Pierre: 1970, *Universitas: Expressions du mouvement communitaire dans le Moyen-Age latin*, Paris.

Michaud-Quantin, Pierre: 1970b, *Etudes sur le vocabulaire philosophique du moyen-âge* (with the collaboration of Michel Lemoine), Rome.

Murdoch, John: 1964, 'Superposition, Congruence and Continuity in the Middle Ages', in I. B. Cohen and R. Taton (1964a) vol. I, pp. 416–441.

Murdoch, John: 1969, 'Mathesis in philosophiam scholasticam introducta', in *Arts libéraux et philosophie au moyen-âge*. Montreal, Paris.

Nakayama, S.: 1972, 'Science and Technology in China', in A. Toynbee (1972).

Needham, Joseph: 1954–1971, *Science and Civilisation in China*, 4 vols. in 7 parts, New York. *In Progress*.

Needham, Joseph: 1963, 'Poverties and Triumphs of the Chinese Scientific Traditions', in Crombie (ed.), 1963, pp. 117–153.

Needham, Joseph: 1969a, *Within the Four Seas: The Dialogue of East and West*, London.

Needham, Joseph: 1969b, *The Grand Titration: Science and Society in East and West*, London.

Needham, Joseph: 1970, *Clerks and Craftsmen in China and the West*, New York.

Nelson, Benjamin: 1961, 'Comments on Edward Grant's Hypotheses in Late Medieval and Early Modern Physics', *Daedalus*, Summer (issued as vol. 91, no. 3, *Proceedings of the American Academy of Arts and Science*), pp. 612–616.

Nelson, Benjamin: 1964, 'Actors, Directors, Roles, Cues, Meanings, Identities: Further

Thoughts on "Anomie"', in *Psychoanalytic Review* **51**:1, pp. 135–160.

Nelson, Benjamin: 1965a, 'Self-Images and Systems of Spiritual Direction in the History of European Civilization', in *The Quest for Self-Control: Classical Philosophies and Scientific Research* (ed. S. Z. Klausner) New York, pp. 49–103.

Nelson, Benjamin: 1965b, '"Probabilists," "Anti-Probabilists" and the Quest for Certitude in the 16th and 17th Centuries', in *Actes du Xme congrès internationale d'histoire des sciences* [*Proceedings of the Xth International Congress for the History of Science*] vol. 1, Paris, pp. 267–273. (Original draft, 1962)

Nelson, Benjamin: 1968a, 'Scholastic *Rationales* of "Conscience", Early Modern Crises of Credibility, and the Scientific-Technocultural Revolutions of the 17th and 20th Centuries', in *Journal of the Society for the Scientific Study of Religion* **VII**:2, pp. 157–177.

Nelson, Benjamin: 1968b, 'The Early Modern Revolution in Science and Philosophy: Fictionalism, Probabilism, Fideism, and Catholic "Prophetism"', in *Boston Studies in the Philosophy of Science* (ed. by R. S. Cohen and M. Wartofsky) vol. 3, Dordrecht, pp. 1–40.

Nelson, Benjamin: 1969a, *The Idea of Usury: From Tribal Brotherhood to Universal Otherhood*, 2nd ed., enlarged, Chicago (originally 1949).

Nelson, Benjamin: 1969b, '*Conscience* and the Making of Early Modern Cultures: *The Protestant Ethic* beyond Max Weber', *Social Research* **36**: 1, pp. 4–21.

Nelson, Benjamin: 1969c, 'Metaphor in Sociology. Review of Robert Nisbet's Social Change and History', in *Science Magazine* **116**, pp. 1498–1500.

Nelson, Benjamin: 1971, 'The Medieval Canon Law of Contracts, Renaissance "Spirit of Capitalism," and the Reformation "Conscience": A Vote *for* Max Weber', in *Philomathes: Studies and Essays in the Humanities in Honor of Philip Merlan* (ed. by R. B. Palmer and R. Hamerton-Kelly), pp. 525–548.

Nelson, Benjamin: 1972a, 'Review of Robert Merton's *Science, Technology and Society in Seventeenth-Century England*', *American Journal of Sociology* **78**:1, pp. 223–231.

Nelson, Benjamin: 1972b, 'Communities, Societies, Civilizations: Post-Millennial Views on the Faces and Masks of Time', in *Social Development* (ed. M. Stanley), pp. 105–133.

Nelson, Benjamin: 1973a, '*Eros, Logos, Nomos, Polis:* Their Shifting Balances in the Vicissitudes of Civilizations', in *Social Research in the Contemporary Study of Religion* (ed. A. Eister). *In Press.*

Nelson, Benjamin: 1973b, 'Copernicus and the Quest for Certitude: East and West', in *Vistas in Astronomy* **15**, Smithsonian Institution Publication. *In Press.*

Nelson, Benjamin: 1973c, 'Weber's *Protestant Ethic*: Its Origins, Wanderings, and Foreseeable Futures', in *Beyond the Classics* (ed. C. Glock and P. Hammond), New York, pp. 71–130.

Nelson, Benjamin: 1973d, 'Civilizational Complexes and Intercivilizational Encounters', in *Sociological Analysis* **34**:2, pp. 79–105.

Nelson, Benjamin: 1973e, 'Civilizational Patterns and Intercivilizational Encounters. A Select Annotated Bibliography' (with the Collaboration of Donald Nielsen), *Bulletin for the International Society for the Comparative Study of Civilizations*, Geneseo, New York, pp. 3–15.

Nelson, Benjamin: 1973–1974a, 'The Work of God's Hand, The Books of Nature and Revelation, and the Machine of the World: 11th–17th Centuries', presented to the *Boston Colloquium for the Philosophy of Science*, Tuesday, April 3, 1973. *Ms. in progress.*

Nelson, Benjamin: 1973–1974b, 'The Quest for Certainty and Certitude, the Books of Revelation, Nature and Conscience: Forgotten Backgrounds of the Early Modern Scientific Revolution', paper presented to a Collegium at the International Symposium,

"The Nature of Scientific Discovery", commemorating the five-hundredth anniversary of the birth of Nicolaus Copernicus (1473–1973), April 25, 1973, sponsored by the Smithsonian Institution and the National Academy of Sciences, Washington, D. C. *In Press.*

Nelson, Benjamin: 1974a, 'On the Shoulders of the Giants of the *Comparative* Historical Sociology of "Science" – in Civilizational Perspective', in *Social Processes of Scientific Development* (ed. by R. Whitley), pp. 13–20.

Nelson, Benjamin: 1975, 'Max Weber's "Author's Introduction" (1920): A Master Clue to his Main Aims', *Forthcoming.*

Nordström, Johann: 1933, *Moyen Age et Renaissance, Essai Historique* (trans. T. Hammer), Paris.

Oakley, Francis P.: 1961, 'Christian Theology and the Newtonian Science: The Rise of the Concept of the Laws of Nature', in *Church History* **30**, pp. 433–457.

Oberman, Heiko: 1963, *The Harvest of Medieval Theology: Gabriel Biel and Late Medieval Nominalism*, Cambridge.

Oberman, Heiko: 1973, 'Reformation and Revolution: Copernicus' Discovery in an Era of Change', *ms.*, Address delivered at Commemoration of Fifth Centenary of the Birth of Copernicus, Co-sponsors: National Academy of Sciences and Smithsonian Institution, April 1973. *In Press.*

Panofsky, Erwin: 1967, *Gothic Architecture and Scholasticism, New York.* (Originally 1957).

Pecham, John: 1970, *John Pecham and the Science of Optics. "Perspectiva Communis"*, (ed. and intro., trans. and notes by D. C. Lindberg), Madison, Wisconsin.

Petrus Hispani: 1947, *Summulae Logicales* (ed. by I. M. Bocheński), Turin.

Petrus Peregrinus (de Maricourt): 1898, *Epistola de magnete* (*ca.* 1269) (ed. by G. Hellmann). Berlin; 1902, (trans. by S. P. Thompson) London.

Pirenne, Henri: 1956, *Economic and Social History of Medieval Europe*, New York.

Ramon de Sabunde: 1966, *Theologia naturalis seu Liber creaturarum* (intro. by Friedrich Stegmüller), Stuttgart.

Ricci, Matthew: 1953, *China in the Sixteenth Century: The Journals of Matthew Ricci: 1583–1610* (trans. from Latin by L. J. Gallagher, S. J.), New York.

Rosen, Edward: 1957, 'The Ramus-Rheticus Correspondence', (ed. by Wiener and Noland) pp. 287–292.

Rosen, Edward: 1959, *Three Copernican Treatises*, New York.

Rosen, Edward: 1964, 'Renaissance Science as Seen by Burckhardt and His Followers', (ed. by T. Helton) pp. 77–103.

Rosen, Edward: 1965, 'The Debt of Classical Physics to Renaissance Astronomers, particularly Kepler', in *Proceedings of the Xth International Congress for the History of Science*, Paris.

Sarton, George: 1955, *Appreciation of Ancient and Medieval Science During the Renaissance (1450–1600)*, Philadelphia.

Sarton, George: 1968, *Introduction to the History of Science*, 3 vols. in 5 parts, Baltimore (Originally 1927–1948).

Schapiro, Herman (ed.): 1964, *Medieval Philosophy: Selected Readings from Augustine to Buridan*, New York.

Schofield, Christine (Jones): 1965, 'The Geoheliocentric Mathematical Hypotheses in Sixteenth Century Planetary Theory', *British Journal for the History of Science* **2**, pp. 291–296.

Singleton, Charles S. (ed.): 1967, *Art, Science and History in the Renaissance*, Baltimore.

Sivin, Nathan: 1968, 'Review of J. Needham, *Science and Civilization in China*, Vol. 4, Part II,' *Journal of Asian Studies* **XXVII**:4, pp. 859–864.

Sivin, Nathan: 1969, *Cosmos and Computation in Early Chinese Mathematical Astronomy*, Leiden.
Sivin, Nathan: 1970, 'Review of A. A. Bennett, *John Fryer: The Introduction of Western Science and Technology into Nineteenth Century China*', *Technology Review*, March, pp. 17–18.
Sivin, Nathan: 1971, 'Review of J. Needham, *The Grand Titration*', *Journal of Asian Studies* **XXX**:4, pp. 970–873.
Sivin, Nathan: 1972, 'Review of J. Needham, *Science and Civilization in China*, Vol. 4, Part III', *Scientific American*, **226**:1, pp. 113–118.
Sivin, Nathan: 1973a, 'Copernicus in China', in *Union internationale d'histoire et de philosophie des sciences: Comité Nicolas Copernic: Colloquia II: Etudes sur l'audience de la théorie heliocentrique. Conférence du Symposium de l'UIHPs. Torún 1973*, Wroclaw, Warsawa.
Sivin, Nathan and Nakayama, H.: 1973b, *Traditional Science in China*, Cambridge, Mass.
Sombart, Werner: 1915, *The Quintessence of Capitalism. A Study of the History and Psychology of the Modern Business Man* (trans. by M. Epstein), New York.
Southern, R. W.: 1953, *The Making of the Middle Ages*, London.
Southern, R. W.: 1970, *Medieval Humanism and Other Studies*, New York.
Spektorsky, E.: 1910–1917, *The Problem of Social Physics in the 17th Century* (Russian), 2 vols., vol. I: Warsaw 1910; vol. II: Kiev 1917.
Sternagel, Peter: 1966, *Die artes mechanicae im Mittelalter. Begriffs- und Bedeutungsgeschichte bis zum Ende des 13. Jahrhunderts*, Källmünz über Regensburg.
Stimson, Dorothy: 1917, *The Gradual Acceptance of the Copernican Hypothesis, Diss.*, Hanover, New Hampshire.
Strong, Edward K.: 1957, 'Newton's "Mathematical Way"', (ed. by P. P. Wiener and A. Noland), pp. 412–432.
Strong, Edward K.: 1959, 'Hypotheses non Fingo', in *Men and Moments in the History of Science* (ed. by H. M. Evans), pp. 162–176.
Sutton, Robert: 1953, 'The phrase *libertas philosophandi*', *Journal of the History of Ideas* **14**, pp. 310–316.
Taton, René (ed.): 1963, *Ancient and Medieval Science*, New York.
Taton, René (ed.): 1964, *The Beginning of Modern Science from 1450 to 1800* (trans. by A. J. Pomerans), New York.
Toynbee, Arnold: 1972, *Half the World. The History and Culture of China and Japan*, London.
Trémontant, C.: 1955, *Études de métaphysique biblique, Paris.*
Trémontant, C.: 1964, *La métaphysique du Christianisme et la crise du XIII siècle*, Paris.
von Gebler, Karl: 1879, *Galileo Galilei and the Roman Curia, London.*
Wallace, William A., O.P.: 1959, *The Scientific Methodology of Theodore of Freiberg*, Fribourg.
Wallace, William A., O.P.: 1962, *The Role of Demonstration in Moral Theology*, Washington, D.C.
Weber, Max: 1920–1, *Gesammelte Aufsätze zur Religionssoziologie*, Tübingen. 3 vols.
Weber, Max: 1928, *General Economic History* (trans. by F. H. Knight), Glencoe, Illinois (Originally 1923).
Weber, Max: 1946, *From Max Weber.* (ed. by H. H. Gerth and C. W. Mills), New York.
Weber, Max: 1949, *Max Weber on the Methodology of the Social Sciences* (trans. and ed. by E. Shils and H. Finch), Glencoe, Illinois.
Weber, Max: 1951, *The Religion of China: Confucianism and Taoism* (trans. by H. H. Gerth), Glencoe, Illinois (Originally 1916–17).

Weber, Max: 1952, *Ancient Judaism* (trans and ed. by H. H. Gerth and D. Martindale), Glencoe, Illinois (Originally 1917–1919).
Weber, Max: 1958a, *The Protestant Ethic and the Spirit of Capitalism* (trans. by T. Parsons; Foreword by R. H. Tawney), New York and London. (Originally 1904–1905).
Weber, Max: 1958b, *The Religion of India* (trans. and ed. by H. H. Gerth and D. Martindale).
Weber, Max: 1962, *The City* (trans. by G. Newirth and D. Martindale), New York (Originally 1921).
Weber, Max: 1964, *The Sociology of Religion* (trans by E. Fischoff; intro. by T. Parsons), Boston (Originally 1922).
Weber, Max: 1968, *Economy and Society*, 3 vols. (trans and ed. by G. Roth and C. Wittich), Totowa, New Jersey.
Westman, Robert S.: 1971, *Kepler's Adoption of the Copernican Hypothesis*, Ann Arbor, Michigan, Univ. Microfilm. Diss.
White, Lynn, Jr.: 1947, 'Natural Science and Naturalistic Art in the Middle Ages', in *American Historical Review* **III**, pp. 421–435.
White, Lynn, Jr.: 1962, *Medieval Technology and Social Change*, Oxford.
White, Lynn, Jr.: 1964, 'Review of Needham's Work', *Isis* **LVIII**, pp. 248–251.
White, Lynn, Jr.: 1969, *Machina ex Deo: Essays in the Dynamism of Western Culture*, Cambridge, Massachusetts.
White, Lynn, Jr.: 1971a, 'Medieval Borrowings from Further Asia,' *Medieval and Renaissance Studies* **5**, pp. 3–26.
White, Lynn, Jr.: 1971b, 'Cultural Climates and Technological Advance in the Middle Ages', in *Viator. Medieval and Renaissance Studies* **2**, pp. 171–201.
Whitehead, Alfred North: 1927, *Symbolism: Its Meaning and Effect*, New York.
Whitehead, Alfred North: 1956, *Adventures of Ideas*, New York. (Originally 1933).
Whitehead, Alfred North: 1968, *Science and the Modern World (Lowell Lectures 1925)*, New York. (Originally 1925).
Whitehead, Alfred North: 1971, *The Concept of Nature (The Tarner Lectures 1919)*.
Wiener, P. P. and A. Noland (eds.): 1957, *Roots of Scientific Thought: A Cultural Perspective*, New York.
Wilson, Curtis: 1956, *William Heytesbury: Medieval Logic and the Rise of Mathematical Physics*, Madison, Wisconsin.
Wittfogel, Karl: 1931, *Wirtschaft und Gesellschaft Chinas*. Leipzig.
Wittfogel, Karl and Fêng Chia-Shêng: 1949, *History of Chinese Society: Liao, 907–1125*, Philadelphia.
Wittfogel, Karl and Fêng Chia-Shêng: 1957, *Oriental Despotism: A Comparative Study of Total Power*, New Haven, Connecticut.
Wolfson, Harry: 1964, 'The Controversy over Causality within the Kalam' (Cohen and Taton, 1964), pp. 602–618.
Wong, George H. S.: 1963a, 'Some Aspects of Chinese Science Before the Arrival of the Jesuits,' in *Chung Chi Journal* **2**, pp. 169–180.
Wong, George H. S.: 1963b, 'China's Opposition to Western Science during Late Ming and Early Ching', in *Isis* **54**, pp. 29–49.
Wrightsman, A. B.: 1970, *Andreas Osiander and Lutheran Contributions to the Copernican Revolution*, University Microfilms (Diss.), Ann Arbor, Michigan.
Zeuthen, H. B.: 1966, *Geschichte der Mathematik im 16. und 17. Jahrhundert*, New York and Stuttgart.
Zilsel, Edgar: 1942a, 'The Sociological Roots of Science', in *American Journal of Sociology* **47**, pp. 544–562.

Zilsel, Edgar: 1942b, 'The Genesis of the Concept of Physical Law', *Philosophical Review* **51**, pp. 245–279.

Zilsel, Edgar: 1942c, 'The Genesis of the Concept of Scientific Progress', in *Journal of the History of Ideas* **6**, pp. 325–349. (Reprinted in Wiener and Noland 1957, pp. 251–275).

Zilsel, Edgar: 1957a, 'The Origins of Gilbert's Scientific Method', in Wiener and Noland (1957), pp. 219–250.

Zilsel, Edgar: 1957b, 'Copernicus and Mechanics', in Wiener and Noland (1957), pp. 276–280.

Zubov, V. P.: 1968, *Leonardo da Vinci* (trans. by David H. Kraus, Foreword by M. P. Gilmore), Cambridge, Mass.

Footnote added in proof. The *full text* of the Anti-Copernican statement by Giovanni Maria Tolosani described above has just come to hand *(13.3.1974)*. The new essay by Eugenio Garin of Florence is entitled 'Alle origini della polemica anticopernicana' and is to be found in *Studia Copernicana* **VI**, Colloquia Copernicana II; Wroclaw / Warszawa / Kraków / Gdánsk: Polska Akademia Nauk, 1973, pp. 31–42. The text of Tolosani's *Opusculum quartum* will be found on pp. 32–42 of Garin's essay.

Only a very few points need to be added here to the description of the work which will be found in footnote 9 above. It will be noted that Tolosani places strong emphasis from the very outset on the *Book of Genesis* and the description of the creation of 'the whole world and its parts'. Scriptural passages believed to describe the creation are set forth throughout the first chapter of the *Opusculum*. As we have already noted, Tolosani acknowledges Copernicus' learning and ability but takes him to task for not explicitly responding to St. Thomas Aquinas' arguments and objections against the Pythagorean view. Tolosani criticized Copernicus also for not truly knowing the '*rationes*' of Aristotle and Ptolomy and for not confuting them in his own work. Toward the close of his third chapter Tolosani takes direct exception to some of Copernicus' physical ideas. The *Opusculum* closes on an accusatory note with Tolosani insisting on Copernicus' 'intolerable errors against divine letters'. He explains that the '*magister Sacri et apostolici Palatii*' had been prevented by death from doing his own confutation of the *De revolutionibus*

PART VII

UNITY OF SCIENCE

(*Chairman:* ERNEST NAGEL)

KENNETH F. SCHAFFNER

THE UNITY OF SCIENCE AND THEORY CONSTRUCTION IN MOLECULAR BIOLOGY*

1. INTRODUCTION

Though philosophers have many times directed their attention at the problems of theory construction, theory testing, and the growth of science, they have been most often preoccupied with these issues in the context of physics.[1] Thus there are a number of important and influential works in the philosophy of science which take actual science seriously,[2] such as P. Duhem's *The Aim and Structure of Physical Theory*, H. Poincaré's *Foundations of Science*, K. R. Popper's *The Logic of Scientific Discovery*, and T. S. Kuhn's *The Structure of Scientific Revolutions*, but which have based their metascientific analyses exclusively on examples of theories drawn from the physical sciences.[3]

It would seem, in the light of the considerable recent progress of biology – especially in the area of molecular genetics – that philosophers ought to take another look at the issue of theory construction in this area, and determine whether or not molecular genetics does construct theories. If this does turn out to be the case, we might well consider the *logic* of theory construction, and the closely associated problems of the logic of theory testing and the development of theories in this science. This paper represents an attempt to do this, and also to assess the implications of theory construction in molecular genetics for various interpretations of the Unity of Science thesis.

What I shall do first is to consider, in some detail, how a highly specific, most influential, and highly corroborated theory emerged in molecular genetics. This is the Jacob-Monod operon theory of gene control. The theory was first proposed in 1960–61 and has, since that time, resulted in both a deeper understanding of gene action and a redirection of research in both molecular biology and embryology.

R. J. Seeger and R. S. Cohen (eds.), Philosophical Foundations of Science, 497–533.

2. The operon theory

Before turning to the history of the development of the operon theory, I wish to discuss the theory as it was proposed in its fully articulated 1961 form, with its interacting control circuit consisting of a regulator gene, a repressor, an inducer or co-repressor, an operator, and the operator's associated structural genes. (The last two elements constitute what was termed the 'operon'.) Then I want to argue that the operon theory, or 'model', is a paradigm case of a molecular biological theory.

2.1. *The 1961 Form of the Jacob-Monod Operon Theory*

The operon theory was proposed to account for the ability of some forms of the bacterium *Escherischia coli*, or *E. coli*, when placed in a lactose environment and lacking other carbon sources, to begin *de novo* synthesis of two enzymes which are essential for the digestion of the lactose. The theory in the form in which it was presented in Jacob and Monod's (1961) now classic review article was articulated as follows:

A convenient method of summarizing the conclusions derived in the preceeding sections of this paper will be to organize them into a model designed to embody the main elements which we were led to recognize as playing a specific role in the control of protein synthesis; namely, the structural, regulator and operator genes, the operon, and the cytoplasmic repressor. Such a model could be as follows:

The molecular structure of proteins is determined by specific elements, the *structural genes*. These act by forming a cytoplasmic 'transcript' of themselves, the structural messenger, [mRNA] which in turn synthesizes the protein. The synthesis of the messenger by the structural gene is a sequential replicative process, which can be initiated only at certain points on the DNA strand, and the cytoplasmic transcription of several, linked, structural genes may depend upon a single initiating point or *operator*. The genes whose activity is thus coordinated form an *operon*.

The operator tends to combine (by virtue of possessing a particular base sequence) specifically and reversibly with a certain (RNA) fraction possessing the proper (complementary) sequence. This combination blocks the initiation of cytoplasmic transcription and therefore the formation of the messenger by the structural genes in the whole operon. The specific 'repressor' (RNA ?), acting with a given operator, is synthesized by a *regulator gene*.

The repressor in certain systems (inducible enzyme systems) tends to combine specifically with certain specific small molecules. The combined repressor has no affinity for the operator, and the combination therefore results in *activation of the operon*.

In other systems (repressible enzyme systems) the repressor by itself is inactive (i.e., it has no affinity for the operator) and is activated only by combining with certain specific small molecules. The combination therefore leads to *inhibition of the operon*.

The structural messenger is an unstable molecule, which is destroyed in the process of information transfer. The rate of messenger synthesis, therefore, in turn controls the rate of protein synthesis.[4]

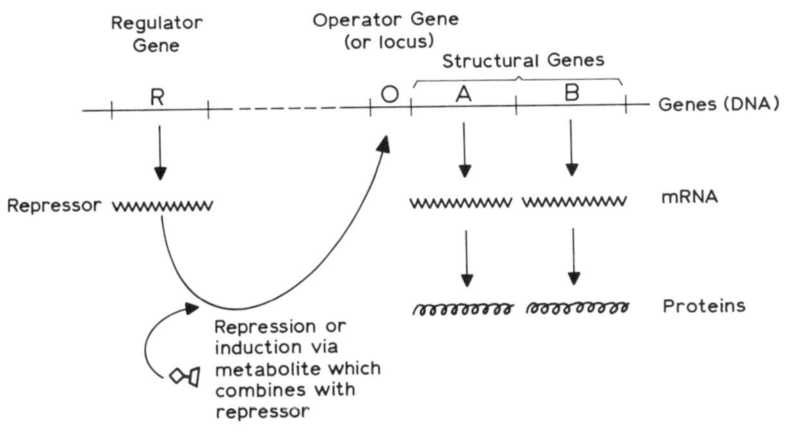

Fig. 1. (After Jacob and Monod, 1961.)

I have provided a diagram (Figure 1) which is based on this form of the operon theory which perhaps shows more clearly how the elements of the model interact.

2.2. *Is the Operon Theory Really a 'Theory'?*

When we think of the term 'scientific theory', we associate with it notions of universality, fairly extensive explanatory power, and somewhat speculative appeals to non-immediately accessible experimental processes or entities. Certainly these features are to be found in theories in physics such as the Bohr theory of the atom. I think it is also quite clear that the operon theory possesses these marks of a theory, and moreover, that it also possesses that 'instrumentalistic' characteristic of a theory of stimulating new experimental and theoretical research in several areas of inquiry. The operon theory also displays other interesting parallels with the theories of physics which have come to be paradigm cases of theory in science, but consideration of these will be deferred until Section 4.

To support my contentions about the theoretical character of the operon theory and its significance, let me quote several distinguished molecular biologists concerning certain features of both the theory and its historical impact on biology. First I shall cite the comments of Gunther Stent, a molecular biologist and historian of molecular biology, which were published as part of an article bearing the title: 'The Operon: On Its Third Anniversary'. Stent wrote:

In the spring of 1961, F. Jacob and J. Monod ... published the review 'Genetic Regulatory Mechanisms in the Synthesis of Proteins' that has exerted a most profound heuristic and dialectic effect on recent physiologicogenetic research ...

....

Whether or not Jacob and Monod's original notion of the interaction of repressor and operator and its effect on messenger formation turns out to be correct in the end, the magnitude of their contribution in putting forward the messenger-operon concept can hardly be overestimated. Not only did the appearance of their review suddenly bring order into what had hitherto been an enormously confused and complex mass of data, but it provided students of the control of protein synthesis with a badly needed new vocabulary for verbalizing interpretations of their experimental results. Thus, whatever future observations may yet reveal about regulation of cellular functions, the promulgation of the operon in 1961, the same year that saw also the discovery of the general nature of the genetic code, is sure to remain one of the principal milestones of molecular biology.[5]

Stent underscored this influence a year and a half later in his *Science* article on the '1965 Nobel Laureates in Medicine or Physiology', this time terming the Jacob and Monod 1961 article "one of the monuments in the literature of molecular biology." Stent went on to note that:

In this review Jacob and Monod proposed two great concepts: the *messenger* RNA and the *operon*. The validity of the messenger concept ... was proved before the review was even in print. When the review *did* appear, it immediately suggested to Marshall Nirenberg the experiments with the synthetic messenger RNA by means of which, in the course of the next two years, Nirenberg and others were able to break the genetic code. The operon concept envisages that genes pertaining to related functions reside in contiguous regions of the bacterial chromosome and form an operon by sharing a common gene of regulation: their operator ... Not only was this regulatory circuit capable of explaining much of the bewildering mass of observations that had accumulated by then on the control of bacterial enzyme synthesis, but it could account also for prophage induction, a process that had remained rather mysterious in the decade since its discovery by Lwoff ... The influence of the work for which Lwoff, Monod, and Jacob are being honored by this prize now far transcends the bounds of molecular biology. Probably its most important aspect has been on developmental biology, a field that, in the last analysis, concerns the understanding of regulation of gene activity in ontogeny.[6]

In their review article on the operon published in 1964, B. N. Ames and R. G. Martin provided a terse summary of the evidence for the operon *theory*:

The study of the regulation of bacterial enzyme synthesis has led to the development of the operon theory. This theory was formulated in order to explain three different kinds of data: (a) the finding by Demerec and his colleagues ... that the genes determining the enzymes of a biochemical pathway were generally in a cluster on the *Salmonella* chromosome; (b) the finding of coordinate repression for the histidine biosynthetic enzymes of the histidine cluster of genes ... in *Salmonella*; and (c) the discovery of regulator ... and operator mutants in the β-galactosidase system of *E. coli* ...

> The crystallization and clearest exposition of the operon theory were presented in the 1961 review of Jacob and Monod.... More recently the theory has been amended and expanded [by Ames and Hartman] to account for the order of genes in an operon, the phenomenon of polarity, and the size of messenger RNA...[7, 8]

The above citations from prominent molecular biologists should indicate both the significance and the theoretical character of Jacob and Monod's contribution. In the later historical and philosophical discussion, other features of the operon theory which emphasize the theoretical aspects of the operon theory will be considered. The reasons for spending this much time to argue that the operon theory is indeed a theory of some scope and influence are (a) because molecular biology is often depicted as almost entirely experimentally oriented to the exclusion of speculative and unifying theoretical aspects,[9] and (b) to establish the significance of the operon theory for philosophers and other scientists who may be unfamiliar with the theory.

3. Experiments and the Construction of the Operon Theory

The roles which experiments and associated or auxiliary theories played in the genesis of the operon theory are most interesting and have potentially important implications for philosophy of biology. The complete historical study is too long to relate in an article of this scope, so I shall have to present several chapters of that story in an attempt to depict in a semi-complete manner the most important features of the story of the construction of the operon theory.[10]

There are three distinguishable developments involved in the formation of the appropriate concepts and the construction of the operon theory on which I wish to focus. These are: (1) the 1959 'Pa-Ja-Mo' experiment (after Pardee, Jacob, and Monod) which led to the notion of the repressor and a new type of gene, then called a 'repressor-making gene', (2) the first clear statement by Jacob and Monod (1959) of the 'double genetic determinism' of 'regulator genes' and 'structural genes', and the tentative proposal of the 'operator' locus, and (3) the corroboration of the operator hypothesis and the discovery of the operon. It is hoped that the analysis of these three developments will disclose interesting aspects of the logic of theory construction in molecular genetics and also admit of potential application to other examples in this important science. Before these specific developments can be considered, however, it is imperative to discuss some of the earlier work of Jacob and Monod in order to understand the background knowledge and techniques appealed to in the construction of the operon theory.

3.1. *The Background of Enzymatic Induction and Bacterial Genetics*

3.1.1. *Enzyme Adaptation.* Jacques Monod became interested in the phenomenon which used to be termed 'enzyme adaptation' at the beginning of World War II. As a result of research conducted under difficult circumstances during the Occupation, and continued afterwards, Monod, together with A. Audureau, showed that the type of enzyme adaptation associated with lactose metabolism in *E. coli* involved a genetic property of the bacterium. After the war, in 1946, Monod began a series of experiments with several coworkers, among them G. Cohen-Bazire and M. Cohn, on the *lac* region of *E. coli* attempting to analyze the causes and kinetics of the *lac* form of enzyme adaptation. Utilizing enzyme kinetic theory, structural chemistry, immunological tests, and isotopic labeling, Monod and his coworkers concluded after a number of experiments that:

(1) the β-galactosidase enzyme that is produced by normal or wild-type *E. coli* in a lactose environment represents *de novo* synthesis of protein molecules (associable with a gene denoted by *z*) rather than an enzyme converted from pre-existing precursors (Monod *et al.*, 1952).

(2) the inducer, which stimulates the production of the enzyme β-galactosidase, does not interact with the enzyme, either as a substrate or through combination with a preformed active enzyme (Monod *et al.*, 1951).[11] The conclusions of these experiments effectively falsified earlier 'classical' hypotheses of enzyme adaptation and warranted a 'rechristening' of the name of the effect, henceforth to be known as 'enzyme induction'.[12] As a consequence of this work Monod (1966) noted that "the conclusion became obvious that the inducer [acted] ... at the level of another specific cellular constituent that would one day have to be identified."[13]

Several years later Monod also discovered that an additional enzyme, different from the β-galactosidase, was also needed to effectively metabolize the lactose in *E. coli* cells. This enzyme, which appeared to aid the lactose in passing through the cell membrane and concentrating therein, was not chemically isolable at the time that it was postulated on the basis of genetic evidence. It was, however, given the name of 'galactoside permease', and assigned a gene as its cause, the *y* gene (Rickenberg *et al.*, 1956; Monod, 1956). The discovery of this second inducible gene was important for at least two reasons. One was that it solved the mystery of the behavior of certain anomalous mutants, termed 'cryptic' mutants; another was the yet to be discovered pleiotropic effect which the *i* gene mutations had on *z* and *y* genes. This will be discussed extensively below. Some time after this discovery, a third protein or enzyme was found to be produced by the stimulated *lac* region. This second new enzyme, after first being mistakenly identified with the permease, was named 'galactoside transacetylase'. It is produced by a gene labeled as *a* (or sometimes as *Ac*).

The discovery of these three proteins (or of the effect in the case of the permease) produced by independent genes was most important in the development of the operon theory, for further research on mutant types of *E. coli* disclosed significant interactions among the *z*, *y*, and *a* genes, which led to the operon hypothesis. Before results of genetic analyses could be linked with the results of the inquiry into enzyme kinetics, however, more careful genetic mappings of *E. coli* had to be carried out.

3.1.2. *Bacterial Genetics*. Bacterial genetics is a fascinating subject and one which has an excellent claim to constitute a major component of the historical foundation of the science of molecular biology.[14] I can only touch in the most cursory manner on ways in which developments in this field, joined with Monod's earlier work, led to the operon theory.

The study of the genetics of bacteria and bacteriophage was initiated by Luria and Delbrück (1943) and later joined by Lederberg and Tatum (1946), Hershey (1946), and a number of other distinguished scientists. Lederberg, in particular, worked on *E. coli* and succeeded in obtaining a variety of mutant strains and in mapping some of the genetic elements of the bacteria. Among Lederberg's mutants was one most interesting strain which synthesized β-galactosidase in the *absence* of any inducer. Such a mutant was termed, in accordance with standard terminology, a *constitutive* mutant.

Monod and his coworkers soon followed Lederberg in his work and isolated some constitutive mutants. It was not until several years after this, however, following the discovery of the independent *z*, *y*, and *a* genes, that Monod made the startling discovery that the constitutive mutation, ascribable to the mutation of *one* gene (*i*) had a *multi*-enzyme or pleiotropic effect. Such a discovery was particularly disconcerting, for as Monod wrote:

We then had to admit that a constitutive mutation, although strongly linked to the loci governing galactosidase, galactoside permease, and transacetylase, had taken place in a gene (*i*) distinct from the other three, (*z*, *y*, and *Ac*), and that the relationship of this gene to the three proteins violated the [one gene-*one* enzyme] postulate of Beadle and Tatum.[15]

François Jacob began his work on bacterial genetics in 1950. In 1952 he began to collaborate with Elie Wollman in an inquiry into the genetic basis of *lysogeny* – the phenomena in which a non-virulent type of bacteriophage, i.e., an *E. coli* virus, becomes, for a time, innocuously incorporated into the genotype of the host bacterium and reproduces with it. One aspect of Jacob and Wollman's inquiry required them to examine bacterial conjugation, a process in which a male *E. coli* bacterium injects genetic material into a female of the same species. Jacob and Wollman discovered that they could interrupt conjugation by placing a mating mixture in a Waring blender, and that as a consequence they could effect the transfer of only a portion of the male chromosome. Results of such interruptions indicated that the transfer of a chromosome followed specific rules for the time rate of injection, and with any particular strain of *E. coli*, that the injection always started at the same point on the chromosome. Thus bacterial conjugation, because of this rule-governed behavior, became a most important tool for examining the interaction of different genes.

Further analysis allowed Jacob and Wollman to map, in some detail, the circular chromosome of the bacterium, and also to discover the existence of extrachromosomal heredity factors, termed *episomes*, which could be added or subtracted from the bacterium. One episome in which we shall be most interested is the sex factor, symbolically represented by the capital letter F, which can be transferred in conjugation, and which can also carry along with itself a small chromosomal fragment from the donor bacterium.

3.2. *The Discovery of the Repressor and the 'Repressor-Making Gene'*

Jacob and Monod began their collaboration in 1957, when, together with Pardee, they conjugated wild-type inducible male bacteria possessing a wild-type galactosidase gene (genotype i^+z^+) with mutant constitutive females possessing deficient galactosidase genes (genotype i^-z^-). The consequence of the conjugation was partially diploid or merozygotic bacteria with the genotype i^-z^-/i^+z^+. Two significant results were forthcoming from this experiment, sometimes referred to as the 'Pa-Ja-Mo' experiment (Pardee *et al.*, 1959).[16] One was the discovery that the galactosidase began to be synthesized very quickly, and at the maximum rate, after the entrance of the z^+ gene into the mutant bacteria. This could be accounted for by assuming that inducibility and constitutivity are expressed in or through the cytoplasm – an idea which later led to the formulation of the messenger (mRNA) hypothesis. Secondly, it was found that the new diploids became inducible two hours after conjugation, indicating that the *inducibility was dominant over constitutivity*, though such dominance was slowly expressed. Again, as in the case of the rapid expression of the injected z^+ gene, the kinetics of the expression of inducibility suggested, because of the rates of expression, that the interaction of the *i* and *z* genes took place via a cytoplasmic messenger synthesized by the *i* gene. (A diagram of the enzyme kinetics of the Pa-Ja-Mo experiment is given in Figure 2.)

In his Nobel lecture, Monod noted that even after the 'Pa-Ja-Mo' experiment had been carried out, the *interpretation* of the experiment was not clear. In a passage which is well worth quoting *in extenso*, Monod wrote:

> Of course I had learned, like any schoolboy, that two negatives are equivalent to a positive statement, and Melvin Cohn and I, without taking it too seriously debated this logical possibility that we called the 'theory of double bluff', recalling the subtle analysis of poker by Edgar Allan Poe.
>
> I see today, however, more clearly than ever, how blind I was in not taking this hypothesis seriously sooner, since several years earlier we had discovered that tryptophan inhibits the synthesis of tryptophan synthetase; also the subsequent work of Vogel, Gorini, Maas, and others showed that repression is not due, as we had thought, to an anti-induction effect. I had always hoped that the regulation of 'constitutive' and inducible systems would be explained one day by a similar mechanism. Why not suppose, then, since the existence of repressible systems and their extreme generality were now proven, that induction could be effected by an anti-repressor rather than by repression by an anti-inducer? This is precisely the thesis that Leo Szilard, while passing through Paris, happened to propose to us during a

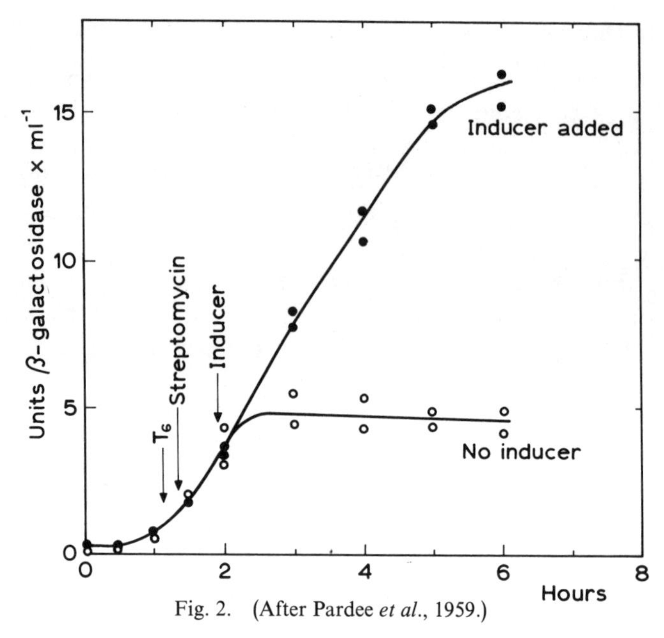

Fig. 2. (After Pardee *et al.*, 1959.)

seminar. We had only recently obtained the first results of the injection experiment, and we were still not sure about its interpretation. I saw that our preliminary observations confirmed Szilard's penetrating intuition, and when he had finished his presentation, my doubts about the 'theory of double bluff' had been removed and my faith established – once again a long time before I would be able to achieve certainty.

These two different but possible interpretations of the 'Pa-Ja-Mo' experiment were presented in the published account of the experiment.[17] As one 'interpretation' or tentative theory of genetic control, Pardee, Jacob, and Monod proposed what they termed an 'inducer model'. This model required the existence of a separate and as yet undiscovered endogenous inducer system in both wild and mutant strains of *E. coli.* In such a model, the i^+ gene would control the synthesis of an enzyme which would destroy or inactivate the endogenous inducer. The *ex*ogenous inducer then, say a galactoside, would still interact with the product of the i gene and prevent it from destroying or inactivating the *en*dogenous inducer.

The second, and preferred, though *prima facie* '*ad hoc* and arbitrary' interpretation offered was the 1959 version of the repressor model. This was the direct ancestor of the operon theory of 1960–61 which was discussed in Section 2. In the 1959 analysis, the i^+ gene was hypothesized to produce a *repressor*, with the *exogenous inducer*, say lactose or some other galactoside, acting as the antagonist of the repressor, preventing the repressor from repressing, and accordingly, permitting protein synthesis to occur. This was the exemplification of Monod's comment that "two negatives [i.e., the repressor and the antagonistic inducer] are equivalent to a positive [i.e., induced protein synthesis]..."

Pardee, Jacob, and Monod believed that the '*ad hoc* and arbitrary' character of the repressor hypothesis was alleviated by the discovery, in the years 1953–57, of repressible biological synthesizing systems.[18] Though they felt that the existence of these systems did little more than justify their proposal of the repressor hypothesis, other considerations did tend to *favor* the repressor model over the inducer account, even though they did not conclusively *falsify* the latter interpretation. These considerations seemed to involve both 'philosophical' and experimental aspects, inasmuch as the repressor model was (1) stated to be 'simpler' than the inducer model, because it did not require the existence of an independent, and undiscovered, inducer synthesizing system,[19] (2) said to imply that "constitutive mutants should... synthesize more enzyme than the induced wild-type"; a fact that was confirmed by already known experiments, and (3) said to imply that "the galactosidase forming system of the mutant should be largely insensitive to the glucose effect [whereby glucose and other carbohydrates inhibit enzyme synthesis], while other inducible systems should retain their sensitivity." The authors noted that "this is precisely the case..."

Pardee, Jacob, and Monod closed their 1959 paper with several most acute and insightful speculations, well worth quoting *in extenso*, which suggested the direction in which later research, both experimental and theoretical, was to go. They wrote:

(5) If adopted and confirmed with other systems, the repressor model may lead to a generalizable picture of the regulation of protein syntheses; according to this scheme, the basic mechanism common to all protein-synthesizing systems would be inhibition by specific repressors formed under the control of particular genes, and antagonized, in some cases, by inducers. Although the wide occurrence of repression effects is certain, the situation revealed with the present system, namely a genetic 'complex' comprising, besides the 'structural' genes (z, y) a repressor-making gene (i) whose function is to block or regulate the expression of the neighboring genes is, so far, unique for enzyme systems. But the formal analogy between this situation and that which is known to exist in the control of immunity and zygotic induction of temperate bacteriophage is so complete as to suggest that the basic

mechanism might be essentially the same. It should be recalled that according to Jacob and Wollman (1956), when a chromosome from a λ-lysogenic ♂ of *E. coli* is injected into a non-lysogenic ♀, the process of vegetative phage development is started, which involves as an essential, probably as a primary, step the synthesis of specific proteins. When the reverse mating (♂ non-lysogenic × ♀ λ-lysogenic) is performed, zygotic induction does not occur; nor does vegetative phage develop when such zygotes are superinfected with λ particles. The λ-lysogenic cell is therefore immune against manifestations of prophage or phage potentialities, and *the immunity is expressed in the cytoplasm* (Jacob, 1958–59). Moreover the immunity is strictly specific, since it does not extend to other, even closely related, phages. The formation, under the control of a phage gene, of a specific repressor, able to block synthesis of proteins determined by other genes of the phage, would account for these findings.

(6) Implicit in the repressor model are two critical questions, which for lack of evidence we have avoided discussing but which should be explicitly stated in conclusion. These questions are:

(a) What is the chemical nature of the repressor? Should it be considered a primary or a secondary product of the gene?

(b) Does the repressor act at the level of the gene itself, or at the level of the cytoplasmic gene-product (enzyme-forming system)?

3.3. *Regulator Genes and the Operator*

In a paper which appeared several months after the Pardee, Jacob, and Monod article, Jacob and Monod (1959) first introduced the term 'regulator gene' (gène régulateur), distinguishing genes of this new type from the ordinary type of gene, the 'structural gene' (gène de structure). Whereas in the account of the 'Pa-Ja-Mo' experiment, the pleiotropic effect of the *i* gene on the *z* and *y* genes had been noted, in this new paper such an effect was extensively discussed and cited as an important distinguishing characteristic of *regulator* genes. It was thus that the Beadle-Tatum hypothesis was rescued by re-evaluating the foundations of genetics and hypothesizing a new class of genes.

The mechanism of the interaction of the regulator gene with the structural genes under its control was not immediately obvious to Jacob and Monod, and in their 1959 paper they considered three possibilities: (1) that the product of the *i* gene, the repressor, acted directly and independently on a particular structure (the repressor target) repeated in each of the genes under the repressor's control, (2) that the repressor acted on a polypeptide chain common to the proteins produced by the controlled genes but determined by an independent cistron, and (3) that the repressor acted on a unique structure, termed the *operator*, about which enough has been said earlier in Section 2 of this paper. Possible consequences of each hypothesis were considered; in particular, Jacob and Monod conjectured as to what types of mutants might be discoverable on the basis of the truth of the alternatives (1), (2), and (3).[20]

3.4. *The Operator and the Operon*

Once the existence of an operator element was deemed worthy of acceptance, in the sense that it seemed worthy of being tested for, attempts were made by Jacob and Monod to look for mutants whose enzyme synthesizing behavior might reveal operator locus mutations. How to do this was apparently not immediately obvious, for Jacob writes that at first:

...it seemed difficult to distinguish mutations affecting the emitter [the *i* gene] from those affecting the receptor [the *o* site] until we realized that the distinction should be relatively easy to make in a diploid.[21]

In a diploid, *loss* of activity of one of the *i* genes ($i^+ \rightarrow i^-$) would be *recessive*, as regards the structural genes under its control, inasmuch as the *other* heterozygous *i* gene would be *dominant*. (Recall the discussion above on the dominance of inducibility.) On the other hand, loss of activity in the *o* site ($o^+ \rightarrow o^c$) would be *dominant* for *its* controlled structural genes (i.e., for those in a *cis* position with respect to the o^c site). Jacob and Monod were guided in important ways in their search for such operator mutants by the striking formal analogy, quoted from the 'Pa-Ja-Mo' paper above, between regulation in the *lac* system and zygotic induction and immunity in the λ phage system. Jacob (1966) noted:

It should be possible in principle, by the use of diploid bacteria, to distinguish, among constitutive mutations, those due to the regulator gene from those due to the receptor. In fact, phage mutants corresponding to one or the other of these types had been known for a long time, although their nature became clear only in the light of this scheme. The existence of such mutations in phages encouraged us to search for analogous bacterial mutations affecting the enzymes of the lactose system. For that purpose, however, diploid bacteria were required.[22]

The partial, and unstable diploids utilized in the 'Pa-Ja-Mo' experiment were difficult to work with, but the new discovery, by Jacob and E. Adelberg, of the *lac* sex factor (F) episome allowed Jacob and Monod, and their coworkers D. Perrin and C. Sanchez (Jacob *et al.*, 1960) to obtain stable variants of *E. coli* diploid for the *lac* region. Some of these, together with their genotypes are given in Table I.

The result of these and later similar experiments corroborated the opera-

TABLE I

Synthesis of galactosidase, protein Cz. (a modified form of galactosidase), and permease by haploid and heterozygous diploid operator mutants

Genotype		Non-induced Bacteria			Induced Bacteria		
Chromosome	F-lac	Galactosidase	Protein Cz.	Permease	Galactosidase	Protein Cz.	Permease
$i^+o^+z^+y^+$		<1	–	nd	100	–	100
$i_3^-o^+z_4^-y^+$	/ $Fi^+o^+z^+y^+$ ...	<1	nd	nd	320	100	100
$i_3^-o^+z_4^-y^+$	/ $Fi^+o^cz^+y^+$ ...	36	nd	33	270	100	100
$i^+o^+z_1^-y^+$	/ $Fi^+o^cz^+y^+$ ...	110	nd	50	330	100	100
$i^+o^+z^+y_R^-$	/ $Fi^+o^cz_1^-y^+$	<1	30	–	100	400	–
$i^+o^+z_1^-y^+$	/ $Fi^+o^cz^+y_R^-$ ...	60	–	nd	300	–	100

Values given are percentages of amounts of proteins observed in induced wild-types. nd = measurement not determined, Alleles z_1^- and z_4^- produce protein Cz. (identified in later articles (Jacob and Monod, 1961) as CRM or cross-reacting material because of immunological similarities with galactosidase). The table discloses that though bacteria heterozygous for *o* and *z* are partially constitutive, only the *z* or *y* alleles located in a *cis* position with respect to o^c are expressed constitutively, the alleles located in a *trans* position remaining strictly inducible. "The constitutive mutation o^c is accordingly pleiotropic, dominant, and manifests its effect only in a *cis* position." (Jacob *et al.*, 1960.) (Permease, and accordingly gene *y*, appears to be misidentified here – cf. p. 503 of this paper and also Stent (1967), p. 157, for reason and evidence of misidentification.)

tor hypothesis, and also the hypothesis that the operator is adjacent to and controls the associated structural genes simultaneously and to the same extent. This discovery of the apparent structural and behavioral unity of the *o-z-y* segment led Jacob *et al.* (1960) to propose that somewhere between the classical or structural gene, as a unit, and the chromosome, as a larger genetic unit, there existed a new unit of coordinated gene action consisting of an operator and the group of structural genes coordinated by it. This new unit of coordinated gene action was termed the *operon*.

Jacob, Monod, and their coworkers remarked that the interaction of regulator gene and operon could be conceived to take place either at the genetic level, or at the cytoplasmic level, and in 1960 they made no attempt to decide between the two modes of activity. Later, in the 1961 review article, Jacob and Monod proposed that the control took place at the genetic level, and, as a consequence of this supposition, they made some most important suggestions concerning protein synthesis that led to the postulation of messenger RNA. They did admit, however, that the choice of the genetic level mode of control was still speculative and lacked specific supporting evidence.

3.5. *Advances in the 1961 Review Article and Beyond*

I have now carried the analysis of the development of the operon theory to the appearance of the 1961 review article discussed in Section 1. That article did make some new information available: it reported the discovery of a new regulator gene mutant, i^s, which effectively falsified the 1959 *inducer* model,[23] and also mentioned a new operator locus mutation, symbolized by o^0. Some negative evidence, based on an experiment performed by Pardee and Prestidge (1959), and implying that the repressor was *not* a protein was also introduced. This inference was based on the fact that the addition of a substance to *E. coli*, such as 5-methyltryptophan, which blocks protein synthesis, still appeared to permit the synthesis of the repressor. The repressor, therefore, was thought to be a non-protein product of the *i* gene – probably an RNA molecule. I have already noted that the messenger or mRNA hypothesis was also first proposed in the 1961 review article.[24]

It seems, however, that the significance of this review article was not a consequence of the new knowledge just cited, but rather was a consequence of the logical presentation and powerful marshalling of evidence supporting the then well-developed operon theory.

The operon theory by no means ceased to develop after 1961. Jacob, Monod, their coworkers, and others have continued their research, and have modified some aspects of the 1961 theory. Though it seemed that the repressor might best be thought of as a polyribonucleotide, it became difficult to conceive how the inducer could possibly interact in a stereospecific manner with the repressor as thus characterized. Monod *et al.* (1963) accordingly proposed that the repressor was an allosteric protein. This has since been confirmed experimentally by Gilbert and Müller-Hill (1966) who, in a most important experiment, chemically isolated the repressor. The operon theory has also been complicated to include a new site, termed the *promoter*, located between the regulator gene and the operon.[25] Perhaps the most dramatic experiment which supported the operon theory forthcoming after 1961 was the fusion experiments which Jacob *et al.* (1965) carried out which fused the *lac* structural genes with a different operator from the purine region of the *E. coli* chromosome, thus bringing the *lac* region under experimentally verifiable control of a control system specifically activated by purine.

The Jacob-Monod operon theory did not go uncriticized, nor lack competitors after its appearance. Alternatives to the Jacob-Monod theory of genetic control in the *lac* region were proposed by Ames and Hartman (1963), by Stent (1964), and others, but the Jacob-Monod approach has been ably defended against these alternative accounts by Beckwith (1967). Recently, Beckwith and his coworkers (Shapiro *et al.*, 1969) have managed to isolate the *lac* operon, by an elegant hybridization technique, and greater knowledge about genetic control is expected to follow from this isolation. Still more recent is the report on direct *in vitro* studies of the repressor with its operator (Bourgeois and Monod, 1970).

4. The Operon Theory and the Unity of Science

There are several implications concerning the Unity of Science thesis which can be drawn from the above account of the structure and development of the operon theory. I shall group these implications under the rubrics of (1) reductionism, (2) system theory, and (3) parallelism in the methodology of theory construction and theory testing.

4.1. *Reductionism*

The operon theory is considered as one of the major contributions of 'molecular biology' to our understanding of processes occurring in living organisms. Now molecular biology is usually understood to concern itself with physico-chemical accounts of biological systems, and to be *the* science which, at least implicitly, constructs reductions of biological phenomena, laws and theories, to physics and chemistry.

One interpretation of the Unity of Science thesis is that it represents an attempt to discover a single basic science to which all other sciences might be reduced. This would be the interpretation of Unity of Science which Carnap (1934) termed a 'unity of laws'. Usually the unifying basic science is taken to be physics, but in the context of a discussion involving biology, it is often taken to be a combination of physics and chemistry, the reduction of chemistry to physics not being held to involve any philosophically interesting issues.[26] Accordingly the question can be raised whether the development of the operon theory supports an interpretation of the Unity of Science which would unify biology by reducing it to physics and chemistry.

It would seem that even though the operon theory was articulated at a level which is very *close* to the physico-chemical level, and even though since 1961 much more has been discovered about the physico-chemical characteristics and interaction of the elements of the operon theory, that it *itself* is not physico-chemical theory. Nevertheless, the operon theory supports a form of the reductionist thesis which I term a 'sophisticated reductionism', in that it constitutes a significant extention of our knowledge of how genes operate which is completely consistent with what is known about the physics and chemistry of gene action. Furthermore, all research to this date suggest that a full physico-chemical account of gene control, at least of the reversible type exemplified in *E. coli*, is likely in the not-too-distant future. As such, however, the operon theory was constructed, formulated, and tested, in terms of and using the techniques of neo-classical genetics, a science which does not *require* a reduction to physics and chemistry.

Another point which perhaps ought to be stressed about the operon theory is that even though it is almost a physico-chemical theory, it is also appropriate to characterize it as 'wholistic' or 'organismic', in that it involves several component parts interacting in ways which result in properties which are ascribable to the whole *lac* region, but which are not properties of the parts of that region taken in isolation. Thus the oftcited opposition between an organismic-wholistic approach to biology and a physico-chemical reductionist approach is, in this case, disclosed to be an untrue opposition. This fact of interaction among parts of a system, however, leads us naturally to the question concerning the *type* of reductionism thesis which the existence of the operon theory supports.

Suppose that all of the entities of the operon theory, e.g., the repressor, operator, and the like, came to be specifiable in terms of physics and chemistry. That is, suppose that *reduction functions*[27] were formulatable which would identify particular repressors with particular allosteric proteins, and particular operator loci with particular DNA sequences. Furthermore, let us suppose that the physico-chemical laws of chemical diffusion, chemical binding, and the conformational folding of proteins complexed with inducers/co-repressors were sufficient to account for genetic control, when the reduction functions were conjoined with such laws. Would we than have the reduction of this sector of biology to physics and chemistry?

In one sense we would not, for we would not really have the *replacement* of biology by physics and chemistry, *unless* we were also to *assume* a considerable amount of organization or chemical systematization which is not apparently explicable on the basis of physics and chemistry. That is to say, we would have to postulate *in addition to* the usual laws and theories of physics and chemistry, sentences which described the complex physico-chemical organizational patterns discovered in living organisms.[28] Though such patterns or structures are most likely explicable as a consequence of chemical-molecular *evolution*,[29] the fact that such patterns have to be accepted as 'initial conditions' in physico-chemical explanations in molecular biology indicates that a simple straightforward or 'naive' reductionism is not warranted. Nevertheless (1) such 'initial conditions' describing biological organization seem generally completely *statable* in physico-chemical terminology, and (2) the laws and theories of physics and chemistry may adequately and without paradox be applied to such biological or physico-chemical structures. Because the operon theory seems to support both of these claims, I conclude that the structure and development of the operon theory supports a type of reductionistic thesis which I term a 'sophisticated reductionism'. Such a 'sophisticated reductionism' accepts organization and systematization as initial conditions in the physico-chemical explanation of biological phenomena, laws, and theories.

The necessity to refer to organization and systematization in such a reductionistic thesis suggests that perhaps another interpretation of the Unity of Science thesis might also apply to my characterization of the operon theory. This would be an approach to the Unity of Science through a discipline termed 'general system theory'.

4.2. *General System Theory*

L. Von Bertalanffy, and a number of other scientists, have professed to see a new approach to unity of science in a subject known as 'general system theory'.[30] In general system theory the emphasis is not on *reducibility* of biology to physics and chemistry but rather on the search for *isomorphisms of systems* in different and perhaps irreducible sciences. Von Bertalanffy writes:

The integrative function of general system theory can perhaps be summarized as follows. So far, the unification of science has been seen in the reduction of all sciences to physics, the final resolution of all phenomena into physical events. From our point of view, unity of science gains a more realistic aspect. A unitary conception of the world may be based, not upon the possibly futile and certainly far-fetched hope finally to reduce all levels of reality to the level of physics, but rather on the isomorphy of laws in different fields. Speaking in what has been called the 'formal' mode, i.e., looking at the conceptual constructs of science, this means structural uniformities of the schemes we are applying. Speaking in 'material' language, it means that the world, i.e., the total of observable events, shows structural uniformities, manifesting themselves by isomorphic traces of order in the different levels or realms.[31]

The possible application of general system theory to reveal the unity between the operon theory and other systems in other sciences is supported by another proponent of general system theory, M. D. Mesarović, who believes that the Jacob-Monod operon theory is an excellent exemplification of the concepts of systems theory.[32] More generally, and in connection with the Jacob-Monod theory and systems like it, Mesarović has expressed the hope that it might be possible to use

...the concepts of systems theory in an indirect way as a conceptual guide to arrive at a detailed description of the structure of a biological system.[33]

One would think, then, that the concepts of system theory would be applicable to our account of the structure and development of the Jacob-Monod operon theory and might show how such an application could constitute an argument for the general system theory interpretation of the Unity of Science thesis.

Unfortunately the *prima facie* possibility that system theory can disclose a new approach to the unity of science does not appear to be supported by our research into the structure and development of the operon theory. The 'system' features of the operon theory, as regards the synthesis of the repressor, the conjunction of the repressor with the inducer/corepressor, and further association of the conjunction with the operator, do not seem to have any analogues or isomorphic counterparts outside the area of *biological* control systems. (Unless, of course, one wishes to trivialize the thesis by stating that Poe's analysis of poker is a relevant logical homology.) Genetic control systems, of the operon variety, are not isomorphic to any standard control system employed in physics or the engineering disciplines. Though one could not deny that some of the mathematical analyses used in physics might be applicable to mathematical idealizations of the more elementary end-point inhibition type of genetic regulatory mechanisms, the kind of mathematics employed seems to possess no features which would distinguish it in such a way so as to constitute a special mathematics of systems theory.[34] These contentions, if true, indicate that there is no non-trivial formulation of 'systems theory' which is applicable to the operon theory in its fully developed 1961 form. Further, it is quite doubtful that the concepts and laws of systems theory, however they might be construed, would have aided Jacob and Monod, and their coworkers, in *developing* the concept of the repressor, the regulator gene, the operator, or the operon. As outlined in the work of von Bertalanffy, and Mesarović then[35] – and these selections seem to be representative samples of the discipline – the concepts and laws of systems theory neither play any explicit or discernably implicit role in the structure and development of the operon theory, nor do they seem in any way to be necessary, sufficient, or of any real assistance, to the understanding of that theory, even in an *ad hoc* fashion.

4.3. *Unity of Science from the Point of View of Parallelism in General Methodology*

Even though I have reached a negative conclusion as regards the suitability of general system theory to clarify the ways in which the structure and development of the operon theory can be related to the unity of science, I do think that there is a type of isomorphism and a unity which reveals itself in the development of the operon theory. This can be termed a *unity of the dynamics of scientific testing and of scientific growth*. It is a unity in that metascientific theories which have been based on the *physical* sciences seem to admit of interesting and insightful application to the case of the development of the operon theory. The account of the dynamics of scientific growth which I will sketch here derives most specifically from the work of Duhem (1954 [1906]) and Popper (1959 [1934]) in the philosophy of the physical sciences, though it is also dependent in important ways on the writings of Hertz (1956 [1894]), Kuhn (1962), Grünbaum (1969), Shapere (1971), and on some of my own work (1970). First I shall consider the applicability of some of Duhem's analyses, especially as concerns theoretical interdependence, to molecular genetics. I shall then turn to analyze the development of the operon theory in modified Popperean terms as a quasi-inductive process of conjectures, refutations, and corroborations. Finally I shall attempt to show, by arguing for the desirability of a number of modifications of both Duhem's and Popper's analyses, that the Duhemian and Popperean aspects of the development of the operon theory represent opposite limiting cases of a single logic of scientific hypothesis evaluation. If this can be shown it will represent another argument for methodological parallelism between physics and biology, as this single logic has recently been utilized to analyze the historical developments in the electrodynamics of moving bodies in physics in the years 1895–1910 (Schaffner, 1970).

4.3.1. *Theoretical interdependence in molecular genetics.* It seems from our earlier analyses of the experiments in molecular physiologicogenetics, that the conclusions of such experiments are expressed in terms of certain symbols which require a knowledge of auxiliary biological and biochemical theories for their proper interpretation. This assertion is a denial of the distinction which Pierre Duhem once drew between experiments in physiology and experiments in physics. Duhem argued that the result of an experiment in physiology was "a recital of concrete and obvious facts," and that "in order to understand it not a word of physiology need be known."[36] On the other hand, the conclusions of experiments in physics

> in no way are ... purely and simply an exposition of certain phenomena; they are abstract propositions to which you can attach no meaning if you do not know the physical theories admitted by the author.[37]

Though the claim by Duhem of this type of distinction between physics and physiology *might* have had some validity in 1906, it is decidedly false as regards contemporary molecular biology. An understanding of the experimental results of the 'Pa-Ja-Mo' experiment, not even including the inducer or repressor hypothesis, but simply, say, the statement of the 'dominance' of 'inducibility' over 'constitutivity', requires a comprehension of bacterial genetics, developed from Mendelian-Morganian genetics. An analysis of the evidence supporting that claim requires an understanding of Monod's earlier work on enzyme induction, which in turn involves enzyme kinetics and structural chemistry. Similarly, the later experiments of Jacob and Monod which determined the existence of the operator locus, and the *cis* mode of action of this genetic entity, also required a sophisticated understanding of bacterial genetics. Jacob's comments regarding the use of diploids to test for the difference in effect between regulator and operator mutations should have made this quite clear.

If the distinction which Duhem upheld between physiology and physics is indeed to be denied in contemporary molecular biology, then we might except that some of Duhem's other claims as regards the difference between physiology and physics might also break down.

Duhem also contended, as a consequence of his claim about the need for auxiliary theories in physics, that:

> the experimental testing of a theory does not have the same logical simplicity in physics as in physiology ... for in physics, an experiment can never condemn an isolated hypothesis but only a whole theoretical group.[38]

Duhem also noted that still another consequence of this organismic character of physical theories was the impossibility of a crucial experiment in physics in which one of a pair of theories is unequivocally falsified.

If we examine the development of the Jacob-Monod operon theory, we can discern several instances which exhibit these Duhemian features. First, I refer to the discovery of the constitutive pleiotropic mutants, whose existence seemed to *falsify* the Beadle-Tatum 'one gene-one enzyme' hypothesis. Such falsification was in actuality not warranted, however, for the difficulty, it was subsequently discovered, was not with the Beadle-Tatum hypothesis, but rather with the auxiliary assumptions of classical genetics, which had assumed that there was only one type of gene – that which we now refer to as a *structural* gene. The Beadle-Tatum hypothesis was thus, as I mentioned earlier, saved from falsification by a re-evaluation of classical genetics and the substitution of a new neo-classical genetics which permitted *two* types of genes to exist, and to interact in such ways that the Beadle-Tatum hypothesis could still be maintained for both types of genes.[39]

Another instance of this Duhemian logical complication, introduced into molecular genetics by the existence of auxiliary theories, was the apparent falsification of the proteinaceous nature of the repressor by the 1959 Pardee and Prestidge experiment. Subsequently, by 1962–63, *other* auxiliary theories required that the repressor *be* a protein, because no other biological molecule had been discovered which could, according to these auxiliary theories of biochemistry, interact with the inducer/co-repressor to the expected degree and also bind with the operator locus.[40] Here, re-evaluation took place at the biochemical level, whereas in the above example, it took place at the genetic or biological level. This double-level *re-evaluation* seems entirely appropriate *vis-à-vis* the operon theory, inasmuch as this theory has elements which are characterized both in terms of genetics and in terms of biochemistry.

There are several lessons which I believe should be drawn from this account of the theoretical interdependence in molecular biology. One is that a new theory, or perhaps simply an experimental discovery expressed in terms of an established theory, might well involve contradiction with other established theories. Secondly, it should be noted that re-evaluation of such a contradiction is fundamental to advance in molecular biology, and that any assessments made of the adequacy of a new experimental discovery or of a new theory might well change as re-evaluation involving the older theories proceeds.

4.3.2. *Conjectures and refutations in the development of the operon theory.* It becomes clear, both from an examination of Monod's comments on the development of the idea of the repressor (quoted above on pp. 505–506), and from the 'Pa-Ja-Mo' paper, that experiment did not *entail* the hypothesis of the repressor, nor was the repressor hypothesis, in any straightforward sense, a *generalization* of the results of experiments. Recall that at least two interpretations were considered: an inducer hypothesis and a repressor hypothesis, with the latter being chosen over the former for various reasons having to do with simplicity, analogy, and the consequences of experimental tests. The inducer hypothesis was not considered to be effectively falsified until i^s mutants were discovered and announced in the 1961 paper of Jacob and Monod. Following Popper's (1959 [1934]) suggestions, I shall accordingly term the proposal of the repressor hypothesis a 'non-inductive conjecture'.[41] By using this locution, however, I do not mean to imply that the experimental evidence did not severely *constrain* the possible conjectures which were available as plausible candidates for explanation of these experiments; I mean only that the experimental evidence did not *entail* a conjecture by means of any deductive or obvious inductive logical machinery.

Another instance of the 'conjectural' element in the development of the operon theory was the hypothesized existence of the operator locus and the manner of action of the repressor. Recall that Jacob and Monod (1959) considered three possibilities of the interaction of the repressor with the structural genes, with only one of these possibilities involving the operator locus in the specialized sense that we now understand it. Consequences of each conjecture were deduced, with the help of auxiliary theories such as hypotheses concerning mutagens, bacterial genetics, and enzyme kinetics, and tests were conducted to determine which consequences were in fact to be observed in the laboratory. As I discussed above, two of the conjectures were falsified, (or at least tentatively falsified in the sense that the Duhemian element often permits later re-evaluation of such falsification), whereas the operator hypothesis was corroborated: it was subjected to a severe test and was not falsified.

That there are also conjectural non-experimental aspects in the 1961 version of the operon theory is supported by Ames and Hartman's (1963) and Stent's (1964) alternative accounts of genetic regulation in the *lac* operon, which are incompatible with the Jacob-Monod theory.

It would take us well beyond the scope of this paper to show that there are *detailed* parallels between the type of hypothesis formation and theory construction which molecular biologists engage in, and that in which physicists, such as Lorentz, Einstein, Bohr, and Schrödinger, say, engaged. This would require extensive historical analysis of the literature of physics.[42] Let it suffice for us to have shown that the *methodology* of conjecture and refutation and, as I shall argue below, the methodology of quasi-inductive advance, both of which have their source in Popper's attempt to characterize the methodology of the physical sciences, are paralleled in molecular biology. The fact that hypothesis, conjecture, and theory are of considerable significance in molecular biology is perhaps only surprising to those biologists and philosophers who might think that contemporary molecular biology is devoid of the theoretical and speculative elements which are so prevalent in theoretical physics.[43]

4.3.3. *The quasi-inductive development of the operon theory.* At the conclusion of his classic, *The Logic of Scientific Discovery*, Popper wrote:

One may discern something like a general direction in the evolution of physics – a direction from theories of a lower level of universality to theories of a higher level. This is usually called the 'inductive' direction; and it might be thought that the fact that physics advances in this 'inductive' direction could be used as an argument in favour of the inductive method.

Yet an advance in the inductive direction does not necessarily consist of a sequence of inductive inferences. Indeed we have shown that it may be explained in quite different terms – in terms of degree of testability and corroborability. For a theory which has been well corroborated can only be superseded by one of a higher level of universality; that is, by a theory which is better testable and which, in addition, *contains* the old, well-corroborated theory – or at least a good approximation to it. It may be better, therefore, to describe that trend – the advance towards theories of an ever higher level of universality – as 'quasi-inductive'.

The quasi-inductive process should be envisaged as follows. Theories of some level of universality are proposed, and deductively tested; after that, theories of a higher level of universality are proposed, and in their turn tested with the help of those of the previous levels of universality, and so on. The methods of testing are invariably based on deductive inferences from the higher to the lower level; on the other hand, the levels of universality are reached, in the order of time, by proceeding from lower to higher levels.

The question may be raised: 'Why not invent theories of the highest level of universality straight away? Why wait for this quasi-inductive evolution? Is it not perhaps because there is after all an inductive element contained in it?' I do not think so. Again and again suggestions are put forward – conjectures, or theories – of all possible levels of universality. Those theories which are on too high a level of universality, as it were (that is, too far removed from the level reached by the testable science of the day) give rise, perhaps, to a 'metaphysical system'. In this case, even if from this system statements should be deducible... which belong to the prevailing scientific system, there will be no *new* testable statement among them; which means that no crucial experiment can be designed to test the system in question. If, on the other hand, a crucial experiment can be designed for it, then the system will contain, as a first approximation, some well-corroborated theory, and at the same time also something new – and something that can be tested.

I believe that the advance toward the operon theory represents such a quasi-inductive advance, if we can also include as part of the quasi-inductive advance, the development of theories which are also *more precise* or *specific*, as well as *more universal*.[44]

The simple repressor hypothesis advanced in the 'Pa-Ja-Mo' paper relied on earlier corroborated hypotheses concerning enzyme induction and bacterial genetics, and these earlier hypotheses are used, as was stressed in Section 3.1.1., as tests of the repressor hypothesis. The later forms of the theory which developed from the 1959 repressor theory indicate that the repressor *hypothesis*, and not simply the 'data' represented by the enzyme kinetic measurements, for example, constituted a constraint on the later theories: they were required to entail the earlier theory or to suggest reasons why such entailment was unnecessary. Thus the late 1959 Jacob-Monod theory of the regulator gene (including the hypothetical operator locus) is more general than, but is constrained by and tested by, the 'Pa-Ja-Mo' repressor hypothesis. Further, the operator hypothesis was tested by searching for operator mutants whose existence could only be revealed by making use of the theories of bacterial genetics, the hypothesis of the *cis* mode of action of the operator, and the quantitative results of enzyme kinetic analyses. The 1960 version of the operon theory was the conjecture proposed to account for the results of genetic mapping order *i o z y* and the simultaneous and quantitatively similar effect of the operator on the structural genes under its control. Finally, the 1961 form of the operon theory, involving the same regulator gene-operator-operon elements as the 1960 form, also included a more precise hypothesis concerning the level of action (at a genetic level) and a specific hypothesis concerning the mechanism of such action, the mRNA hypothesis. Clearly the later post-1961 versions of the operon theory include the 1961 theory, or give specific reasons why they do not do so, e.g., as regards the case of the chemical nature of the repressor.

4.3.4. *Criticism and modification of the Duhemian and Popperean aspects of parallelism in the logic of testing and growth of science.* I have now argued that the development of the operon theory is quasi-inductive in the Popperean sense. But many philosophers of science feel that Popper's emphasis on falsifiability and on simple falsification is incompatible both with the Duhemian thesis that falsification is a very complex process and with the historically demonstrated resistance of actual scientific theories to falsification.[45] The question that might well be raised is whether the joint Duhem-Popper account I sketched above is completely consistent. I believe that it is not – or at least that it is not consistent without further modifications of both Duhem and Popper. I also believe that with modification their views can be incorporated into a synthesis which is strongly analogous to what I have termed elsewhere the logic and epistemology of comparative theory evaluation. Let me begin a discussion of modifications with Duhem, and then turn to Popper's analysis.

There are several elements of Duhem's analysis of the flexibility of theory *vis-à-vis* experience which need modification if his analysis is to be an accurate reconstruction of scientific *practice*. First it should be noted that most scientists act as if there were certain types of *simplicity* restrictions on the possible alteration of auxiliary theories in the face of falsifying evidence. Duhem suggested that a crucial experiment was impossible because of the interconnectedness of scientific theory: if there was disagreement between theory and experiment it was unclear that the theory purportedly being tested was to blame. The Duhemian thesis, or the D-thesis to use Grünbaum's terminology, follows on this view.[46] The D-thesis maintains that *if* we are willing to make the appropriate adjustments in the auxiliary theories, that any specific theory can be saved from any falsifying data. This assumes, though, that there is no restriction in the number of theoretical modifications, e.g., the introduction of *ad hoc* hypotheses or the appeal to statements representing the action of special entities with peculiar behaviors, which can be conjoined to the corpus of scientific theories so as to insulate the theory being tested from falsification. If there is a criterion of simplicity which rules against such complication, and which imputes a lower value of acceptability to theories possessing such complications, then the retainability of the apparently falsified theory over an against a *rejection* of that theory plus no additional complications is *not* guaranteed.

Secondly, it should be pointed out that auxiliary theories when altered so as to permit the retention of some prima-facie falsified theory, might well then conflict with still *other* well confirmed or corroborated theories or accepted experiments. For example, if we alter our theory about possible tertiary structures and the binding abilities with small molecules of RNA, so as to save the 1961 Jacob-Monod hypothesis that the repressor is a polyribonucleotide, this modification of this auxiliary hypothesis might well conflict with the accepted quantum-mechanical theory of chemical binding. Accordingly, the freedom which a scientist has in altering his auxiliary theories is constrained, not only by considerations of what I would term *relative simplicity*, but also for reasons of what might be termed *theoretical context sufficiency* and *experimental adequacy*. If we incorporate the effects of these three types of constraints into the Duhemian analysis of science, we can see that falsification, or at least a *relative falsification*, is more likely to be expected.[47, 48]

Thus modified, the Duhemian analysis can be introduced into Popper's account of the quasi-inductive path of the growth of science. But Popper, like Duhem, also requires modification if his analysis is to function as an adequate explication of the historical development of science.

Popper's account is both too weak as regards the type of constraints which earlier science places on later scientific theories, and too restrictive as regards the asserted deductive relation of the later theories to the earlier theories. Popper's analysis is weak in the same sense as Duhem's: he places too much emphasis on the directly related theoretical and experiential base of science and not enough on the *inter*theoretical and *meta*theoretical relations. More specifically, Popper does not take into account, at least explicitly, the richness and variability of the interaction of new theories with the accepted theoretical corpus of the time. He talks as if the extant theories are either totally consistent with the new theory, or as if the extant theories are simply *aids* to the derivation of basic statements – roughly 'observational' reports – which can then serve as a test of the new theory. Popper does not consider explicitly in his analysis of the quasi-inductive path of science the role which the theoretical context exerts on the acceptability of any new theory, nor does he consider ways in which a new theory might force a fundamental revision in some related and at present mutually contradictory theory.

Popper is more adequate as regards the importance of *simplicity* as a principle of constraint on new scientific theories. For Popper, simplicity is an important consideration in the comparative evaluation of hypotheses, if only because he identifies it with *falsifiability*. For Popper, the more falsifiable a hypothesis or theory is, the greater is its empirical content, and thus the greater its virtue as a scientific theory. But if one is not disposed to identify simplicity with falsifiability, then Popper's analysis is not sufficient, as he does not offer any *independent* analysis of simplicity as a constraint other than as falsifiability. Barker (1961) has, I believe, successfully shown that simplicity and falsifiability are not in all circumstances identical. Barker argues that if H_2 entails H_1, and if H_2 is a detailed elaboration of H_1 – suppose it is H_1 and H_a – then although H_2 may well be more falsifiable, it will not, according to our intuition, be simpler than H_1. Accordingly I believe that Popper's analysis must be supplemented with an analysis of simplicity as an independent desideratum.

What I do accept as suggestive and valuable in Popper's analysis is his stress on the element of theoretical and empirical constraint provided by previously established theories and their evidential bases. The previous theory which has been accepted in a given science or a branch of a science sets itself up as a standard of adequacy.[49] It tells us which experiments are important and should be explained by any new theory. It is necessary for a new theory to do as well as or better than the older theory in that area, or else fail to establish itself as worthy of consideration as a scientific theory for that area.

What is too restrictive in Popper's account is the other aspect of this same type of constraint, i.e., that provided by the previously accepted theories and their evidential bases. There are several cases in the history of science in which the previous theory (or even an approximation to it) in a given discipline does *not* follow from the later theory proposed to account for the same, or roughly the same, sector of experience. One has only to think of Newtonian mechanics and Aristotelian mechanics, or, perhaps better, of Lavosier's oxygen theory and the theory of phlogiston.[50] It is not the case that every earlier theory must follow as a limiting case or an approximation of a later theory in the same branch of science. Acceptance of this possibility that the later theory may vary considerably from the earlier one amounts to a relaxation of the quasi-inductive constraint, though it does not eliminate it completely. For even in the case of significant differences between earlier and later theories, the *'experimental results'* accounted for by the earlier theory still must be explained by the later theory. Where to draw the line of demarcation between theory and experiment is relative to the degree of change, but I believe that it can be said in general that even in extreme cases of theory change in a branch of science, that some of the earlier theories are preserved as part of the interpretation of the experimental findings. This inseparable union of experiment *and* theory indicates that there is much merit to D. Shapere's suggestion that it is a 'domain of adequacy' – a *mixture* of experiment and theory, which must be accounted for by a new theory or a modification of a theory.[51]

4.3.5. In summary, then, I want to propose, somewhat schematically, an analysis of the assessment of new scientific hypotheses and of theories which involves three aspects: (1) an evaluation of the *theoretical context sufficiency*, i.e., of the degree to which a new hypothesis or theory accords with the relevant connected hypotheses and theories of established science. An assessment made in terms of this category recognizes theoretical interdependence; it also allows that an established hypothesis or theory may be wrong and the new hypothesis or theory correct; (2) an evaluation of the *experimental adequacy*, i.e., of the degree to which the new hypothesis or theory accounts for the experiments, and perhaps part of the theory, which the previous hypothesis or theory in this branch of science did. One would also evaluate, in terms of this category, whether the new hypothesis or theory accommodates the data better, i.e., with more precision, and predicts new data in addition; (3) an evaluation of the *relative simplicity* of the new hypothesis or theory in its dimensions of (a) a relative minimization of previously untested and independently uncorroborated elements, (b) a relative minimization of the ontology and the terminology, and (c) a relative minimization of the number of independent component hypotheses and the degree of complexity of these hypotheses. The total assessment of a new hypothesis' or theory's worth will be a weighted sum of the three assessments, with experimental adequacy ranking highest and relative simplicity lowest.[52]

Assessments made within these three categories will vary with new discoveries, new proofs, and new formulation of hypotheses and new axiomatizations of a theory, and depend as well on what is occurring in connected branches of science. This changeable character of assessments I view not as a potential criticism of this logic of comparative theory evaluation or of hypothesis evaluation, rather I see it as a virtue of the logic. Science is a changing subject and a dynamic logic of comparative theory evaluation, though it *itself* may not change, should reflect the changes in science by changing values of its assessments.

Such a logic together with its associated underlying epistemology – not to be discussed here but available elsewhere – provides a synthetic philosophical unification of the accounts of Duhem and Popper – and also articulates what I hope is the appropriate total framework in terms of which the history of the operon theory and other theories in biology can best be analyzed.

I should like to close this essay by making a 'practical suggestion' which is based, in particular, on the importance of falsification and the way in which tests of the operon theory appeared to develop in the quasi-inductive manner described. I should like to suggest that the theoretical biologist ought to be most careful to consider the quasi-inductive aspects of the advance of biological theory in addition to searching for far-reaching and general 'bold conjectures'. If the quasi-inductive approach is disregarded, theoretical biologists, such as general systems theorists, 'mathematical biologists', and disciples of C. H. Waddington, are likely to find themselves developing a biological metaphysics rather than a biological science.[53]

University of Pittsburgh

NOTES

* This work has been aided by a grant from the National Science Foundation.

[1] There are of course some exceptions to this generally true rule. See, for example, Beckner's (1959) excellent monograph, which discusses the logic of concept formation in biology, albeit from a classical, non-molecular, point of view.

[2] I do not consider, for example, that Carnap's masterful *Logical Foundations of Probability* (1950) takes actual science seriously. It does not use specific scientific hypotheses as examples, nor does it attempt to formulate a logic which does justice to science's concern with theories and universals. For an extensive discussion of Carnap *vis-à-vis* these difficulties, see Lakatos (1968a), but also see Hintikka (1965) for an important extension of Carnap's approach which mitigates some of these difficulties.

[3] The actual situation is even more restrictive than I have indicated, in that only Kuhn (1962) concerns himself with theories of chemistry as well as theories of physics.

[4] Jacob and Monod (1961), p. 352.

[5] Stent (1964), pp. 816, 820.

[6] Stent (1965), pp. 463, 464. But see note 24 below for a qualification of Stent's comments about the origin of the messenger concept.

[7] Ames and Martin (1964), p. 235.

[8] It should be mentioned that the amendments and extensions of the theory noted by Ames and Hartman constituted an *alternative* to the Jacob and Monod analysis. See Beckwith (1967) for a discussion of this and other such alternatives.

[9] See Waddington (1968), p. 103.

[10] There is as yet, to my knowledge, no detailed historical analysis of the development of the operon theory. Stent's papers (1965, 1967), the Nobel lectures of Jacob (1966) and Monod (1966), and Jacob and Monod's recent (1970) brief historical paper were the basic secondary sources used for this article. Carlson (1966), Ch. 24, was also very helpful. Since this essay was completed in 1970, I have developed a more extensive analysis of the background of the operon theory. This research, based on extensive interviews with the scientists involved in the development of the theory, is presented in a paper of mine, 'Logic of Discovery and Justification in Regulatory Genetics', in *Studies in History and Philosophy of Science*, in press.

[11] It should be pointed out here that the disassociation of the enzyme activity from enzyme induction was immensely simplified by the discovery of a number of galactoside analogues. Cf. Jacob and Monod (1970).

[12] See Cohn *et al.* (1953), and also Stent's (1967), p. 153 account of this episode for more detail.

[13] Cf. Monod (1966), p.477.

[14] Cf. Stent (1968) and Cairns, Stent and Watson (1966) for support of this thesis. For an excellent introduction to this material, see Ravin (1965), Ch. 2.

[15] Monod (1966), pp. 478–479.

[16] Cf. Stent (1965), p. 463 for this locution.

[17] Cf. Pardee, Jacob, and Monod (1959), esp. pp. 175–177.
[18] Cf. Monod's interesting discussion of Vogel's introduction of the term 'repression', in Monod (1966), pp. 477–478. Also see my forthcoming paper cited in note 10 above.
[19] The sense of 'simplicity' employed by Pardee, Jacob, and Monod seems to be a kind of 'inductive simplicity' or perhaps a 'corroborative simplicity', which I have characterized elsewhere as 'fitness'. See my (1970), Section IV C. Also see Section 4.3.4. in the present paper.
[20] Cf. Jacob and Monod (1959), pp. 1283–1284.
[21] Jacob (1966), p. 1472.
[22] Jacob and Monod in their recent (1970) paper on the history of the operon theory note, in further support of the importance of the *lac*-λ analogy, that "systematic comparisons of the two systems proved invaluable. The o^c mutation, for instance, was discovered because the assumed symmetry between the two systems predicted their existence ..., similarly i^s mutants could be interpreted by analogy with the super-repressed lambda mutants ...".
[23] Cf. Jacob and Monod (1961), pp. 331–332.
[24] I must qualify somewhat this assertion that the mRNA hypothesis was first proposed in the 1961 review article, even in the face of Stent's comments quoted above on p. 6. Robert C. Olby (1970) has in an important article recently argued that the messenger hypothesis had its origin in Jacob's verbal account of improved versions of the 'Pa-Ja-Mo' experiment and the Riley *et al.* experiment. Thus the messenger idea was known within a small community before the Jacob and Monod (1961) article, though both articles (Brenner *et al.*, 1961 and Gros *et al.*, 1961) which reported experimental support for a messenger hypothesis did refer to the forthcoming Jacob and Monod (1961) article.
[25] Cf. Beckwith and Zipser (1970) for the results of the most recent mappings.
[26] This is not quite the case, as I tried to point out in my (1967c), Ch. III, but such issues are not crucial to any of the problems discussed in this present paper.
[27] Cf. my (1967a) and also my (1969d). 'Reduction functions' represent a synthesis of Nagel's (1961) 'Connectability Assumptions', Feigl's (1967 [1958]) and Sklar's (1964) 'Synthetic Identities', and Quine's (1964) 'Proxy Function'.
[28] See my (1967b), and also my (1969a) for a discussion of this issue. Also see Polanyi (1968) for a 'vitalistic' interpretation of the existence of such organization.
[29] I have discussed several theories of chemical evolution in connection with this thesis in my (1969c).
[30] Cf. Von Bertalanffy (1951).
[31] Von Bertalanffy (1968), pp. 48–49.
[32] Cf. Mesarović (1968), pp. 71–72, 74.
[33] Mesarović (1968), p. 71.
[34] See for example the interesting work of Goodwin (1963), and also Maynard-Smith's (1968) esp. pp. 107–115, for mathematical analyses of simple biological control systems.
[35] See Mesarović (1968), pp. 59–65, and Von Bertalanffy (1968), esp. Chs. 3 and 7.
[36] Duhem (1954 [1906]), p. 147.
[37] Duhem (1954 [1906]), pp. 147–148.
[38] Duhem (1954 [1906]), pp. 180, 183. Also see A. Grünbaum (1966 and 1969) for penetrating and lucid discussions of the complexities involved in the falsifiability of a scientific hypothesis *vis-à-vis* Duhem's arguments. Grünbaum's latter work is especially interesting as it also considers such questions in connection with a biological example, specifically the 'falsification' of the hypothesis of spontaneous generation by Pasteur, and its revival or the re-evaluation of its falsification by Oparin, Miller, and Urey.

For a somewhat different analysis of the interdependency of scientific theories which

is based on a post-positivistic account of correspondence rules see my (1969b) and (1969d)
[39] I am using the term 'neo-classical genetics' to represent the modifications and clarifications of classical Mendelian-Morganian genetics which arose as a consequence of studies of the genetics of phage and bacteria, e.g., the ideas of genetic fine structure and the 'cistron' locution of Benzer (1957).
[40] The reasons for supposing in 1962–63 that the repressor was a protein were somewhat more extensive than I have indicated in the body of the paper. In their paper proposing an allosteric enzyme model of the repressor, however, Monod, Changeux, and Jacob did not give specific reasons for the change from a polyribonucleotide to a protein repressor other than to note that (a) the hypothesis that the repressor was a polyribonucleotide "did not by itself account for repressor-inducer interaction," (b) that this hypothesis "has met with further serious difficulties while several lines of indirect experimental evidence suggest that the active product of a regulator gene is a protein, present in exceedingly minute amount in cells." (Monod *et al.*, 1963, p. 327). The further serious difficulties noted in (b) apparently were the discovery of temperature-sensitive and effector-insensitive repressor mutations, and the phenotypic reversal of R^- mutations by extra-genic suppressors (see Stent, 1964, for references). It seems to be clear, however, that (a) is of considerable weight, and Stent, commenting on this situation in the Spring of 1964, noted that the belief that the repressor is a protein "is embraced in the remainder of this article, although it must be admitted that the reasons for so doing still remain more doctrinal than empirical." (Stent, 1964). It was only with the work of Bourgeois *et al.* (1965), demonstrating that i^- mutations could be suppressed by well-characterized suppressors of chain-terminating mutations, that clear evidence that the repressor was a protein first became available.
[41] It might be asked why I do not simply term the proposal of the repressor hypothesis, and the other hypotheses constituting the operon theory, hypotheses which are advanced in accordance with the standard hypothetico-deductive characterization of science. Such an account would license the 'conjectural' element discussed, but without analyzing the development of the operon theory in terms of the Popperean approach. The reason for using the Popperean mode of analysis was to be philosophically faithful to certain aspects of the mode of presentation, argument, and testing, which is discernible in Jacob, Monod, and their coworkers' scientific papers – in particular (1) to the continual advancement of sometimes two, sometimes three, alternative and mutually inconsistent hypotheses, and their attempts to *falsify* as well as confirm or corroborate such hypotheses, and (2) to the 'quasi-inductive' elements of the operon theory's development to be discussed in the next section. These aspects accord quite well with Popper's (1959 [1934] and 1962) approach and justify its use. They do not, however, require an *uncritical* acceptance of Popper's analyses, as see Section 4.3.4. I would also like to point out that a deeper analysis of the discovery situation associated with the repressor concept indicates that the Popperean account is too one-sided, and that the constraints involved in the context of discovery can be sufficiently strong to allow *inference* to new hypotheses. See my forthcoming paper cited in note 10 for an extensive discussion of this point.
[42] For philosophical developments of Popper's analyses and an application of these developments to theory construction and evolution in physics see Lakatos (1970).
[43] Cf. Waddington (1968), esp. pp. 103–104. Waddington believes that biology needs a 'theoretical biology' just as physics needs a 'theoretical physics'.
[44] This slight modification of Popper's conception of quasi-inductive advance is, I believe, supported by Popper's remarks concerning 'degrees of testability'. See his (1959 [1934]) Ch. 6, esp. pp. 121–123.

[45] The possibility of a fundamental contradiction between the Popperean and the Duhemian aspects of the philosophical analysis of the development of the operon theory presented in this paper was suggested by A. Grünbaum in personal conversation. Difficulties with the Popperean falsification model have been discussed in the literature by Lakatos (1968a, 1970), and the historically demonstrated resistance of actual scientific theories to falsification, and the implications of this resistance *vis-à-vis* Popper's analyses, have been discussed by Kuhn (1962).

[46] The D-thesis has also been defended in the context of a discussion of the analytic-synthetic distinction by W. V. Quine in his influential article 'Two Dogmas of Empiricism' (Quine, 1961) and in his book (1960).

[47] I have discussed in my (1970) how constraints like these are employed in scientific revolutions (specifically in connection with the revolution which replaced Lorentzian electrodynamics with an Einsteinian electrodynamics) to insure progress in science.

[48] I am going to assume that different theories holding for some subject area in science can conflict both with each other and with certain auxiliary theories, for example, that Lorentz's and Einstein's theories of the electrodynamics of moving bodies conflicted with each other and also with certain aspects of classical mechanics. Further I am going to assume that a common 'observational' base can be specified so as to require explanation of accepted sentences describing this observational base by different theories purporting to hold for a specific scientific subject area. Both of these assumptions have been questioned by proponents of the 'historical school' of philosophy of science, e.g., Kuhn (1962) and Feyerabend (1962), but I have defended them extensively in my (1970) essay, and here must assume them without further proof.

[49] In addition to the long quote from Popper (1959 [1934]) given in the text above which supports the point that the previous theories and their experimental base constitute important constraints on a new theory which might replace the earlier theory holding in a specific subject area(s), it might also be useful to refer to Popper's comments about revised theories *vis-à-vis* the experimental base of older theories enunciated in connection with a postulated Humean change in the laws of nature. Popper wrote:
"Let us suppose that the sun will not rise tomorrow (and that we shall nevertheless continue to live, and also to pursue our scientific interests). Should such a thing occur, science would have to try to *explain* it, *i.e.*, to derive from it laws. Existing theories would presumably require to be drastically revised. But the revised theories would not merely have to account for the new state of affairs: *our older experiences would also have to be derivable from them.*" (Popper (1959 [1934]) p. 253.)

[50] Such cases have been stressed by Kuhn (1962) and by Feyerabend (1962).

[51] See Shapere (1971) for an explication of the concept of a domain.

[52] These categories of comparative theory or hypothesis evaluation are in these pages characterized primarily as criteria of *justification.* (I owe this observation to D. Shapere.) I believe, however, that such criteria do function, together perhaps with other criteria, in the context of *discovery* as general constraints on the thought of scientific discoverers. Sometimes the criteria are explicit, sometimes implicit. For such an interesting explicit instance of the workings in physics of the three criteria discussed in the body of this paper see M. Planck's (1900) paper in which he first proposed the correct black-body radiation law (Ter Haar, 1967, paper 1), and in which he explains what the steps in his thinking were. For some of the 'other criteria' which might be involved in a logic of discovery, see Hanson (1961), Hesse (1970), Shapere (1971), and Schaffner (in press, see note 10).

[53] Such a 'suggestion' ought to be viewed more as a *caveat* than as a criticism. Waddington's conviction that more theory is needed in biology could have extremely saluary conse-

quences. Recently Goodwin and Cohen (1969) have developed a theory of embryological differentiation which would seem to meet Waddington's criteria for a theory. It remains to be seen, however, whether such a theory, which seems quite unfalsifiable in its present form, represents as viable an approach as the different type of attack on the differentiation problem suggested by Bonner (1965) and, more recently, by Zubay (1968), pp. 466–467, and by Britten and Davidson (1969).

BIBLIOGRAPHY

Ames, B. N. and Hartman, P. E., *Cold Spring Harbor Symp. Quant. Biol.* **28** (1963) 349–356.

Ames, B. N. and Martin, R. G., *Ann. Rev. Biochem.* **33** (1964) 235–258.

Barker, S. F., 1961, *Phil. Sci.* **28** (1961) 162.

Beckner, M., *The Biological Way of Thought*, Columbia University Press, New York, 1959.

Beckwith, J., *Science* **156** (1967) 597–604.

Beckwith, J. and Zipser, D. (eds.), *The Lactose Operon*. Cold Spring Harbor, New York, Cold Spring Harbor Laboratory of Quant. Biol., 1970.

Benzer, S., in *The Chemical Basis of Heredity* (ed. by W. D. McElroy and B. Glass), Johns Hopkins, Baltimore, 1957.

Bonner, J., *The Molecular Biology of Development*, Clarendon Press, Oxford, 1965.

Bourgeois, S., Cohn, M., and Orgel, L., *J. Mol. Biol.* **14** (1965) 300–302.

Bourgeois, S. and Monod, J., *Ciba Symposium on Control Processes in Multicellular Organisms* (ed. by G. E. W. Wolstenholme and J. Knight), J. and A. Churchill, London, 1970, p. 3.

Brenner, S., Jacob, F., and Meselson, M., *Nature* **190** (1961) 576.

Britten, R. J. and Davidson, E. H., *Science* **165** (1969) 349–357.

Cairns, J., Stent, G., and Watson, J. (eds.), *Phage and the Origins of Molecular Biology*, Cold Spring Harbor Laboratory of Quant. Biol., Cold Spring Harbor, New York, 1966.

Carlson, E. A., *The Gene: A Critical History*, Saunders, Philadelphia, 1966.

Carnap, R., *The Unity of Science* (transl. by M. Black), K. Paul Trench, Trubner, & Co., London, 1934.

Carnap, R., *The Logical Foundations of Probability*, The University of Chicago Press, Chicago, 1950.

Cohn, M., Monod, J., Pollock, M. R., Spiegelman, S., and Stanier, R. Y., *Nature* **172** (1953) 1096.

Duhem, P., *The Aim and Structure of Physical Theory* (transl. by R. P. Wiener), Princeton University Press, Princeton, 1954 [1906].

Feigl, H., *The 'Mental' and the 'Physical'*, University of Minnesota Press, Minneapolis, 1967 [1958].

Feyerabend, P. K., 1962, in *Minnesota Studies in The Philosophy of Science, III* (ed. by H. Feigl and G. Maxwell), Univ. of Minnesota Press, Minneapolis, pp. 28–97.

Gilbert, W. and Müller-Hill, B., *Proc. Natl. Acad. Sci. U.S.* **56** (1966) 1891–1898.

Goodwin, B., *Temporal Organization in Cells*, Academic Press, New York, 1963.

Goodwin, B. and Cohen, M., *J. Theoret. Biol.* **25** (1969) 49–107.

Gros, F., Hiatt, H., Gilbert, W., Kurland, C. G., Risebrough, R. W., and Watson, J. D., *Nature* **190** (1961) 581.

Grünbaum, A., *Mind, Matter, and Method: Essays in Philosophy in Honor of Herbert Feigl* (ed. by P. K. Feyerabend and G. Maxwell), University of Minnesota Press, Minneapolis, 1966, pp. 276–293.

Grünbaum, A., *Studium Generale* **22** (1969) 1061.

Hanson, N. R., *Current Issues in the Philosophy of Science* (ed. by H. Feigl and G. Maxwell), Holt, Rinehart, and Winston, New York, pp. 20–35.

Hershey, A., *Cold Spring Harbor Symp. Quant. Biol.* **11** (1946) 67–76.

Hertz, H., 1956 (1894), *The Principles of Mechanics* (transl. by D. E. Jones and J. T. Walley), Dover, New York.

Hesse, M. B., 'Logic of Discovery in Maxwell's Electromagnetic Theory', (1970) (mimeo), privately circulated.

Hintikka, J., *Proceedings of the 1964 International Congress for Logic, Methodology and Philosophy of Science* (ed. by Y. Bar-Hillel), Amsterdam, North-Holland, 1965, pp. 274–288.

Jacob, F., *Harvey Lectures, 1958–1959* **54** (1958–59) 1.

Jacob, F., *Science* **152** (1966) 1470.

Jacob, F. and Monod, J., *C. R. Acad. Sci., Paris* **249** (1959) 1282–1284.

Jacob, F. and Monod, J., *J. Mol. Biol.* **3** (1961) 318–356.

Jacob, F. and Monod, J., in Beckwith and Zipser (1970) pp. 1–4.

Jacob, F., Perrin, D., Sanchez, C., and Monod, J., *C. R. Acad. Sci., Paris* **250** (1960) 1727–1729.

Jacob, F., Ullmann, A., and Monod, J., *J. Mol. Biol.* **13** (1965) 704–719.

Jacob, F. and Wollman, E., *Ann. Inst. Pasteur* **91** (1956) 486.

Kuhn, T. S., *The Structure of Scientific Revolutions*, University of Chicago Press, Chicago, 1962.

Lakatos, I., *The Problem of Inductive Logic* (ed. by I. Lakatos), North-Holland, Amsterdam, 1968a, pp. 315–417.

Lakatos, I., *Proc. Aristot. Soc.* **69** (1968b) 149–186.

Lakatos, I., *Criticism and the Growth of Knowledge* (ed. by I. Lakatos and A. Musgrave), Cambridge University Press, Cambridge, 1970, pp. 91–196.

Lederberg, J. and Tatum, E. L., *Cold Spring Harbor Symp. Quant. Biol.* **11** (1946) 113–114.

Luria, S. E. and Delbrück, M., *Genetics* **28** (1943) 491–511.

Maynard Smith, J., *Mathematical Ideas in Biology*, Cambridge University Press, Cambridge, (1968).

Mesarović, M. D., *Systems Theory and Biology*, (ed. by M. D. Mesarović), Springer-Verlag, New York, 1968, pp. 59–87.

Monod, J., in *Enzymes: Units of Biological Structure and Function* (ed. by O. H. Gaebler), Academic Press, New York, 1956, p. 7.

Monod, J., *Science* **154** (1966) 475–483.

Monod, J. and Audureau, A., *Ann. Inst. Pasteur* **72** (1946) 530.

Monod, J., Cohen-Bazire, G., and Cohn, M., *Biochim. Biophys. Acta* **7** (1951) 585–599.

Monod, J., Changeux, J., and Jacob, F., *J. Mol. Biol.* **6** (1963) 306–329.

Monod, J., Pappenheimer, A., and Cohen-Bazire, G., *Biochim. Biophys. Acta* **9** (1952) 648–660.

Nagel, E., *The Structure of Science*, Harcourt, Brace, and World, New York, 1961.

Olby, R. C., *Daedalus* **99** (1970) 938–987.

Pardee, A., Jacob, F., and Monod, J., *J. Mol. Biol.* **1** (1959) 165–178.

Pardee, A. and Prestidge, L., *Biochim. Biophys. Acta* **36** (1959) 545–547.
Planck, M., see Ter Haar 1967, paper 1.
Poincaré, H., *Foundations of Science* (transl. by G. B. Halsted), The Science Press, Lancaster, Pa., 1946 [1901, 1905, 1908].
Polanyi, M., *Science* **160** (1968) 1308–1312.
Popper, K. R., *The Logic of Scientific Discovery*, Basic Books, New York, 1959 [1934].
Popper, K. R., *Confectures and Refutations: The Growth of Scientific Knowledge*, Basic Books, New York, 1962, pp. 215–250.
Quine, W. V., *Word and Object*, MIT Press, Cambridge, Mass. 1960.
Quine, W. V., *From a Logical Point of View*, 2nd rev. ed., Harper and Row, New York, 1961, pp. 20–46.
Quine, W. V., *J. Phil.* **61** (1964) 209–216.
Ravin, A., *The Evolution of Genetics*, Academic Press, New York, 1965.
Rickenberg, H. V., Cohen, G. N., Buttin, G., and Monod, J., *Ann. Inst. Pasteur* **91** (1956) 829–857.
Riley, M., Pardee, A., Jacob, F., and Monod, J., *J. Mol. Biol.* **2** (1960) 225.
Schaffner, K., *Phil. Sci.* **34** (1967a) 137–147.
Schaffner, K., *Science* **157** (1967b) 644–647.
Schaffner, K., 'The Logic and Methodology of Reduction in the Physical and Biological Sciences', unpublished Ph.D. dissertation, Columbia University, 1967c.
Schaffner, K., *J. Hist. Biol.* **2** (1969a) 19–33.
Schaffner, K., *Phil. Sci.* **36** (1969b) 280–290.
Schaffner, K., *American Scientist* **57** (1969c) 410–420.
Schaffner, K., *Brit. J. Phil. Sci.* **20** (1969) 325–348.
Schaffner, K., *Minnesota Studies in the Philosophy of Science* (ed. by R. Stuewer) University of Minnesota Press, Minneapolis, 1970, pp. 311–354, 365–373.
Shapere, D., 'Explanation and Scientific Progress', in *Boston Studies in the Philosophy of Science*, in press (Mimeo, 1971).
Shapiro, J., MacHattie, L., Eron, L., Ihler, G., Ippen, K., Beckwith, J., Arditti, R., Reznikoff, W., and MacGillivray, R., *Nature* **224** (1969) 768.
Sklar, L., 'Intertheoretic Reduction in the Natural Sciences', unpublished Ph.D. dissertation, Princeton University, 1964.
Stent, G. S., *Science* **144** (1964) 816–820.
Stent, G. S., *Science* **150** (1965) 462–464.
Stent, G. S., *The Neurosciences* (ed. by G. Quarton, T. Melnechuk, and F. Schmitt), Rockefeller University Press, New York, 1967, pp. 152–161.
Stent, G. S., *Science* **160** (1968) 390–395.
Ter Haar, D., *The Old Quantum Theory*, Pergamon Press, Oxford, 1967.
Von Bertalanffy, L., *Human Biology* **23** (1951) 302–312.
Von Bertalanffy, L., *General System Theory*, G. Braziller, New York, 1968.
Waddington, C. H., *Towards a Theoretical Biology: I. Prolegomena*, (ed. by C. H. Waddington), Aldine Pub. Co., Chicago, 1968, pp. 103–108.
Zubay, G. L., *Papers in Biochemical Genetics*, Holt, Rinehart, and Winston, New York, 1968.

LAWRENCE SKLAR

THE EVOLUTION OF THE PROBLEM OF THE UNITY OF SCIENCE

A serious part of the excitement of science is coming upon the unexcepted. Novel facts, unanticipated experimental results, and surprising consequences of theories with which we thought we were thoroughly familiar, provide the sense of recurring 'newness' which is part of the fascination of the scientific pursuit. Philosophers usually aren't as lucky, and philosophy usually lacks this quality of eternal freshness.

Sometimes, however, we do find a feature of serious, programmatic philosophy something like the continual surprise of science. When this occurs I think there is an element of deep satisfaction. It is as though, for once, philosophy has taken hold upon us, rather than our leading it, tamely, where we will. I believe this happens most frequently when, in the course of a longterm 'program' of philosophical investigation, the 'argument' develops in ways we would never have anticipated when we started, leading us to realize just how little we really knew, even 'in principle', about the premisses from which we started. No new *facts*, perhaps, to amaze us, but certainly astonishing *conclusions*. The effect is most rewarding when the conclusions to which we are apparently forced are not only unexpected, but also wildly at variance with what we *hoped* we would be led to believe at the ultimate end of our investigations.

Consider, for example, the strange evolution of empiricism from mechanistic materialism through representative realism and Berkelian idealism to the phenomenalism of Hume, Mill and Mach. Custom has, perhaps, staled us to the peculiar and totally unexpected direction forced upon philosophy by the epistemological critique of materialism which arises almost inevitably out of the application of empiristic principles; but something of the wonder can be recovered by watching the faces of students in courses in elementary philosophy when they are led for the first time through Berkeley's *First Dialogue*. Starting from an attempt to provide a firm epistemological foundation to the new materialist science, and applying our empiricist canon with rigor and consistency, we end up with a metaphysic which, if not idealist, is at least far removed from

R. J. Seeger and R. S. Cohen (eds.), Philosophical Foundations of Science, 535–545.

the naive materialist ontology at which we started. I am not claiming, of course, that this development of the empiricist argument really is inevitable, nor that its terminal conclusions in phenomenalism are correct.

I believe that a similar feature can be noted in the evolution of recent discussion regarding the unity of science thesis, its meaning and its truth. In anticipation, let me reveal the main thread of my argument. In the nineteenth century the 'unity of science' theme was most commonly found in association with 'crude' materialism, a not very well formulated but plainly ontological or metaphysical thesis. The rebirth of interest in the problem of the unity of science in the twentieth century occurred in the context of logical positivism, an orientation to philosophy a primary feature of which was the radical eschewal of metaphysics. Finally human discourse was to be cleared of the tissue of confusion resulting from the unfortunate propensity of man to take literal nonsense as sense. In so far as philosophical discourse was intelligible, it was metadiscourse, about words, languages or theories. And in so far as the unity of science question was a genuine question, its study was the study of theories, not of things. But the evolution of the examination of the problem of the unity of science, an evolution stemming directly from the efforts of the logical positivists, has led us back to the once despised metaphysical and ontological questions, and, indeed, has served as a major impetus to the now current interest among analytical philosophers of questions as metaphysical as any that perplexed the minds of Aristotle, Leibniz or the younger Wittgenstein.

When the problem of the unity of science began to receive renewed attention in the thirties two main approaches seem to have predominated. First, there was the attempted development, utilizing the new-found resources of formal logic, of a rigorous and precise 'linguistic' version of the old empiricist program of reducing all assertions to assertions about phenomena or appearances. Second there was the attempt to carry over into the study of physical theory the concept of extensional isomorphism which had been found so very useful in the study of 'reductionism' in mathematics. Let us very briefly examine these two approaches to see why the search for a principle of unity now leads in other directions.

The 'philosophical' reduction of all the sciences to lawlike statements framed in a language of phenomena, sense-data or appearences, and, hence, the 'unification' of all science to the science of appearances, proved both too hard and too easy to serve as a methodological account of the unity of science.

The project was 'too hard' because the alleged reduction of all language in science to sense-datum language remained elusive, if not impossible, of accomplishment. The unpacking of the difficulties inherent in this project constitutes a good portion of recent work in epistemology, but this is hardly the place to review the dialectic of this question. For our purposes it is more important to note the failure of this program to account for the 'unity of science' in the other direction; that is, the fact that 'philosophical' reduction is 'too easy' to serve the role intended for it.

What can I mean by the claim that philosophical reductionism is 'too easy' a solution to the problem of the unity of science, when I acknowledge the overwhelming difficulties in carrying out the philosophical reductionist program? Only this – even if philosophical reductionism were to be successfully completed, it is not at all clear that it would serve as an adequate account of what is meant by the unity of science. It is almost universally agreed, now, that the unity of science is, if it exists at all, an accomplishment to be achieved within the *practice* of science. That is, if science really is a 'unity', this unity must be demonstrated as one consequence of our scientific discovery and theoretical systematization. A 'unity' imposed on science as a consequence of philosophical reductionism is not this kind of unity at all, and, hence, philosophical reductionism is not an appropriate grounding, even if it is correct, for what is usually meant by the claim that all science can be unified into a single all encompassing science *by the progress of science itself*.

A similar objection holds, I believe, against attempts to found the unity of science on the notion of extensional isomorphism, a notion of great value in discussing 'reductionism' in mathematics. The technique of demonstrating a reduction by means of extensional isomorphism is well known. Take a theory. Formalize it. Find a re-interpretation of the axioms in some new domain. Take the re-interpretation as showing a reduction of the original theory to the theory of the new domain. Consider, for example, the axiomatization of complex numbers. Show that by reinterpreting the complex numbers as ordered pairs of reals, and addition and multiplication on the complexes as a new pair of operations on ordered pairs, the axioms remain true in the new interpretation. Then claim that complex numbers *are* ordered pairs, or more conservatively, that one has unified the theory of complex numbers into the theory of ordered pairs of reals.

No matter how useful such a procedure may be in the unification of mathematics, it is surely not at the foundation of any unity of physical science. Once again, the 'unification' obtained is too easy for our purposes. Suppose we did have a formalized, first-order fragment of an ordinary science, say biology. We find a re-interpretation of the axioms which makes them all come out true in the domain of the integers, with the predicates and relations involved interpreted as predicates and relations among the integers, a reinterpretation guaranteed to exist by the Löwenheim-Skolem Theorem. Does this reinterpretation 'reduce a fragment of biology to arithmetic'? Well, if it does, who cares about *that* kind of reduction? *Real* unity of science must be the result of scientific discovery relating previously unrelated branches of natural science. Let me, then, proceed on the assumption that there is agreement that the unity of science, if it exists at all, is to be found as the end product of scientific practice, not philosophical speculation or formal manipulation of theories. It is still the philosopher's task, however, to tell us just what the scientist does when he achieves a unification within the ongoing practice of science.

It is interesting to note that almost all who have looked for unity of science in the direction I propose have assumed, almost without question, that the unity is to be obtained by a series of successful reductions of theories the one to another, the end product being a hierarchical, asymmetric structure in which upper-level theories are reduced to lower-level ones, the whole structure, at any specific time, resting upon some most basic reducing theory. Apparently unity cannot come as a result of federalism among the sciences, but only as a by-product of the imperialism of physics. I, for one, believe that this assumption is correct, although I also believe that we have yet to fully comprehend the rationale behind our desire for an 'inverted pyramid' model of scientific unity. Nor do we yet have clear and detailed accounts of the nature of the asymmetry of the relation, the quality which makes some theories 'more basic' than others.

Not that we have no ideas on this subject at all. We take *generality* to be a mark of basicness. Or, again, we take it to be a mark of basicness that a theory deals with entities which are spatial parts of the wholes which constitute the subject matter of a less basic theory, so that the hierarchy of reduction, at least in part, parallels the hierarchy of the whole-part relation. Further, we are convinced that the hierarchy of reduction is matched by an asymmetry of explanatory power – more basic theories explain the less basic ones. But exactly how the various asymmetries fit together, and whether the ones noted fully capture the asymmetry of the reduction hierarchy, are questions not yet fully explored by any means.

To summarize, our present picture of the unity of science looks something like this: (1) Unification of the sciences is a scientific task, whose analysis may be left to philosophers, but whose accomplishment is the task of working scientists, *qua* scientists. (2) The unification is to be obtained, if it is obtainable at all, by a process of concatenation of reductions, less basic theories and sciences being reduced to more basic, until, finally, unity is obtained by the ultimate reduction of all theory to some most basic theory. (3) The hierarchical structure so obtained will be the structure of explanation in science, the explanation of less basic theories relying upon reference to the more basic with, perhaps, at any time there being some most basic unexplained explainer at the base of the structure, at any time.

But, we must now ask, just what is it to reduce one theory to another? Only by answering this question to our satisfaction can we begin to speculate upon the possibility of the attainment of unification, and it is just this question to which philosophers have devoted extensive attention in recent years.

What have we discovered about the nature of the reduction of one theory to another so far? A great deal, but not nearly enough.

At least we are now aware, as philosophers were not always aware, of just what a varied, diverse and wideranging group of scientific accomplishments reductions are, and of just how little the various reductions have in common methodologically.

In some reductions the reduced theory is simply derivable from the reducing. In some cases the reduced theory is only approximately derivable from the reducing. Sometimes we say that a reduction has taken when only a *fragment* of the reduced theory is derivable, or even approximately derivable, from the reducing theory. Frequently we allege a reduction to have taken place when the reduced theory is shown, by the very act of reduction, to be an incorrect theory due for elimination from science. In other cases the reduction serves to add additional confirmation to the reduced theory, significantly increasing our confidence in its correctness. Sometimes this last case holds with the modification that it is a somewhat changed version of the reduced theory whose place in science is secured by the reduction.

The rich structure of reduction has been explored more than once in the literature, and it is not my intention here to try one more exhaustive characterization of the nature of inter-theoretic reduction. Rather I wish to focus my remaining attention on three questions: (1) To what extent do models of reduction force straightforwardly metaphysical questions upon us, reluctant as we may be to grant them scientific 'meaningfulness'? (2) Do there appear to be any systematic approaches by means of which we might hope to avoid falling into the clutches of metaphysics, while at the same doing doing full justice to all the features of reduction? (3) Are there certain features of the reductive process, noted in very special cases of reduction, to which insufficient attention has yet been paid?

Where in the study of reduction does metaphysics raise its fearsome head? The cases which give us trouble can be characterized as follows: (1) The reduced theory is not eliminated or replaced in the act of reduction, but, rather, has its place in our scientific worldpicture made more secure by the reductive process. (2) But the reduced theory contains basic concepts not present in the reducing theory. The standard examples are the reduction of the theory of macroscopic matter to the theory of its atomic constituents and the reduction of physical optics to the electromagnetic theory.

First we note that any simple derivation of the reduced theory from the reducing is out of the question because of the concepts in the reduced theory not present in the reducing. But, our intuition tells us, the reducing theory does *explain* the reduced, and the explanation does seem to fit the deductive-nomological pattern. Well, then, all we need is some auxiliary premisse (or premisses) sufficient to allow us to deduce the reduced theory from the conjunction of the reducing theory and this new auxiliary statement.

But we cannot stop at this point, for such an auxiliary statement

(Nagel's 'bridging hypothesis') is all too easy to come by, unless severe restrictions are put on its nature. Just take, for example, the conditional formed of reducing theory as antecedent and reduced theory as consequent. Such a bridging law would allow us to reduce any theory to any other. Again, we frequently find examples of correlating laws which we intuitively reject as sufficient for establishing reductions (e.g. Wiedemann-Franz law).

The 'bridge' between the theories must, then, have something special about it for it to constitute the means to a reduction. The move most frequently taken at this point is to suggest that the bridge constitutes a reduction when it is not a mere lawlike correlation between phenomena, but, instead, a lawlike *identification*. For example, physical optics reduces to the electromagnetic theory because light-rays *are* electromagnetic waves, and are not merely correlated with them. Combine this thesis with the observation that, intuitively, identifications, unlike correlations, seem not to require explanation, and the move looks even more enticing. We have achieved the derivability of the reduced theory from the reducing at the cost of an additional hypothesis, the identification, but have not had to pay the price of adding to our body of science another idenpendent piece itself requiring explanation. Identifications work for us but ask no wages in return.

Beautiful. But Pandora's Box has been opened and the dark spirits of metaphysics are a flutter about the room, mocking our empiricist predispositions. What is an identification? What is a lawlike identification? How do you tell when two classes of entities are merely correlated and when there is, properly speaking, only one class previously encountered under two aspects or two descriptions? Why don't identifications require explanation? Is it because they are not discovered but simply 'invented'? Is the difference between a correlation and an identification a difference in fact? Or a decision? Or a convention? If there really are no observational consequences which would resolve a dispute between a correlationist and an identificationist, should we really allow that there are two different theories at all? Aren't they the same theory, merely conventionally transcribed in misleadingly different ways?

This is surely not the place even to begin to plumb the depths of these questions. I only wish to reiterate the theme on which I opened. After several decades of retreat to meta-language, linguistic mode, talk about theory rather than talk about the world, etc., the question of the unity of science (among other questions) has now dragged us, willy-nilly, back to the foggy realms of metaphysics. If the unity of science is to be found in the unity of theory, apparently the unity of theory must be discerned as a linguistic reflection of unity 'in the world'.

Let me look at some other ways in which 'unity' themes raise metaphysical issues. Much of the recent interest in the unity of science was inspired by those interested in the traditional problem of the relationship of the mental to the physical, of mind to brain. The argument went something like this: Traditionally the mind–body problem has been treated in isolation. But if looked upon as the problem of the relationship of psychology to neuro-physiology, it appears as one reduction problem among many. Perhaps, then, this traditional philosopher's problem can be solved by treating it as just one instance of the general problem of the reduction of theories.

The type of reduction in science relevant to the mind/body problem is immediately seen to be the kind of 'identificatory' reduction we have just been discussing. Just as quickly, not surprisingly, the mind/body 'identificatory reduction' is seen to have very special features differentiating it, if, indeed, it is intelligible at all, from the more mundane identificatory reductions of physical science.

Let me just rehearse one of these special issues, of particular interest to my general theme here. What is the relation of mind to body? Well, mental entities *are* certain physical entities, and the identification is one of lawlike form, etc. For example? Well, sensations are certain brain processes. But now the cat is out of the bag, for it is processes (events, states-of-affairs, happenings?), not *things* whose identity is asserted in this particular special identificatory reduction. Puzzles are piled upon puzzles. In addition to the plethora of metaphysical questions enumerated above, we now have to face up to all the traditional issues surrounding the existence, nature, identity, etc. of those mysterious entities, halfway things and halfway propositions, whose nature has traditionally been so perplexing to metaphysicians.

It may be interesting to note, in passing, that these very same metaphysical issues just touched upon have arisen in the main-stream of philosophy once again, for a related but somewhat different reason – again a reason deeply embedded in the general problem of the unity of science.

It has frequently been claimed that the sciences of human behavior can never be reduced to the physical sciences. The arguments with which I am now concerned are not those which posit irreducibility because of a special ontological committment of human sciences, but rather those which locate the autonomy of the human sciences in their (allegedly) distinctive pattern of explanation. It goes something like this: Scientists in the natural sciences explain by reference to causes. Causes of events are events bound to the event to be explained non-analytically but only by means of some natural law. Thus to explain an action by a *want* is to explain behavior in terms of a condition (wanting) which is bound to the behaving in an analytic way. It is an analytic truth that, *ceteris paribus*, people do what they want, whereas the causes of events never analytically entail, even *ceteris paribus*, the occurrence of the effect event.

But the attack on this argument is clear, and has been followed up in some detail. The presence, or absence, of an analytic connection is a function of the description of the events employed, not of the events themselves. It is analytic that Caesar's fatal stabbing was followed by Caesar's death, but, surely, not analytic that Caesar's being punctured, vigorously, by senator-wielded knives was followed by the dictator's death. Yet the fatal stabbing was the vigorous puncture wounding, and that one self-same event, under any of the infinity of possible descriptions truly applicable to it, was, for sure, the cause of Caesar's death.

Very well. But, once again, it would be sheer self-deception not to recognize the congeries of metaphysical issues we have opened up. For once we allow events to be the kinds of things in our ontology which can be described and redescribed in ways not analytically, or even necessarily, equivalent, all our old problems about the existence and identity of quasi-propositional entities are once more to the fore.

I will turn now to much briefer treatments of the other two questions of which I promised to treat. First, are there any plausible suggestions that can be made as to how we might have avoided getting bogged down in metaphysics in the first place? That is, are there any reasonable routes we might think of, to do programmatic justice to our intuitions about the nature of scientific reduction, which do not raise all the subtle issues carried along by the identity thesis in its train?

My own intuition tells me that sooner or later the metaphysical issues, no matter how cleverly disguised, will appear at some point in our philosophical argument. But explorations as to their avoidability are worth pursuing, nonetheless. Consider, for example, the following line of thought: Why did we invoke identification in the first place? Because it was not sufficient for a hypothesis to allow the deduction of the reduced theory from the reducing for it to constitute a reducing hypothesis. Other constraints on the 'bridge' are surely necessary. To be sure, but is the requirement that it be an identification the only thing we can think of? What of all the other constraints one might apply: syntactical – requiring the bridging hypothesis to take on special form; semantic – requiring a special status to the conditional which appears in the bridging law, etc.; epistemic – requiring a special confirmatory basis for the bridging law, for example that it have evidential support independent of the evidential support for either reduced or reducing theory; pragmatic – requiring that the bridging law play some special role in directing scientific investigation or speculation, say that we resolve to preserve it immune from revision and so modify the direction our future science takes to allow us this liberty? I don't know whether any such moves would prove satisfactory. As I have said, my own guess is that they would just postpone the day of metaphysical reckoning. But it might take us longer to get to the metaphysical mysteries, and they might not appear as mysterious when we arrived at their abode.

Finally, a few remarks *a propos* my third question: Are there special cases of reduction to which insufficient attention has been paid, which fail to fit any of the well-developed known patterns, and whose examination is well worthy of the philosopher's time?

To partially answer the question let me just note my favorite example of a thoroughly un-understood reduction, the reduction of thermodynamics to statistical mechanics. I think this case of reduction is badly understood for two reasons: (1) Our present day understanding of theories in science fails radically to do justice to the variety of theories and their special natures. Since thermodynamics is a very special theory, indeed quite unlike any other, our understanding of its nature is particularly bad. (2) We have as yet almost no real understanding of the way in which a statistical theory functions in science. In particular, the crude models of statistical explanation so far explicated by philosophers do not even come close to elucidating the pervasive but puzzling role played in science by statistical mechanics. Recently, there has appeared in the literature some debate about just this reduction. My own feeling is that the crude dialectic we have seen concerning whether or not temperature *is* mean kinetic energy of molecules is a debate framed in too clumsy concepts to provide any real insight into this very special case of 'unification' at all.

Why should philosophers care about such anomalous cases of reduction? Aren't they curiousities to be noted and then ignored? I think not. Once again, let me argue my case by focussing on a single example.

The reduction of thermodynamics and the laws of ideal gases to statistical mechanics is a 'parasitical' reduction. Before the whole program ever gets off the ground we must first accept the reduction of the theory of macroscopic objects to their molecular constituents. Only if we believe that blobs of gases are collections of molecules could we ever entertain the hypothesis that some properties of the gas are intimately related in a statistical way to some properties of the underlying molecular structure. Further, the reductions are asymmetrically related – we could know that gases were of molecular constituency without knowing the work of Boltzmann and Gibbs, but the theory of the statistical mechanical reduction of thermodynamics presupposes knowledge and acceptance of the theory of molecular constitution.

What is the importance of all this to philosophers? Well, consider the version of the reduction of the mental to the physical which views mentality as the possession by the body of certain mentalistic properties, attributable to the body as a whole. Notice how the reduction of the mental to the physical appears, in this version, parasitical on the reduction of the body of the person to its constituents and their structure. It's as though we say: "Well, the body is just a bunch of elementary particles arranged thus-and-so, but what about that funny property of it, its paininess." Is the 'left-over' pain-property in this case just like the 'left-over' temperature, of a gas, once we know that the gas itself is just a collection of molecules with a particularly weak structure imposed upon them? 'Just-alike' pains and temperatures certainly are not, but alike enough to pursuade me, if no one else, that philosophers of mind have as much to gain as philosophers of physics in getting at least one special and anomalous case of reduction worked out in detail.

University of Michigan

SYNTHESE LIBRARY

Monographs on Epistemology, Logic, Methodology,
Philosophy of Science, Sociology of Science and of Knowledge, and on the
Mathematical Methods of Social and Behavioral Sciences

1. J. M. BOCHEŃSKI, *A Precis of Mathematical Logic*. 1959, X + 100 pp.
2. P. L. GUIRAUD, *Problèmes et méthodes de la statistique linguistique*. 1960, VI + 146 pp.
3. HANS FREUDENTHAL (ed.), *The Concept and the Role of the Model in Mathematics and Natural and Social Sciences. Proceedings of a Colloquium held at Utrecht, The Netherlands, January 1960*. 1961, VI + 194 pp.
4. EVERT W. BETH, *Formal Methods. An Introduction to Symbolic Logic and the Study of Effective Operations in Arithmetic and Logic*. 1962, XIV + 170 pp.
5. B. H. KAZEMIER and D. VUYSJE (eds.), *Logic and Language. Studies dedicated to Professor Rudolf Carnap on the Occasion of his Seventieth Birthday*. 1962, VI + 256 pp.
6. MARX W. WARTOFSKY (ed.), *Proceedings of the Boston Colloquium for the Philosophy of Science, 1961–1962*, Boston Studies in the Philosophy of Science (ed. by Robert S. Cohen and Marx W. Wartofsky), Volume I. 1963, VIII + 212 pp.
7. A. A. ZINOV'EV, *Philosophical Problems of Many-Valued Logic*. 1963, XIV + 155 pp.
8. GEORGES GURVITCH, *The Spectrum of Social Time*. 1964, XXVI + 152 pp.
9. PAUL LORENZEN, *Formal Logic*. 1965, VIII + 123 pp.
10. ROBERT S. COHEN and MARX W. WARTOFSKY (eds.), *In Honor of Philipp Frank*, Boston Studies in the Philosophy of Science (ed. by Robert S. Cohen and Marx W. Wartofsky), Volume II. 1965, XXXIV + 475 pp.
11. EVERT W. BETH, *Mathematical Thought. An Introduction to the Philosophy of Mathematics*. 1965, XII + 208 pp.
12. EVERT W. BETH and JEAN PIAGET, *Mathematical Epistemology and Psychology*. 1966, XXII + 326 pp.
13. GUIDO KÜNG, *Ontology and the Logistic Analysis of Language. An Enquiry into the Contemporary Views on Universals*. 1967, XI + 210 pp.
14. ROBERT S. COHEN and MARX W. WARTOFSKY (eds.), *Proceedings of the Boston Colloquium for the Philosophy of Science 1964–1966, in Memory of Norwood Russell Hanson*, Boston Studies in the Philosophy of Science (ed. by Robert S. Cohen and Marx W. Wartofsky), Volume III. 1967, XLIX + 489 pp.
15. C. D. BROAD, *Induction, Probability, and Causation. Selected Papers*. 1968, XI + 296 pp.
16. GÜNTHER PATZIG, *Aristotle's Theory of the Syllogism. A Logical-Philosophical Study of Book A of the Prior Analytics*. 1968, XVII + 215 pp.
17. NICHOLAS RESCHER, *Topics in Philosophical Logic*. 1968, XIV + 347 pp.

18. ROBERT S. COHEN and MARX W. WARTOFSKY (eds.), *Proceedings of the Boston Colloquium for the Philosophy of Science 1966–1968*, Boston Studies in the Philosophy of Science (ed. by Robert S. Cohen and Marx W. Wartofsky), Volume IV. 1969, VIII+537 pp.
19. ROBERT S. COHEN and MARX W. WARTOFSKY (eds.), *Proceedings of the Boston Colloquium for the Philosophy of Science 1966–1968*, Boston Studies in the Philosophy of Science (ed. by Robert S. Cohen and Marx W. Wartofsky), Volume V. 1969, VIII+482 pp.
20. J. W. DAVIS, D. J. HOCKNEY, and W. K. WILSON (eds.), *Philosophical Logic*. 1969, VIII+277 pp.
21. D. DAVIDSON and J. HINTIKKA (eds.), *Words and Objections: Essays on the Work of W. V. Quine*, 1969, VIII+366 pp.
22. PATRICK SUPPES, *Studies in the Methodology and Foundations of Science. Selected Papers from 1911 to 1969*. 1969, XII+473 pp.
23. JAAKKO HINTIKKA, *Models for Modalities. Selected Essays*. 1969, IX+220 pp.
24. NICHOLAS RESCHER *et al.* (eds.), *Essay in Honor of Carl G. Hempel. A Tribute on the Occasion of his Sixty-Fifth Birthday*. 1969, VII+272 pp.
25. P. V. TAVANEC (ed.), *Problems of the Logic of Scientific Knowledge*. 1969, XII+429 pp.
26. MARSHALL SWAIN (ed.), *Induction, Acceptance, and Rational Belief*. 1970, VII+232 pp.
27. ROBERT S. COHEN and RAYMOND J. SEEGER (eds.), *Ernst Mach: Physicist and Philosopher*, Boston Studies in the Philosophy of Science (ed. by Robert S. Cohen and Marx W. Wartofsky), Volume VI. 1970, VIII+295 pp.
28. JAAKKO HINTIKKA and PATRICK SUPPES, *Information and Inference*. 1970, X+336 pp.
29. KAREL LAMBERT, *Philosophical Problems in Logic. Some Recent Developments*. 1970, VII+176 pp.
30. ROLF A. EBERLE, *Nominalistic Systems*. 1970, IX+217 pp.
31. PAUL WEINGARTNER and GERHARD ZECHA (eds.), *Induction, Physics, and Ethics, Proceedings and Discussions of the 1968 Salzburg Colloquium in the Philosophy of Science*. 1970, X+382 pp.
32. EVERT W. BETH, *Aspects of Modern Logic*. 1970, XI+176 pp.
33. RISTO HILPINEN (ed.), *Deontic Logic: Introductory and Systematic Readings*. 1971, VII+182 pp.
34. JEAN-LOUIS KRIVINE, *Introduction to Axiomatic Set Theory*. 1971, VII+98 pp.
35. JOSEPH D. SNEED, *The Logical Structure of Mathematical Physics*. 1971, XV+311 pp.
36. CARL R. KORDIG, *The Justification of Scientific Change*. 1971, XIV+119 pp.
37. MILIČ ČAPEK, *Bergson and Modern Physics*, Boston Studies in the Philosophy of Science (ed. by Robert S. Cohen and Marx W. Wartofsky), Volume VII. 1971, XV+414 pp.
38. NORWOOD RUSSELL HANSON, *What I do not Believe, and other Essays* (ed. by Stephen Toulmin and Harry Woolf). 1971, XII+390 pp.
39. ROGER C. BUCK and ROBERT S. COHEN (eds.), *PSA 1970. In Memory of Rudolf Carnap*, Boston Studies in the Philosophy of Science (ed. by Robert S. Cohen and Marx W. Wartofsky), Volume VIII. 1971, LXVI+615 pp. Also available as a paperback.
40. DONALD DAVIDSON and GILBERT HARMAN (eds.), *Semantics of Natural Language*. 1972, X+769 pp. Also available as a paperback.
41. YEHOSUA BAR-HILLEL (ed.), *Pragmatics of Natural Languages*. 1971, VII+231 pp.
42. SÖREN STENLUND, *Combinators, λ-Terms and Proof Theory*. 1972, 184 pp.
43. MARTIN STRAUSS, *Modern Physics and Its Philosophy. Selected Papers in the Logic, History, and Philosophy of Science*. 1972, X+297 pp.

44. MARIO BUNGE, *Method, Model and Matter.* 1973, VII+196 pp.
45. MARIO BUNGE, *Philosophy of Physics.* 1973, IX+248 pp.
46. A. A. ZINOV'EV, *Foundations of the Logical Theory of Scientific Knowledge (Complex Logic)*, Boston Studies in the Philosophy of Science (ed. by Robert S. Cohen and Marx W. Wartofsky), Volume IX. Revised and enlarged English edition with an appendix, by G. A. Smirnov, E. A. Sidorenka, A. M. Fedina, and L. A. Bobrova. 1973, XXII+301 pp. Also available as a paperback.
47. LADISLAV TONDL, *Scientific Procedures*, Boston Studies in the Philosophy of Science (ed. by Robert S. Cohen and Marx W. Wartofsky), Volume X. 1973, XII+268 pp. Also available as a paperback.
48. NORWOOD RUSSELL HANSON, *Constellations and Conjectures* (ed. by Willard C. Humphreys, Jr.). 1973, X+282 pp.
49. K. J. J. HINTIKKA, J. M. E. MORAVCSIK, and P. SUPPES (eds.), *Approaches to Natural Language. Proceedings of the 1970 Stanford Workshop on Grammar and Semantics.* 1973, VIII+526 pp. Also available as a paperback.
50. MARIO BUNGE (ed.), *Exact Philosophy – Problems, Tools, and Goals.* 1973, X+214 pp.
51. RADU J. BOGDAN and ILKKA NIINILUOTO (eds.), *Logic, Language, and Probability.* A selection of papers contributed to Sections IV, VI, and XI of the Fourth International Congress for Logic, Methodology, and Philosophy of Science, Bucharest, September 1971. 1973, X+323 pp.
52. GLENN PEARCE and PATRICK MAYNARD (eds.), *Conceptual Change.* 1973, XII+282 pp.
53. ILKKA NIINILUOTO and RAIMO TUOMELA, *Theoretical Concepts and Hypothetico-Inductive Inference.* 1973, VII+264 pp.
54. ROLAND FRAÏSSÉ, *Course of Mathematical Logic* – Volume I: *Relation and Logical Formula.* 1973, XVI+186 pp. Also available as a paperback.
55. ADOLF GRÜNBAUM, *Philosophical Problems of Space and Time.* Second, enlarged edition, Boston Studies in the Philosophy of Science (ed. by Robert S. Cohen and Marx W. Wartofsky), Volume XII. 1973, XXIII+884 pp. Also available as a paperback.
56. PATRICK SUPPES (ed.), *Space, Time, and Geometry.* 1973, XI+424 pp.
57. HANS KELSEN, *Essays in Legal and Moral Philosophy*, selected and introduced by Ota Weinberger, 1973, XXVIII+300 pp.
58. R. J. SEEGER and ROBERT S. COHEN (eds.), *Philosophical Foundations of Science, Proceedings of an AAAS Program, 1969.* Boston Studies in the Philosophy of Science (ed. by Robert S. Cohen and Marx W. Wartofsky), Volume XI. 1974, IX+545 pp. Also available as paperback.
59. ROBERT S. COHEN and MARX W. WARTOFSKY (eds.), *Logical and Epistemological Studies in Contemporary Physics*, Boston Studies in the Philosophy of Science (ed. by Robert S. Cohen and Marx W. Wartofsky), Volume XIII. 1973, VIII+462 pp. Also available as a paperback.
60. ROBERT S. COHEN and MARX W. WARTOFSKY (eds.), *Methodological and Historical Essays in the Natural and Social Sciences. Proceedings of the Boston Colloquium for the Philosophy of Science, 1969–1972*, Boston Studies in the Philosophy of Science (ed. by Robert S. Cohen and Marx W. Wartofsky), Volume XIV. 1974, VIII+405 pp. Also available as a paperback.
63. SÖREN STENLUND (ed.), *Logical Theory and Semantic Analysis. Essays Dedicated to Stig Kanger on His Fiftieth Birthday.* 1974, V+217 pp.
64. KENNETH SCHAFFNER and ROBERT S. COHEN (eds.), *Proceedings of the 1972 Biennial Meeting, Philosophy of Science Association*, Boston Studies in the Philosophy of Science (ed. by Robert S. Cohen and Marx W. Wartofsky), Volume XX. 1974, VIII+445 pp. Also available as a paperback.

65. HENRY E. KYBURG, JR., *The Logical Foundations of Statistical Inference*. 1974, IX + 421 pp.
66. MARJORIE GRENE, *The Understanding of Nature: Essays in the Philosophy of Biology*, Boston Studies in the Philosophy of Science (ed. by Robert S. Cohen and Marx W. Wartofsky), Volume XXIII. 1974, XII + 360 pp. Also available as a paperback.

In Preparation

61. ROBERT S. COHEN and MARX W. WARTOFSKY (eds.), *For Dirk Struik. Scientific, Historical, and Political Essays in Honor of Dirk J. Struik*, Boston Studies in the Philosophy of Science (ed. by Robert S. Cohen and Marx W. Wartofsky), Volume XV. Also available as a paperback.
62. KAZIMIERZ AJDUKIEWICZ, *Pragmatic Logic*, transl. from the Polish by Olgierd Wojtasiewicz.
67. JAN M. BROEKMAN, *Structuralism: Moscow, Prague, Paris.*
68. NORMAN GESCHWIND, *Selected Papers on Language and the Brain*, Boston Studies in the Philosophy of Science (ed. by Robert S. Cohen and Marx W. Wartofsky) Volume XVI. Also available as a paperback.
69. ROLAND FRAÏSSÉ, *Course of Mathematical Logic* – Volume II: *Model Theory.*

SYNTHESE HISTORICAL LIBRARY

Texts and Studies
in the History of Logic and Philosophy

Editors:

1. M. T. BEONIO-BROCCHIERI FUMAGALLI, *The Logic of Abelard*. Translated from the Italian. 1969, IX + 101 pp.
2. GOTTFRIED WILHELM LEIBNITZ, *Philosophical Papers and Letters*. A selection translated and edited, with an introduction by Leroy E. Loemker. 1969, XII + 736 pp.
3. ERNST MALLY, *Logische Schriften* (ed. by Karl Wolf and Paul Weingartner). 1971, X + 340 pp.
4. LEWIS WHITE BECK (ed.), *Proceedings of the Third International Kant Congress*. 1972, XI + 718 pp.
5. BERNARD BOLZANO, *Theory of Science* (ed. by Jan Berg). 1973, XV + 398 pp.
6. J. M. E. MORAVCSIK (ed.), *Patterns in Plato's Thought. Papers arising out of the 1971 West Coast Greek Philosophy Conference*, 1973, VIII + 212 pp.
7. NABIL SHEHABY, *The Propositional Logic of Avicenna: A Translation from al-Shifā': al-Qiyās*, with Introduction, Commentary and Glossary. 1973, XIII + 296 pp.
8. DESMOND PAUL HENRY, *Commentary on De Grammatico: The Historical-Logical Dimensions of a Dialogue of St. Anselm's*. 1974, IX + 345 pp.
9. JOHN CORCORAN, *Ancient Logic and Its Modern Interpretations*. 1974, X + 208 pp.

In Preparation

10. E. M. BARTH, *The Logic of the Articles in Traditional Philosophy*.
11. JAAKKO HINTIKKA, *Knowledge and the Known*.
12. E. J. ASHWORTH, *Language and Logic in the Post-Medieval Period*.